HERBICIDES

CHEMISTRY, DEGRADATION, AND MODE OF ACTION

Second Edition, Revised and Expanded

Volume 1

Edited by

P. C. KEARNEY and D. D. KAUFMAN

Pesticide Degradation Laboratory
United States Department of Agriculture
Agricultural Research Center
Beltsville, Maryland

MARCEL DEKKER, INC.   New York and Basel

The introduction and subsequent destruction and/or accumulation of foreign substances in environmental systems has formed the basis for many new and expanded phases of modern science. Drugs, detergents, pesticides, pollutants, and a myriad of other substances have received major attention. Part of this flurry of activity can be traced to the publication of Silent Spring, by Rachel Carson, and the subsequent debates on the real and/or imaginary hazards of pesticides. The continuing environmental interest in the fate of all organic substances in the environment has stimulated much new and original research on the fate of pesticides in soil, water, air, and animal systems. The more stringent guidelines required for registration of new pesticides have added immensely to the body of literature existing on the environmental parameters of pesticides. While the production and use of insecticides have remained fairly static over the last several years, there has been a dramatic increase in the use of herbicides in plant production programs. As the world food situation becomes more critical, experts agree that the most rapid rates of expansion in pesticide production will be in the field of herbicides. Because of the continuing environmental interest and the current and projected production figures for herbicide production, the need for a more comprehensive understanding of their environmental fate and behavior is readily apparent.

The subject matter circumscribed by Degradation of Herbicides, published in 1969, has expanded tremendously in detail. Although Degradation of Herbicides has filled a need, the subject has expanded sufficiently to warrant revision and the inclusion of topics omitted almost entirely from the first edition. Thus, Herbicides: Chemistry, Degradation, and Mode of Action is an updated and expanded revision of our original book. The field of herbicide chemistry has expanded so rapidly since the previous edition that the new book appears in two volumes. These volumes represent the most comprehensive coverage available on the organic herbicides.

The first volume is comprised of nine chapters on important classes of herbicides: the phenoxyalkanoic acids, s-triazines, phenylureas, substituted uracils, thiocarbamates, chloroacetamides, amitrole, chlorinated aliphatic acids, and dinitroaniline. The second volume continues to examine individual classes of herbicides, i. e. , diquat and paraquat, benzoic acids, carbamates, phenols, diphenyl ethers, organic arsenicals, picloram herbicides, and then considers two important processes affecting herbicides, photodecomposition and volatilization. The scope of each chapter varies slightly but covers important topics of physical and chemical properties, synthesis, formulation, degradation in plants, soils, and animals, mode of action, and selectivity. The author and subject indexes to both volumes appear at the end of Volume 2.

The new edition should serve as an excellent resource for scientists currently engaged in various phases of pesticide research. It is also a valuable resource to the student for introducing the concepts and literature pertinent to the field. The book is subdivided into various sections which should enhance its continued use as a teaching text for the beginning or more advanced student. For regulatory or government agencies involved in policy-making decisions regarding pesticides, the book serves an important need in providing background information on the most important classes of herbicides, as well as important processes affecting environmental fate and behavior of all chemicals.

Just a word about the use of the term "degradation." All too often, alterations of organic substances in biological systems have been erroneously ascribed to "metabolism" in the sense of enzyme-catalyzed reactions. Numerous examples of pesticide transformations are now documented in which purely chemical systems are involved. We have attempted to avoid this pitfall and use the more generalized term "degradation" to cover all transformations of organic herbicides without particularly trying to ascribe these to enzyme or particulate systems except where deemed appropriate.

Beltsville, Maryland   Philip C. Kearney
                                                                 Donald D. Kaufman

CONTRIBUTORS TO VOLUME 1

MASON C. CARTER, Department of Forestry, Auburn University, Auburn, Alabama[*]

GERARD DUPUIS, Agrochemicals Division, Ciba-Geigy Ltd., Basel, Switzerland

EDITH EBERT, Agrochemicals Division, Ciba-Geigy Ltd., Basel, Switzerland

HERBERT O. ESSER, Agrochemicals Division, Ciba-Geigy Ltd., Basel, Switzerland

S. C. FANG, Department of Agricultural Chemistry, Oregon State University, Corvallis, Oregon

CHESTER L. FOY, Department of Plant Pathology and Physiology, Virginia Polytechnic Institute and State University, Blacksburg, Virginia

JOHN A. GARDINER, Biochemicals Department, Research Division, E. I. du Pont de Nemours & Co., Wilmington, Delaware

HANS GEISSBÜHLER, Agrochemicals Division, Ciba-Geigy Ltd., Basel, Switzerland

TOMASZ GOLAB, Lilly Research Laboratories, Division of Eli Lilly and Company, Indianapolis, Indiana

ERNEST G. JAWORSKI, Monsanto Agricultural Products Co., Monsanto Company, St. Louis, Missouri

MICHAEL A. LOOS, Department of Plant Pathology and Microbiology, University of Natal, Pietermaritzburg, South Africa[†]

GINO J. MARCO, Agricultural Division, Ciba-Geigy Corporation, Greensboro, North Carolina

---

[*]Present address: Department of Forestry and Natural Resources, Purdue University, West Lafayette, Indiana.

[†]Present address: Department of Microbiology and Virology, University of Stellenbosch, Stellenbosch, South Africa.

HENRY MARTIN, Agrochemicals Division, Ciba-Geigy Ltd., Basel, Switzerland

GERALD W. PROBST, Lilly Research Laboratories, Division of Eli Lilly and Company, Indianapolis, Indiana

CHRISTIAN VOGEL, Agrochemicals Division, Ciba-Geigy Ltd., Basel, Switzerland

GÜNTHER VOSS, Agrochemicals Division, Ciba-Geigy Ltd., Basel, Switzerland

WILLIAM L. WRIGHT, Lilly Research Laboratories, Division of Eli Lilly and Company, Indianapolis, Indiana

CONTENTS OF VOLUME 1

CONTENTS OF VOLUME 2

Chapter 1

PHENOXYALKANOIC ACIDS

MICHAEL A. LOOS

Department of Plant Pathology and Microbiology
University of Natal
Pietermaritzburg, South Africa[*]

[*]Present address: Department of Microbiology and Virology, University of Stellenbosch, Stellenbosch, South Africa.

1

I. INTRODUCTION: DEVELOPMENT AND CHEMISTRY OF PHENOXY HERBICIDES

A. Development

The chlorine-substituted phenoxyacetic acids, 2,4-D, MCPA, and 2,4,5-T (Table 1), were introduced as selective weedkillers at the end of World War II, following the publication of secret wartime research on their growth-regulating and herbicidal activities [1, 2]. They have high activity against many broadleaved weeds but not against graminaceous species; thus, 2,4-D and MCPA are commonly used for the control of weeds in cereal crops, grass pastures, and lawns, as well as in many other situations [3-6]. Many woody broadleaved plants are effectively controlled with 2,4,5-T [3-6]. In 1969 U.S. production of 2,4-D, the most widely used phenoxy herbicide, was 47,077,000 lb, and that of 2,4,5-T was 4,999,000 lb [7]. During the five preceding years, and especially from 1966 to 1968 when these two herbicides were used as defoliants in Vietnam [8-11], production was considerably higher [7].

The 2-phenoxypropionic acid herbicides were introduced mainly to control certain weed species that are not controlled by the phenoxyacetic acid herbicides [4-6]. The 4-phenoxybutyric acid herbicides, on the other hand, are useful because of their low or negligible toxicity to certain crop plants, including legume species, that are damaged by low concentrations of the phenoxyacetic acids [3-6]. The selectivity of these 4-phenoxybutyric acid herbicides is related in part to the ability of plants to β-oxidize them to the corresponding phenoxyacetic acids.

Related to the phenoxyalkanoic acids are several phenoxyethyl ester herbicides (Table 2). Sesone was developed in the early 1950s as a soil-applied herbicide to prevent germination of weed seeds and growth of weed seedlings in horticultural crops. It is active only when applied to soil, where it is converted to 2,4-dichlorophenoxyethanol and 2,4-D. It is suitable for use on deep-rooted crops where there is little chance of the crop roots absorbing 2,4-D from the surface layers of the treated soil [3-5, 12]. More recently introduced phenoxyethyl herbicides are MCPES, 2,4,5-TES, 2,4-DEP, 2,4-DEB, and erbon [4-6]. Erbon, which has the herbicidal properties of the 2,4,5-trichlorophenoxy compounds and dalapon, was developed as a nonselective, persistent soil-applied herbicide for use in noncrop areas.

TABLE 1

Phenoxyalkanoic Acid Herbicides: Structural Formulas, Names, and Properties[a]

Basic structure

CH₂—COOH → O with ring substituents X, Y, Z

$CH_2\text{---}COOH$

Common names / Chemical name / Ring substituents	Mol wt / SA per molecule[b] (Å²)	Melting point (°C)	Water ppm (°C)	Organic (g/100 ml)	pK (°C)	Acute oral LD₅₀ (mg/kg)
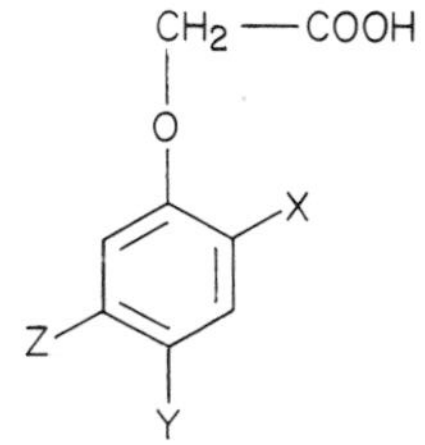 MCPA [(4-Chloro-o-tolyl)-oxy]acetic acid X = CH₃, Y = Cl, Z = H	200.6 ――― 56.4	118-119	640 825 (25) 1174 (25) 1600 (25) 550 (20) 1300 (17)	n-Heptane, 0.5 Xylene, 4.9 Toluene, 6.2 Ether, 77 Ethanol, 153	2.90 3.40	700
2,4-D (2,4-Dichlorophenoxy)-acetic acid X, Y = Cl, Z = H	221.0 ――― 56.1	140.5 139-139.5	400 500 550 620 (25) 725 (25) 900 (25) 50 (20) 530 (17)	CCl₄, 0.1 Ether, 27 Acetone, 85 Ethanol, 130	2.64 2.80 3.22 (60) 3.31	375 500 300-1000
2,4,5-T (2,4,5-Trichloro-phenoxy)acetic acid X, Y, Z = Cl	255.5 ――― 60.5	153.5-155 154-157 158	200 251 (25) 268 (25) 280 (25) 238 (20) 238 (30)	Xylene, 0.61 Toluene, 0.73 Ether, 23.4 Ethanol, 54.8	3.14 3.46 (60)	300 375 500

(continued)

TABLE 1 (continued)

Basic structure	Common names Chemical name Ring substituents	Mol wt SA per molecule [b] ($Å^2$)	Melting point ($°C$)	Solubility		pK ($°C$)	Acute oral LD_{50} (mg/kg)
				Water ppm ($°C$)	Organic (g/100 ml)		
$CH_3—CH—COOH$	MCPP Mecoprop 2-[(4-Chloro-o-tolyl)- oxy]propionic acid X = CH_3, Y = Cl, Z = H	$\dfrac{214.6}{55.2}$	94–95	600 (20) 620 (20) 895 (25)		3.20 3.38 (60) 3.75	700 1500
	2,4-DP Dichlorprop 2-(2,4-Dichlorophenoxy)- propionic acid X, Y = Cl, Z = H	$\dfrac{235.1}{54.6}$	117.5–118.1 118–119	180 350 (20) 710	Kerosene, 0.21 Xylene, 5.1 Toluene, 6.9 Benzene, 8.5 Isopro- panol, 51.0 Acetone, 59.5	3.0 3.28 (60)	
	Silvex 2-(2,4,5-Trichloro- phenoxy)propionic acid X, Y, Z = Cl	$\dfrac{269.5}{59.1}$	179–181	150 140 (25)	CCl_4, 0.095 Benzene, 0.47 Ether, 9.76 Methanol, 13.4 Acetone, 18.0	3.10 (60) 4.41	

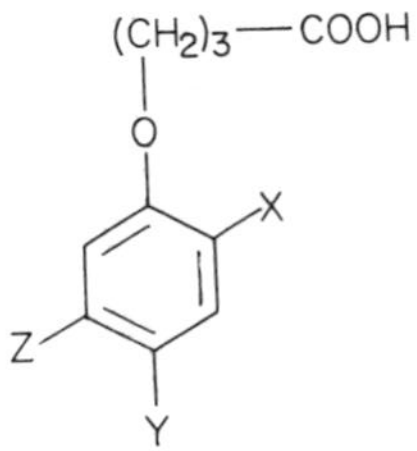

(CH2)3—COOH

(structure: phenoxy ring with substituents X, Y, Z and O-linkage)

	MCPB 4-[(4-Chloro-o-tolyl)-oxy]butyric acid X = CH₃, Y = Cl, Z = H	2,4-DB 4-(2,4-Dichloro-phenoxy)butyric acid X, Y = Cl, Z = H	2,4,5-TB 4-(2,4,5-Trichloro-phenoxy)butyric acid X, Y, Z = Cl
	$\dfrac{228.7}{65.8}$	$\dfrac{249.1}{65.4}$	$\dfrac{283.6}{69.7}$
	99–100 102–103	117–118 119–120.5 121–122	114–115
	44 44–48	53 46 (25)	42
	Ether, sol Ethanol, 15 Acetone, 20	Ethanol, sol Acetone, 10	
	4.86 (60) 6.21	4.58 (60) 5.95	4.78
	1500 375– 1200	700 1500 300– 1000	

---

[a] Data from Refs. 6 and 24–27. Different values for the same property are results from different sources.

[b] SA, surface area.

TABLE 2

Phenoxyethyl Ester Herbicides: Structural Formulas, Names, and Properties[a]

Phenoxyethyl moiety	Acid moiety	Common name / Chemical name	Mol wt / Melting point (°C)	Solubility (% by weight)	Acute oral LD_{50} (mg/kg)
CH_2-CH_2- (2-chloro-4-methylphenoxy ring)	$-O-SO_2-O-Na$	MCPES 2-[(4-Chloro-o-tolyl)-oxy]ethyl sodium sulfate	288.7		
CH_2-CH_2- (2,4-dichlorophenoxy ring)	$-O-SO_2-O-Na$	Sesone 2-(2,4-Dichloro-phenoxy)ethyl sodium sulfate	309.1 170	Benzene, 0.05 Acetone, 0.64 Heptane, 9.4 Water, 26.5 (25°C)	730[b] 1400[b]
(2,4-dichlorophenoxy ring)	P	2,4-DEP Tris[2-(2,4-dichloro-phenoxy)ethyl]-phosphite	649.1	Water, insol Kerosene ⎱ Diesel oil ⎰ 1 Xylene ⎱ Aromatic ⎰ miscible naphthas	850

Structure	Name	Value	Solubility	
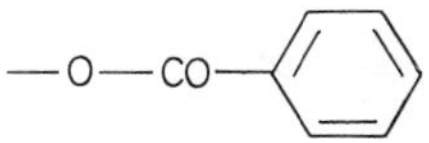 —O—CO— (phenyl)	2,4–DEB 2-(2,4-Dichloro-phenoxy)ethyl benzoate	$\dfrac{311.2}{74}$	Water, 0.0048	1700
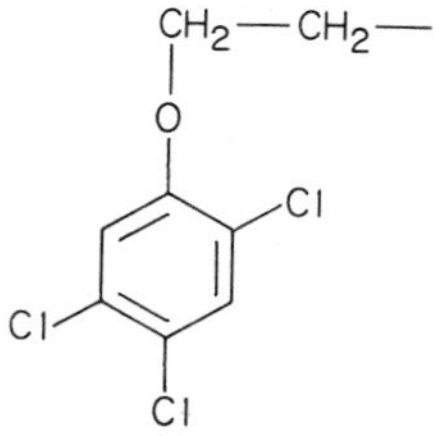 —O—SO₂—O—Na ($-O-SO_2-O-Na$)	2,4,5–TES 2-(2,4,5-Trichloro-phenoxy)ethyl sodium sulfate	343.6	Water, 6	720
—O—CO—C(Cl)(Cl)—CH₃	Erbon 2-(2,4,5-Trichloro-phenoxy)ethyl 2,2-dichloro-propionate	$\dfrac{366.5}{49-50}$	Water, insol Xylene Kerosene } sol Ethanol Acetone	1120

(Structures at left: benzene ring with —CH₂—CH₂— linkage to a 2,4,5-trichlorophenoxy group; —O—CO—Ċ(Cl)(Cl)—CH₃ with Cl substituents)

The phenoxy herbicides, which are readily degraded in the environment, were for many years not considered an environmental or public
health hazard. However, since 1969 the U.S. government has curtailed
the use of 2,4,5-T on food crops [13, 14], following reports of teratogenic
(fetus-deforming) effects of this herbicide and the isooctyl ester of 2,4-D
in laboratory experiments with rats and mice [13, 15]. The 2,4,5-T used
in these experiments was contaminated with approximately 30 ppm of
2,3,7,8-tetrachlorodibenzo-p-dioxin (TCDD) [15], a highly toxic, embryotoxic, and teratogenic compound [16, 17] produced in the manufacture of
2,4,5-T during the high-temperature hydrolytic conversion of 1,2,4,5-
tetrachlorobenzene to 2,4,5-trichlorophenol [18, 19]:

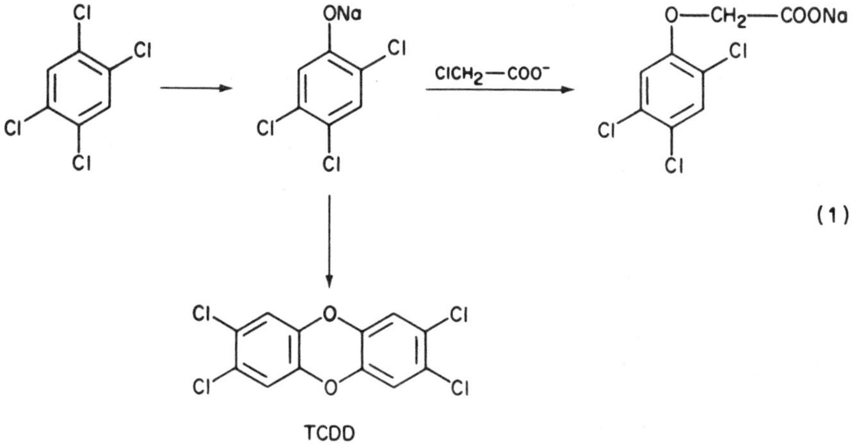

(1)

TCDD

The observed teratogenicity of the 2,4,5-T may, at least in part, have
resulted from its high dioxin content [17]. The dioxin is slowly degraded
by photodecomposition in aqueous suspension and on wet or dry soil [18].
Its accumulation in plants growing in treated soils is unlikely [20]. However, residues may persist for some time on treated leaves [20]. The
dioxin problem seems to have been overcome by the use of 2,4,5-T-manu-
facturing processes that hold the dioxin content below 0.5 ppm [18, 19],
but the 2,4,5-T itself may be teratogenic at high dosage levels. Thus, Roll
[21] has reported that 2,4,5-T with a dioxin content of less than 0.1 ppm
produced an increase in cleft palate, in addition to embryotoxic effects,
when administered orally to mice (Mus musculus) from days 6-15 of pregnancy at doses of 35-130 mg/kg body weight. A dose of 20 mg/kg/day was
established as the "teratogenic no-effect level." However, fetuses of rats
(Rattus norvegicus) given oral doses of up to 24 mg 2,4,5-T (containing
0.5 ppm dioxin)/kg on days 6-15 of pregnancy, and of rabbits (Oryctolagus
cuniculus) given doses of up to 40 mg 2,4,5-T/kg on days 6-18 of pregnancy
showed no evidence of teratogenicity or embryotoxicity [22]. The spraying

of gamebird eggs with 2,4-D at rates not exceeding normal field applications increased the mortality and physical abnormalities of the chick embryos in one study [23], but these effects were not confirmed in subsequent investigations [618-621]. However, the possibility of hazards has been causing public concern, and in the United States the registration of herbicide uses has been transferred to the Pesticide Regulation Division of the Environmental Protection Agency (EPA). The potential hazards of phenoxy herbicides to public health and the quality of the environment have therefore become important considerations to be weighed carefully against the undoubted great benefits of these compounds to agriculture and other human activities.

B.   Physical and Chemical Properties

The chemical and common names, structures, and properties of phenoxy herbicides are given in Tables 1 and 2. Review references other than those indicated below the table are Freed [28, 29] and the Encyclopedia of Chemical Technology [30].

The solubilities of the phenoxyalkanoic acids in water are low, as might be expected of compounds with a considerable lipophilic component in the molecule [31]. Heats of solution (ΔH) for MCPA, 2,4-D, and 2,4,5-T are 8.2, 6.1, and 8.9 kcal/mole, respectively [29]. Surface activity, which increased with increasing chlorine substitution, was demonstrated with 2- and 4-chlorophenoxyacetic acid, 2,4-D, and 2,4,5-T [31].

The phenoxyacetic and phenoxypropionic acid herbicides are stronger acids than the parent acetic and propionic acids, which have pK (25°C) values of 4.75 and 4.87, respectively [32]. This is apparently not the case with the phenoxybutyric acids, as the parent n-butyric acid has a pK (20°C) of 4.81 [32].

The phenoxyalkanoic acids absorb strongly in the ultraviolet region of the spectrum, showing several absorption maxima [28]. Maxima in the region 270-290 nm (aqueous solution) have proved useful in spectrophotometric analysis [33], and several quantitative analytical procedures have been developed [34-38]. Analytical wavelengths for the major herbicidal phenoxy acids in aqueous solution are [28, 33] as follows: MCPA, MCPP, and MCPB, 279 nm; 2,4-D, 2,4-DP, and 2,4-DB, 283 nm; and 2,4,5-T, silvex, and 2,4,5-TB, 288-289 nm. The molar extinction coefficients for MCPA, 2,4-D, and 2,4,5-T at these wavelengths are 1500, 1900, and 1200, respectively [28].

Infrared spectra are extremely valuable for qualitative identification of isolated phenoxy acids. Infrared spectrophotometry has been used for the quantitative determination of phenoxybutyric acid herbicides, but recoveries were poor, necessitating correction by an isotope-dilution procedure

[39]. Infrared absorption bands of possible value for the analysis of phenoxy-acetic acids have been suggested [28], but quantitative infrared analysis is now of little interest following the introduction of gas-chromatographic methods for the phenoxy herbicides.

The 2-phenoxypropionic acids have an asymmetric carbon in the mole-cule and may be resolved into (+) and (-) optical isomers. The (+) isomers, which are responsible for the herbicidal or auxin activity of the compounds (see Sec. III, B), have the D absolute configuration as in D-(+)-glyceralde-hyde and D-(-)-lactic acid [40]:

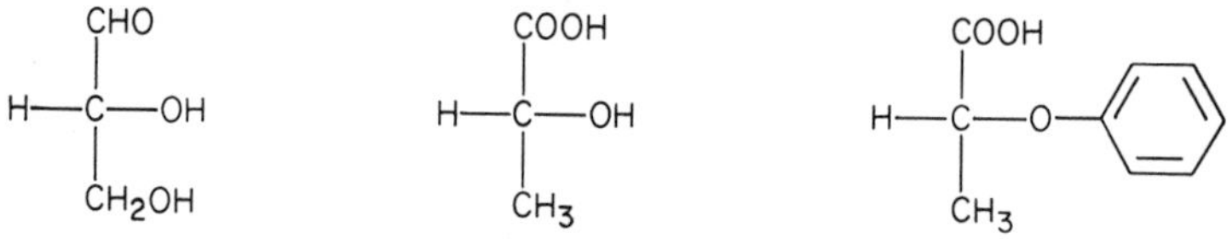

CHO
|
H—C—OH
|
CH_2OH

D-(+)-
Glyceraldehyde

COOH
|
H—C—OH
|
CH_3

D-(-)-
Lactic acid

COOH
|
H—C—O—⬡
|
CH_3

D-(+)-2-Phenoxy-
propionic acid

C.   Synthesis

The usual procedure for the synthesis of the phenoxyacetic and 2-phe-noxypropionic acid herbicides involves reaction of the appropriate phenol with chloroacetic or 2-chloropropionic acid in alkaline aqueous medium [6, 26, 30]:

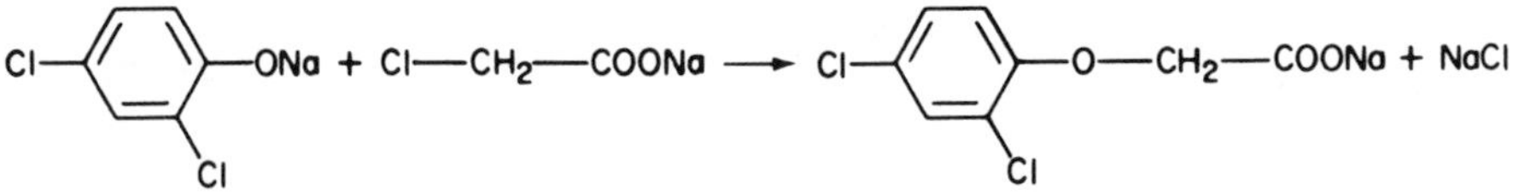

Cl—⬡—ONa + Cl—CH_2—COONa → Cl—⬡—O—CH_2—COONa + NaCl
                                                                                                        Cl

(2)

Cl—⬡—ONa + Cl—CH—COONa → Cl—⬡—O—CH—COONa + NaCl
                              | |
                              CH_3 CH_3

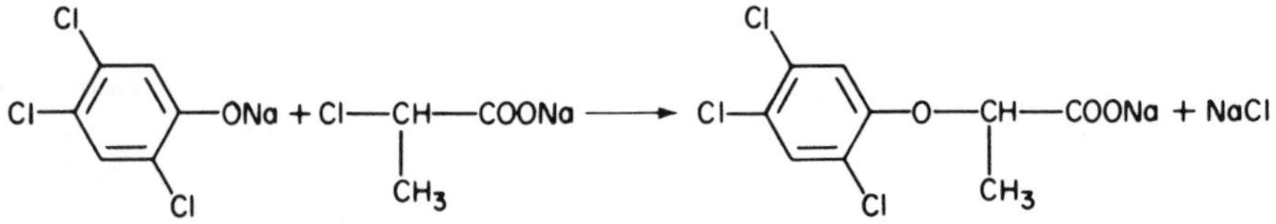

(3)

A pH of 10-12 and a temperature of about 105°C are favorable reaction conditions [26].

The phenoxybutyric acids are synthesized using γ-butyrolactone in place of the chlorinated acid [6, 26, 30]:

$$Cl-\underset{Cl}{\underset{|}{C_6H_3}}-ONa + \underset{O\quad O}{\text{(butyrolactone)}} \longrightarrow Cl-\underset{Cl}{\underset{|}{C_6H_3}}-O-CH_2-CH_2-CH_2-COONa \qquad (4)$$

Other procedures may be used for the synthesis of these compounds, for example, chlorination of phenoxyacetic acid or its esters, or 2-methyl-phenoxyacetic acid [6, 26, 41, 42]:

$$C_6H_5-O-CH_2COOH + 2Cl_2 \longrightarrow Cl-\underset{Cl}{\underset{|}{C_6H_3}}-O-CH_2COOH + 2HCl \qquad (5)$$

$$\underset{CH_3}{\underset{|}{C_6H_4}}-O-CH_2COOH + Cl_2 \longrightarrow Cl-\underset{CH_3}{\underset{|}{C_6H_3}}-O-CH_2COOH + HCl \qquad (6)$$

For further information on the above and other processes see Melnikov [26].

Sesone and the other phenoxyethyl sodium sulfates are produced from the reaction of the appropriate phenoxyethanol with chlorosulfonic acid, and subsequent salt formation [6,43]:

$$Cl-\underset{Cl}{\underset{|}{C_6H_3}}-O-CH_2-CH_2-OH \xrightarrow{+\ ClSO_3H}$$

$$Cl-\underset{Cl}{\underset{|}{C_6H_3}}-O-CH_2-CH_2-O-SO_3H \xrightarrow{+NaOH}$$
$$+ HCl$$

$$Cl-\underset{Cl}{\underset{|}{C_6H_3}}-O-CH_2-CH_2-O-SO_3Na \qquad (7)$$

The synthesis of 2,4-DEP involves the reaction of 2-(2,4-dichloro-phenoxy)ethanol with PCl_3 in benzene in the presence of pyridine [44]:

$$(8)$$

D.   Chemical Reactions

The phenoxyalkanoic acids undergo reactions of alkyl carboxylic acids, aromatic compounds, and ethers. Reactions that are important for analysis of the phenoxy herbicides and the preparation of formulations are considered here. Other reactions are mentioned in the article by Melnikov [26].

Salts are formed with metal cations, ammonia, and organic amines. The low water solubility of the parent phenoxy acids is greatly increased by salt formation. Salt formation is important in the production of water-soluble formulations [3, 4]. Solubility data for metal salts of phenoxy herbicides have been reviewed [25, 26]; however, the figures for 2,4-D, where a comparison is possible, show disturbing discrepancies, suggesting the need for a comprehensive, confirmatory study. Nevertheless, the data do indicate that the magnesium and calcium salts of MCPA and 2,4-D are considerably less water soluble than the sodium and potassium salts.

Ester formation is important in the manufacture of many phenoxyalkanoic acid herbicide formulations. It is also important in the analysis of phenoxy acids by gas chromatography, which requires the production of volatile derivatives. For the latter purpose, methyl esters are produced rapidly and essentially quantitatively using as agents diazomethane [45-49], boron trifluoride-methanol [48, 50-52], or dimethyl sulfate [53]. The acid-catalyzed reaction of acid with alcohol, although relatively slow [53], was used for the formation of butoxyethanol esters of 2,4-D and 2,4,5-T for the determination of these herbicides in citrus [54]. Esterification of phenoxy acids with 2-chloroethanol has been used to increase sensitivity and overcome interference in gas-chromatographic analyses of herbicides extracted from soil [55, 56]. Esters of 2,4-D and 2,4,5-T in commercial formulations were transesterified to the methyl ester by boron trichloride-methanol [57].

Reaction with thionyl chloride converts phenoxyacetic acids to phenoxy-acetyl chlorides [58-61]. These were subsequently reacted with amino acids in a Schotten-Baumann reaction, which forms amide linkages between the phenoxyacetyl and amino acid moieties:

$$\text{Cl—C}_6\text{H}_3(\text{Cl})\text{—O—CH}_2\text{—COOH} + \text{SOCl}_2 \longrightarrow \text{Cl—C}_6\text{H}_3(\text{Cl})\text{—O—CH}_2\text{—CO·Cl}$$

$$+ \text{HCl} + \text{SO}_2$$

$$\text{Cl—C}_6\text{H}_3(\text{Cl})\text{—O—CH}_2\text{—CO·Cl} + \text{H}_2\text{N—CH}\overset{\text{COONa}}{\underset{\text{R}}{|}} \xrightarrow{+\text{NaOH}}$$

$$\text{Cl—C}_6\text{H}_3(\text{Cl})\text{—O—CH}_2\text{—CO—NH—CH}\overset{\text{COONa}}{\underset{\text{R}}{|}} \quad + \text{NaCl} + \text{H}_2\text{O} \qquad (9)$$

Similar amino acid derivatives of dichlorprop and silvex were prepared in this way [62, 63].

Degradation of the side chain of phenoxyacetic acids when heated in concentrated sulfuric acid at 150°C for 2 min is the basis of several colori-metric procedures for these compounds. The reaction, which yields form-aldehyde as one of the products, is believed to proceed as follows [64]:

$$\text{C}_6\text{H}_5\text{—O—CH}_2\text{—COOH} + \text{H}_2\text{O} \longrightarrow \text{C}_6\text{H}_5\text{—OH} + \text{CH}_2(\text{OH})\text{—COOH}$$

$$\text{CH}_2(\text{OH})\text{—COOH} \longrightarrow \text{CO} + \text{H}_2\text{O} + \text{HCHO} \qquad (10)$$

The liberated formaldehyde, and hence the parent phenoxyacetic acid, is determined by its color reaction with chromotropic acid (1,8-dihydroxy-naphthalene-3,6-disulfonic acid) [64-68], J-acid (6-amino-1-naphthol-3-sulfonic acid), or phenyl-J-acid (6-anilino-1-naphthol-3-sulfonic acid) [69,70].

The ether bond of phenoxyacetic acids is readily and quantitatively cleaved by concentrated hydrogen iodide or pyridine hydrochloride to yield

the corresponding phenol [26, 28, 71-73]. Pyridine hydrochloride cleavage
has been used in quantitative analytical methods for phenoxyacetic acid
herbicides, the phenolic product being determined colorimetrically [71] or
by gas chromatography [72, 73].

Nitration of the aromatic nucleus of 2,4-D yields mainly 2,4-dichloro-
5-nitrophenoxyacetic acid, and traces of 2,4-dichloro-6-nitrophenoxyacetic
acid [26]. Nitration of MCPA and MCPB was used to increase the sensitivity
of electron affinity gas-chromatographic analysis of these compounds; in
each case two products, presumably the 5- and 6-nitro derivatives, were
formed [74]. Nitro derivatives prepared from 2,4-D, MCPP, and 2,4-DB
were detected on thin-layer plates by reducing the nitro group with stannous
chloride in hydrochloric acid, diazotizing with sodium nitrite, and coupling
with N-(1-naphthyl)ethylenediamine dihydrochloride to yield colored derivatives [75].

The detection of phenoxy herbicides on paper and thin-layer chromatograms using silver nitrate as the chromogenic reagent [76-80] is based on
the presence of chlorine in the molecule. Ultraviolet irradiation of the
chromatograms results in the development of dark-colored spots. Color
development presumably involves the release of chloride by photolytic reactions (see Crosby and Li [81]), and the subsequent formation of silver
chloride, which darkens on exposure to light as a result of decomposition
to silver and chlorine [82].

E.   Formulations

Most commercial formulations of phenoxyalkanoic acids contain the
herbicide in the salt or ester form. Among the salts, the amines are particularly important, although the sodium, potassium, and ammonium salts
are also used [5, 6]. Examples of amine salts (the cationic moiety only)
are shown in Table 3; the di- and trisubstituted amines are the important
cations in amine salt formulations of the phenoxy herbicides [3, 4]. The
phenoxyalkanoic acid amine salts, with the exception of those of N-oleyl-
1,3-propylenediamine, are highly soluble in water [25, 26] and are used
in the formulation of aqueous concentrates. Their high water solubility
permits high rates of application using low-gallonage equipment [3]. The
N-oleyl-1,3-propylenediamine salts of 2,4-D and 2,4,5-T were developed
as oil-soluble herbicides [6].

Phenoxyalkanoic acid esters are oil soluble, but they may be applied
as emulsions in water if a suitable emulsifying agent is included in the
formulation [3]. Esters normally exhibit greater herbicidal activity than
the parent acids, because of improved absorption by the target plants (see
Sec. III,A). The esters of low-molecular-weight alcohols are volatile
(Table 4) and may be a hazard to nontarget crops, but this problem was

TABLE 3

Amine Cations of Phenoxyalkanoic Acid Amine Salts

Organic group (R)	Cation structure and name		

	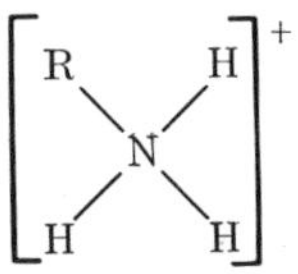 $\begin{bmatrix} R & \diagdown & \diagup & H \\ & N & \\ H & \diagup & \diagdown & H \end{bmatrix}^+$	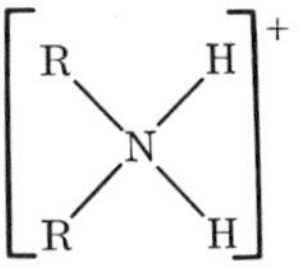 $\begin{bmatrix} R & \diagdown & \diagup & H \\ & N & \\ R & \diagup & \diagdown & H \end{bmatrix}^+$	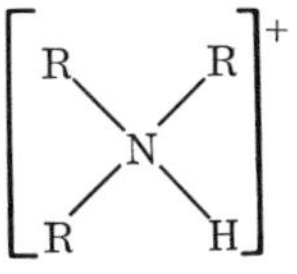 $\begin{bmatrix} R & \diagdown & \diagup & R \\ & N & \\ R & \diagup & \diagdown & H \end{bmatrix}^+$
CH_3-	Methylamine	Dimethylamine	Trimethylamine
$CH_3 \cdot CH_2-$	Ethylamine	Diethylamine	Triethylamine
$HOCH_2 \cdot CH_2-$	Ethanolamine (2-amino-ethanol)	Diethanolamine (2,2'-imino-diethanol)	Triethanolamine (2,2',2''-nitrilo-triethanol)
$CH_3 \cdot CHOH \cdot CH_2-$	Isopropanol-amine (1-amino-2-propanol)	Diisopropanol-amine (1,1'-imino-di-2-propanol)	Triisopropanol-amine (1,1',1''-nitrilo-tri-2-propanol)

$$CH_3(CH_2)_7 CH=CH(CH_2)_8 \overset{+}{N}H_2 (CH_2)_3 \overset{+}{N}H_3$$

N-Oleyl-1,3-propylenediamine

TABLE 4

Vapor Pressures of Low-Molecular-Weight Alcohol Esters
of 2,4-D and 2,4,5-T at 25°C[a]

2,4-D				2,4,5-T	
Branched-chain esters		Straight-chain esters		Straight-chain esters	
Ester	mm Hg	Ester	mm Hg	Ester	mm Hg
		Methyl	23×10^{-4}	Methyl	9.8×10^{-4}
		Ethyl	11×10^{-4}	Ethyl	6.5×10^{-4}
2-Propyl	14×10^{-4}	n-Propyl	6.7×10^{-4}	n-Propyl	2.1×10^{-4}
2-Butyl	5.7×10^{-4}	n-Butyl	3.9×10^{-4}	n-Butyl	Low volatile
2-Pentyl	2.4×10^{-4}	n-Pentyl	3.0×10^{-4}		
2-Hexyl	2.2×10^{-4}	n-Hexyl	Low volatile		
4-Heptyl	Low volatile				

[a]From Jensen and Schall [83, Tables III and IV, and Fig. 1].

overcome by the introduction of "low-volatile" esters [3]. Important low-volatile esters of the phenoxyalkanoic acids are [4, 84, 85] the following:

2-Ethylhexyl
$$CH_3-CH_2-CH_2-CH_2-\underset{\underset{\displaystyle CH_3-CH_2}{|}}{CH}-CH_2-$$

Isooctyl
$$CH_3-\underset{\underset{\displaystyle CH_3}{|}}{CH}-CH_2-\underset{\underset{\displaystyle CH_3}{|}}{CH}-CH_2-CH_2-$$
and other isomers

Butoxyethyl
$$CH_3-CH_2-CH_2-CH_2-O-CH_2-CH_2-$$

Propylene glycol butyl ether
$$CH_3-CH_2-CH_2-CH_2-(O-CH_2-\underset{\underset{\displaystyle CH_3}{|}}{CH}-)_x$$

Tetrahydrofurfuryl
$$\underset{\underset{\displaystyle H_2C}{|}}{H_2C}\underset{\underset{\displaystyle \diagdown_{O}\diagup}{}}{\text{———}}\underset{\underset{\displaystyle CH-}{|}}{CH_2}$$

The low-volatile esters have estimated vapor pressures at 25°C of less than 1.5×10^{-4} mm Hg [83, 84]. Although the n-butyl ester of 2,4,5-T falls in this group, plant tests indicate a volatility hazard [85], probably because of the high toxicity of the compound [83].

II. DEGRADATION

A. Degradation in Plants

1. Phenoxyacetic Acids

The ability of plants to degrade phenoxyacetic acid herbicides has been known since 1950, when Holley et al. [86] and Weintraub et al. [87] reported the metabolism of ^{14}C-labeled 2,4-D by the bean Phaseolus vulgaris L. As indicated in periodic reviews [88-97], information on degradation processes has steadily accumulated.

a. Side-chain degradation. Degradation of the side chain of phenoxy-acetic acid herbicides has been observed in many plants [86, 87, 98-112] but it appears to play a major role in herbicide breakdown in only a few species or varieties. Such plants include the red currant (Ribes sativum

Syme) [99], certain varieties of apple (Malus sylvestris Mill.) [100, 101], the strawberry (Fragaria sp.), and garden lilac (Syringa vulgaris L.) [100], all of which showed high rates of $^{14}CO_2$ release from 2,4-D labeled with ^{14}C in the side chain. Thus, excised leaves or leafy shoots of these plants released 7-33% of the ^{14}C label from [1-^{14}C]2,4-D as $^{14}CO_2$ in 20-24 hr [99-101]. In experiments lasting several days, up to 50% of the [1-^{14}C]2,4-D label and up to 20% of the [2-^{14}C]2,4-D label were released as $^{14}CO_2$ by red currant and strawberry leaves [99, 100]. Red currant, strawberry, and Cox's Orange Pippin apple leaves also decarboxylated [1-^{14}C]2,4,5-T, [1-^{14}C]4-chlorophenoxyacetate, and [1-^{14}C]MCPA at rates comparable to those observed for the decarboxylation of [1-^{14}C]2,4-D, but little or no decarboxylation of [1-^{14}C]2-chlorophenoxyacetate was observed [99, 100].

Leafe [102] showed that side-chain degradation was a major mechanism in the metabolism of MCPA by bedstraw (Galium aparine L.), although only 7% of the radioactivity of [1-^{14}C]MCPA or [2-^{14}C]MCPA supplied to the plants was evolved as $^{14}CO_2$. However, most of the radioactive carbon of the MCPA side chain was cleaved from the aromatic portion of the molecule, some of it being incorporated into cell constituents such as starch, proteins, and nucleic acids. The extent of side-chain degradation was thus greater than the decarboxylation data indicated. Weintraub et al. [113] reported that ^{14}C released from [1-^{14}C]2,4-D and [2-^{14}C]2,4-D in the bean was incorporated into plant acids, sugar, dextrins, starch, pectin, protein, and cell-wall substances. It is not known whether the ^{14}C was released from the MCPA or 2,4-D side chain as $^{14}CO_2$ and incorporated into the cell components by CO_2 fixation, or whether a radioactive metabolite of MCPA or 2,4-D was directly incorporated into the cell material.

Many plants are capable of a slow or limited degradation of the side chain of phenoxyacetic acid herbicides, which is probably not of any great significance in herbicide metabolism. For example, many of the plants studied by Luckwill and Lloyd-Jones [100], including several apple varieties, liberated less than 8% of the ^{14}C from [1-^{14}C]2,4-D as $^{14}CO_2$ in 20 or 21 hr. A daily release of only 1-2% of the [1-^{14}C]2,4-D label as $^{14}CO_2$ was reported for the bean [98, 103, 112], corn (Zea mays L.) [103, 112], and cotton (Gossypium hirsutum L.) [104], while the release of less than 1% per day was observed with Stayman apples [101], sorghum (Sorghum vulgare Pers.) [104], blackjack oak (Quercus marilandica Muenchh.), persimmon (Diospryos virginiana L.), green ash (Fraxinus pennsylvanica Marsh.), sweetgum (Liquidambar styraciflua L.), winged elm (Ulmus alata Michx.) [105], and big leaf maple (Acer macrophyllum Pursh.) [106]. The results of Fang et al. [114] with the bean, indicating release of 17.5% of the radioactivity of [2-^{14}C]2,4-D as $^{14}CO_2$ in 3 days, were criticized by Weintraub et al. [98], who believed that this rate of $^{14}CO_2$ release was much too high. Other plants exhibiting a slow decarboxylation of 2,4-D are the black currant (Ribes nigrum L.) [99], tick bean (Vicia faba L. var. minor) [107], cocklebur (Xanthium sp.), smartweed (Polygonum pennsylvanicum L.), jimsonweed (Datura stramonium L.), and bur cucumber (Sicyos angulatus L.) [108], and the cultivated cucumber (Cucumis sativus L.) [109]. The bur and cultivated cucumbers can also slowly decarboxylate 2,4,5-T [109], as can the bean [103] and big leaf maple [106].

Fluorine-containing phenoxyacetic acids, like their chlorine-containing counterparts, also undergo side-chain degradation in plants. The rate of degradation may be similar to that of the corresponding chlorine-containing acid, as in the case of 4-fluorophenoxyacetic acid and 2,4-difluorophenoxy-acetic acid decarboxylation in the bean [103], or it may be quite different. Thus, in contrast to their rapid decarboxylation of 2,4-D, McIntosh apple shoots decarboxylated 2-chloro-4-fluorophenoxyacetic acid only very slowly [101]. On the other hand, blackjack oak decarboxylated this compound con-siderably faster than 2,4-D, although the amount of radioactivity released as $^{14}CO_2$ from the carboxyl-labeled 2-chloro-4-fluorophenoxyacetic acid amounted in 22 hr to only 7% of the applied dose [105].

Loss of the side chain from a phenoxyacetic acid without further meta-bolic changes to the molecule would yield the corresponding phenol (Fig. 1). Luckwill and Lloyd-Jones [100], investigating 2,4-D metabolism in the strawberry, showed the formation of a phenol which they assumed was 2,4-dichlorophenol. The amount of phenol formed, assuming that it was 2,4-dichlorophenol, agreed closely with the amount of [2-^{14}C]2,4-D that lost its label as $^{14}CO_2$. Moreover, the symptoms that developed in 2,4,5-T-treated strawberries suggested that 2,4,5-trichlorophenol, which would be formed by loss of the side chain from 2,4,5-T, was damaging the plants rather than 2,4,5-T itself. Chkanikov et al. [115], using paper chromatog-raphy, demonstrated the production of 2,4-dichlorophenol from 2,4-D in the bean, sunflower (Helianthus sp.), corn, and barley (Hordeum vulgare L.). Using red raspberry (Rubus sp.) homogenates, they showed that 2,4-dichlorophenol and 2,4,5-trichlorophenol could be metabolized in plants. The n-butyl ester of 2,4,5-T was metabolized to 2,4,5-trichloro-phenol by sweetgum and southern red oak (Quercus falcata Michx.) [116].

Different mechanisms have been suggested for degradation of the side chain of phenoxyacetic acids in plants. Tick beans metabolizing 2,4-D [107] and bedstraw metabolizing MCPA [102] released both side-chain carbons as CO_2 at the same rate, suggesting removal of the 2,4-D or MCPA side chain as a two-carbon unit or release of the methylene carbon immediately after the carboxyl carbon. By contrast, a stepwise degradation of the side chain seemed to occur in the bean [98], red currant [99], strawberry [100], cotton, and sorghum [104], with the carboxyl carbon being released as CO_2 about twice as fast as the methylene carbon. However, even in these plants the phenoxyacetic acid molecule might be degraded by cleavage at the ether linkage with liberation of the side chain as a two-carbon unit. This

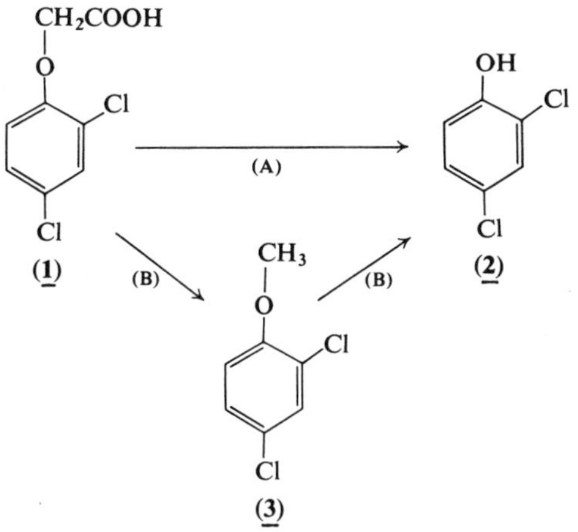

CH₂COOH — rendered as chemical figure

FIG. 1. Conversion of 2,4-D (1) in plants to 2,4-dichlorophenol (2). Hypothetical pathways involving (A) cleavage of the 2,4-D at the ether linkage and (B) the intermediary formation of 2,4-dichloroanisole (3). (Reprinted from Loos [94].)

possibility is suggested by the course of acetate metabolism in certain plants, which liberate the carboxyl carbon as carbon dioxide considerably faster than the methyl carbon [117]. Fleeker [592] obtained evidence that the 2,4-D side chain is liberated in the red currant as glyoxylic acid or a similar two-carbon compound.

Stepwise removal of the phenoxyacetic acid side chain would involve the formation of an intermediate containing only one of the carbons of the original side chain. Luckwill and Lloyd-Jones [99, 100] obtained evidence that such an intermediate was formed in 2,4-D-treated red currant and strawberry leaves but they were unable to extract it from the leaf residue for identification. They suggested that it might be a bound form of 2,4-dichloroanisole, which would be formed by the decarboxylation of 2,4-D (Fig. 1, pathway B), but the existence of this pathway in plants has still to be established. There was no evidence of 2,4,5-trichloroanisole production in the conversion of 2,4,5-T to 2,4,5-trichlorophenol in sweetgum and southern red oak [116].

b. Side-chain lengthening. In experiments of Bach [118], beans metabolized 2,4-D to unidentified compounds containing aliphatic hydroxyl groups, suggesting that the 2,4-D side chain might undergo lengthening in a manner similar to that involved in fatty acid biosynthesis. The gas-chromatographic studies of Linscott and co-workers [119, 120] confirmed

this suggestion. Alfalfa (Medicago sativa L.) lengthened the 2,4-D side
chain by two carbon units to yield 2,4-DB and 6-(2,4-dichlorophenoxy)-
caproic acid [119]. The side-chain lengthening probably occurred in the
epidermal cells, as a result of the activity of enzymes responsible for the
synthesis of surface waxes [121]. Resistant bromegrass (Bromus inermis
Leyss.), timothy (Phleum pratense L.), and orchardgrass (Dactylis glom-
erata L.) lengthened the side chain of 2,4-D by a single carbon to yield
3-(2,4-dichlorophenoxy)propionic acid [120]. However, Steen et al. [593]
could not confirm the one-carbon lengthening of the 2,4-D side chain in
these grasses, nor could they detect it in other monocotyledons and
dicotyledons.

c. Ring hydroxylation. Hydroxylation of the ring of phenoxyacetic
acids, suggested but not proved by Holley [122] in 1952, is now a well-
established metabolic reaction of plants. Fawcett et al. [123], using paper
chromatography, showed that wheat (Triticum sativum L.) and pea (Pisum
sativum L.) tissue hydroxylated unsubstituted phenoxyalkanoic acids, including phenoxyacetic acid, in the 4 position (Fig. 2, pathway A). In addition, the higher phenoxyalkanoic acids were β-oxidized. Similar reactions
were observed in oats (Avena sativa L.), barley, and corn, but not in peanuts (Arachis hypogaea L.), soybeans [Glycine max (L.) Merr.], and alfalfa
[124]. In the oats, barley, and corn, the higher phenoxyalkanoic acids with
an even number of carbons in the side chain appeared to be first β-oxidized
to phenoxyacetic acid, then hydroxylated to yield 4-hydroxyphenoxyacetic
acid. The 4 hydroxylation of phenoxyacetic acid in oats was confirmed by
Thomas et al. [125].

Thomas and co-workers [126] also investigated the hydroxylation of
chlorine-substituted phenoxyacetic acids in oats. If the 4 position of the
ring was unsubstituted, the phenoxyacetic acid was hydroxylated in that
position (Fig. 2, pathway A); the hydroxylated phenoxy acids were then
conjugated to glucose to form the 4-O-β-D-glucosides (Fig. 3, 4). Thus,
2-chlorophenoxyacetic acid was converted to the glucoside of 2-chloro-4-
hydroxyphenoxyacetic acid, and 2,6-dichlorophenoxyacetic acid to the
glucoside of 2,6-dichloro-4-hydroxyphenoxyacetic acid. Chlorine-substituted
phenoxyacetic acids with no unsubstituted 4 position, for example 2,4-D
and 4-chlorophenoxyacetic acid, were not hydroxylated by the oats. The
fate of 2,4,6-trichlorophenoxyacetic acid was exceptional, with hydroxylation
occurring in the 3 position (Fig. 2, pathway B). The 3-hydroxy-2,4,6-
trichlorophenoxyacetic acid then combined with glucose to form the
glucoside.

The identity of the principal 2,4-D metabolite in the bean, which was
believed by Holley [122] and Crosby [127] to be a hydroxyphenoxyacetic acid,
was reinvestigated [112, 128, 129]. The ether-insoluble fraction of alcohol
extracts from 2,4-D-treated beans contained conjugated 2,4-D metabolites,
which on treatment with β-glucosidase or acid hydrolysis, yielded two

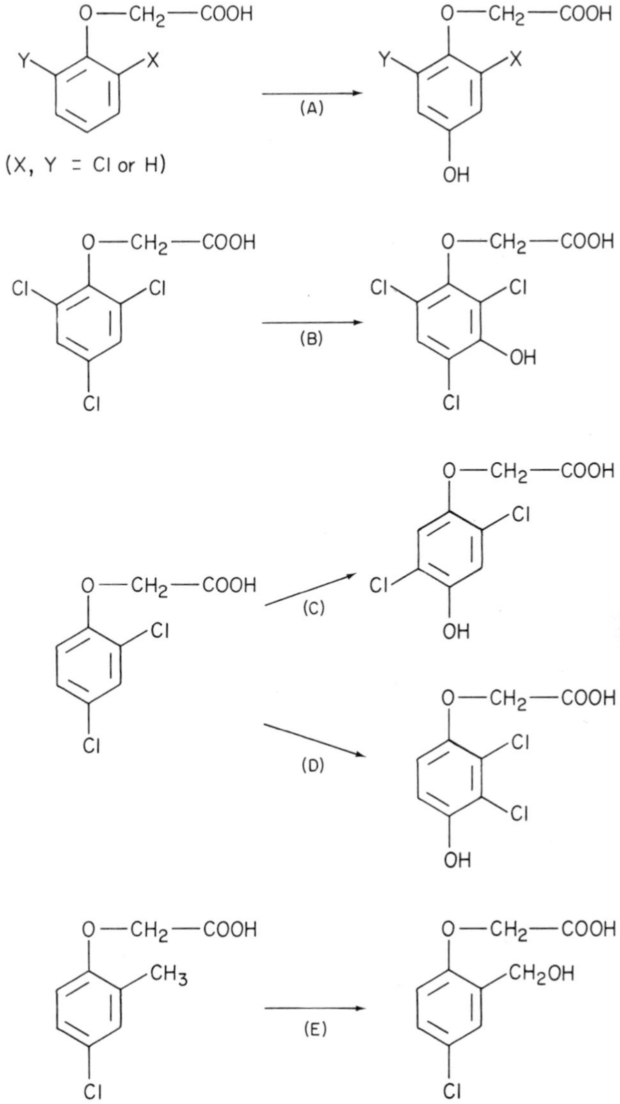

FIG. 2.   Hydroxylation of phenoxyacetic acids by plants. (A) 4 Hydroxylation of phenoxyacetic acid, 2-chloro-, and 2,6-dichlorophenoxyacetic acid;
(B) 3 hydroxylation of 2,4,6-trichlorophenoxyacetic acid; (C) and (D) 4 hydroxylation of 2,4-D with chlorine shift to the 5 or 3 position; (E) hydroxylation of the 2-methyl group of MCPA. (Adapted from Loos [94].)

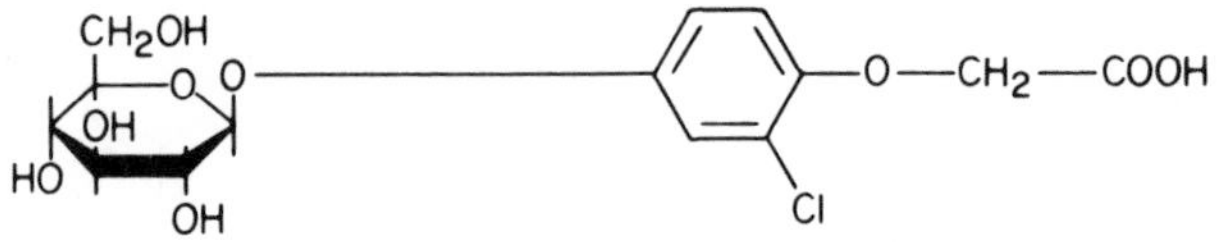

FIG. 3.   Conjugated forms of phenoxyacetic acids detected in plants.
(4) β-D-Glucoside of 4-hydroxy-2-chlorophenoxyacetic acid; (5) β-D-glucoside of 4-chloro-2-hydroxymethylphenoxyacetic acid; (6) β-D-glucose ester
of 2,4-D; and (7) 2,4-dichlorophenoxyacetylaspartic acid. (Adapted from
Loos [94].)

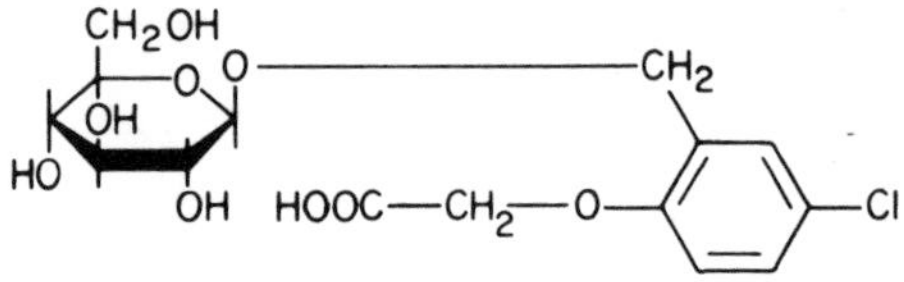

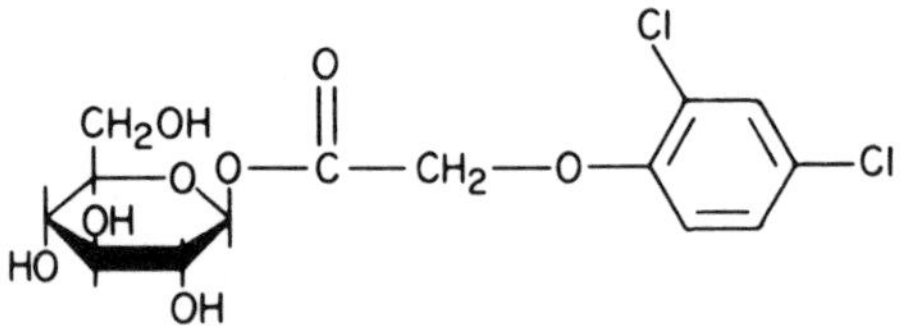

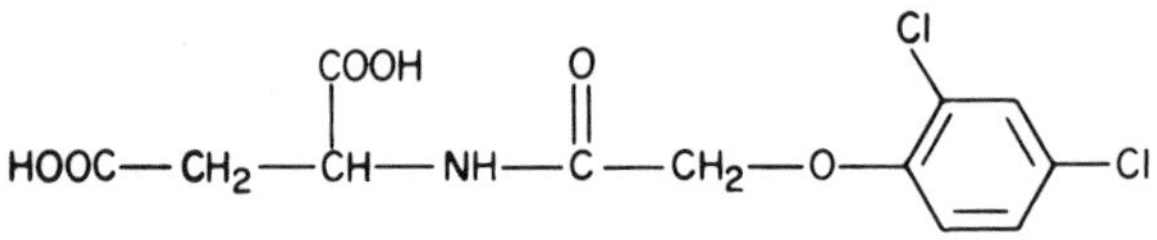

phenolic acids that were separated by thin-layer or paper chromatography.
The major phenolic acid was 2,5-dichloro-4-hydroxyphenoxyacetic acid and
the minor 2,3-dichloro-4-hydroxyphenoxyacetic acid. These two acids were
also present as aglycones in the ether-soluble fraction of the alcohol extracts
[129]. The 4 hydroxylation of 2,4-D in the bean was thus accompanied by a

chlorine shift from the 4 to the 5 or 3 position (Fig. 2, pathways C and D).
These reactions are examples of hydroxylation-induced intramolecular
migration, or the "NIH shift," which is now recognized as a general mechanism of aromatic metabolism [130]. Hydroxylation of 2,4-D involving the
NIH shift also occurred in soybean [129, 131, 594, 595], wheat, barley
[129], oats [129, in contrast to 126], corn [112, in contrast to 129], blue-
grass (Poa pratensis L.) [112], wild oat (Avena fatua L.), wild buckwheat
(Polygonum convolvulus L.), leafy spurge (Euphorbia esula L.), and yellow
foxtail [Setaria glauca (L.) Beauv.] [132], but not in buckwheat (Fagopyrum
esculentum Moench) [129]. It occurred to a very small extent in wild mustard [Brassica kaber (DC.) L. C. Wheeler var. pinnatifida (Stokes) L. C.
Wheeler], perennial sowthistle (Sonchus arvensis L.), and kochia [Kochia
scoparia (L.) Roth] [132].

Small amounts of 2-chloro-4-hydroxyphenoxyacetic acid were produced
from 2,4-D by wild buckwheat, wild oat, leafy spurge, and yellow foxtail
[132]. Beans produced small quantities of 2,5-dichloro-4-hydroxyphenoxy-
acetic acid from 2,4,5-T [129]. These conversions are examples of 4
hydroxylation linked to chlorine elimination as an alternative to the chlorine
shift.

d. Hydroxylation of MCPA methyl group. Hydroxylation of the 2-
methyl group of MCPA to yield 4-chloro-2-hydroxymethylphenoxyacetic
acid (Fig. 2, pathway E) was demonstrated in peas [133], rape (Brassica
napus L. var. arvensis), and red campion (Melandrium rubrum) [134].
This metabolite formed a glycoside, tentatively identified as 4-chloro-2-
(β-D-glucopyranosidomethyl)phenoxyacetic acid (Fig. 3, 5), in all three
species [134].

e. Conjugation with plant constituents. The formation of glucosides
of hydroxyphenoxyacetic and hydroxymethylphenoxyacetic acids, discussed
in the previous sections, constitutes some examples of conjugation with
plant constituents. Phenoxyacetic acids with no hydroxy group may undergo
esterification with glucose. Thus, Klämbt [135] and Ojima and Gamborg
[136] reported the formation of the glucose ester of 2,4-D in wheat and soybean cell suspension cultures, respectively, while Thomas et al. [126]
reported that oats converted 2,4-D, 4-chloro-, and 2,6-dichlorophenoxy-
acetic acid to their β-D-glucose esters (Fig. 3, 6). Rape converted MCPA
to a β-sugar ester, tentatively identified as 4-chloro-2-methylphenoxy-
acetyl-β-D-glucose [134].

Aspartic acid combined with 2,4-D to form 2,4-dichlorophenoxyacetyl-
aspartic acid (Fig. 3, 7) in wheat [135], peas [137], red and black currants
[99], and probably also in bur and cultivated cucumbers [109], according
to paper-chromatographic evidence. Similarly, MCPA was converted to
N-(4-chloro-2-methylphenoxyacetyl)-L-aspartic acid in peas [133], rape,

and red campion [134]. Soybean cotyledon callus tissue metabolized 2,4-D
to the N^α-(2,4-dichlorophenoxyacetyl) derivatives of glutamic and aspartic
acids, alanine, valine, leucine, phenylalanine, and tryptophan [131, 594,
595].

The identity of the conjugates discovered by Butts and his colleagues
[138, 139], which were designated "unknowns 1 and 3," is uncertain.
Unknown 1 is a major metabolite of 2,4-D in beans [138, 140] and peas
during the early stages of 2,4-D metabolism [141]. Unknown 3 is the major
2,4-D metabolite in corn and wheat [139], as well as in tomatoes (Lycopersicon esculentum Mill.) and peas during the later stages of 2,4-D metabolism [141]. Unknowns 1 and 3 were distinguished from 2,4-D and from
each other by their R_F values on paper chromatograms. Both were hydrolyzed by acid or emulsin and unknown 1 was hydrolyzed by takadiastase to
yield 2,4-D [138, 139]. These observations suggested that the two conjugates were glycosides of 2,4-D [138, 139]. However, Butts and Fang [142]
subsequently suggested that the plant component of the complexes was protein. From beans treated with 2,4-D or 2,4,6-trichlorophenoxyacetic acid
they isolated 2,4-D- and 2,4,6-trichlorophenoxyacetic acid-"protein" complexes which on hydrolysis were found to contain at least 12 amino acids.
The relative amounts of the amino acids from each complex were rather
similar. When the 2,4-D-protein complex was injected into the stems of
beans, 2,4-D was decarboxylated about three times as fast as when free
2,4-D was injected, suggesting that the complex was a product of a detoxication process and perhaps an intermediate in the metabolism of 2,4-D
[142].

Work on 2,4-D degradation in the bean by Bach and Fellig [118, 143-
145] did not result in the positive identification of any 2,4-D metabolites,
but did cast suspicion on the formation of 2,4-D-protein complexes. The
2,4-D metabolites detected in these studies must have included 2,3- and
2,5-dichloro-4-hydroxyphenoxyacetic acid and their 4-O-β-D-glucosides,
and apparently also metabolites with lengthened side chains [118]. Bach
[118] did not believe that the ether-insoluble, radioactive products of
labeled 2,4-D metabolism were polypeptide or protein conjugates of 2,4-D
as proposed by Butts and Fang [142], as their behavior during ion-exchange
and paper chromatography and extraction into organic solvents could not be
reconciled with a polypeptide structure. He proposed that these 2,4-D
metabolites yielded amino acids on hydrolysis because of contamination
with relatively large amounts of "bound" amino acids, for example amino
acid amides of miscellaneous plant metabolites.

The work of Bach [118] also raises doubts as to whether the unknown 1
complex in beans contained 2,4-D, as reported by Jaworski and Butts [138],
or a metabolite of 2,4-D that is not readily distinguishable from the parent
acid by paper chromatography. However, in cotton there is evidence that
a conjugate resembling the unknown 1 of beans does contain 2,4-D [146].

This conjugate, when injected into cotton seedlings, produced symptoms similar to those produced by 2,4-D, suggesting that the conjugate was hydrolyzed in the plant with the release of free 2,4-D. The conjugate had no effect when applied externally to the cotton seedlings, eliminating the possibility that it was contaminated with free 2,4-D.

Meagher [147] has reported evidence for the conjugation of 2,4-D to pectic acid in citrus peel. A portion of the 2,4-D taken up by citrus peel was bound in the acetone-insoluble fraction. The herbicide was released from this fraction in a conjugated form by prolonged heating. Pectic acid isolated from the acetone-insoluble fraction of the 2,4-D-treated citrus peel and subjected to the same heat treatment, liberated a conjugate of 2,4-D, thus suggesting conjugation of at least some of the "bound" 2,4-D with pectic acid.

   f. Ring cleavage. A radioactive metabolite in cultivated cucumber treated with ring-labeled 2,4-D was identified by thin-layer chromatography using three different solvent systems as monochloroacetic acid [148]. The production of this compound indicates cleavage of the 2,4-D ring and degradation of the cleavage product.

   g. Formation of unidentified metabolites. In addition to the metabolites discussed in the foregoing sections, many unidentified radioactive metabolites originating from ^{14}C-carboxyl- or methylene-labeled phenoxyacetic acids have been reported in plants. The unidentified metabolites have usually been distinguished from the parent phenoxyacetic acids by their solubility characteristics and paper chromatography. Examples of such compounds are the metabolites found in 2,4-D-treated cotton [104, 146] and 2,4-D- or 2,4,5-T-treated big leaf maple [106] that were similar in their chromatographic behavior to the unknown 1 complex of Jaworski and Butts [138], and the metabolite found in 2,4-D-treated sorghum [104] that resembled the unknown 3 complex of Fang and Butts [139]. An unidentified metabolite of 2,4-D detected in the bean by Weintraub et al. [113] was a relatively volatile or unstable, ether-soluble, acidic material. Bermuda buttercup (Oxalis pes-caprae L.) also produced unstable metabolites and probably volatile metabolites from 2,4-D [111]. Metabolites of 2,4-D or silvex, which were possibly sugar esters, were detected by Crosby [127] in the bean and by Meagher [147] in citrus peel. Erickson et al. [149] reported the formation of an ester-like complex from 2,4-D in lemons. Other plants in which unidentified metabolites of 2,4-D have been detected are young cherry trees (Prunus avium L.) [150], red and black currants [99], apples, strawberries [100], tick beans [107], ironweed (Vernonia baldwinii Torr.) [151], jimsonweed [152], bur and cultivated cucumbers [109], blackjack oak [153], birdsfoot trefoil (Lotus corniculatus L.) [154], nightflowering catchfly (Silene noctiflora L.) [155], and honeyvine milkweed [Ampelamas albidus (Nutt.) Britt.] [156].

Several plants that produced unidentified metabolites from 2,4-D, also produced metabolites from other phenoxyacetic acids. For example, red and black currants produced unidentified metabolites from 2,4,5-T and 2- and 4-chlorophenoxyacetic acid [99]. Metabolites of 2,4,5-T were detected in bur and cultivated cucumbers [109], honey mesquite seedlings [Prosopsis juliflora var. glandulosa (Torr.) Cockerell] [157], and blackjack oak [153]. Blackjack oak also produced several unidentified compounds from 2-chloro-4-fluorophenoxyacetic acid [153]. MCPA metabolites were detected in bedstraw [102].

The unidentified metabolites discussed in this section were probably formed by the known degradation mechanisms outlined in the foregoing sections. However, some may perhaps be the products of degradation mechanisms that have not yet been described.

2.   Higher ω-Phenoxyalkanoic Acids

The degradation of ω-phenoxyalkanoic acids with a side chain containing three or more carbon atoms has been the subject of detailed investigation (for reviews see Loos [94, 97], Casida and Lykken [95], Robertson and Kirkwood [96], Wain et al. [158-162], and Linscott [163]), especially by Wain and his colleagues [123, 164-166], who clearly demonstrated the importance of β oxidation of the side chain of these compounds in plants. Indeed, with the exception of certain 3-phenoxypropionic acids, the higher ω-phenoxyalkanoic acids are nonherbicidal in the absence of β oxidation. Other degradation reactions are analogous to those of the phenoxyacetic acids.

a. β Oxidation of the side chain.   Synerholm and Zimmerman [167] first obtained evidence that plants metabolize ω-phenoxyalkanoic acids by β oxidation of the side chain. In a study of the growth-regulating activities of the homologous series of 2,4-dichlorophenoxyalkanoic acids from the acetic to the caprylic acid, they found that only the 2,4-dichlorophenoxy acids with an even number of carbon atoms in the side chain were active on tomato. They postulated that only the 2,4-dichlorophenoxyacetic acid was active per se and that the butyric, caproic, and caprylic acids owed their activity to their β oxidation in the plant to the acetic acid homolog. The acids with an odd number of carbon atoms in the side chain would be converted to the half-ester of 2,4-dichlorophenol with carbonic acid. This compound would decompose to CO_2 and 2,4-dichlorophenol, which is inactive as a growth regulator [168]. The scheme proposed for the β oxidation of the two groups of 2,4-dichlorophenoxyalkanoic acids is shown in Fig. 4.

The formation of phenoxyacetic acids from higher ω-phenoxyalkanoic acids with an even number of carbons in the side chain and of phenols from ω-phenoxyalkanoic acids with an odd number of carbons in the side chain was demonstrated by chemical and chromatographic procedures, for example,

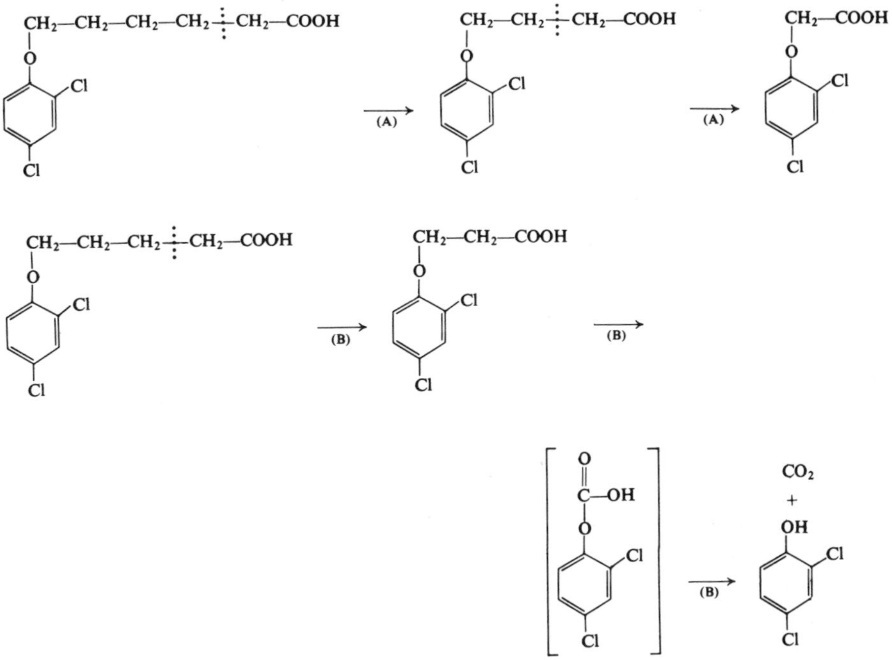

FIG. 4. β Oxidation of 2,4-dichlorophenoxyalkanoic acids in plants. (A) Formation of 2,4-D from acids with an even number of carbons in the side chain and (B) formation of 2,4-dichlorophenol from acids with an odd number of carbons in the side chain. (After Synerholm and Zimmerman [167]. Reprinted from Loos [94].)

in flax (<u>Linum usitatissimum</u> L.) seedlings metabolizing unsubstituted phenoxyalkanoic acids [164], wheat and pea tissue metabolizing 2,4-dichlorophenoxyalkanoic acids [166], and wheat tissue metabolizing 2,4,5-trichlorophenoxyalkanoic acids [165]. However, flax seedlings metabolizing 10-phenoxy-n-decanoic acid unexpectedly produced considerable amounts of phenol [164]. Fawcett et al. [164] proposed that the phenoxydecanoic acid side chain was degraded by ω oxidation as well as by β oxidation.

Pea tissue failed to produce 2,5-dichlorophenoxyacetic acid and 2,4,5-T when treated with higher homologs with an even number of carbons in the side chain [123, 165]. Paper ionophoresis and bioassay with wheat coleoptile cylinders showed that 2,5-dichlorophenoxycaproic acid was β-oxidized by the pea only to the butyric acid stage [123]. The ability of certain substituents on the phenoxy ring to block β oxidation of the aliphatic acid side chain

TABLE 5

Plant Growth Responses to Homologous Series of
Phenoxyalkanoic Acids and β-Oxidation Reactions
Causing the Responses[a]

Response type	Plant response to homologs[b]						β-Oxidation reactions in plant[b]					
1	H	$\underline{C}^{c}$	V	$\underline{B}^{c}$	P	$\underline{A}^{c}$	H	C	V	B	P	A[d]
2	H	C	V	B	P	$\underline{A}^{c}$	H	C	V	B	P	A[d]
3	$\underline{H}^{c}$	$\underline{C}^{c}$	$\underline{V}^{c}$	$\underline{B}^{c}$	$\underline{P}^{c}$	$\underline{A}^{c}$	H	C	V	B	P[d]	A[d]
4	$\underline{H}^{c}$	C	$\underline{V}^{c}$	B	$\underline{P}^{c}$	$\underline{A}^{c}$	H	C	V	B	P[d]	A[d]

[a] Compiled from Fawcett et al. [123]. Reprinted from Loos [94].

[b] H, C, V, B, P, A: heptanoic, caproic, valeric, butyric, propionic, and acetic homologs, respectively.

[c] Positive response of test plant; other homologs gave no growth responses.

[d] Active as plant growth regulator per se.

had a profound influence on the growth-regulating activity of ω-phenoxyalka-noic acids in the pea [123, 165].

The most comprehensive investigation of β oxidation among ω-phenoxy-alkanoic acids, as indicated by growth-regulating activity in homologous series, is that of Fawcett et al. [123]. These workers used three test plants to study growth-regulating activity in 18 homologous series of phe-noxyalkanoic acids with no substituents or different combinations of chlorine and methyl substituents on the phenoxy ring. With all series the pattern of growth-regulating activity on the test plants could be explained in terms of β oxidation of the phenoxyaliphatic acid side chain. The comprehensive study [123] confirmed the results of earlier studies [164-166] and indicated additional conclusions.

Four response patterns were observed in wheat cylinder and pea curva-ture or pea segment tests with the 18 series of phenoxyaliphatic acids [123]. These response patterns and the β-oxidation reactions that were proposed to explain them are indicated in Table 5. Type 1 and 3 responses are ob-tained where β oxidation results in the production of the acetic or propionic acid homolog. If only the acetic acid homolog is active as a growth regu-lator, a type 1 response is observed; if both the acetic acid and propionic acid homologs are active, a type 3 response results. The wheat cylinder tests gave type 1 responses with all of the phenoxyaliphatic acids tested, except the unsubstituted series and the 2-chloro series, which produced type 3 responses. In the wheat tests, none of the combinations of substituents on the phenoxy ring prevented β oxidation of the long-chain phenoxyaliphatic acids to the active phenoxyacetic or phenoxypropionic acid.

In the pea tests, type 1, 2, and 4 responses were observed, depending on the combination of substituents on the phenoxy ring. In the type 2 and 4 responses, β oxidation of the long-chain phenoxyaliphatic acids appears to be inhibited at the butyric acid or propionic acid stage. If the acetic but not the propionic acid homolog is a growth regulator for the pea, a type 2 response results; if both the acetic and propionic homologs are active, then a type 4 response is produced.

The various combinations of substituents on the phenoxy ring produced types 1, 2, and 4 responses in the pea tests as follows [123]:

Type 1	Type 2	Type 4
3-Chloro-	2,5-Dichloro-	Unsubstituted
4-Chloro-	2,3,4-Trichloro-	2-Chloro-
3,4-Dichloro-	2,4,5-Trichloro-	2,3-Dichloro-
2,4-Dichloro-	2-Chloro-5-methyl-	
3-Chloro-4-methyl-	5-Chloro-2-methyl-	
4-Chloro-2-methyl-	2,4-Dichloro-5-methyl-	
4-Chloro-3-methyl-	2,5-Dichloro-4-methyl-	
	4,5-Dichloro-2-methyl-	

Thus, the homologous series with a 2-chloro or 2-methyl substituent group on the phenoxy ring usually produced a type 2 or 4 response, indicating that these substituent groups hindered β oxidation of long-chain phenoxyaliphatic acids at the butyric or propionic acid stage. However, the blocking effect of the 2-chloro and 2-methyl substituents was eliminated by the introduction of a chlorine atom in the 4 position, provided there was no additional chlorine or methyl substituent in the 3 position. Fawcett et al. [123] concluded that both steric and electronic effects of the substituent groups were involved in the hindering of β oxidation of the phenoxyaliphatic acid side chain. Type 1 responses were produced by substituent combinations that did not hinder

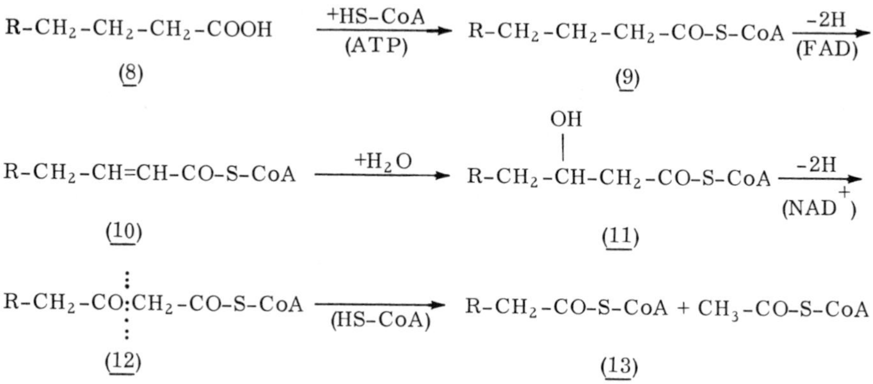

$$R-CH_2-CH_2-CH_2-COOH \xrightarrow[\text{(ATP)}]{+HS-CoA} R-CH_2-CH_2-CH_2-CO-S-CoA \xrightarrow[\text{(FAD)}]{-2H}$$

$$\underline{(8)} \qquad\qquad\qquad\qquad\qquad\qquad \underline{(9)}$$

$$R-CH_2-CH=CH-CO-S-CoA \xrightarrow{+H_2O} R-CH_2-\overset{\displaystyle OH}{\underset{\displaystyle |}{C}}H-CH_2-CO-S-CoA \xrightarrow[\text{(NAD}^+\text{)}]{-2H}$$

$$\underline{(10)} \qquad\qquad\qquad\qquad\qquad\qquad \underline{(11)}$$

$$R-CH_2-\overset{\vdots}{\underset{\vdots}{C}}O{:}CH_2-CO-S-CoA \xrightarrow{\text{(HS-CoA)}} R-CH_2-CO-S-CoA + CH_3-CO-S-CoA$$

$$\underline{(12)} \qquad\qquad\qquad\qquad\qquad\qquad \underline{(13)}$$

FIG. 5. β Oxidation of unsubstituted and substituted alkanoic acids. Cofactors: HS-CoA, coenzyme A; ATP, adenosine triphosphate; FAD, flavin-adenine dinucleotide; and NAD$^+$, nicotinamide-adenine dinucleotide. Intermediates in the degradation of butyric acid (8, R = H) would be butyryl CoA (9), crotonyl CoA (10), 3-hydroxybutyryl CoA (11), 3-ketobutyryl CoA (12), and acetyl CoA (13). (Adapted from Loos [94].)

β oxidation. The unsubstituted phenoxyaliphatic acids, which had no substituents to hinder β oxidation, unexpectedly showed a type 4 response. However, in this series the activity of the acetic acid homolog was low and it was assumed that too little phenoxyacetic acid was produced by β oxidation of the caproic and butyric homologs to give a growth response.

The responses of tomato to the various substituted phenoxyaliphatic acids were in most cases similar to those of the pea [123].

The sequence of reactions involved in the β oxidation of aliphatic acids is indicated in Fig. 5. Evidence for this pathway in the degradation of phenoxyalkanoic acids in plants is the production of small amounts of 2,4-dichlorophenoxycrotonic acid and 2,4-D in 2,4-DB-treated alfalfa [121], soybeans, and cocklebur [169]. In the hindered β oxidation of phenoxybutyric acids in the pea, the oxidation sequence appears to be blocked after the formation of the 3-hydroxy-4-phenoxybutyric acid derivative. Thus, Fawcett et al. [123] obtained evidence that 2,5-dichlorophenoxybutyric acid was converted in the pea to 4-(2,5-dichlorophenoxy)-3-hydroxybutyric acid, but this intermediate was not metabolized by the pea tissue to 2,5-dichlorophenoxyacetic acid. Wheat tissue, on the other hand, carried out this conversion. Both wheat and pea tissue metabolized 4-(2,4-dichlorophenoxy)-3-hydroxybutyric acid to 2,4-D.

Gas chromatography has proved a useful method for demonstrating the metabolism of phenoxybutyric acid herbicides in plants by β oxidation, for

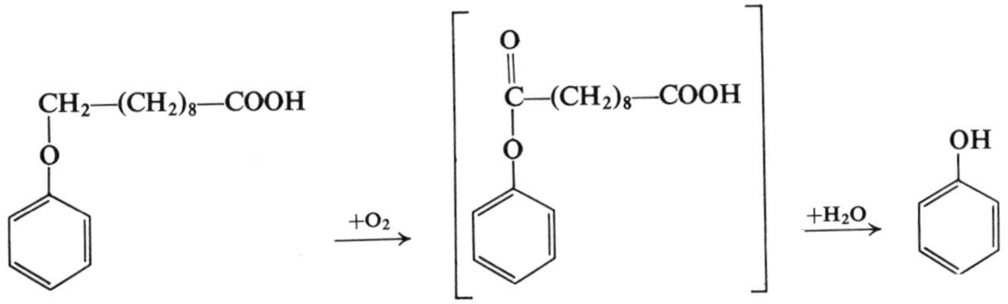

FIG. 6.   ω Oxidation of 10-phenoxy-n-decanoic acid in flax. (After Fawcett et al. [164]. Reprinted from Loos [94].)

example, the metabolism of 2,4,5-TB to 2,4,5-T in wheat, but not in peas [170], MCPB to MCPA in beans [174], and 2,4-DB to 2,4-D in the big leaf maple [171], timothy grass, birdsfoot trefoil, and peas [172]. In all of these studies the conversion was demonstrated where herbicide was applied externally to intact plants, in contrast to many of the earlier chemical studies [123, 165, 166] where plant segments were used and absorption and translocation of the herbicide played no role. In the 2,4-DB experiments with the pea, the plants were grown under sterile conditions to exclude the possibility of β oxidation by microorganisms on the plant surface [172]. The work of Norris and Freed [171] and Fertig et al. [172] demonstrates the application of gas chromatography to quantitative studies of β oxidation.

b. ω Oxidation of the side chain.   Fawcett et al. [164], in their chemical studies of β oxidation, found that 10-phenoxy-n-decanoic acid unexpectedly produced large amounts of phenol during its breakdown in flax seedlings. Other phenoxyalkanoic acids with an even number of side-chain carbons produced little or no phenol. It was proposed that the 10-phenoxy-n-decanoic acid was broken down by ω oxidation and cleavage of the molecule at the bridging oxygen, as well as by β oxidation. The proposed ω oxidation pathway is indicated in Fig. 6.

c. Side-chain lengthening.   The alfalfa studies of Linscott and colleagues [119, 121] showed lengthening of the side chain of 2,4-DB by two carbon units to yield 6-(2,4-dichlorophenoxy)caproic acid, 10-(2,4-dichlorophenoxy)decanoic acid, and apparently other 2,4-dichlorophenoxyalkanoic acid homologs. At the same time, some of the 2,4-DB was degraded to 2,4-D. Esters of 4-(2,4-dichlorophenoxy)crotonic acid and 4-(2,4-dichlorophenoxy)-3-hydroxybutyric acid, which are intermediates in the breakdown of 2,4-DB to 2,4-D, were converted to 2,4-DB and also appeared to undergo side-chain lengthening. Side-chain lengthening and β oxidation are competing processes in the metabolism of 2,4-DB by alfalfa. Similar reactions were subsequently reported for soybean and cocklebur [169].

d. Ring hydroxylation. Unsubstituted phenoxyalkanoic acids in the
tissues of pea, wheat, oats, barley, and corn were metabolized by ring
hydroxylation as well as β oxidation [123, 124]. The hydroxylation of the
ring either preceded [123] or followed [124] β oxidation of the phenoxyalkanoic acid side chain.

e. 2,4-DB degradation in silage. Linscott and his colleagues [163,
173, 174] studied the degradation of ^{14}C-carboxyl-labeled 2,4-DB in forage
plants "ensiled" on a laboratory scale. Loss of radioactivity from the silage
indicated the extent of degradation. Silage prepared from alfalfa, birdsfoot
trefoil, timothy, orchardgrass, or bromegrass lost about 30-60% of the
applied radioactivity in 1-3 months. The three grass silages decomposed
2,4-D and silvex to about the same extent as 2,4-DB. Most of the ^{14}C loss
from the 2,4-DB-treated alfalfa silage occurred within the first 15 days,
and considerable losses occurred during the first 8 hr. It is not known
whether the plant enzymes or microorganisms in the silage were responsible
for the 2,4-DB degradation, but the rapid breakdown in the early stages of
the silage fermentation suggested that the degradation was mainly an aerobic
process. However, losses of radioactivity still continued when conditions
were almost certainly highly anaerobic. The pathway of degradation of the
2,4-DB was not investigated.

3. Higher α-Phenoxyalkanoic Acids

The metabolism by plants of α-phenoxyalkanoic acids with three or
more carbon atoms in the side chain has been little studied, in spite of the
importance of 2-phenoxypropionic acids as weedicides. On the basis of
available evidence the 2-phenoxypropionic acids tend to be persistent in
plants, which may be advantageous in weed control, for example in the control of prickly pear (Opuntia polyacantha Haw.) with silvex [110] and bedstraw
with mecoprop [102]. However, silvex in the prickly pear was slowly decarboxylated and metabolized to unidentified metabolites [110], and silvex and
dichlorprop in the big leaf maple were slowly decarboxylated [106].

4. Ester, Amide, and Nitrile Derivatives of Phenoxyalkanoic Acids

Most ester, amide, and nitrile derivatives of phenoxyalkanoic acids
are probably not active as plant growth regulators per se but require conversion to an active acid as a prerequisite for activity [160, 161]. Degradation of these compounds in plants is therefore, in many cases, an
activation mechanism.

a. Esters. Hydrolysis of 2,4-D esters has been reported in barley
[175] and lemons [149] treated with the isopropyl ester, and in corn and
beans treated with the butoxyethanol and propylene glycol butyl ether esters
[176]. Meal from castor beans (Ricinus communis L.) contained an esterase
that hydrolyzed the butyl ester of 2,4-D [177], while the octyl ester was

hydrolyzed by crude protein preparations or extracts from cucumber, spinach, corn, and seeds of pumpkin (Cucurbita pepo L.) [178]. Wheat seeds and seedlings of corn, cucumber, and soybean contain enzymes able to hydrolyze the 2-naphthyl ester of phenoxyacetic acid [179]. The butoxy-ethanol ester of 2,4,5-T was hydrolyzed in red maple (Acer rubrum L.) and white ash (Fraxinus americana L.) [180], and the 2-ethylhexyl ester in live oak (Quercus virginiana Mill.) and mixed grasses (Andropogon scoparius Michx., Paspalum plicatulum Michx., and Sorghastrum spp.) [182]. The ability to hydrolyze esters of phenoxyacetic acids thus appears to be wide-spread among plants, and is presumably an important factor in weed control involving ester formulations of the phenoxyalkanoic acid herbicides.

The formation of the ethyl ester of 2,4-D was detected by gas chroma-tography in forage treated with the 2,4-D butyl ester [182]. The transfor-mation mechanism is unknown but transesterification and a type of β oxidation were suggested as possible mechanisms [182].

b. Amides. Fawcett et al. [166] obtained evidence that amides of ω-2,4-dichlorophenoxyalkanoic acids, from the acetic to the heptanoic homolog, were hydrolyzed by wheat or pea tissue. The resulting acids were β-oxidized to 2,4-D or 2,4-dichlorophenol, depending on the number of carbons in the aliphatic acid side chain. The homologous series of 2,4-dichlorophenoxyalkanoic acid amides thus showed the same alternation of growth-regulating activity in the wheat cylinder, pea curvature, pea seg-ment, and tomato leaf epinasty tests as the free acids.

Wood and Fontaine [58] and Krewson and his colleagues [59-63] investi-gated the growth-regulating activity of many amino acid derivatives of phenoxyacetic and 2-phenoxypropionic acids. The amino group of the amino acids was linked to the carboxyl group of the phenoxy acids. In studies with the amino acid derivatives of 2,4-D [58], MCPA [59], and 4-chlorophenoxy-acetic acid [60], it was found that the L- and DL-amino acid derivatives usually had about the same growth-regulating activity as the free parent phenoxy acids, whereas the D-amino acid derivatives were inactive or had little activity. It was postulated that the test plants were able to hydrolyze the derivatives of the natural L-amino acids but not those of the unnatural D-amino acids to yield free phenoxyacetic acids, which were assumed to be necessary for activity [59-61]. However, it was subsequently established [62, 63] that certain amino acid derivatives of dichlorprop and silvex differed in their growth-regulating activity from the parent phenoxypropionic acids. These results suggested that hydrolysis of the amide bond might not be essential for activity, i.e., that the amino acid derivatives were active per se [62, 63]. This important suggestion needs further investigation.

c. Nitriles. The work of Fawcett and his colleagues [166, 183] indi-cates that ω-(2,4-dichlorophenoxy)alkane nitriles are degraded in wheat tissue by two different mechanisms. One mechanism involves hydrolysis of

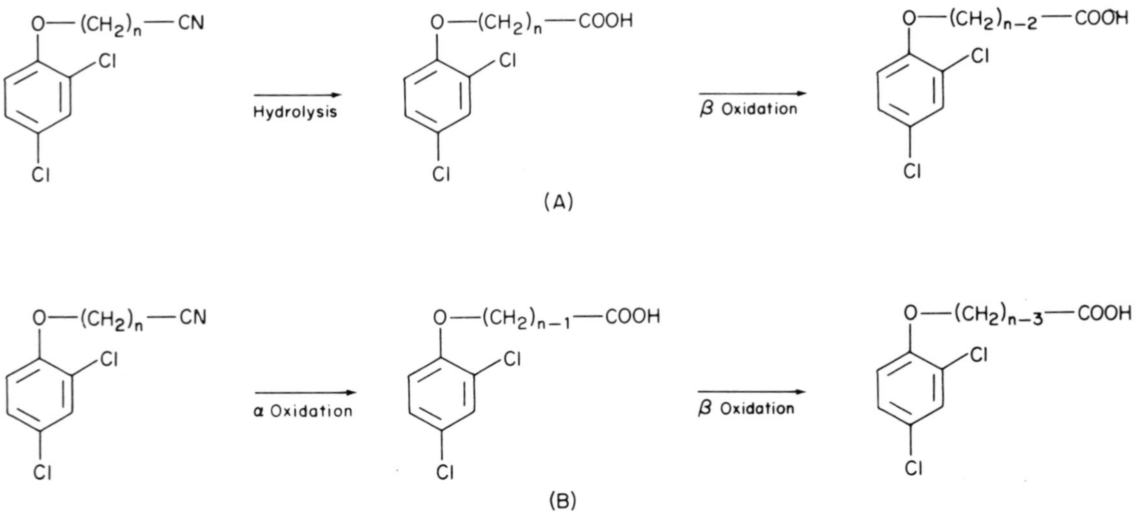

FIG. 7. Degradation of ω–(2,4-dichlorophenoxy)alkane nitriles in wheat tissue. Alternative pathways involving (A) hydrolysis followed by β oxidation and (B) α oxidation followed by β oxidation. (After Fawcett et al. [166]. Reprinted from Loos [94].)

the nitrile group to a carboxyl group followed by β oxidation of the resulting 2,4-dichlorophenoxyalkanoic acid (Fig. 7, pathway A). The second mechanism involves the oxidative removal of the nitrile carbon from the side chain (α oxidation) followed by β oxidation of the resulting acid (Fig. 7, pathway B). These mechanisms were proposed to explain the formation of both 2,4-D and 2,4-dichlorophenol from all the ω-(2,4-dichlorophenoxy)-alkane nitriles supplied to the wheat tissue, with the exception of the propionitrile. The propionitrile produced no detectable 2,4-D and showed little growth-regulating activity in the wheat cylinder test.

With peas and tomatoes only 2,4-dichlorophenoxyacetonitrile showed growth-regulating activity [166, 183]. Only this nitrile produced 2,4-D on incubation with pea tissue. None of the nitriles, except possibly the propionitrile, produced 2,4-dichlorophenol. The higher ω-(2,4-dichlorophenoxy)-alkanoic acids and amides all produced either 2,4-D or 2,4-dichlorophenol. The pea tissue therefore was unable to convert the higher ω-(2,4-dichloro-phenoxy)alkane nitriles to the corresponding amides or acids, or to α-oxidize them to an acid with a shorter side chain.

B.   Degradation by Microorganisms

1.   Phenoxyacetic Acids

The introduction of MCPA, 2,4-D, and 2,4,5-T as weedkillers was accompanied by reports of their detoxication in soils [184-192]. Under conditions favorable for its degradation, 2,4-D disappears within about two to three weeks [33, 193, 194]. The other two herbicides disappear more slowly, for example, MCPA often persists for about six weeks or longer [33, 193, 194], although much shorter persistence has been noted [195]. The persistence of 2,4,5-T, in studies where the molecule was degraded, varied from 45 to 270 days [192, 194]. The stimulation of herbicide degradation by warm, moist conditions and the addition of organic matter to the soil [186-190], a correlation between the rate of 2,4-D degradation and numbers of aerobic soil bacteria [196], and inhibition of herbicide degradation in air-dried and autoclaved soils [186-190] provided early evidence that degradation was microbiological. The effects of soil pH [196] and liming [187] on 2,4-D degradation were consistent with this hypothesis. The optimum pH of 5.3 for 2,4-D degradation in the organic soils studied by Corbin and Upchurch [197] is unexpectedly low, but can be explained by assuming that fungi were the agents of 2,4-D degradation in these highly acid soils of pH 3.6-3.8.

The role of microorganisms in the degradation of phenoxyacetic acid herbicides in soil was conclusively demonstrated by Audus [198-201; see also 89, 193, 194]. The kinetics of 2,4-D, MCPA, and 2,4,5-T detoxication in soil perfusion experiments were exactly what would be expected of a microbial process [198, 200, 201], and a 2,4-D-degrading bacterium was subsequently isolated from 2,4-D-perfused soil [199, 200]. A variety of bacteria and actinomycetes capable of degrading phenoxyacetic acids have since been isolated from soil and used to study degradation pathways (for reviews see Audus [89, 193, 194], Freed and Montgomery [91], and Loos [94, 97]).

a. Organisms degrading phenoxyacetic acids.   The bacteria and actinomycetes known to decompose phenoxyacetic acids are listed in Table 6. The table also indicates the phenoxyacetic acids that were metabolized by the various organisms. All of the intensively studied species metabolized several phenoxyacetic acids, although the rate and extent of metabolism were not the same for all substrates [206, 212-215, 223, 225]. Many phenoxyacetic acids were degraded completely, or almost completely, with loss of their aromatic structure and release of their chlorine as chloride [204, 206, 208, 210, 212, 214, 218, 220, 223, 225].

The fungus Aspergillus niger metabolizes phenoxyacetic acids by intro-ducing a hydroxyl group into the aromatic ring [233-238]. The hydroxylated phenoxyacetic acids are not further degraded by the fungus.

b. Adaptation of organisms to phenoxyacetic acid substrates.   Most of the phenoxyacetic acid-decomposing bacteria and actinomycetes listed in Table 6 were isolated from soil by the enrichment culture technique using herbicide substrates, i.e., their numbers in the soil were increased to a level at which they could be isolated by supplying them with the appropriate herbicide (indicated in the table by the superscript "a") as a nutrient source. The soil perfusion experiments of Audus [193, 194, 198, 200, 201] show the course of substrate disappearance in such enrichment cultures. Initially a small amount of substrate is removed by adsorption on soil colloids. This phase is followed by a lag period, during which no appreciable change in herbicide concentration is observed. The length of the lag period varies according to the substrate; thus, the lag phases of MCPA and 2,4,5-T are considerably longer than that of 2,4-D. Finally, there is the phase of rapid substrate disappearance. Subsequent additions of substrate disappear rapidly without a lag. Observations of 2,4-D decomposition in pots of soil [191] and of 2,4-D and MCPA decomposition in field experiments [195, 239] confirm the shortening of the detoxication period, i.e., lessening or even disappear-ance of the lag period, with repeated application of herbicide to the soil. In untreated soils the numbers of organisms with the potential to degrade phenoxy herbicides are low, for example, seven Natal soils contained, by most probable number count, from 1 to 245 2,4-D-decomposing organisms and from 0.34 to 1377 MCPA-degrading organisms per gram of soil [240]. The populations of 2,4,5-T-degrading organisms were apparently very much smaller, as 2,4,5-T degradation in 5-g soil samples occurred consistently with only two of the soils [241]. Six of the soils showed consistent 2,4,5-T degradation in 50-g samples, and all soils showed it in 500-g samples. The populations of herbicide-degrading organisms in soils presumably have to attain a critical size to degrade the herbicide at a detectable rate; hence at least a part of the lag phase in soil perfusion experiments would be the time required for the population to multiply to this level [193, 194, 201]. Once the critical level is reached, further additions of herbicide are degraded without a lag. However, herbicide metabolism does not necessarily imply that the compound is used as a growth substrate by the responsible organism, for example, 2,4,5-T cometabolism by benzoate-grown Brevibacterium sp. [219].

The potential of soil organisms to degrade phenoxy herbicides must result from the possession of enzyme systems that normally metabolize compounds structurally related to the herbicides. These enzymes may be sufficiently nonspecific to degrade the phenoxy compounds and their breakdown products, but they are likely to require induction [212-215, 223, 242]. On the other hand, mutations may be required to give the enzymes the necessary specificity to metabolize the herbicides and their metabolites. Either mechanism would contribute to the observed lags in herbicide breakdown.

The roles of the postulated induction and mutation mechanisms in the development of phenoxy herbicide-degrading microbial populations in soil have still be to established. However, indirect evidence seems to favor induction as the most significant mechanism. Soils tend to show very consistent and reproducible lag periods for the degradation of 2,4-D and other phenoxy acids; if the responsible organisms developed as a result of mutations, a much greater variability would be expected [193, 194]. Furthermore, samples of any one soil, when enriched quite independently, seem to produce active populations of the same organism [193]. This result can be reconciled with the mutation hypothesis only by assuming that a particular bacterial species in the soil is more prone than other species to undergo the mutation(s) leading to activity [194]. The tendency for enzymes degrading phenoxyacetic acids to have a wide specificity, as demanded by the induction hypothesis, is evident from Table 6. The inducible nature of the enzymes has been established for many of the organisms [212-215, 223, 242]. On the other hand, the possible role of mutation should not be disregarded. The reproducibility of the lag period for different samples of a soil often no longer applies as the samples become smaller, for example, near the end point in most-probable-number dilution series [240]. It seems possible that long lags in such samples might result from late mutations to herbicide-degrading ability. Such mutations would not be detected in the presence of organisms that rapidly degrade the herbicide following enzyme induction. Mutation may well play a role in the adaptation of phenoxy herbicide-degrading organisms to new phenoxyacetic acid substrates, for example, the increased ability of the 2,4-D-degrading Arthrobacter sp. of Loos et al. [223] to metabolize MCPA following serial culture in MCPA medium [218] may have developed by this mechanism.

TABLE 6

Bacteria and Actinomycetes That Degrade Phenoxyacetic Acids

Organism	References	Phenoxyacetic acids metabolized											
		Phenoxyacetic acid	2-Chlorophenoxyacetic acid	4-Chlorophenoxyacetic acid	2,4-D	2,6-Dichlorophenoxy-acetic acid	2,4-Dibromophenoxy-acetic acid	4-Bromo-2-chlorophenoxy-acetic acid	MCPA	4-Hydroxyphenoxyacetic acid	4-Chloro-2-hydroxy-phenoxyacetic acid	2,4-Dichloro-6-hydroxy-phenoxyacetic acid	2,4,5-T
Bacteria													
Pseudomonad	202-204			+[a]							+		
Pseudomonas sp. (P. subcreta - P. pictorum)	202, 205, 206		+	+	+[a]				+		+		
Pseudomonas sp.	206-210	+	+	+	+				+[a]		+		+
Mycoplana sp.	211	+	+	+	+[a]				+		+		
Achromobacter sp.	212			+[a]					+[a]				
Achromobacter sp.	212, 213	+	+	+	+		+	+	+		+		+
Achromobacter sp.	214, 215		+	+	+[a]	+			+			+	+

Organism	Ref.								
Flavobacterium peregrinum	212, 213				+[a]				
F. peregrinum	216, 217			+[a]		+			
F. peregrinum	218		+[a]				+[a]		
Brevibacterium sp.	219							+[a]	
Corynebacterium sp.	220				+				
Corynebacterium-like organism	221, 222				+[a]				
Arthrobacter globiformis (Bacterium globiforme)	199, 200			+[a]					
Arthrobacter sp.	218, 223–230				+[a]		+		
Sporocytophaga congregata (Flavobacterium aquatile)	221, 222, 231				+[a]			+[a]	
Actinomycetes									
Nocardia sp.	194			+	+				
Streptomyces viridochromogenes	232			+	+				+

[a]Substrate in enrichment and isolation media. (Adapted from Loos [94].)

39

Soil that has been "enriched" in bacteria able to decompose a particular phenoxyacetic acid, retains for long periods the ability to degrade the enrichment substrate rapidly [191, 193-195]. Thus, the ability to degrade 2,4-D rapidly without a long lag persisted in soil in the absence of 2,4-D for at least a year after the initial 2,4-D treatment. The perfusion of 2,4-D-treated soil with distilled water for periods up to 60 days, resulted in little loss of 2,4-D-decomposing ability [193, 194]. These results provide additional evidence that bacterial proliferation and not just enzyme induction is involved in the development of herbicide-degrading populations of soil bacteria [193]. However, the results do not support or contradict either the mutation hypothesis or the enzyme induction hypothesis for the development of herbicide-degrading ability.

Enrichment populations of bacteria that develop in soil in response to a particular phenoxyacetic acid substrate are often adapted to degrade other phenoxyacetic acid substrates [191, 193, 194, 200, 201, 212]. Soils perfused with 2,4-D until degradation was complete, rapidly detoxicated MCPA and vice versa [200, 201, 212]. Both MCPA- and 2,4-D-treated soils degraded 4-chlorophenoxyacetic acid, which in turn stimulated the development of a microbial population that degraded 2,4-D and MCPA [212]. In all cases the rate of degradation of the second molecule was slower than its rate of degradation in soil in which it had induced its own enrichment microflora. Audus [193, 194, 200, 201] postulated that 2,4-D and MCPA each induced its own specific degradation enzymes, either in the same bacterial species or in different species, and that these enzymes could incidentally degrade the other molecule. The efficiency of degradation was greatest, in each case, with the substrate that induced the formation of the enzymes. The observation that 2,4,5-T was degraded in soil enriched with MCPA, but not in soil enriched with 2,4-D, confirmed differences in the enzyme systems induced by 2,4-D and MCPA [193, 194, 201].

Patterns of "cross-adaptation" in soils treated with phenoxyacetic acids were extensively studied by Brownbridge (see Audus [193, 194]). The most striking feature of these experiments was the ability of MCPA-enriched soil to degrade rapidly all phenoxyacetic acids supplied to it. Degradation was often as rapid in the MCPA-enriched soils as in soils enriched with the test compound itself, and in the case of 2,5-dichlorophenoxyacetic acid, was more rapid in MCPA-enriched soil. Soil enriched with MCPA-degrading organisms readily degraded 2-chlorophenoxyacetic acid, which was unable to induce its own enrichment microflora when perfused through soil for almost one year. Soils perfused with 2,4-D were less efficient than the MCPA-enriched soils in degrading other phenoxyacetic acids. Thus, replacement of the 2-chlorine of 2,4-D with a methyl group decreased the biodegradability of the molecule as indicated by the increased lag phase in the development of enrichment populations [193, 194, 200, 201], but once the MCPA enrichment population had developed, it had a greater ability than the 2,4-D enrichment population to metabolize other phenoxyacetic acids. Brownbridge (see Audus [193, 194]) suggested that high activation energies were required for enzymatic attack on 2-methyl-substituted molecules. An enzyme generating these energies could, therefore, act on homologous molecules requiring similar or smaller energies.

The ability to metabolize more than one phenoxyacetic acid is common among bacteria isolated from 2,4-D or MCPA enrichment cultures (Table 6). The 2,4-D-decomposing bacteria generally degraded 2- and 4-chlorophenoxyacetic acid and MCPA [212-215, 223, 225], in agreement with the soil perfusion results of Brownbridge, except for the metabolism of 2-chlorophenoxyacetic acid. However, 2-chlorophenoxyacetic acid metabolism by the isolated bacteria was often slow in comparison with the degradation of 2,4-D, 4-chlorophenoxyacetic acid, and MCPA [213, 215, 223]. Phenoxyacetic acid and 2,4,5-T were also commonly oxidized by 2,4-D-decomposing bacteria [211, 214, 215, 223], but the consumption of oxygen during the metabolism of these substrates was small. In the case of Arthrobacter sp. metabolizing phenoxyacetic acid, the low oxygen uptake was due to incomplete metabolism of the molecule to the corresponding phenol [223]. The MCPA-decomposing Achromobacter strain of Steenson and Walker [212, 213] degraded 2,4-D, 2- and 4-chloro-, 4-bromo-2-chloro-, and 2,4-dibromophenoxyacetic acid; but if it was grown on 2,4-D instead of MCPA, it lost the ability to degrade MCPA. This result agrees with the conclusion of Audus [193, 194, 200, 201] that different enzyme systems are responsible for the degradation of 2,4-D and MCPA.

The oxidation of 4-chloro-2-hydroxy- and 2,4-dichloro-6-hydroxyphenoxyacetic acid by strains of Achromobacter that degraded 4-chlorophenoxyacetic acid and 2,4-D, respectively [212, 215], is of particular interest, as these hydroxy acids were proposed as intermediates in the bacterial degradation of 4-chlorophenoxyacetic acid and 2,4-D [202]. However, only 4-chlorophenoxyacetic acid hydroxylation has since been substantiated as a reaction of a main degradation pathway (see the next section).

Steenson and Walker [242] investigated the induction of 2,4-D- and MCPA-degrading enzymes in Flavobacterium peregrinum and Achromobacter sp. Flavobacterium peregrinum was adapted to metabolize 2,4-D when grown on 2,4-D, MCPA, or 2-chloro-4-methylphenoxyacetic acid, but it did not decompose MCPA. MCPA was therefore an inducer but not a substrate of the 2,4-D-degrading enzyme system. Enzyme induction by compounds that are not substrates of the enzymes they induce, is well known in bacteria (see Pardee [243]). In the Achromobacter, 2,4-dichlorophenol and 5-chloro-2-cresol were inducers of both the 2,4-D- and MCPA-degrading enzyme systems. These phenols are probably early intermediates in the degradation of 2,4-D and MCPA, respectively, by the bacterium [213]. The induction of bacterial enzymes of aromatic metabolism by

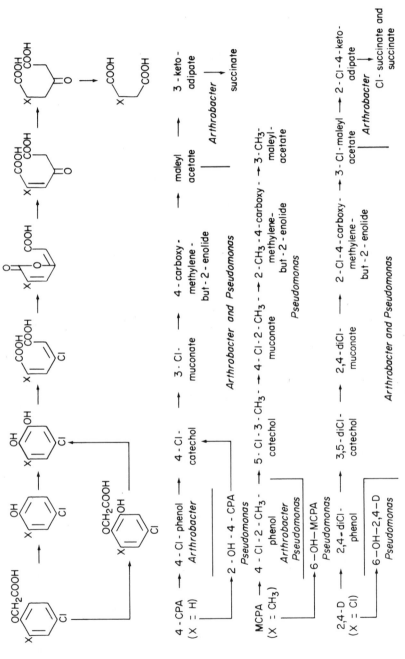

FIG. 8. Metabolism of 4-chlorophenoxyacetic acid (4-CPA), MCPA, and 2,4-D by Arthrobacter and Pseudomonas spp. (Adapted from Loos [97] by permission of the International Union of Pure and Applied Chemistry.)

intermediates in the degradation pathways rather than by the supplied substrates is now a well-established phenomenon [244].

c. Degradation pathways. Published research prior to 1967 (reviewed by Loos [94]) suggested pathways of degradation of phenoxyacetic acids by bacteria, but provided little unequivocal or detailed evidence in support of the pathways. Details of the pathways in an Arthrobacter sp. and pseudomonads are now available following intensive studies in the laboratories of Alexander [218, 223-230, 245] and Evans [202-210].

Figure 8 shows the pathways for the degradation of 4-chlorophenoxyacetic acid, MCPA, and 2,4-D demonstrated in Arthrobacter sp. and pseudomonads. The pathways were established by studies with growing or resting whole cells and crude or partially purified enzyme preparations. There is convincing, and in many cases unequivocal, chemical evidence for the formation of the various intermediates. The most important mechanism, demonstrated in the breakdown of all three compounds by the arthrobacter and in the breakdown of MCPA and 2,4-D by the pseudomonads, involved removal of the acetic acid side chain to yield the corresponding phenol [206, 208, 218, 223-225], followed by ortho hydroxylation of the phenol to produce a catechol [206, 208, 227]. The catechols were in all cases converted to a muconic acid by ortho cleavage of the aromatic ring [206, 208, 210, 228, 229]. Lactonization of the muconic acids involved dehydrochlorination [206, 208, 210, 228, 229], and in this respect it differed from the lactonization of 3-carboxy- or unsubstituted muconic acid formed during the breakdown of nonchlorinated aromatic compounds [244]. The resulting butenolides were probably delactonized to 3-hydroxymuconic acids in their conversion to maleylacetic acids [208], a step that is not indicated in Fig. 8. The final steps of degradation produce acids, for example succinic and presumably acetic, that can be metabolized by way of the tricarboxylic acid cycle [229, 230].

The metabolism of 4-chlorophenoxyacetic acid by a pseudomonad isolated from a soil under conifers involved ortho hydroxylation of the acid to 4-chloro-2-hydroxyphenoxyacetic acid, followed by loss of the side chain to yield 4-chlorocatechol [202-204]. Further metabolism of the 4-chlorocatechol was similar to its metabolism in the Arthrobacter sp. The production of 4-chloro-6-hydroxy-2-methylphenoxyacetic acid and 2,4-dichloro-6-hydroxyphenoxyacetic acid during the metabolism of MCPA and 2,4-D, respectively, by Pseudomonas N.C.I.B. 9340 seemed to result from side reactions, as the organism was unable to oxidize these hydroxy acids [206, 208].

Other possible pathways for the metabolism of 2,4-D and possibly MCPA are apparent from the studies with Pseudomonas sp. [206, 208]. These involve elimination of the 4-chlorine from the aromatic ring and its replacement by a hydrogen. Possible pathways leading from 2,4-D to

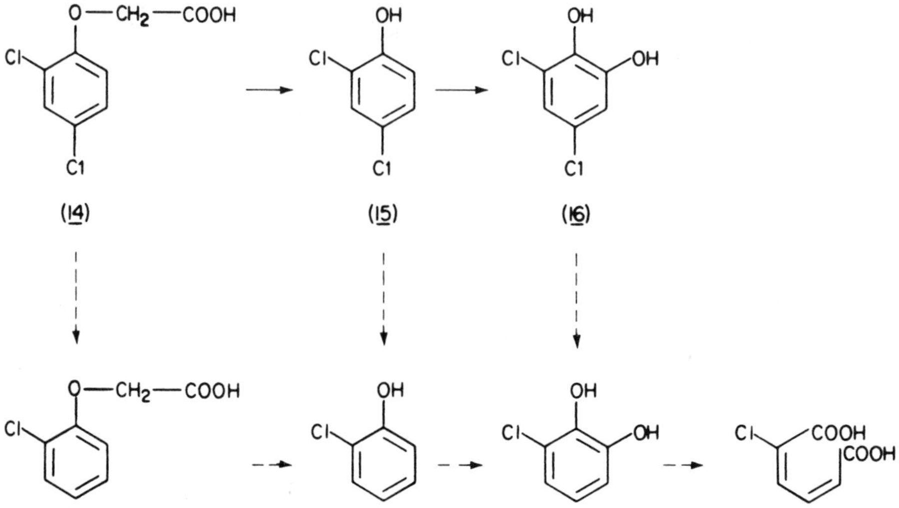

FIG. 9. Possible pathways for the metabolism of 2,4-D (14) by
Pseudomonas sp. resulting in the production of 2-chlorophenol (18) and
2-chloro-cis,cis-muconic acid (20). Postulated intermediates are
2,4-dichlorophenol (15), 3,5-dichlorocatechol (16), 2-chlorophenoxyacetic
acid (17), and 3-chlorocatechol (19). (After Evans et al. [206].)

2-chlorophenol and 2-chloromuconic acid, which were produced by bacteria
growing on 2,4-D [205, 206], are indicated in Fig. 9; 2-methylphenoxy-
acetic acid and ortho-cresol might be produced from MCPA by analogous
reactions [208]. The presence of a monochloro- and unsubstituted phenol
in cultures of a Nocardia sp. metabolizing 2,4-D [194] might reflect a
similar mechanism of chlorine elimination in this organism, whereas the
presence of chlorohydroquinone [194] might reflect removal of chlorine
linked to ring hydroxylation as observed in certain plants [129, 132].

Crude extracts of Pseudomonas N.C.I.B. 9340 catalyzed removal of
the MCPA side chain as glyoxylate [209]. The glyoxylate is possibly
oxidized to oxalate in intact cells. Indirect evidence indicated that the side
chain of 2,4-D also was probably removed as glyoxylate by Arthrobacter
sp. [245]. In this organism, the released glyoxylate undergoes condensation
with decarboxylation and incorporation of ammonia, to form alanine
(Fig. 10). Resting cells and cell-free extracts of MCPA-grown Arthrobacter
sp. quantitatively metabolized [^{18}O]phenoxyacetic acid to [^{18}O]phenol, which
retained all the label of the parent acid [226]. Enzymatic cleavage of the

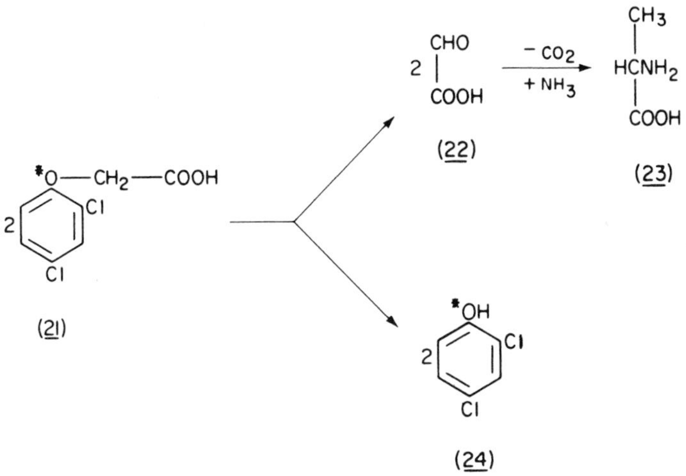

FIG. 10.   Proposed metabolism of 2,4-D (21) side chain by Arthrobacter sp., yielding glyoxylate (22), alanine (23), and 2,4-dichlorophenol
(24). (After Tiedje and Alexander [245]. Adapted from Loos [97] by
permission of the International Union of Pure and Applied Chemistry.)

phenoxyacetic acid molecule therefore occurred between the ether oxygen
and the aliphatic side chain.

There was no evidence with either Pseudomonas N.C.I.B. 9340 or
Arthrobacter sp. of significant degradation of phenoxyacetic acids via the
corresponding anisole, i.e., involving an initial decarboxylation [208, 224,
225], although traces of 2,4-dichloroanisole were produced by cultures of
the arthrobacter growing on 2,4-D [224].

The cometabolism of 2,4,5-T by benzoate-grown Brevibacterium sp.
produced 3,5-dichlorocatechol [219].

Aspergillus niger hydroxylates phenoxyacetic acids without metabolizing
the resulting hydroxy acids [233-238]. Phenoxyacetic acid is converted to
2-, 3-, or 4-hydroxyphenoxyacetic acid [233, 235, 236], 4-chlorophenoxy-
acetic acid to 2- or 3-hydroxy-4-chlorophenoxyacetic acid, and 2-chloro-
phenoxyacetic acid to 3-, 4-, 5-, or 6-hydroxy-2-chlorophenoxyacetic acid
[234]. With 2-chlorophenoxyacetic acid, the chlorine can be replaced by a
hydroxyl group to give 2-hydroxyphenoxyacetic acid [234]. The fungus
hydroxylates 2,4-D and MCPA in the 5 position, but 2,4-D is also metabolized, as in higher plants, by 4-hydroxylation coupled with a chlorine shift
to form 2,5-dichloro-4-hydroxyphenoxyacetic acid [237, 238].

d.   Resistance to microbial degradation.   Extensive studies have been
conducted to determine the effect of molecular structure on the breakdown

of phenoxyacetic acids in soil [33, 193, 194, 246]. The following effects of ring substitution on the degradation of the phenoxyacetic acid molecule were reported by Audus [193, 194]:

1. Chlorine substitution in the para position made the molecule much more labile. Thus, 4-chlorophenoxyacetic acid, at a concentration of 100 ppm, was 80% detoxified in 12.1 days, compared to 34.5 days for phenoxyacetic acid.

2. Chlorine substitution in the ortho position deactivated the molecule, so that 2-chlorophenoxyacetic acid was not degraded in one year.

3. Meta substitution also deactivated the molecule, but 10 ppm of 3-chlorophenoxyacetic acid was detoxified in about 60 days.

4. In the di- and trichlorophenoxyacetic acids the activating effect of the 4 substitution largely overcame the deactivating effect of other substituents. In contrast to 2-chlorophenoxyacetic acid, 2,4-D was rapidly detoxified (16.0 days), while 2,4,5-T was more easily degraded than 2,5-dichlorophenoxyacetic acid. However, none of the phenoxyacetic acids with a meta-chlorine in the molecule was rapidly degraded.

5. The methyl group was more deactivating than the chloro group. The periods required for 80% detoxication of 100 ppm MCPA and 2,4-dimethylphenoxyacetic acid were 85.7 and 102.5 days, respectively, compared with 16.0 days for 2,4-D.

A similar pattern of resistance to degradation was observed by Alexander and his collaborators [33, 246] who particularly emphasized the high degree of resistance of phenoxyacetic acids with a meta-chlorine substituent. Such molecules were termed recalcitrant [247-249]. However, the absence of degradation of some of the compounds investigated by Alexander and Aleem [33], for example 2,4,5-T, may have resulted from the small soil inocula (4 g) used by these workers [97].

2. Higher ω-Phenoxyalkanoic Acids

    a. β Oxidation. In 1955 Wain [158, 159] suggested that soil microorganisms probably β-oxidized ω-phenoxyalkanoic acids in a manner similar to that of plants. This mechanism has since been demonstrated in soil bacteria such as Nocardia [250-253], Pseudomonas, and Micrococcus [253], and the fungus Aspergillus niger [233]. According to the evidence currently available [250-253], soil bacteria β-oxidize ω-phenoxyalkanoic acids in the same way as plants do (Fig. 4). Substituents on the phenoxy ring that hinder β oxidation in plants also tend to hinder β oxidation in microorganisms, for example, an ortho-chloro or ortho-methyl group [251, 253]. In agreement with the plant studies [123], the hindrance appeared to be reduced in

the case of Nocardia coeliaca by the introduction of a 4-chloro group into
the 2-chlorophenoxyalkanoic acid molecule [253] . By contrast, Nocardia
opaca β-oxidized MCPB with greater difficulty than 2-methylphenoxybutyric
acid, and 3,4-dichlorophenoxybutyric acid with greater difficulty than either
3- or 4-monochlorophenoxybutyric acid, indicating a greater hindrance to
β oxidation by disubstitution than by monosubstitution [251] . The introduction
of a methyl substituent group on the 4-carbon of phenoxybutyric acid or
MCPB considerably retarded β oxidation of these compounds by Nocardia
opaca; however, lengthening the side chain of the methyl-substituted MCPB
to give 6-(4-chloro-2-methylphenoxy)-n-heptanoic acid reduced the retarding
effect of the methyl substitution [252] . Lengthening of the side chain also
facilitated the β oxidation of ω-phenoxyalkanoic acids by Nocardia coeliaca
[253] . Phenoxyalkanoic acids with 10 or 11 carbons in the side chain
appeared to undergo α oxidation as well as β oxidation in Nocardia coeliaca
[253] . Thus, in the initial stage of side-chain oxidation one carbon atom
is lost instead of two, so that short-chain phenoxy acids that would not be
produced by β oxidation alone were detected as metabolic products.

   As indicated in Fig. 5, 3-hydroxyacyl-coenzyme A thioesters are key
intermediates in the β oxidation of aliphatic acids. The metabolism of
phenoxybutyric acids to their corresponding 3-hydroxy derivatives was
demonstrated with Nocardia opaca, especially where ring substituent groups
hindered β oxidation to the phenoxyacetic acids [250, 251] . Blockage of
β oxidation in plants also appears to occur at the 3-hydroxyphenoxybutyric
acid stage.

   Gutenmann et al. [254] , using gas chromatography, showed that β oxidation of 2,4-dichlorophenoxyalkanoic acids occurred in a natural soil. The
metabolism of 2,4-DB via 2,4-D was also indicated by the sequential induction technique [255] . The degradation of higher phenoxyalkanoic acids by
the mixed natural soil microflora was thus similar to their degradation by
the pure cultures of soil bacteria studied by Webley et al. [250-252] and
Taylor and Wain [253] . However, the mixed population was able to further
degrade the products of β oxidation, namely 2,4-D and 2,4-dichlorophenol,
so that these compounds did not accumulate in high concentrations. Gutenmann and Lisk [256] subsequently obtained evidence that the β oxidation of
2,4-DB in soil proceeded by way of the expected first intermediate, the
unsaturated 2,4-dichlorophenoxycrotonic acid.

   b. β Oxidation and ring hydroxylation. Aspergillus niger β-oxidized
the side chain and hydroxylated the ring of 4-phenoxybutyric and 5-phenoxy-
valeric acid to form 4-hydroxyphenoxyacetic and 3-(4-hydroxyphenoxy)-
propionic acid, respectively [233] . These conversions resemble the
metabolism of unsubstituted phenoxyalkanoic acids in certain higher plants
[123, 124] . The fungus also produced small amounts of 2-hydroxyphenoxy-
acetic acid and 3-(2-hydroxyphenoxy)propionic acid from the phenoxybutyric
and phenoxyvaleric acids, respectively [233] .

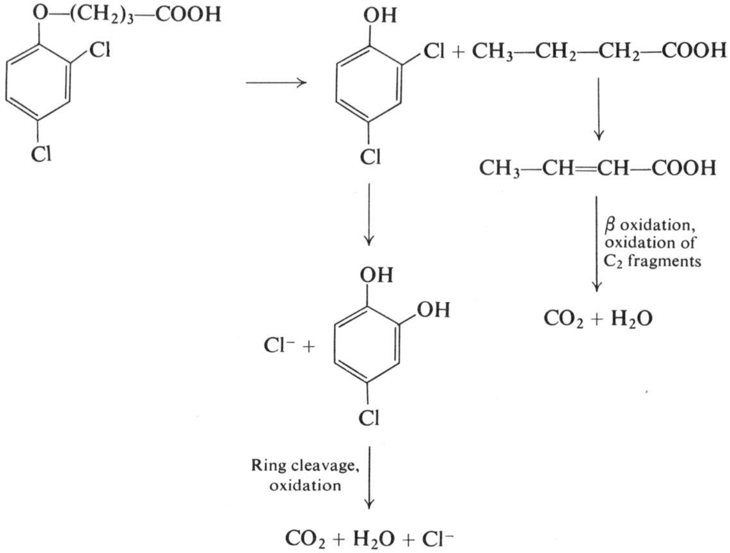

FIG. 11.   2,4-DB degradation by Flavobacterium sp., involving
cleavage of the ether linkage. (After MacRae et al. [257]. Reprinted from
Loos [94].)

c. Cleavage of the ether linkage.   MacRae et al. [257] demonstrated
that a 2,4-DB-decomposing Flavobacterium sp. [246] produced 2,4-di-
chlorophenol, 4-chlorocatechol, butyric, and crotonic acid when grown in
the presence of 2,4-DB. These results suggested that the 2,4-DB was
cleaved at the ether linkage and the products metabolized as shown in
Fig. 11. In agreement with the ether cleavage hypothesis, the bacterium
produced phenolic material and the expected unsubstituted aliphatic acids
when incubated with ω-2,4-dichlorophenoxyalkanoic acids from the propionic
to the undecanoic homolog [258, 259]. The cleavage of 2,4-DB at the ether
linkage, unlike β oxidation, would lead to immediate detoxication of the
herbicide. This detoxication was shown experimentally when alfalfa seed
inoculated with the flavobacterium was grown in 2,4-DB-treated sterilized
soil [260]. However, in nonsterilized soil, the inoculated organism could
not establish itself well enough to effect a noticeable detoxication of the
2,4-DB.

d. Ring cleavage.   The results of Alexander and Aleem [33] showed
that the ring structure of several ω-phenoxyalkanoic acids was lost during
their degradation in soil. Loss of aromatic structure was indicated by
disappearance of the ultraviolet absorption spectra of the compounds. Many

or all of the compounds were probably metabolized prior to ring cleavage by the mechanisms outlined above.

e. Resistance to microbial degradation. The effect of ring and side-chain substitution on the β oxidation of ω-phenoxyalkanoic acids by two Nocardia species has been discussed. Alexander and his co-workers [33, 246] investigated the degradation of ω-phenoxyalkanoic acids in soils and soil suspensions. The ω-phenoxypropionic and butyric acids with no meta substituent in the molecule were readily degraded, in many cases more rapidly than the corresponding phenoxyacetic acids. However, in agreement with the observations on phenoxyacetic acid degradation, the meta-substituted ω-phenoxyalkanoic acids were highly resistant to microbial attack, although in one experiment 2,4,5-TB was degraded in 103 days [246].

Field soil that had received nine spring and fall treatments of MCPA detoxified MCPB more rapidly than untreated soil [261]. Degradation of the MCPB by an MCPA-degrading enrichment microflora is suggested.

3. Higher α-Phenoxyalkanoic Acids

The results of Alexander and co-workers [33, 246] and Altom and Stritzke [596] demonstrate the ability of soils to degrade certain α-phenoxyalkanoic acids. The pathways of degradation are unknown, but the aromatic ring was destroyed, as indicated by disappearance of the ultraviolet absorption spectrum of the compound [33]. Degradation of the α-phenoxypropionic, -butyric, and -valeric acids was retarded in comparison with the corresponding phenoxyacetic and ω-phenoxyalkanoic acids. When the side chain was lengthened to give the 2-phenoxycaproic acids, the 4-chloro- and 2,4-dichlorophenoxycaproic acids were readily degraded, but consistent with the trend among phenoxyacetic and ω-phenoxyalkanoic acids, meta-substituted 2-phenoxycaproic acids resisted degradation. Silvex was detoxified by a strain of Streptomyces viridochromogenes isolated from soil [232].

A field soil adapted to metabolize MCPA and MCPB at a rapid rate following repeated applications of MCPA, showed no improvement over untreated soil in the detoxication of mecoprop, dichlorprop, and silvex [261].

4. Phenoxyalkanoic Acid Esters

Aly and Faust [262] obtained manometric evidence that sewage micro-organisms oxidized the alcohol moiety of the isopropyl and butyl esters of 2,4-D. The 2,4-D was not decomposed during the 9-day incubation period. It was assumed that hydrolysis of the ester preceded oxidation of the alcohol moiety. Bailey et al. [263] reported the hydrolysis of the propylene glycol butyl ether ester of silvex in pond waters, and showed that it followed first-order reaction kinetics. The hydrolysis was 90% complete in 16-24 hr and 99% complete in 33-49 hr. Adsorption of both the ester of silvex and the free acid appeared to occur on the sediment; essentially complete

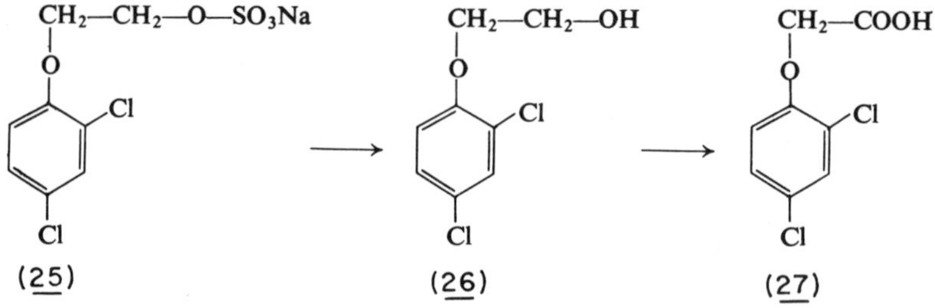

FIG. 12.   Degradation of sesone (25) in soil to 2,4-dichlorophenoxy-ethanol (26) and 2,4-D (27). (After Vlitos [264]. Reprinted from Loos [94].)

disappearance of both occurred by the fifth week following treatment. The report emphasized the possible roles of physicochemical factors in determining the fate of the herbicides. The results of Smith [597] suggest that the isopropyl and n-butyl esters of 2,4-D are rapidly hydrolyzed nonbiologically in soils; however, the isooctyl ester is more stable and can possibly undergo some biological hydrolysis.

5.   Phenoxyethyl Esters

The phenoxyethyl ester herbicides are active against susceptible plant species only when applied through the soil [4, 12]. The activation of sesone has been intensively studied [264-269]. Sesone shows little toxicity in sterilized soils of pH 5.5 and higher, but in nonsterile soil it is rapidly converted to 2,4-dichlorophenoxyethanol and 2,4-D [264]. Figure 12 shows the proposed degradation sequence. A strain of Bacillus cereus var. mycoides that degraded sesone to 2,4-dichlorophenoxyethanol was isolated [264, 265]. The formation of 2,4-D, which was also demonstrated by Audus [266], is apparently microbiological, but the responsible organisms were not isolated [264, 266]. The contribution of 2,4-dichlorophenoxy-ethanol to the phytotoxicity of sesone-treated soil is controversial [267, 268]. In soils with a pH of 4.0 or lower, 2,4-dichlorophenoxyethanol is formed nonbiologically from sesone [269].

Conversion to the corresponding phenoxyethanol or phenoxyacetic acid is apparently also the mechanism of activation of the other herbicidal phenoxyethyl esters [4].

C.   Degradation and Elimination by Animals

1.   Phenoxyacetic and 2-Phenoxypropionic Acids

Most studies on the fate of phenoxyacetic and 2-phenoxypropionic acids in animals have shown a rapid and essentially complete elimination of these compounds, without metabolism, in the urine. This was the fate of 2- and 4-chloro- and unsubstituted phenoxyacetic acid in rabbits [270] , 2,4-D in rats [271] , sheep (Ovis aries) [272] , and steers (Bos taurus) [273] , and MCPA, 2,4,5-T, and silvex in cows [274, 275] . Excretion rate depends on the compound; for example, rabbits fed phenoxyacetic acids at rates of 100 or 125 mg/kg body weight excreted in 6 hr 44-72% of the dose of unsubstituted and 2-chlorophenoxyacetic acid, but only 13-15% of the 4-chloro-phenoxyacetic acid [270] . However, by 24 hr more than 70% of all three acids had been excreted. Most of the 2,4-D fed to sheep [272] and the 2,4-D, 2,4,5-T, and silvex fed to cattle [273, 275] was eliminated within 1 day posttreatment but MCPA fed to cows was mainly excreted on the third day posttreatment [274] . The dose of herbicide may also influence the excretion pattern; for example, rats that received 1-10 mg of 2,4-D eliminated more than 90% within 24 hr, whereas those that received 60-100 mg eliminated only 40-60% in 24 hr and a further 16-36% from 24 to 38 hr posttreatment [271] .

Consistent with the excretion patterns just given, blood and other tissues normally show a transient accumulation of herbicide following absorption, and subsequent decreases in concentration as the herbicide is eliminated from the animal body. Radioactive 2,4-D fed to sheep at a dose of 4 mg/kg showed a peak concentration in the blood at 1.25 hr, followed by a rapid decrease within the next hour [272] . Erne [276] observed maximum concentrations of 2,4-D in the plasma of rats, pigs (Sus scrofa), calves, and chickens (Gallus domesticus) from 2 to 7 hr after oral application of 50-200 mg 2,4-D/kg, the peak concentrations for a dose of 100 mg/kg reaching 100-200 μg 2,4-D/ml plasma. Little 2,4-D remained in the plasma at 24 or 36 hr posttreatment. Low levels of 2,4-D (less than 0.4 μg/g) appeared in eggs from the 2,4-D-treated chickens, especially in the yolks at 24 and 48 hr posttreatment. The kidneys, livers, lungs, and spleens of the chickens, pigs, and rats accumulated from 60 to 260 μg 2,4-D/g fresh weight of tissue at 6 hr posttreatment, but the 2,4-D concentrations were greatly reduced at 24 hr and were almost eliminated at 48 or 72 hr posttreatment. Lower levels of 2,4-D accumulated transiently in other tissues.

A similar pattern is evident in the rat experiments of Khanna and Fang [271] , who showed that 2,4-D concentration reached a peak in most tissues at about 8 hr posttreatment. However, stomach tissue absorbed and accumulated a high concentration of 2,4-D within the first hour. A high dose of 2,4-D (80 mg/rat) was apparently absorbed over a much longer period than a low dose (1 mg/rat), resulting in high concentrations of 2,4-D persisting in the tissues for longer periods than in the case of the low dose. In six different tissues, most of the 2,4-D (56-88%) accumulated in the soluble fraction of the cells, followed by the nuclear (8-38%), microsomal (1-12%), and mitochondrial (1-11%) fractions.

Edible tissues of a sheep sampled 4 days after oral application of 4 mg radioactive 2,4-D/kg showed only very low levels of radioactive material [272]. By contrast, a sheep poisoned by four successive daily doses of 250 mg 2,4,5-T/kg accumulated from 20 to 60 ppm of the herbicide in its liver, muscle, omental, and renal fat, and 176 ppm in its kidneys [277]. Similarly, pigs poisoned by repeated daily oral doses of 50 or 100 mg 2,4-D/kg showed high levels of 2,4-D in their tissues [276].

Mice injected with 2,4-D eliminated most of the herbicide from their bodies within 24 hr [279]. The disappearance of 2,4,5-T was slower. The mechanisms of elimination were not investigated. No 2,4-dichlorophenol was detected in the bodies of the 2,4-D-treated mice.

Most of the 2,4-D fed to a dairy cow disappeared from the rumen within 23 hr [280]. The concentration in the rumen soon after feeding was 3.5 ppm; this had declined to 1 ppm by 8-12 hr, and to less than 0.5 ppm by 23 hr posttreatment. 2,4-D incubated with metabolically active rumen fluid in vitro for 12 hr showed no decrease in concentration. The conclusion was that the 2,4-D concentration decreased in the rumen in vivo as a result of dilution by imbibed water, movement out of the rumen with the feed residues, and possibly by absorption on the walls of the rumen.

Early investigations of 2,4-D in the milk of cows receiving the herbicide in their feed indicated no residues [280-282]. In two of the studies, high 2,4-D doses, namely 5.5 g [281] and 1.15 g, i.e., 50 ppm in 50 lb of feed [282], were administered daily. The analytical methods were bioassay and gas chromatography, respectively, the latter being sensitive to 0.1 ppm 2,4-D in the milk. A similar result was obtained with MCPA, using a gas-chromatographic analytical method sensitive to 0.4 ppm herbicide [282]. However, methods sensitive to 0.01 ppm 2,4-D have since shown residues of the herbicide in the milk of cows grazing pastures sprayed with 2 lb/acre of the isopropyl, 2-ethylhexyl, or butyl ester of 2,4-D [283]. The highest residue in the milk was 0.06 ppm 2,4-D. Residues were maximal 1-2 days after spraying and were still detectable after 7 days but not after 14 days. Methods sensitive to 0.05 ppm herbicide failed to detect residues in the milk of cows receiving in their feed 1000 ppm MCPA, 300 ppm 2,4-D or silvex, or 100 ppm 2,4,5-T (although at this rate cream contained detectable 2,4,5-trichlorophenol), but low concentrations of the last three herbicides were observed in the milk at the higher feeding rates of 1000, 1000, and 300 ppm, respectively [278].

The few reports of metabolism of phenoxyacetic acids in animals and the low levels of metabolites detected, seem to indicate the rarity of this phenomenon. Small amounts of an acid-hydrolyzable conjugate of 2,4-D appeared in the urine of pigs [284], and traces of an unidentified metabolite of radioactive 2,4-D were found in rats, especially in the liver and urine [271]. A metabolite of 2,4,5-T excreted with the unchanged acid in the urine

of rats was identified as N-(2,4,5-trichlorophenoxyacetyl)glycine [285].
Low levels of the corresponding phenol were detected in milk from cows
receiving 100 ppm 2,4,5-T in the feed, in the milk of one cow receiving
1000 ppm 2,4-D or MCPA, and in the tissues of sheep or cattle receiving
2,4-D or 2,4,5-T at rates of 1000 ppm or greater [278].

2.   Phenoxyacetic and 2-Phenoxypropionic Acid Esters

Hydrolysis of phenoxyacetic and 2-phenoxypropionic acid esters to the
free acids apparently occurs readily in animals. The butyl ester of 2,4-D
fed orally to pigs and rats was recovered from urine, plasma, red blood
cells, or liver almost entirely in the free acid form [284]. The rate of
absorption of ester by the pigs, rats, and calves was low compared with the
rate of absorption of free acid [276]. Cows fed Kuron, the propylene glycol
butyl ether esters of silvex, over a 4-day period, eliminated 67.5% of the
dose as silvex in the urine during the treatment period and two subsequent
days [275]. No unhydrolyzed Kuron was excreted in the urine, and no Kuron
or silvex appeared in the milk. Only low concentrations of 2,4,5-T and its
propylene glycol butyl ether esters were found in the omental fat of cattle
fed the esters at rates of 0.15 or 0.75 mg/kg for up to 32 weeks [277].

Ester hydrolysis in sheep was variable and possibly influenced by the
dose of herbicide. Sheep dosed on four successive days with 0.15 or 0.75
mg/kg of the propylene glycol butyl ether esters of 2,4,5-T hydrolyzed a
portion of the esters to 2,4,5-T, particularly at the higher dose [277].
Esters and 2,4,5-T rapidly appeared in the blood and urine, usually within
1-3 hr posttreatment. Only low concentrations of 2,4,5-T (0.08 ppm)
accumulated in the omental fat. A sheep that received a single dose of
25 mg esters/kg hydrolyzed little of the esters to 2,4,5-T, and 86% of the
dose was recovered as esters in the urine. By contrast, sheep that received
a toxic dose of 250 mg esters/kg for four or six successive days showed
high concentrations of 2,4,5-T in their omental and renal fat, liver, and
muscle (33-113 ppm), and particularly in their kidneys (240-368 ppm).
Ester concentrations in these tissues were about 1 ppm or less. Conversion
of the esters to 2,4,5-T occurs quite rapidly in raw muscle and other tissues, necessitating analysis for this metabolite as well as the esters in
residue studies [286].

The butyl and isooctyl esters of 2,4-D disappeared from the bodies of
injected mice more rapidly than the 2,4-D acid [279]. Disappearance of the
butyl ester was particularly rapid, and the rate appeared to be increased
by pretreatment injections of herbicide for 5 days preceding the test. Mechanisms of herbicide disappearance were not investigated.

Fish (Lepomis gibbus) and blue crabs (Callinectes sapidus) held for
3 days in water treated with the butoxyethanol ester of 2,4-D at a rate of
30 lb acid equivalent per acre, accumulated relatively low concentrations

of 2,4-D and/or ester in their bodies [287]. The residues calculated as
ester were less than or equal to 0.3 ppm in the fish, and less than 0.8 ppm
in the crabs. By contrast, oysters (Crassostrea virginica) and clams (Mya
arenaria) accumulated in their tissues from 3.5 to 3.8 ppm of the herbicide,
predominantly in the ester form. Further studies with bluegills (Lepomis
machrochirus R.), rainbow trout (Salmo gairdneri R.), and channel catfish
(Ictalurus punctatus R.) held in water containing 0.3 or 1.0 ppm of 2,4-D
butoxyethanol ester, showed maximum accumulations of herbicide in fed
fish after 1-2 hr and in fasted fish after 1-8 hr [288]. The ester, but not
free 2,4-D, was readily absorbed by the fish through the gill membranes.
The absorbed ester was hydrolyzed in the fish tissues to 2,4-D, which was
rapidly excreted.

3. Higher ω-Phenoxyalkanoic Acids

ω-Phenoxyalkanoic acids undergo a limited β oxidation in animals. Thus,
rabbits fed 4-phenoxybutyric or 6-phenoxycaproic acid eliminated a portion
of the dose (13-38% and 20-23%, respectively) as phenoxyacetic acid in the
urine [270]. Much of the 4-phenoxybutyric acid was eliminated unchanged
at a rate similar to that of phenoxyacetic acid.

In 2,4-DB-fed dairy cows, 2,4-DB and 2,4-D, together equivalent to
12.7% of the administered dose, were eliminated in the urine [273]. As
2,4-D is excreted quantitatively in the urine, it could be calculated that
9.3% of the administered 2,4-DB had been β-oxidized. Similar results were
obtained with cows fed MCPB; with doses equivalent to 2.5 and 5 ppm in the
feed the recoveries of MCPA in the urine indicated 9.2 and 7.2% β oxidation,
respectively [274]. All metabolically produced 2,4-D and MCPA was eliminated within 1 day of the 2,4-DB or MCPB feeding.

A rapid disappearance of 2,4-DB from the rumen of a dairy cow was
observed [51]. A concentration of 5 ppm 15-20 min after feeding was
reduced to 0.5 ppm about 5 hr later. This rate of disappearance was many
times greater than could be accounted for by movement of water into the
rumen or passage of material out of it, and therefore presumably resulted
from the activities of the rumen microflora. However, there was no evidence of β oxidation of the 2,4-DB to 2,4-D in the rumen. No 2,4-DB was
eliminated in the feces.

No 2,4-DB or 2,4-D residues were detected in the milk of cows
receiving a daily dose of 2,4-DB equivalent to 50 ppm in 50 lb of feed [282].
The analytical method would have detected 0.1 ppm of either herbicide.

4. Erbon

The metabolism of erbon has been investigated in sheep [289, 290].
Erbon incubated with blood, urine, or feces for at least 1 hr was converted
completely to 2-(2,4,5-trichlorophenoxy)ethanol (Fig. 13); the same

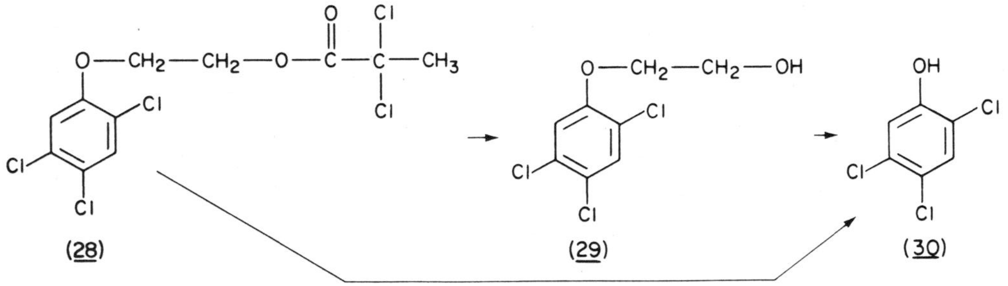

FIG. 13. Possible pathways for the metabolism of erbon (28) in sheep via the identified metabolites 2-(2, 4, 5-trichlorophenoxy)ethanol (29) and 2, 4, 5-trichlorophenol (30).

conversion took place in other sheep tissues, but was not as rapid or complete [289]. However, in further experiments with fortified liver homogenate or rumen fluid the conversion was almost complete in 20 min, with most of the erbon converted to the trichlorophenoxyethanol within 10 min [290]. Small amounts of 2,4,5-trichlorophenol were produced as a second metabolite in feces and brain tissue [289].

Erbon administered orally to a living sheep was mainly eliminated in the urine as the metabolites, 2-(2,4,5-trichlorophenoxy)ethanol and 2,4,5-trichlorophenol, which were present in approximately equal molar concentrations [290]. Almost 70% of the dose was eliminated by this route in 96 hr, as opposed to less than 2% in the feces. Most of the elimination (54%) occurred from 7 to 23 hr posttreatment, following a sharp rise and subsequent fall of 2-(2,4,5-trichlorophenoxy)ethanol in the blood during the first 6 hr posttreatment. A sheep that received daily oral doses of 100 mg erbon/kg and which died on the eighth day accumulated 5-6 ppm of 2-(2,4,5-trichlorophenoxy)ethanol and 2-6 ppm of 2,4,5-trichlorophenol in its kidneys, liver, and omental fat, but 1 ppm or less of the metabolites in its brain and muscle tissue. Sheep dosed daily with 50 or 25 mg of erbon/kg accumulated less than 1 ppm of the trichlorophenoxyethanol in their kidneys, liver, and omental fat and showed no ill effects during the 11-day experimental period.

III.   MODE OF ACTION

A.   Uptake and Translocation

Phenoxy herbicides may be applied to leaves, stems, or, by way of the soil, roots of plants. To exert a herbicidal effect, they must be absorbed into the plant and translocated to the sites of action. Herbicide uptake and translocation have been well reviewed [4, 93, 96, 291-304]; these reviews should be consulted for general information on plant structure and physiology, which is presented here without references. The role of uptake and translocation in determining the susceptibility or resistance of plants to phenoxy herbicides is considered in Sec. III, C.

1.   Foliar-Applied Herbicides

Herbicides applied to leaves must traverse the cuticle covering the surface of these organs to enter the plant. The material of the cuticle is lipoidal (cuticular waxes) and semilipoidal (cutin). Immediately below the cuticle are the hydrophilic cellulose and pectic materials of the epidermal cell walls. The cellulose is organized in micelles and microfibrils. Interfibrillar spaces or channels designated ectodesmata traverse the epidermal cell walls and cuticle to just below the surface of the latter [304]. Some of the ectodesmata apparently serve as extrusion channels for wax, while others seem to be filled with an aqueous phase. Thus, there appear to be both lipid and aqueous routes for penetration of herbicides into the plant once the surface of the cuticle has been penetrated. Even stomatal penetration of herbicides involves passage through an invaginated cuticle [301, 303, 304]; hence, entry into and departure from this structure control the rate of herbicide uptake into the plant.

Herbicide formulation has a significant influence on absorption. Sodium and potassium salts of the phenoxy herbicides have a low lipid solubility and penetrate poorly, particularly if the application medium has a high pH [4, 291, 294, 305, 306]. As the pH is lowered the herbicide is increasingly less dissociated, and the free acid, owing to its lower polarity, penetrates the cuticle more readily [4, 291, 294, 305, 306-308]. The low pH also lowers the polarity of the cuticle by suppressing the ionization of carboxyl groups [293]. Ammonium and amine salts of the phenoxyalkanoic acids, i.e., salts derived from relatively weak bases, penetrate the cuticle more readily than the sodium and potassium salts [305, 306, 309]. Esters penetrate the cuticle even more readily [291, 305, 309], but rapid penetration and high contact toxicity of the short-chain alkyl esters seem to be disadvantages in weed control, possibly because of damage to the translocation mechanism [175, 310]. Heavy esters formed from long-chain alcohols with lipophilic and hydrophilic properties, for example butoxyethanol esters, have proved very effective in respect to both absorption and translocation [291]. The suggestion [291, 294, 310] that the short-chain alkyl esters have little tendency to leave the cuticle and enter the living protoplasm of the underlying cells, i.e., the symplast system, is not supported by work with barley showing hydrolysis of the isopropyl ester of 2,4-D during penetration [175]. This hydrolysis should enhance herbicidal activity by producing a hydrophilic carboxylic acid with increased tendency to partition into the symplast. However, the work of Wathana et al. [311] on 2,4-DB absorption by soybean and cocklebur shows clearly that very large amounts of applied herbicide can be immobilized in the cuticle. This immobilization is likely to be intensified where elongation of the phenoxy acid side chain takes place.

The extent to which the molecular structure of the phenoxyalkanoic acid affects absorption is not clear. In many studies, 2,4-D has been absorbed more rapidly into leaves than the more lipophilic 2,4,5-T [109, 171, 312-315] and 2,4-DB [171], while MCPA penetrated sunflower leaves, but not flax (linseed) cotyledons, more readily than 2,4,5-T [316]. On the other hand, Sargent [317, cited in 302] reported that penetration of phenoxyacetic acids into bean leaf discs was favored by increased chlorination.

The inclusion of surface-active compounds as emulsifying agents in formulations of phenoxy acids and esters often increases toxicity, apparently through increased absorption of the herbicide [305, 318-324]. In addition to their role as wetting agents and possibly humectants [300], the surfactants dissolve surface waxes [325] and may loosen the structure of the cuticle [301, 326]. The carrier in oil formulations, which are used particularly against woody weed species [291, 327], will also interact with the lipoidal cuticle and might be expected to assist absorption. However, the postulated benefits of oil additives [291, 301] often fail to enhance herbicide action [327]. The surfactant and oil additives should be free of rapid phytotoxic effects [291, 321, 322, 328-330] which might reduce herbicide uptake and translocation. Inorganic ions, such as ammonium, nitrate, and phosphate, may be useful additives for promoting phenoxy herbicide absorption [331, 332], but dimethyl sulfoxide seems to be of limited value or even detrimental [333].

Temperature increases in the physiological range, i.e., below 30°-35°C, increase the absorption of phenoxy herbicides [157, 307, 334-337], presumably by increasing disorganization of the lipid materials arranged in micelles in the cuticle and plasma membrane, with consequent increased cuticle and membrane permeability [293]. Low humidity or growth under conditions of moisture stress have been reported to decrease phenoxy herbicide absorption [4, 305, 336] or to have little or no effect [157, 338-340]. Increased herbicide uptake at high humidity possibly involves opening of the stomata [336], which may also be a factor in the stimulation of uptake as light intensity is increased [307, 337, 341]. The role of the stomata is uncertain, but both physical and metabolic factors appear to be involved in 2,4-D absorption [337]. Herbicide uptake through stomata does not avoid the barrier of the cuticle, but this structure seems to be thinner and more readily penetrated inside the stomatal chamber, where favorable moisture conditions for absorption should occur [303, 304, 342].

After a herbicide has traversed the cuticle, it has two possible paths for movement through the plant, namely the living symplast continuum consisting of the cell protoplasts connected by plasmodesmata and including the sieve-tube system of the phloem, or the nonliving apoplast consisting of the continuous cell-wall system and including the xylem conducting elements. Foliar-applied phenoxy herbicides characteristically move from the leaves by way of the symplast, moving with the stream of assimilates from "source to sink," for example from photosynthesizing leaves to sites of active growth in roots, shoots, flowers, and fruits [109, 310, 313, 321, 343-345]. The autoradiographs of Radwan et al. [346] show clearly the movement of foliar-applied 2,4-D in the phloem of the petioles. No 2,4-D is translocated from the leaves of dark-grown plants until sugar is made available, either by photosynthesis or direct addition, for movement by way of the phloem [347-352]. However, in the aspen (Populus tremula L.), transport of 2,4-D and 2,4,5-T was predominantly acropetal in the xylem, even when movement started in the phloem [352]. A powerful lateral movement of herbicide from phloem to xylem was therefore indicated.

The path of herbicide movement after penetration of the cuticle is through the epidermis and mesophyll cells to the cells of the phloem. In the bean, the shortest path involves at least one palisade parenchyma cell and one or more cells of the bundle sheath, the minimum distance from leaf surface to phloem being about 25-30 μm [353]. This distance was traversed by 2,4-D, including passage through the cuticle, in about 1 hr, compared to a translocation rate in the phloem of 10-100 cm/hr [353]. According to Crafts [4, 294, 298], free 2,4-D acid applied in aqueous solution to the leaf will enter the cuticle at a pH of about 3.3, but will be at a pH of about 5.5 in the epidermis and 7.0 in the phloem. At these higher pH levels the acid will be dissociated, which will benefit transport in the aqueous phase. However, compared with certain other herbicides, the movement of phenoxy herbicides such as 2,4-D and 2,4,5-T through the plant by way of the symplast is restricted [171, 344, 354-356], i.e., they tend to be retained by living cells [344, 356]. Possible barriers to herbicide movement are discussed by Robertson and Kirkwood [96]. In the phloem, phenoxy herbicides tend to be accumulated by the phloem parenchyma, and hence to move laterally out of the mainstream of assimilate transport [4]. Retention of the phenoxy herbicides seems to be minimized by rapid translocation processes; thus, weed control by foliar application is best achieved during periods of vigorous root growth when the herbicide will be drawn with assimilates into this sink [4, 344, 357]. Some plants may even leak herbicide into the surrounding medium following its effective translocation to the roots, for example 2,4-D-treated cotton [4], jimsonweed [152], night-flowering catchfly, common lambsquarters [155], and honeyvine milkweed [Ampelamus albidus (Nutt.) Britt.] [357], and 2,4-D- or 2,4,5-T-treated beans [4, 358].

Soil moisture and atmospheric humidity have marked effects on the translocation of phenoxy herbicides. Beans subjected to moisture stress by growth in soils of low moisture content, translocated greatly reduced amounts of 2,4-D or 2,4,5-T from treated leaves to the roots, stems, and stem apices [338-340, 359]. High atmospheric humidities stimulated the translocation of 2,4-D or 2,4,5-T, particularly in the downward direction, in beans [336, 358], honey mesquite [157], and wolftail (Carex cherokeensis Schwein) [333]. Under conditions of 100% relative humidity, 2,4-D applied to cotyledons or leaves of cotton and several oak species (Quercus douglasii Hook. & Arn., Q. suber L., and Q. wislizenii A. DC.) readily entered the xylem, and was translocated at a high rate by both phloem and xylem [360-362]. Translocation in these plants at lower humidities normally occurs mainly in the phloem. Low atmospheric humidities, presumably associated with high transpiration rates, stimulated xylem transport of 2,4,5-T in the bean [358] and honey mesquite [157].

Translocation of 2,4-D in the bean increased as the temperature was raised from 20° to 30°C [336]. Temperature affected the pattern of 2,4,5-T translocation in honey mesquite; thus, movement was primarily basipetal to the roots at 21°C, basipetal to the roots and acropetal to the growing tip at 30°C, and acropetal for a short distance in the upper stems at 38°C [157]. These movements apparently reflected translocation patterns, for example, no sink at the shoot tip at low temperature, and possibly contact injury to the vascular tissue by the 2,4,5-T at 38°C.

The effect of light on herbicide translocation through its effect on photosynthesis and the translocation of assimilates has been indicated previously. Increasing the light intensity from 40 to 4000 ft-c decreased the translocation of 2,4,5-T in post oak (Quercus stellata Wangenh.), but in three other woody plants had little influence on the percentage of absorbed 2,4,5-T that was translocated [341].

2.   Stem-Applied Herbicides

Absorption of phenoxy herbicides into green or succulent stems is similar to absorption into leaves [93]. Stems covered by bark are normally not penetrated by herbicides in aqueous solution, and require an oil carrier such as diesel oil when the herbicide is applied as a basal spray. However, aqueous solutions of herbicide are suitable where a cut is first made through the bark to the outer sap wood.

Translocation of stem-applied 2,4,5-T in red maple, white ash, and marabu (Dichrostachys nutans Benth.) was mainly acropetal in the xylem, with little or no downward movement in the phloem [180, 363]. For the control of marabu, Hay [363] proposed that sufficient spray should be applied at ground line to ensure that phytotoxic quantities of herbicide reach the exterior of the roots. This is similar to the recommendation of Crafts [93] that excess spray should be applied to stems so that liquid flows down from the crown of the plant and kills the buds from which shoots arise.

3.   Root-Applied Herbicides

Phenoxy herbicides are readily absorbed into the roots of plants [347, 364-371] but translocation to the tops tends to be very restricted [364-366, 370]. The route, as outlined by Crafts [93, 298], is essentially aqueous, through the external apoplast in the root hair region of absorbing roots and into the symplast. Physical adsorption and metabolic absorption are apparently involved in herbicide uptake, both being promoted by a low external pH [365]. Alternatively, herbicide uptake may essentially involve adsorption on cell surfaces that require metabolic energy to maintain their integrity [598]. The herbicide may be accumulated within the symplast to higher concentrations than in the external medium [371]. Within the symplast the herbicide migrates from the epidermis cells through the cortex and endodermis into the stele, where it leaks into the apoplast, probably as a result of the high carbon dioxide and low oxygen tension [93, 298]. Translocation out of the roots takes place in the transpiration stream of the xylem [346, 347]. The poor translocation of root-applied phenoxy herbicides apparently results from their strong retention by living cells in their passage through the symplast, as indicated by increased translocation in the presence of metabolic inhibitors such as dinitrophenol [365, 370].

B.   Biochemical Basis of Activity

1.   Auxin and Antiauxin Activity of Phenoxyalkanoic Acids

The phenoxyalkanoic acid herbicides belong to the large group of plant growth-regulating compounds designated auxins. Auxins may be defined as organic substances which at low concentrations (less than 10^{-3} M) promote growth along the longitudinal axis of shoots and inhibit elongation of roots [372]. Natural auxins include indole-3-acetic acid (IAA) and related compounds such as 3-(indole-3-)propionic, 4-(indole-3-)n-butyric, and 3-(indole-3-)pyruvic acid, and indole-3-acetaldehyde and indole-3-acetonitrile [373, 374]. Synthetic auxins, such as the herbicidal phenoxyalkanoic acids, show effects similar to those of the natural auxins in standard plant and plant tissue tests, for example the oat and wheat coleoptile elongation tests, split pea stem curvature test, growth inhibition of wheat and flax roots, induction of root formation, parthenocarpic fruits and morphological responses, the tomato leaf epinasty test, and others [374, 375]. Some of these tests, for example the coleoptile elongation and epinasty tests, indicate effects on cell elongation, whereas others indicate effects on both cell division and elongation [376, 377]. For detailed reviews of the effects of auxin herbicides on plants see Audus [378].

Examples of compounds with auxin activity are shown in Fig. 14 [40, 160, 372, 374, 379-383]. The synthetic auxins include, besides the phenoxy herbicides, important herbicides of other chemical classes, for example, the phenylacetic and benzoic acids and 4-amino-3,5,6-trichloropicolinic acid (picloram). A mode of action depending on auxin activity is therefore suggested for these herbicides. It is generally accepted that the phenoxy and other auxin herbicides are analogs of indoleacetic acid, with the ability to take the place of this natural auxin in reactions which control the growth of the plant. However, normal control of growth is lost in the case of the herbicides, which do not respond to the mechanisms controlling IAA concentrations at different sites in the plant.

While it is logical to postulate that synthetic auxins owe their activity to their structural resemblance to IAA, it has proved a difficult task to determine from a study of these compounds the minimum molecular requirements for auxin activity. The tortuous progress may be traced through a succession of reviews [31, 40, 160, 372, 374, 379-381, 384-388]. The requirements proposed in 1938 by Koepfli et al. [389], namely a ring

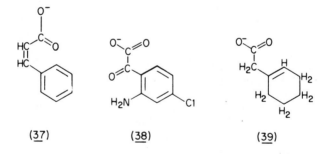

FIG. 14.   Examples of compounds with auxin activity. Anions of (31)
IAA, (32) 2,4-D, (33) naphthalene-2-oxyacetic, (34) 2,3,6-trichlorophenyl-
acetic, (35) naphthalene-1-acetic, (36) naphthalene-2-acetic, (37) cis-
cinnamic, (38) 4-chloroisatic, (39) cyclohex-1-eneacetic acid.

FIG. 14 (cont.). (40) 1,2,3,4-tetrahydronaphthalene-1-carboxylic, (41) indane-1-carboxylic, (42) acenaphthene-1-carboxylic, (43) fluorene-9-carboxylic, (44) 9,10-dihydrophenanthrene-9-carboxylic, (45) naphthalene-1-carboxylic, (46) 2,3,6-trichlorobenzoic, (47) 4-amino-3,5,6-trichloropicolinic acid, and (48) N,N-dimethyl-S-carboxymethyldithiocarbamate.

containing one or more double bonds, a side chain with a carboxyl group (or
a group which is convertible to a carboxyl) situated at least one carbon atom
away from the ring, and a particular spatial relationship between the acid
group and the ring, were proved inadequate by the discovery of the auxin
activity of compounds (40)-(48) (Fig. 14) and their relatives. It is clear
from the activity of (48) and certain other dithio- and thiocarbamates [379,
390, 391] that a ring structure is not essential, and from the activity of
(40)-(47) that a carbon between the carboxyl and the ring is not a requirement for auxin activity. Furthermore, many compounds with the same basic
structure as the auxins in Fig. 14, but with different ring or side-chain
substitutions, show little or no auxin activity and act as antiauxins [392,
393]. Such observations seem to complicate the picture, but, in fact, they
have played an important role in the development of theories of the molecular
requirements for auxin activity and auxin reaction with receptor sites in the
plant. As a basis for considering these theories, experimental data on the
auxin and antiauxin activity of phenoxyalkanoic acids will be presented.

Antiauxin effects must first be explained. Antiauxins are defined as
competitive inhibitors of auxins [381]. The competitive inhibition of auxin-
stimulated elongation of oat coleoptile segments or curvature of split pea
stems by antiauxins is readily observed and recognized as an antiauxin
effect [392]. Inhibitory effects of antiauxins on nonauxin-induced elongation
of shoot parts can be ascribed to competitive inhibition by the antiauxins of
endogenous auxin activity, which is at suboptimal levels in shoots and
coleoptiles [393]. Antiauxins stimulate root growth by counteracting the
inhibitory effects of endogenous auxin which is present in roots at supraoptimal concentrations; as a corollary they counteract the intensified inhibition of root growth resulting from the addition of exogenous auxin [393].
The stimulation of root growth by antiauxins may be very marked in certain
plants, such as wheat, and the term "root auxins" has been used by some
authors [381, 394].

Auxin and antiauxin activities of many different phenoxyalkanoic acids
have been intensively studied by Åberg [395-402] and Wain and his colleagues
[376, 377, 392, 403-409]. The compounds exhibited activities ranging from
purely auxinic to purely antiauxinic; Åberg [410, 411] has used the term
"intermediate regulator" for compounds with both auxinic and antiauxinic
character. Effects of selected chlorine-substituted phenoxyacetic acids in
plant tests, representing the entire range of activities, are indicated in
Fig. 15 and Table 7. The last column of the table shows the system of Åberg
[411] for summarizing the auxinic and antiauxinic character of phenoxyalkanoic acids. These assessments of auxin and antiauxin activity are based
on the growth effects shown in Fig. 15, and additional confirmatory tests
[395-402]. Thus, auxin activity was confirmed by the ability of the antiauxin,
2-(1-naphthylmethylthio)propionic acid, to alleviate the growth inhibition
caused by the test compound in the flax root test, and antiauxin activity by

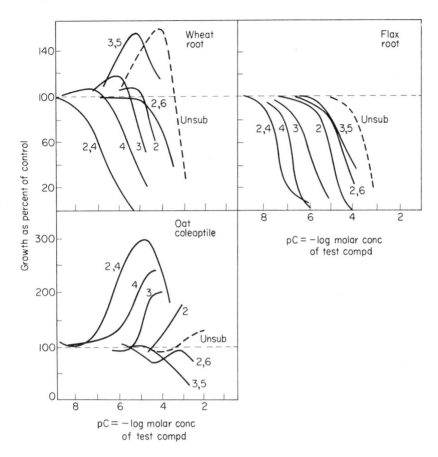

FIG. 15. Activity of chlorine-substituted phenoxyacetic acids in auxin assay tests. The numbers beside the curves indicate the positions of the chlorine substituents. Unsub is unsubstituted phenoxyacetic acid. (Compiled from Åberg [397, 399, 400] by permission of The Agricultural College of Sweden and B. Åberg.)

the ability of the test compound to antagonize the stimulation of oat coleoptile elongation caused by IAA, or the inhibition of flax root growth caused by 2,4-D. The possibility of an incorrect assessment of auxin or antiauxin activity was minimized by these procedures, but not completely eliminated, for example, in the case of 3,5-dichlorophenoxyacetic acid, which is possibly incorrectly classified as an antiauxin [412]. The results of the auxin activity tests in Wain's laboratory (Table 7), with the exception of the phenoxyacetic acid results, support the auxin ratings of Åberg [411].

TABLE 7

Activity of Selected Phenoxyacetic Acids in Auxin Assay Tests[a]

Phenoxyacetic acid	Minimum active concentration in tests[b]										Activity[d] in tomato epinasty test at indicated concentration (% in lanolin) [405]	Activity rating[d] according to Åberg [411]	
	Wheat coleoptile cylinder[c]		Oat coleoptile cylinder		Maize root segment	Cress root	Pea segment	Pea curvature				Auxin	Antiauxin
	pC [407]	ppm [405]	pC [377]	ppm [404]	pC [408]	pC [408]	pC [407]	pC [407]	ppm [405]	ppm [404]			
2,4-Dichloro-	7[e]	0.1[e]	5[e]	0.1[e]	5[e]	6[e]	7[e]	7[e]	0.1[e]	0.1[e]	+++ at 1	+++	0
4-Chloro-	7[e]	0.1[e]	4[e]	0.1[e]	6[e]	6	5[e]	5[e]	0.1[e]	0.1[e]	++ at 1	+++	(+)
3-Chloro-	4	10		1	5	5	5	5	10	1	+++ at 5	++	+
2-Chloro-	3	100	4	1	4	4	3	3	100	1	+ at 1	+	+
2,6-Dichloro-	3	I at 100	I at 25				4	4	10	5	0 at 5	(+)	+
Unsubstituted	I	I	I	I					I	25	0 at 5	+	++
3,5-Dichloro-	I	I	I	I			I	I	I	I	0 at 5	0	++

[a]Numbers in brackets are references.

[b]I = inactive.

[c]pC = -log molar concentration.

[d]Decreasing activity from +++ to (+); 0 = inactive.

[e]Lowest concentration tested.

Other phenoxyalkanoic acids can be similarly rated according to their auxin and antiauxin activity by comparing their growth-regulating effects with those recorded in Fig. 15 and Table 7. This has been done in Table 8, which summarizes the auxin and antiauxin activities of the complete range of α-phenoxyalkanoic acids studied by Åberg [395-402] and Wain and his colleagues [376, 377, 392, 403-409]. Results from Thimann [413], Muir and Hansch [414], and Matell [415] are also included. The table does not indicate the responses of individual test plants to the phenoxy acids, but for many of the compounds this information is available in Tables 5, 7, and 11 of Jönsson [381]. The activity of ω-phenoxyalkanoic acids has been discussed in Sec. II.

    a. Activity of D- and L-enantiomers. Table 8 shows quite clearly that auxin activity in the 2-phenoxypropionic, -n-butyric, and -n-valeric acids is associated with the D configuration, i.e., the configuration corresponding to that of D(-)-lactic acid. With the exception of 2-iodophenoxypropionic acid, all of the D enantiomers are dextrorotatory [40].

The L enantiomers, in all cases, displayed considerable antiauxin activity in the experiments of Åberg [397, 399, 401, 402, 411]. However, some of the iodo- and dichlorophenoxypropionic acids also showed evidence of a weak auxin activity [401, 411]. The antiauxin activity of L(-)-2-(2,4-dichlorophenoxy)- and L(-)-2-(2,4,5-trichlorophenoxy)propionic acid is confirmed by the results of Wain and his colleagues [392, 403].

    b. Effect of side-chain length. In the α-phenoxy-n-alkanoic acids, auxin activity increases as the side chain increases in length from the acetic to the propionic acid (D enantiomer) (Table 8). The 2-phenoxy-n-butyric acids have activity similar to that of the corresponding 2-phenoxy-n-propionic acids, according to activity data for the DL racemates [377, 405, 408], and the D and L enantiomers in the case of the unsubstituted and 2,4-dichlorophenoxy acids [397, 415]. Unsubstituted D(+)-2-phenoxy-n-valeric acid shows a decrease in activity compared with D(+)-2-phenoxy-n-butyric acid, and further lengthening of the side chain to give D(+)-2-phenoxy-n-caproic acid results in the complete loss of auxin activity [397]. The 2-phenoxyisobutyric acids, which lack an α-hydrogen, are mainly inactive as auxins (see the following section), as is DL-2-phenoxyisovaleric acid, which does not lack an α-hydrogen [397].

Among the ω-phenoxyalkanoic acids with three or more carbons in the side chain, only a few 3-phenoxypropionic acids are active per se; the activity of other homologs depends on β oxidation to one of these compounds. Active 3-phenoxypropionic acids include the unsubstituted and 2-chlorophenoxypropionic acids, which were active in wheat coleoptile and pea stem tissue, and the 2,3-dichlorophenoxypropionic acid, which was active in pea tissue [123].

TABLE 8

Auxin and Antiauxin Activity of α-Phenoxyalkanoic Acids

Class of acid	Auxin activity[a]						References
	0	Slight	(+)	+	++	+++	
Phenoxyacetic							
Unsubstituted			Unsub ++				396, 404, 411, 414
Fluoro		2-F ++	3-F ++		4-F 0		398, 411
Chloro				2-Cl +	3-Cl +	4-Cl (+)	Table 7, 396, 399, 414
Bromo				2-Br +	3-Br (+) 4-Br +		396, 411, 414
Iodo			4-I +++	2-I +	3-I (+)		396, 398, 411, 414
Dihalogeno	3,5-Cl₂ ++ 2,6-Br₂		2,6-Cl₂ + 2,4-I₂ +	2,3-Cl₂ ++	2,5-Cl₂ (+)	2,4-Cl₂ 0 3,4-Cl₂ +	400, 404–408, 411, 414

2,4-F₂ →
← 2,4-Br₂

Substituent						References
Trihalogeno	2,4,6-Br$_3$ 2,4,6-I$_3$ 6-Br-2,4-Cl$_2$ 2,4-Cl$_2$-6-I 2,4-Br$_2$-6-Cl 2,4-Br$_2$-6-I 2,6-Br$_2$-4-Cl	2,4,6-Cl$_3$ ++	2,3,6-Cl$_3$	←2,3,4-Cl$_3$→ 2,3,5-Cl$_3$ 3,4,5-Cl$_3$	2,4,5-Cl$_3$ 0 2,4-Cl$_2$-6-F 2,4-Br$_2$-6-F	404–408, 411, 414
Tetrahalogeno				←2,3,4,5-Cl$_4$→		406
Pentahalogeno	2,3,4,5,6-Cl$_5$					406
Methyl			3-Me ++ 4-Me ++		2-Me (+)	396, 399, 407, 411, 414
Ethyl	2-Et		4-Et ++			396, 411
Isopropyl	4-iP +++ 2-iP					396, 411, 414
Isopropenyl			2-iPe +	3-iPe (+)		398
tert-Butyl	4-tB ++	←2-tB→ +	3-tB +			396, 398, 411
Trifluoromethyl					3-CF$_3$ (+)	398

(continued)

TABLE 8 (continued)

Class of acid	0	Slight	(+)	+	++	+++	References
				Auxin activity[a]			
[Phenoxyacetic] **Dialkyl**	3,5-Me$_2$ ++ 3-Et-5-Me	2,6-Me$_2$ +		2,3-Me$_2$ (+) 2,4-Me$_2$ (+) 2,5-Me$_2$ (+) 3,4-Me$_2$ (+)			400, 402, 406, 407 414
Trialkyl	2,4,6-Me$_3$		2,3,5-Me$_3$				406, 414
Halogeno/Alkyl	3-Cl-5-Me 4-Cl-2-Et		2-Cl-3-Me 2-Cl-5-Me	2-Cl-4-Me 2-Cl-6-Me 3-Cl-2-Me 3-Cl-4-Me 5-Cl-2-Me	4-Cl-3-Me	4-Cl-2-Me	405, 408, 409, 414
Dihalogeno/Alkyl	2,4-Cl$_2$-6-Me 2,6-Cl$_2$-4-Me 2,6-I$_2$-4-Me		2,5-Cl$_2$-4-Me	2,4-Cl$_2$-5-Me		4,5-Cl$_2$-2-Me	406, 408, 414
Halogeno/Dialkyl	4-Cl-2,6-Me$_2$ 4-Cl-2iP-5-Me ++		4-Cl-3,5-Me$_2$	4-Cl-3,5-Me$_2$			406, 411, 414

Methoxy	2-MeO ++			3-MeO + 4-MeO +	396, 398, 411, 414
Dihalogeno/Methoxy		2,4-Cl₂-6-MeO			413
Methylthio		4-MeS ++		3-MeS (+)	398
Nitro	2-NO₂ ++			3-NO₂ +	396, 411, 414
		←————— 4-NO₂ —————→ ++			
Dinitro	2,4-(NO₂)₂				414
Dinitro/Methyl	2,4-(NO₂)₂-6-Me				406
Dichloro/Nitro			2,4-Cl₂-5-NO₂		414
2-Phenoxypropionic					
Unsubstituted	L(-)-Unsub +			D(+)-Unsub 0	397, 411
Fluoro	L(-)-4-F +			D(+)-4-F	411
Chloro	L(-)-2-Cl ++			D(+)-2-Cl 0	399, 411
	L(-)-3-Cl ++			D(+)-4-Cl 0	
	L(-)-4-Cl ++			D(+)-3-Cl (+)	

(continued)

TABLE 8 (continued)

Class of acid	Auxin activity[a]						References
	0	Slight	(+)	+	++	+++	
[2-Phenoxypropionic]							
Bromo	L(−)-4-Br ++				D(+)-4-Br +		411
Iodo	L(−)-4-I +++		L(+)-2-I + L(−)-3-I +	D(+)-4-I +	D(−)-2-I 0	D(+)-3-I (+)	411
Dihalogeno	L(−)-2,5-Cl$_2$ (+) L(−)-3,5-Cl$_2$ + L(−)-2,4-Cl$_2$ ++	L(−)-2,3-Cl$_2$ + L(−)-2,6-Cl$_2$ + L(−)-3,4-Cl$_2$ ++	D(+)-3,5-Cl$_2$ +	D(+)-2,3-Cl$_2$ (+) D(+)-2,6-Cl$_2$ (+)		D(+)-2,4-Cl$_2$ 0 D(+)-2,5-Cl$_2$ 0 D(+)-3,4-Cl$_2$ 0 DL-2,4-Br$_2$	401, 411
Trihalogeno	L(−)-2,4,5-Cl$_3$ ++ D(+)-2,4,6-Cl$_3$ +++ L(−)-2,4,6-Cl$_3$ +++ DL-2,4,6-Br$_3$ DL-2,4-Br$_2$-6-I	DL-2,4,6-Cl$_3$ DL-2,4,6-I$_3$ DL-2,4-Cl$_2$--6-I DL-2,4-Br$_2$--6-Cl		DL-2,6-Br$_2$--4-Cl	DL-2,3,4-Cl$_3$ DL-6-Br--2,4-Cl$_2$	D(+)-2,4,5-Cl$_3$ 0	405, 408, 411 405, 406, 408

Type							Ref
Tetrahalogeno	D(+)-2,3,4,6-Cl$_4$ +++ L(-)-2,3,4,6-Cl$_4$ +++						←— DL-2,3,4,5-Cl$_4$ —→	406, 411
Methyl	L(-)-2-Me ++ L(-)-3-Me ++ L(-)-4-Me ++		D(+)-4-Me +	D(+)-3-Me +	D(+)-2-Me 0			399, 411
Dialkyl	L(-)-2,3-Me$_2$ ++ L(-)-2,4-Me$_2$ ++ L(-)-2,5-Me$_2$ ++ L(-)-2,6-Me$_2$ ++ L(-)-3,4-Me$_2$ ++ L(-)-3,5-Me$_2$ ++ DL-3-Et-5-Me	D(+)-2,6-Me$_2$ + D(+)-3,5-Me$_2$ +	D(+)-3,4-Me$_2$ (+)	D(+)-2,3-Me$_2$ (+) D(+)-2,4-Me$_2$ (+)	D(+)-2,5-Me$_2$ 0			402, 406
Halogeno/Methyl	L(-)-4-Cl-2-Me +++				D(+)-4-Cl-2-Me			415
Dihalogeno/Methyl	DL-2,6-Cl$_2$-4-Me DL-2,4-Cl$_2$-6-Me							406

(continued)

TABLE 8 (continued)

Class of acid	Auxin activity[a]						References
	0	Slight	(+)	+	++	+++	
[2-Phenoxypropionic] Halogeno/Dialkyl	L(−)-4-Cl--2-iP-5-Me +++	DL-4-Cl--2,6-Me$_2$	D(+)-4-Cl--2-iP-5-Me +++				406, 411
2-Phenoxy-n-butyric							
Unsubstituted	L(−)-Unsub +				D(+)-Unsub 0		397, 411
Chloro				DL-3-Cl	DL-2-Cl	DL-4-Cl	377, 405
Dichloro	L(−)-2,4-Cl$_2$ DL-3,5-Cl$_2$			DL-2,3-Cl$_2$ DL-2,6-Cl$_2$	DL-2,5-Cl$_2$ DL-3,4-Cl$_2$	D(+)-2,4-Cl$_2$ DL-2,4-Cl$_2$	377, 395, 405, 408
Trichloro		DL-2,4,6-Cl$_3$			DL-2,3,4-Cl$_3$	DL-2,4,5-Cl$_3$	377, 405, 408
Chloro/Methyl						DL-4-Cl-2-Me	377, 405
2-Phenoxy-isobutyric							
Unsubstituted	Unsub +++						377, 397, 405
Chloro	2-Cl +++ 3-Cl	4-Cl +++					377, 399, 405

			References
Dichloro	2,5-Cl$_2$	2,3-Cl$_2$	377, 401, 405
	2,6-Cl$_2$	2,4-Cl$_2$ ++	
		3,4-Cl$_2$	
		3,5-Cl$_2$ ++	
Trichloro	2,4,6-Cl$_3$	2,3,4-Cl$_3$	376, 377, 405, 408
		2,4,5-Cl$_3$	
Chloro/Methyl	4-Cl-2-Me		376, 377
2-Phenoxy-n-valeric	L(-)-Unsub ++	D(+)-Unsub (+)	397
2-Phenoxy-isovaleric	DL-Unsub ++		397
2-Phenoxy-n-caproic	D(+)-Unsub ++		397
	L(-)-Unsub ++		

[a] Antiauxin activity, if known, is indicated beneath the acid. Auxin or antiauxin activity decreases from +++ to (+) or slight; 0 = inactive as auxin or antiauxin.

c. α-Hydrogen effect. The α-hydrogen effect is shown in the phenoxyalkanoic acids by the lack of auxin activity of 2-phenoxyisobutyric acids in
tests involving cell elongation, for example the tomato leaf epinasty, oat
curvature, oat cylinder, and wheat cylinder tests [376, 377, 397, 399, 401,
405] . By contrast, most of the corresponding 2-phenoxy-n-butyric acids
(D enantiomers or DL racemates) showed considerable activity in these tests
[376, 377, 397, 405] .

| 2-Phenoxyisobutyric acid | D(+)-2-Phenoxy-n-butyric acid |

Some 2-phenoxyisobutyric acids showed auxin activity in tests which
depend on cell division as well as elongation, for example the tomato
parthenocarpy tests, the tomato leaf-rooting test, and the production of
morphogenic responses in tomato, as well as in the pea curvature test,
which does not assess simple cell elongation or division [376, 377, 405] .
However, in many cases the activity was lower than that of the corresponding
2-phenoxy-n-butyric acids tested in the same way. Fawcett et al. [405]
obtained no indications from chromatographic studies that 2-(2,4,5-tri-
chlorophenoxy)isobutyric acid was converted in pea to the corresponding
propionic or acetic acid. Wain and his colleagues [160, 374, 376, 377, 405]
concluded that a hydrogen on the α-carbon plays an important role in determining the auxin activity of the phenoxyalkanoic acids and is essential for
high auxin activity. Replacement of the two α-hydrogens of 2,4-D by
deuterium did not reduce auxin activity [416] , in contrast to their replacement by fluorines which greatly reduced activity [417] , in spite of the
van der Waals radius of fluorine being close to that of hydrogen [32] .
Replacement of one hydrogen by fluorine gave a (+) enantiomer, which was
less active than 2,4-D on pea, and a (-) enantiomer with negligible activity
[417] .

The 2-phenoxyisobutyric acids which are inactive as auxins have a
marked antiauxin activity which may result in considerable stimulation of
root growth [392, 394, 397, 399, 401, 418-420] .

d. Replacement of ether oxygen by sulfur or nitrogen. The effect of
replacing the ether oxygen of phenoxyalkanoic acids by sulfur or nitrogen is
indicated by the growth-regulating activities of S-phenylthioglycolic acids
and their α-alkyl homologs [381, 405, 421, 422] , N-phenylglycines [423-
425] , and N-phenylalanines [395, 423, 426] . Many of these compounds are
effective as auxins, but they generally show a lower activity than the corre-

sponding phenoxy acids. D(+)-N-(2,4,5-Trichlorophenyl)alanine is a much
stronger auxin than the L(-) enantiomer [423]. The sulfoxide (aryl-SO-CH$_2$-
COOH) corresponding to MCPA showed low auxin activity which disappeared
in the sulfone (aryl-SO$_2$-CH$_2$-COOH) [422]; the sulfoxide and sulfone corresponding to 2,4-D were both inactive [421]. Conversion of active phenyl-
glycines to their amides had little effect on activity [424]. Methyl substitution
of the bridging nitrogen reduced the auxin activity of 4-chloro- and 3,4-
dichlorophenylglycinamide [427], while the activity of N-phenylglycine and
N-(4-chlorophenyl)glycine was eliminated by acetylation but not by carboxy-
methylation of the bridging nitrogen [425, 427].

   e. Alternatives to the carboxyl group. Replacement or substitution
of the carboxyl group of phenoxyalkanoic acids with groups that can be converted to it in the plant, for example nitrile, amide, and ester groups, has
been considered in Sec. II. Phenoxythioacetic acids, i.e., with a -COSH
in place of the -COOH, showed activity comparable to, higher or lower than
that of the corresponding phenoxyacetic acids [428]. The phenoxymethane-
phosphonous acid analogs of 4-chlorophenoxyacetic acid, 2,4-D, and 2,4,5-
T are antiauxins or inactive, but 2-chlorophenoxymethanephosphonous acid
has a weak auxin activity [429]. Among the corresponding phosphonic and
phosphonate analogs, ethyl 2,4-dichlorophenoxymethanephosphonate showed
slight auxin activity in the split pea and pea epicotyl tests; it also increased
the dry weight of tops of wheat seedlings, as did the 2-chloro analog [430].
Toxicity was increased in the 2,4-dichlorophenoxymethanephosphonate
diethyl ester [429]. By contrast, 2,4-dichlorophenoxymethanesulfonic acid
was inactive [431].

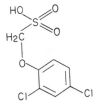

2-Chlorophenoxy-   2,4-Dichlorophenoxy-   2,4-Dichlorophenoxy-
methanephosphonous acid   methanephosphonic acid   methanesulfonic acid

   f. Effects of ring substitution. From the data of Table 8 and the
supporting references, the effects of ring substitution on the growth-
regulating activity of α-phenoxyalkanoic acids may be summarized as
follows:

   1. The auxin activity of unsubstituted acids, particularly phenoxy-
       acetic acid, is relatively low.

2. Introduction of a single halogen into the ring in most cases
    increases auxin activity. The position effect on activity is
    4- > 3- > 2-, with the exception of 4-I and to a lesser extent, 4-Br.
    The decreasing activity of the 4-halogen-substituted phenoxyacetic
    and D(+)-2-phenoxypropionic acids in the direction 4-Cl > 4-Br >
    4-I is associated with an increasing van der Waals radius of the
    substituent atom (Cl 1.80; Br, 1.95; I, 2.15 Å) [410].

3. An activating effect of para substitution by F, Cl, and Br, is also
    evident in the dihalogenophenoxy acids. Dimeta and diortho halo-
    genation, particularly the former, are deactivating.

4. Para halogenation again shows its activating effect in the trihalo-
    geno and tetrahalogeno compounds, except where the product is a
    2,4,6-trisubstituted phenoxy acid. The latter compounds, and
    other diortho-substituted trihalogenophenoxy acids, tend to have
    low or no auxin activity. Exceptions to this rule are the 2,4-
    dichloro-6-fluoro- and 2,4-dibromo-6-fluorophenoxyacetic acids,
    and 2-(6-bromo-2,4-dichlorophenoxy)propionic acid. The deacti-
    vating effect of dimeta substitution is counteracted by additional
    ortho and para substitution (see the 2,3,5- and 3,4,5-trichloro
    and the 2,3,4,5-tetrachloro compounds).

5. The 2,3,4,5,6-pentachlorophenoxyacetic acid, with all ring posi-
    tions substituted, is inactive.

6. Monomethyl substitution shows the opposite position effect to halogen
    substitution, i.e., 2- > 3- > 4-, particularly in the D(+)-2-phenoxy-
    propionic acid series. The 2-methylphenoxyacetic and D-2-(2-
    methylphenoxy)propionic acids are more active auxins than the
    corresponding 2-chloro compounds. Replacement of the methyl
    hydrogens of 3-methylphenoxyacetic acid by fluorines increases
    auxin activity from a low to a very high level.

7. In the dimethylphenoxy acids, as in the dihalogeno compounds,
    dimeta substitution is more deactivating than diortho substitution.
    A methyl group in one of the ortho positions is activating, but a
    para-methyl tends to be deactivating.

8. Increasing the size of the alkyl group in the 2 and 4 positions results
    in loss of auxin activity, for example the 2-ethyl, 2- and 4-isopropyl,
    and 2- and 4-tert-butyl substitutions. Substitution in the 3 position
    does not appear to be influenced in this way, as indicated by the
    activity of 3-isopropenyl- and 3-tert-butylphenoxyacetic acid.

9. The trends outlined above for halogeno and methyl substitution can
    be detected in the phenoxy acids substituted with both halogen and
    methyl groups. The 4-chloro-2-methyl combination is particularly
    effective.

10. Activity in the methoxyphenoxyacetic acids shows the same trend as in the halogenophenoxyacetic acids, i.e., decreasing auxin activity from the 4-substituted to the 2-substituted acid, which is completely inactive as an auxin. The trend in the methylthiophenoxyacetic acids cannot be discerned from Table 8.

11. Position effects in the nitrophenoxyacetic acids seem to resemble those in the bromo- and iodophenoxyacetic acids, i.e., an increase in auxin activity from 2 to 3 substitution, and weak or zero auxin activity associated with the large 4 substituent.

12. There is a general inverse relationship between auxin and antiauxin activity, i.e., the weaker the auxin activity of a compound, the stronger is its antiauxin activity. The most powerful auxins are free of antiauxin activity, and vice versa.

13. Ring substituents other than those discussed under 2-11 have proved of little or no significance in the promotion of auxin activity.

2. Concepts of Auxin and Antiauxin Structure and Interaction with Cell Receptor Molecules

a. Auxins. Concepts of the molecular requirements for auxin activity have undergone considerable modification since the proposals of Koepfli et al. [389], as a result of the continued discovery of new compounds with auxin activity.

The requirement for an acidic group is still valid, but may be fulfilled by a carboxyl or, usually less efficiently, by a thiocarboxyl (-CO-SH), sulfonic acid (-SO$_2$-OH), hydrogen sulfate (-O-SO$_2$-OH), phosphonous acid [-PO(H)-OH], phosphonic acid [-PO(OH)$_2$] group, or even a tetrazole group (-C=N-N-N-NH) or the aci form of the nitromethane group (-CH=NO-OH) [374, 379, 381]. The auxinic phenols [432, 433] have a weakly acidic hydroxyl group.

The requirement for a ring with one or more double bonds has had to be modified in view of the auxin activity of the thio- and dithiocarbamates [379, 390, 391]. These compounds may be analogous to the auxins with a benzene ring in that they can assume a planar structure through internal electron shifts [379, 390, 391, 434]:

Further analogies with the auxins containing an aryl group are the auxin
activity of (-)- and antiauxin activity of (+)-N,N-dimethyl-O-(1-carboxy-
ethyl)thiocarbamate [435], and the activity of N,N-dimethyl-(ω-carboxy-
alkyl)dithiocarbamates only if they are β-oxidized by the plant tissue to the
carboxymethyl homolog [436]. The modified requirement for activity
therefore becomes a planar structure which may or may not be a ring.

There is general agreement that the acidic group and the planar part
of the molecule are required for reaction with a specific receptor site or
sites in the cell. As will be discussed later, there is now direct evidence
for the protein nature of the receptor(s). There is agreement that the
steric, electronic, and hydrophilic-lipophilic properties of the auxins control their activity, although details are not completely clear.

The role of steric features of the side chain is indicated by the α-hydrogen effect, and the auxin activity of the D and inactivity of the L isomers of
α-phenoxy and other α-arylpropionic, -butyric, and -valeric acids [40,
160, 374, 381]. Wain and colleagues [160, 374, 392, 403] have postulated
a three-point attachment to the receptor site through the carboxyl group,
planar group, and the α-hydrogen, which must be in the correct configuration. However, this hypothesis ignores the active benzoic, naphthoic,
α-methylene- and α-isopropylidenephenylacetic acids, which have no
α-hydrogen [166, 374, 379, 381]. To overcome this problem, Jönsson [381]

α-Methylene-
phenylacetic acid

α-Isopropylidene-
phenylacetic acid

proposed that the α-hydrogen in the correct configuration is essential, not
for reaction with a receptor site, but because a larger atom or group would
sterically hinder the correct approach of the auxin to the receptor. This proposal is less restrictive than the three-point attachment hypothesis and permits an explanation of the auxin activity of 2-(indole-3)-, 2-(2,4,5-trichloro-
phenoxy)-, and possibly 2-(2,3,4-trichlorophenoxy)isobutyric acid [405] and
the small difference in the activity of the enantiomers of 2-(indole-3)- and
2-(naphthalene-1)propionic acid [438, 439]. In these compounds the ring
system is particularly effective and the rather small buckling of the side
chain caused by the α-methyl group is perhaps not critical [381].

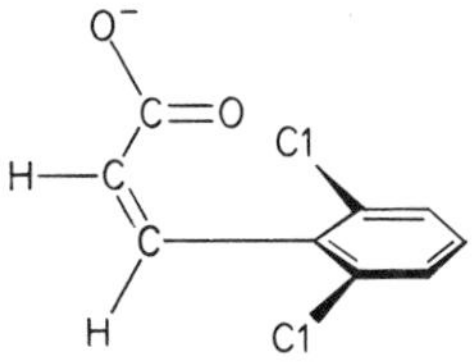

FIG. 16.   Auxins with a fixed spatial relationship between the carboxyl group and planar part of the molecule. (49) 2,6-Dichloro-cis-cinnamic acid, (50) 2,6-dichlorobenzoic acid, (51) 2-chloronaphthalene-1-carboxylic acid, (52) 1,4-dihydronaphthalene-1-carboxylic acid, (53) acenaphthene-1-carboxylic acid, (54) benzodioxole-2-carboxylic acid.

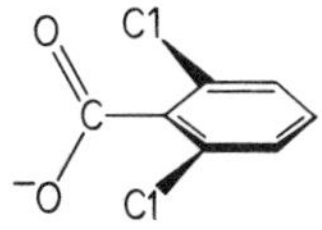

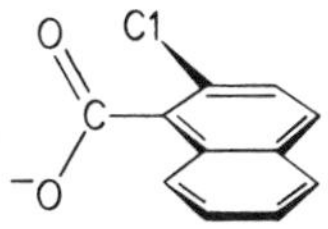

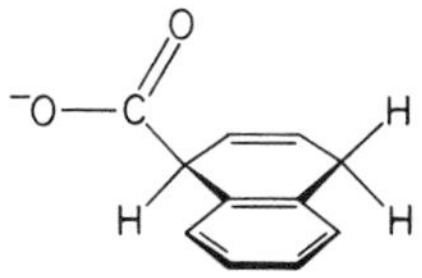

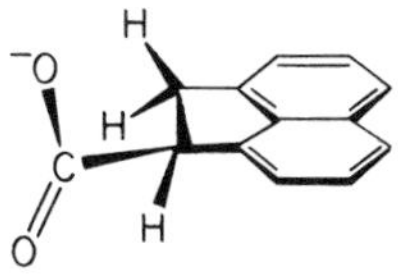

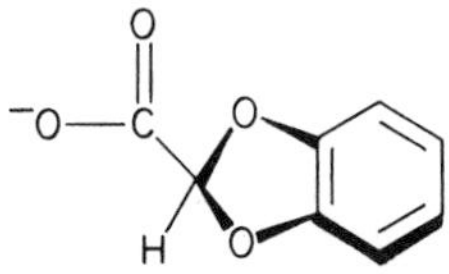

The spatial relationship between the carboxyl group and the planar part of the molecule has received considerable attention. In compounds where the relationship can be defined, i.e., where free rotation of the side chain is restricted, activity of aromatic auxins is associated with a carboxyl oxygen or oxygens out of the plane of the ring. Examples are 2,6-dichloro-cis-cinnamic acid [440], 2,6-dichlorobenzoic acid [379, 384], 2- and 8-methyl or halogenonaphthalene-1-carboxylic acid [379], 1,4-dihydro- and 1,2,3,4-tetrahydronaphthalene-1-carboxylic acid [441], acenaphthene-1-carboxylic acid [441], and unsubstituted, 5-chloro-, and 4,6-dichloro-1,3-benzodioxole-2-carboxylic acid [442] (Fig. 16). Arrangements such as those in Fig. 16 are obviously possible in the arylacetic and aryloxyacetic acid auxins where free rotation of the side chain can occur, as are other arrangements which may result in higher activity. Thus, the inactivity of 2,4,6-tribromophenoxyacetic acid as an auxin in contrast to the high activity of 2,4-dichloro-6-fluoro- or 2,4-dibromo-6-fluorophenoxyacetic acid has been ascribed to the restriction of free rotation of the side chain by the two ortho-bromines but not by a bromine or chlorine and the small fluorine atom [160, 374, 406]. The activity of the compounds illustrated in Fig. 16 does not support the theory of Jönsson [381, 443] concerning the requirements for appreciable auxin activity, specifically the proposal that the side chain, including the carboxyl carbon, should be situated almost in the plane of the ring, forming a pseudoring with the carboxyl group close to the original ring and near the center of the extended ring system thus formed. However, their activity does support the structure proposed by Chamberlain and Wain [444] for highly active diortho-substituted phenylacetic acids, namely, the carboxyl above the plane of the ring with its axis of rotation approximately perpendicular to the ring surface (Ugochukwu and Wain [440], Fig. 4). Large ortho and para substituents on the phenoxyalkanoic acids are associated with auxin inactivity (Table 8) [396, 398, 410, 411, 445].

Electronic effects have proved difficult to interpret, but the effects of ring substitution on auxin activity of the phenoxyalkanoic acids, benzoic acids, and phenols indicate their importance. With the exception of the weakly electron-donating methyl group, the substituents that increase activity are electron withdrawing, for example, nitro and halogen substituents. In the phenoxyalkanoic acids, a free ortho position with a relative deficit of negative charge, for example, resulting from ortho- and/or para-halogen, meta-nitro, or meta-trifluoromethyl substitution, is associated with high activity (Table 8). Methyl substituents, although weakly electron donating, resemble the halogens in being ortho-para directing in respect to electrophilic substitution, and can replace halogen substituents without too much decrease in activity. Activity in the benzoic acids [407, 414, 446] and phenols [432, 433] seems to be associated with a vacant para position with a relative deficit of negative charge, as in the 2,6-dihalogeno compounds.

On the basis of such results, Porter and Thimann [372, 447, 448] have suggested that auxin activity is associated with a fractional positive charge on the planar part of the molecule at a distance of about 5.5 Å from the negative charge on the ionized carboxyl group. The 5.5 Å is an approximate value; for example, in the phenols the distance between the oxygen of the ionized hydroxyl group and the para-carbon is only 4.2 Å. The proposals are applicable to IAA and its benzothiophene and benzofuran analogs, in which the heteroatom carries a δ^+ charge as a result of the donation of a lone pair of electrons into the ring system [447, 448], the thio- and dithio-carbamates, and 4-amino-3,5,6-trichloropicolinic acid (picloram) which would have the δ^+ on the 4-amino group [382]. Thimann [372] suggested that 3-amino-2,5-dichlorobenzoic acid is an exception to the theory, but this compound might act in the same way as picloram which also has the amino group meta to the carboxyl. Thimann [447] and Wain and his colleagues [374, 392, 444] postulated an electrostatic attraction between the acidic group of the auxin and a positively charged receptor group, for example amino, followed by an attraction of the planar part of the auxin to a complementary planar site on the receptor molecule, presumably involving van der Waals forces. Covalent bonding was not visualized.

Possible chemical reactions between auxin and receptor involving covalent bond formation have been suggested by Hansch, Muir, and colleagues [387, 414, 446, 449]. Proposed reactions of 2,4-D and 2,6-dichlorobenzoic acid are indicated in Fig. 17. Amide bond formation might succeed the formation of the salt linkage between the carboxyl group of the auxin and the amino group of the receptor protein. The $-X^-$ attacking group would be a nucleophile, possibly the $-S^-$ of cysteine. There is evidence for the postulated halide elimination in 2,6-dichlorobenzoic acid-treated oat coleoptile tissue [449], and in reaction mixtures of ortho-halogenobenzoic acids, nucleophiles, and metal ions in vitro [450]. Nucleophilic attack with elimination of fluoride might occur with the active 2,4-dichloro-6-fluoro-and 2,4-dibromo-6-fluorophenoxyacetic acids as a result of the increased stability of the fluoride ion over that of chloride [450], but the theory of ortho-nucleophilic attack has had to be extended to account for the activity of certain other diortho-halogen-substituted phenoxy acids [387]. According to the extended theory, the carboxyl of the side chain of these acids can, if necessary, assume a position quite close to the para position of the ring which, if unsubstituted, would be subject to nucleophilic attack. A similar mechanism was proposed for the phenylacetic acids [387] but subsequent work indicates that the decrease in electron density in the meta rather than the para position, as required by this hypothesis, is the important electronic factor determining auxin activity of these compounds [451]. Also the extended theory does not explain the rather high activity of DL-2-(6-bromo-2,4-dichlorophenoxy)propionic acid [406].

Molecular orbital calculations suggest that interaction of the ortho position of benzoic acids with a nucleophile plays a role in determining auxin activity [452, 453]. Similarly, interaction of the 8 position of 1-naphthoic acids with a nucleophile or a radical appears to be important for activity [454]. However, a chemical reaction was not visualized, but rather the formation of a molecular complex of the charge-transfer type [454]. In the phenoxyacetic and S-phenylthioglycolic auxins, the meta position seemed to be the interacting position (Fukui et al., cited by Koshimuzu et al. [454]), in contrast to the ortho position in the Hansch-Muir hypothesis. The calculations of Cocordano and Ricard [455], who used linear combination of atomic orbitals procedures, suggest that there may not be a dilemma in these results, as growth activity of chlorinated

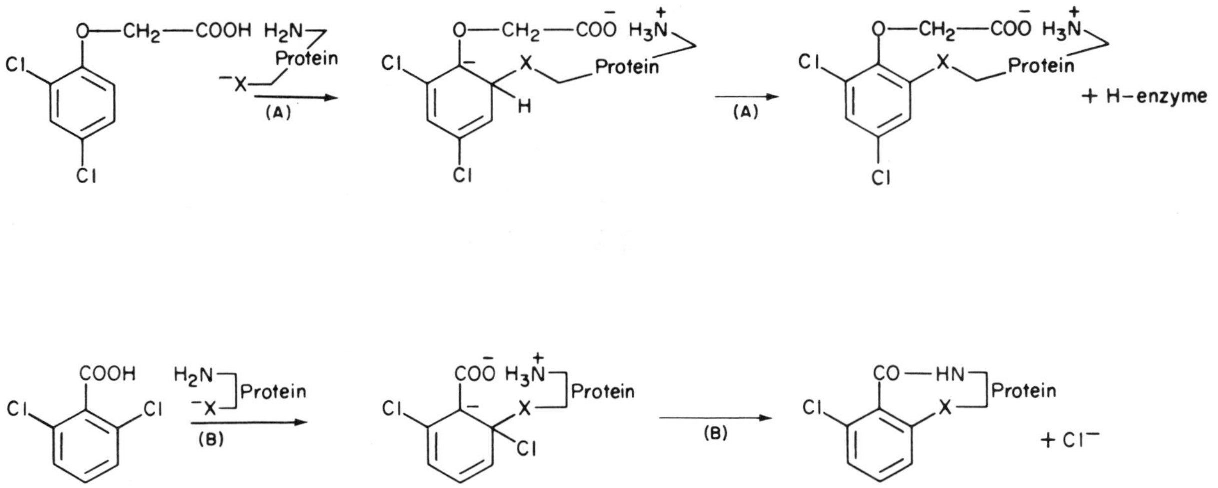

FIG. 17.   Postulated reactions of auxins with cell receptor proteins involving covalent bond formation. (A) 2,4–D type, without halogen elimination, and (B) 2,6–dichlorobenzoic acid type, with halogen elimination. (After Hansch et al. [449] and Muir and Hansch [414].)

phenoxyacetic acids seemed to be associated with the ability to react with a cell component through the 3 and 6 positions, simultaneously. A sulfur atom is theoretically capable of bonding with these positions. Unfortunately, this approach was not applied to such "problem" compounds as the active 2,4-dichloro-6-fluoro-, 2,4-dibromo-6-fluoro-, and 2,3,4,5-tetrachloro-phenoxyacetic acids.

The hydrophilic-lipophilic properties of the auxins seem to be of considerable importance in determining activity. An early hypothesis of Veldstra (see Veldstra and Booij [31]) attributed the activity of auxins to their interaction with cell membranes, resulting in increased permeability to nutrients and, consequently, increased cell growth rates. Although this view is no longer held, the hydrophilic-lipophilic balance in a molecule appears to affect its auxin activity, probably by regulating its rate of movement to the primary site of action [456, 457]. Thus, in the statistical studies of Hansch, Muir, and co-workers [452, 456, 457], Hammett sigma functions could be used to predict activity of phenoxyacetic and phenylacetic acids on the basis of electron density at the ortho and meta positions, respectively, if functions reflecting degree of lipophilicity of the molecules were also introduced into the equations.

b. Antiauxins. The competitive nature of the inhibition of auxin action by antiauxins is reminiscent of the well-known competitive inhibition of enzyme activity by substrate analogs. The action of antiauxins has therefore been visualized as a reaction with the auxin receptor site that does not lead to a growth response but which prevents competing auxin molecules from reacting in the manner required to trigger a growth response [392, 393, 403, 410, 420].

The most prominent antiauxins are close analogs of auxins, but possessing no or very little auxin activity. Powerful antiauxins among the phenoxyalkanoic acids include most 2-phenoxyisobutyric acids [392, 394, 397, 399, 401, 418-420], L enantiomers of the 2-phenoxypropionic, n-butyric, and n-valeric acids [392, 397, 399, 401, 403, 411], D- and L-2-phenoxy-n-caproic acid [397, 411], many diortho-substituted phenoxy-alkanoic acids [392, 400, 411, 458, 459], phenoxyalkanoic acids with large ortho or para substituent groups [396, 398, 410, 411], and unsubstituted phenoxyacetic acid [394, 396, 397, 411, 460]. The 3,5-disubstituted phenoxyalkanoic acids have also been regarded as powerful antiauxins [392, 400, 401, 411], but the antiauxin status of 3,5-dichlorophenoxyacetic acid was not confirmed by Firn and Wain [412]. Between the high-activity auxins and antiauxins are many compounds with intermediate activity, i.e., which act both as weak auxins and weak antiauxins (Table 8). The theory indicated in the preceding paragraph can explain, and hence is supported by, these structure-activity relationships.

The action of antiauxins was explained by Wain and his colleagues [392, 403] as attachment to one or two of the three points to which, according to their hypothesis, auxins must attach to elicit a growth response. The 2-phenoxyisobutyric acids have the ring and carboxyl group for attachment to two of the points, but lack the essential α-hydrogen. The L enantiomers of the 2-phenoxypropionic, n-butyric, and n-valeric acids have the α-hydrogen in the wrong configuration for attachment. An alternative explanation, indicated previously, is that the α-hydrogen in the correct configuration is not required for attachment, but if it is replaced by an alkyl group the correct approach of the auxin to the receptor is prevented. In the case of 2,6-dichlorophenoxyacetic acid, the ring system is unable to undergo the correct attachment for high auxin activity. Benzoic acids and cyclohexanecarboxylic acid acted as antiauxins against a range of auxins, but showed poor or negligible activity against IAA, as did indole-3-carboxylic acid [392]. However, these examples of inactivity against IAA are exceptions to the more general observation (see references in the preceding paragraph) that antiauxins are nonspecific in respect to the auxins that they antagonize, which supports the idea of a common site or sites of action for the many different auxins. Propionic and palmitic acid are unable to function as antiauxins; thus, both a carboxyl group and a saturated or unsaturated ring seem to be minimum requirements for antiauxin activity [392]. Åberg [410] has concluded that antiauxin activity of a compound is determined both by its lack of intrinsic auxin activity and its affinity for the auxin receptor site.

Bonner and his colleagues [420, 458, 461-463, summarized in 464, 465] interpreted the results of their kinetic experiments as indicating two-point attachment between receptor and auxin at suboptimal concentrations, and one-point competing attachment of antiauxins to the receptor. Michaelis-Menten and Lineweaver-Burke kinetics for enzyme-substrate systems seemed to apply to the auxin-receptor system. Inhibitory effects of anti-auxins on auxin action were typical competitive inhibitions. These results were supported by the experiments of Ingestadt [460], but deviations from the theoretical kinetics were observed. The competitive inhibition of auxin-stimulated growth of oat coleoptile segments by 2,4-dichloroanisole, 2-phenoxyisobutyric acids, 2,6-dichloro-, and 2,4,6-trichlorophenoxyacetic acid was taken as evidence of one-point attachment by antiauxins, as these compounds have either a ring or side chain, but not both, suitable for attachment [420, 458]. However, the experiments of Bonner and co-workers have not been free of criticisms (see Housley [465] and Galston and Purves [466]), including serious doubts that 2,4-dichloroanisole has more than a very slight antiauxin activity [467, 468].

c. Summary. Structure-activity relationships among the many different auxins seem to support the concept of a common mode of action and a common receptor site, although the details of the primary reaction are still obscure. The stereospecificity of the auxin-receptor interaction indicates

a protein-effector type of complexing. This conclusion is strengthened
by the competitive antagonism of auxin action by structurally related antiauxins.

3. Auxin-induced Stimulation of Nucleic Acid and Protein Synthesis

Much of the mystery surrounding the mode of action of natural and synthetic auxins disappeared when it was discovered that they stimulate nucleic
acid and protein synthesis (see Refs. 469-472 for reviews). Stimulation of
these syntheses leads naturally to a stimulation of cell growth, which may
be abnormal.

Early work showed the stimulated synthesis of ribonucleic acid (RNA)
and deoxyribonucleic acid (DNA) in IAA-treated tobacco pith tissue cultures
[473] , and of nucleic acids and protein in stems of 2,4-D-treated bean [474,
475] and cucumber seedlings [476] , cucumber hypocotyl, and corn mesocotyl tissues [476] . The stimulation of nucleic acid and protein synthesis
in relation to cell division and elongation was studied in detail in 2,4-D-
treated soybeans [477-479] . Treatment of seedlings with 2,4-D 60 hr after
germination inhibited root and shoot growth by inhibition of both cell division
and elongation. Normal meristematic activity resumed above the cotyledons
and in the apex of the hypocotyl, but not in the roots, 2-3 days after the
2,4-D application. Abnormal proliferation of fully elongated cells at the
base of the hypocotyl began within 12 hr, and led to root initiation between
24 and 48 hr. Hypocotyl tissue 5-10 mm below the apex showed an increase
in the diameter of cortical and pith cells within 12 hr and division within
36 hr; this region of the hypocotyl was characterized by abnormal radial
enlargement. The synthesis of RNA, DNA, and protein in the meristematic
hypocotyl apex was temporarily inhibited by the 2,4-D, but synthesis
resumed prior to the resumption of growth. In the lower regions of the
hypocotyl, the 2,4-D treatment induced an increased RNA synthesis after
12 hr, followed by an increased DNA and protein synthesis, so that a
marked accumulation of these materials coincided with the massive proliferation of tissue. The high RNA/protein ratio which developed at this
time subsequently declined. The increase in RNA was greatest in the
ribosomal fraction but was also considerable in the soluble fraction. It was
postulated [478] that 2,4-D stimulated a renewal of nuclear activity which
resulted in the production of ribosomal and soluble RNA and subsequently
protein to support massive tissue proliferation. This abnormal growth,
which would presumably result in destruction of the phloem [480] , might be
the main factor responsible for death of the plants [479] .

Subsequent work with cocklebur (Xanthium sp.) led to a similar conclusion [481] . Treatment with 2,4-D inhibited growth of the roots and
leaves, and promoted abnormal enlargement of the axis (apex, stem, and
tap root). The RNA, DNA, and protein content of the axis was stimulated,
the RNA content increasing more rapidly than the protein content. The RNA,

DNA, and protein of the leaves and, to a lesser extent, the cotyledons and lateral roots, were mobilized and translocated to support the growth of the axis, especially as the capacity for photosynthesis was lost after about 2 days. The induced senescence of the leaves, cotyledons, and lateral roots, i.e., the organs responsible for obtaining the plant's energy and nutrient requirements from the environment, must be a major factor contributing to the death of the plant.

Corn seedlings likewise showed a stimulation of protein and nucleic acid content when sprayed with growth-stimulating or herbicidal concentrations of 2,4-D [482]. It was postulated that the herbicidal action probably resulted from an excessive and exhaustive nucleic acid and protein synthesis which would preclude normal cell development and function. Wheat was unusual in that 2,4-D, 2,4,5-T, and other synthetic auxins caused increases in DNA and protein in the roots but decreases in RNA [599].

Excised soybean hypocotyl tissue incorporated [8-^{14}C]ADP into RNA at an increased rate in the presence of 5-100 ppm 2,4-D, i.e., concentrations which stimulated growth [483]. The RNA content of mature, but not meristematic or rapidly elongating tissue, increased by 25-30% in response to the 2,4-D. After a 1.5-hr exposure to 25 ppm 2,4-D, cells contained almost half the incorporated radioactivity in the nuclei-rich fraction and 25% in the ribosome-rich fraction. With further incubation, radioactivity was transferred from the nuclei-rich fraction to the ribosomes. The RNA synthesis was inhibited by actinomycin D, indicating that it was DNA dependent. These results provided evidence that the stimulation of RNA synthesis in response to auxin occurred in the nucleus.

The DNA and RNA of soybean hypocotyl tissue were purified and characterized by sucrose density gradient centrifugation and methylated albumin-kieselguhr (MAK) column chromatography [484]. A species of RNA distinct from soluble and ribosomal RNA was detected in the hypocotyl tissue [484] and intact soybean roots [485]. This RNA had a high AMP:GMP (adenosine monophosphate:guanosine monophosphate) ratio among its nucleotides similar to that of the soybean DNA, and was designated DNA-like RNA or D-RNA. It was synthesized at approximately twice the rate of ribosomal RNA, and had a mean life of about 2 hr. On the basis of these properties and its association with polyribosomes [486], the D-RNA appeared to be messenger RNA.

The importance of D-RNA synthesis in the stimulation of cell growth by 2,4-D was demonstrated by inhibitor studies [487-490]. Actinomycin D inhibited protein synthesis and normal or 2,4-D-induced cell elongation by inhibiting the synthesis of all species of RNA. By contrast, 5-fluorouracil selectively inhibited ribosomal and soluble RNA synthesis to a large degree, but did not significantly affect the synthesis of D-RNA or cell elongation. In the presence or absence of this inhibitor, 2,4-D stimulated the production of AMP-rich RNA, i.e., D-RNA and RNA which bound tenaciously to the MAK column. Tissue treated for 4 hr with 5-fluorouracil and varying actinomycin D concentrations, then induced to grow by 2,4-D treatment, showed a good agreement between the degree of inhibition of D-RNA and protein synthesis and the inhibition of growth. The antiauxin trans-cinnamic acid inhibited both the 2,4-D-stimulated elongation of etiolated pea internode segments and RNA synthesis, including that of the presumed messenger RNA [491, 492]. It appears, therefore, that the stimulatory effect of 2,4-D on plant cell growth is largely determined by the stimulation of messenger RNA synthesis, although the 2,4-D also stimulates the synthesis of other RNA, particularly ribosomal RNA. The stimulated messenger RNA synthesis, supported if necessary by the stimulated synthesis of ribosomal and soluble RNA, stimulates the production of proteins, including enzymes that play key roles in the growth processes. These are considered in Sec. III,B,4.

In agreement with the concept that auxins have a common mechanism of action, IAA and other auxins have shown effects on nucleic acid and protein synthesis which are similar to those described for 2,4-D (see Key [472]).

Experiments with isolated nuclei [493-496] confirmed that the nucleus is the site of auxin-stimulated RNA and DNA synthesis. Subsequently, O'Brien et al. [497, 498] showed that RNA synthesis by isolated chromatin-RNA polymerase preparations from soybean hypocotyls was stimulated by 2,4-D treatment of the plants prior to harvest, although not by 2,4-D treatment of the chromatin-enzyme preparations in vitro. The soluble RNA polymerase of corn behaved similarly [600]. The RNA synthesized by the chromatin-RNA polymerase preparations from control and 2,4-D-treated soybeans differed with respect to their elution profiles on MAK columns and the rate at which they incorporated [^{14}C]leucine into protein, the ratio of the activities being about 1:2 [497, 498]. When Escherichia coli RNA polymerase was added to the two systems, they saturated at about the same level of polymerase, suggesting that an equal quantity of template was available in both cases. The increased RNA synthesis as a result of 2,4-D treatment of the soybeans seemed, on the basis of these experiments, to result from the induction or activation of RNA polymerase rather than from more template being made available. The increased polymerase activity was expressed in the production of longer RNA chains, with only a 20-50% increase in the number of RNA chains [601]. However, in experiments with pea chromatin it appeared that auxin stimulated RNA synthesis by causing more DNA template to be made available [499]. In cucumber, both RNA polymerase activity and the availability of template seemed to be increased by auxin treatment of seedlings [602]. The loss of specific nucleohistones from 2,4-D-treated cucumber roots perhaps indicates a mechanism for the uncovering of template [603]. Further details of the auxin-induced stimulation of RNA synthesis by isolated chromatin-RNA polymerase systems are discussed in Sec. III, B, 5.

Chromatin preparations from soybean hypocotyls also showed stimulated DNA synthesis when the plants were treated with 2, 4-D prior to harvest [500]. No stimulation of DNA polymerase activity was evident during the first 6 hr after application of the auxin. As 2, 4-D-stimulated RNA synthesis can occur during that time, it appears that 2, 4-D, at concentrations which can induce cell division in soybean hypocotyl tissue, stimulates RNA polymerase activity prior to activation of DNA polymerase activity and cell division.

4.   Stimulation of Other Growth-Controlling Processes

a.   Cell-wall extension.   The elongation of cells following the treatment of certain plant tissues with auxins involves an increase in the plastic deformability (extensibility) of the cell walls, i.e., a loosening or plasticization of the wall structure, followed by extension of the walls under the influence of the cell turgor pressure [501-505]. Studies with a variety of plant tissues have shown that auxin-induced cell expansion and increased cell-wall extensibility are inhibited by inhibitors of RNA and protein synthesis [506-518], but not by 5-fluorouracil at concentrations that are noninhibitory for the synthesis of D-RNA [487, 517, 518]. Such results indicate that the continuing extension of cell walls during auxin-stimulated cell elongation depends on the stimulated synthesis of RNA, particularly messenger RNA, and protein, which could include or consist of wall-extending enzymes.

Postulated enzyme reactions include degradation reactions which would cleave bonds in the wall materials to render the structure more extensible, and synthetic reactions which would incorporate new material in the expanded wall [471, 505]. Degradative enzymes reported to be stimulated by auxins include hemicellulase, β-1,3- and/or β-1,6-glucanase in barley coleoptiles [492, 519], dextranase (α-1,6- and/or α-1,3-glucanase) in oat coleoptiles [520], hemicellulase, but not cellulase, in experiments of a few hours duration with pea epicotyls [492, 519], and cellulase, β-1,3-glucanase, and pectin-esterase in longer experiments with decapitated pea seedlings [521-524, 604, 605]. In the pea seedling experiments, the cellulase was preferentially stimulated (induced), with activity increasing rapidly from 1 to 3 days after application of the auxin. However, the significance of the stimulation of these degradative activities for cell enlargement is by no means clear, as attempts to establish a role of the enzymes by adding them from external sources have yielded conflicting or negative results [492, 519, 525, compared to 526, 527], even with tissues where increased cell-wall extensibility could be demonstrated [527, 528]. Also, the reversibility of wall loosening in the presence of respiratory inhibitors makes it questionable whether the process is mediated by polysaccharide-hydrolyzing enzymes which promote essentially irreversible reactions [502, 504, 526].

The role of auxin-stimulated synthesis of cell-wall materials is also uncertain, although it will obviously be important when auxin-induced elongation continues over a long period. The stimulated synthesis of pectin materials is well established [501, 529], but the auxin-stimulated incorporation of methyl groups, at least, is not a prerequisite for auxin-induced cell elongation [530]. The stimulated synthesis of cellulose in oat coleoptiles [530, 531] is apparently a consequence of auxin-stimulated cell elongation rather than a direct result of auxin-stimulated enzyme synthesis [529], but that of matrix hemicelluloses, including the dominant β-glucan and glucuronoarabinoxylan polysaccharides, appears to be a direct auxin effect as it occurred when cell elongation was inhibited by calcium ions [529]. In pea stem tissue, the synthesis of both cellulose and matrix polysaccharides was stimulated by auxin in the presence of growth-inhibiting concentrations of calcium [532] but, in contrast to the oat tissue, the stimulation of cell-wall synthesis in the pea stems was little affected by inhibitors of RNA and protein synthesis [533]. Auxin treatment of the pea tissue, besides stimulating cell-wall synthesis at the synthetase level, indirectly stimulated hexokinase activity, and hence the synthesis of wall precursors, by increasing uptake of the glucose substrate of hexokinase [532, 533]. The rapid stimulation of pea β-glucan synthetase (uridine-diphosphate glucose:β-1,4-glucan glucosyl transferase) by auxin appeared to involve reactivation of previously existing, reversibly deactivated enzyme; the process was energy requiring and did not occur when auxin and sugars were added to isolated synthetase particles [606]. The increase in β-glucan synthetase activity accompanied, but could not have been responsible for, the rapid stimulation of cell elongation by auxin described in the following paragraph [607]. Stimulated synthesis of an apparently different β-glucan synthetase (guanosine diphosphate glucose: β-1,4-glucan glucosyl transferase) was observed in decapitated pea seedlings from 1 to 4 days after application of auxin; protein synthesis was necessary for this stimulation [608]. However, protein synthesis in the absence of auxin was detrimental, apparently because of the production of a protein inactivator of the glucan synthetase [608].

Auxin effects on cell elongation other than through the stimulated synthesis of enzymes involved in cell-wall extension are suggested by kinetic experiments. Plant tissues showed stimulated elongation within 10-15 min of treatment with 10^{-6}-10^{-5} M auxin [534-537] and in the case of corn coleoptiles treated with 2×10^{-7} M IAA, within 2-3 min of treatment [537]. With oat coleoptiles, the lag in IAA-stimulated elongation could be eliminated completely by raising the auxin concentration to 5×10^{-3} M and the temperature to $40°$C [538]. Studies with inhibitors of RNA and protein synthesis indicate that the initial stimulation of elongation by auxin does not depend on RNA and protein synthesis [534, 609] but requires the presence of a pool of "growth limiting protein" [539]. Protein synthesis is required subsequently to maintain the pool [539], and one effect of auxin is to cause its rapid expansion [540]. Details of the reactions involved are unknown.

Rapid wall extension has been observed in the tissues of several plants held under tension and exposed to low pH [610-614]. Hager et al. [610] suggested that auxin causes rapid cell-wall loosening by promoting hydrogen ion liberation from the plasma membrane; increase of these ions on or in the wall was believed to stimulate a wall-loosening enzyme. The experiments of Rayle et al. [612, 613] suggested the possibility that an increased hydrogen ion concentration promoted the nonenzymatic breakage of acid-labile bonds in the wall structure. However, subsequent work [614] has indicated that in green pea stem segments the rapid extension caused by auxin is not mediated by an increase in hydrogen ion concentration.

b. Ethylene production. The stimulation of ethylene production in auxin-treated plants, first indicated in 1935 [541] and well established during the past decade, is of great significance owing to the role of this compound as an endogenous plant growth regulator (see Pratt and Goeschl [542]). Auxin-stimulated ethylene production is similar to other auxin-stimulated processes in that it is inhibited by inhibitors of RNA and protein synthesis, such as actinomycin D, puromycin, 2-thiouracil, and 4-fluorophenylalanine [543]. There is evidence that in etiolated pea shoots IAA induces the formation of a short-lived RNA required for the synthesis of a highly labile protein, which controls the rate of ethylene production in vegetative tissue [544].

Ethylene production by plant tissues in response to auxin treatment was studied in detail by Burg and Burg [545, 546]. Ethylene was produced by etiolated and light-grown pea stem and sunflower (Helianthus annuus L.) hypocotyl tissues incubated in nutrient solutions containing more than 10^{-6} M IAA. The production rate increased with increasing IAA concentration, attaining high levels at auxin concentrations in excess of 10^{-5} or 10^{-4} M. At these concentrations the auxin-stimulated elongation of the tissues was inhibited, except that of light-grown pea tissue, which was insensitive to the effects of ethylene. This and further evidence indicated that the inhibition of elongation of pea stem and sunflower hypocotyl tissues at supraoptimal concentrations of auxin is caused by ethylene. Auxin-induced fresh weight increases were similarly inhibited with increasing ethylene production in the sunflower tissue, but not in the etiolated pea stems, which showed an abnormal, isodiametric expansion of cortex cells similar to that produced by treatment with ethylene in the presence of 10^{-6} M IAA. The ethylene apparently did not interfere with auxin-induced cell enlargement, but diverted it from a longitudinal to an isodiametric expansion, resulting in swelling rather than elongation of the pea stems. Inhibition of elongation occurred about 3 hr after high concentrations of IAA or ethylene were applied to the stems. Carbon dioxide, an analog of allene and hence of ethylene, reversed the inhibitory effect of ethylene on elongation; the action of CO_2 was competitive, possibly involving displacement of ethylene from a metal-containing receptor site in the cell [547].

Oat and corn coleoptiles also produced ethylene at IAA concentrations greater than 10^{-7}-10^{-6} M [545, 546]. However, in contrast to the pea and sunflower responses, rates of auxin-induced growth and ethylene production both increased to a maximum at about 10^{-5}-10^{-4} M IAA, then decreased sharply at higher IAA concentrations. Growth inhibition at supraoptimal auxin concentrations in the coleoptiles was obviously not determined by ethylene production as it was in the etiolated pea stems and light- and dark-grown sunflower hypocotyls.

The inhibitory effect of auxins on root growth in peas was ascribed mainly to ethylene production [548]. Pea root sections produced ethylene within 15-30 min in response to IAA at a concentration of 10^{-8} M or greater, the ethylene production increasing with increasing IAA concentration. Growth of the sections, whether determined as elongation or increase in weight, was increasingly inhibited as ethylene production increased. Elongation was inhibited more rapidly than the weight increase, resulting in swelling of the root tissue. An inhibitory effect of auxin additional to the ethylene effect was observed at IAA concentrations greater than about 5×10^{-6} M. However, ethylene production seemed to account for more than 90% of the inhibition caused by 10^{-4} M IAA. Carbon dioxide antagonized the ethylene- or auxin-induced inhibition of growth. Ethylene apparently also mediated the inhibition by IAA of tomato seedling root elongation and elongation of root sections of oat, sunflower, mung bean (Phaseolus aureus Roxb.), and castor bean. Both auxin and ethylene, at concentrations which resulted in root swelling, stimulated the production of root hairs on the surface of the swollen tissue.

The claim that ethylene is responsible for most of the IAA inhibition of root growth in peas was challenged by Andreae et al. [549]. These workers showed differences in the effects of IAA and ethylene on the growth of pea root sections which suggested that ethylene played only a minor role in mediating the inhibition of root growth by IAA. However, subsequent detailed studies by Chadwick and Burg [550] seemed to confirm their earlier conclusions [548] of a mainly ethylene-mediated inhibition of root growth at low IAA concentrations, and an additional nonspecific acid toxicity at high auxin levels.

Plants treated with 2,4-D have shown stimulated ethylene production similar to that induced by IAA. The response of soybean seedlings to 2,4-D [551, 552] resembled that of pea stems to IAA [545, 546] in that both etiolated and light-grown seedlings showed 2,4-D-stimulated ethylene production, with the latter being less sensitive than the former to growth inhibition by ethylene. Carbon dioxide reversed the inhibitory effect of supraoptimal 2,4-D concentrations on the growth of the etiolated soybeans, indicating that it was mediated by ethylene production [551], but this reversal could not be shown with the light-grown seedlings on account of carbon dioxide toxicity [552]. Light-grown soybeans evolved ethylene about eight times faster than light-grown corn seedlings, but the corn was far more

sensitive to growth inhibition by ethylene [552]. The selective action of
2,4-D as a herbicide could not be ascribed to differences in the ability of
the two plant species to produce ethylene, a possibility suggested by the
2,4-D-induced production of ethylene in cotton but not in sorghum [553].
The effects of ethylene on soybeans indicate that the gas probably does not
account for the herbicidal action of 2,4-D, as its application did not result
in death of the seedlings [552]. On the other hand, it probably accounts for
a number of the effects of 2,4-D and other auxins on plants, for example
epinastic responses and the stimulation of adventitious root formation [541,
554], as well as inhibitory and swelling effects as discussed in the preceding
paragraphs [615]. However, in the subhook region of etiolated pea seedlings
the effects of ethylene and 2,4-D were distinct, with ethylene inhibiting cell
and tissue elongation, presumably by inhibiting the polar transport of auxin
from the hook to the subhook region, and 2,4-D reversing these effects by
making up the auxin deficit [616]. Ethylene inhibited DNA synthesis in the
nonmeristematic subhook cells, as well as in the cells of the shoot, lateral
bud, and root primary meristems, where it strongly inhibits cell division
[615]. Applications of IAA which stimulated ethylene production inhibited
DNA synthesis in the hook and subhook regions of etiolated peas, whereas
most lower concentrations were noninhibitory or stimulatory [617].

Although ethylene-stimulated RNA and protein synthesis has been
observed in certain plant tissues, for example abscission zone explants of
bean [555, 556], soybean hypocotyls [551, 557], apple slices [558], and
fig (Ficus carica L.) fruits [559], the stimulation of these processes by
auxins is apparently not mediated by ethylene. Thus, the elongating section
of the hypocotyls of soybean seedlings showed only a small increase in RNA
following ethylene treatment, compared to the marked accumulation of RNA
and DNA following treatment with 2,4-D [557]. The 2,4-D effect was
reduced when ethylene was applied at the same time. Furthermore,
chromatin isolated from the elongating and basal hypocotyl tissue of the
2,4-D-treated soybeans showed a greatly increased capacity for RNA synthesis compared with chromatin from control and ethylene-treated soybeans, although the latter chromatin was more active than chromatin from
the control [557].

5.   Primary Actions of Auxins

a.   Stimulation of RNA polymerase.   Experiments with isolated
chromatin-RNA polymerase systems have shown that the enzymatic synthesis of RNA is not stimulated directly by auxins, but through "mediator"
molecules, which appear to be proteins. A labile "factor" isolated from
the cotyledons of control soybeans, i.e., seedlings not exposed to 2,4-D,
stimulated activity of the chromatin-bound RNA polymerase from control
but not from 2,4-D-treated plants; however, the factor stimulated the
activity of saturating amounts of E. coli RNA polymerase in the presence

of chromatin from either source [560] . The factor was subsequently fractionated into two components: a high-molecular-weight material which
stimulated the endogenous RNA polymerase, and a fraction of lower molecular weight which stimulated the E. coli RNA polymerase. It was suggested
that auxin might form a complex with the factor in the cytoplasm which
would enable it to be transported across the nuclear membrane, possibly in
a modified form. An interaction of the complex or modified complex with
RNA polymerase allowing transcription of hormone-specific genes, not
previously expressed, would permit physiological expression of the auxin.
The chromatin-RNA polymerase preparations from 2,4-D-treated plants
were presumably fully stimulated by factor transported in this manner and
hence incapable of further response.

The stimulation of E. coli RNA polymerase activity by the low-molecular-
weight material was similar to that mediated by "hormone reactive protein"
from pea bud, soybean, or tobacco nuclei in a reaction mixture containing
the protein, E. coli RNA polymerase, pea bud chromatin, and auxin [499] .
The hormone-reactive protein chromatographed into heavy and light material, the former possibly being a multimer of the latter. It was suggested
that the protein plus auxin interacted with chromatin, thus making more
template available for the transcription. The hormone-reactive protein was
lost from soybean or tobacco nuclei by leakage during isolation of the nuclei
in the absence of auxin, but not if auxin was present. The protein isolated
from pea roots was more effective in stimulating RNA polymerase activity
in the presence of 2,4-D and pea root chromatin than in the presence of
2,4-D and pea bud chromatin [561] . Likewise, the pea bud hormone-reactive
protein was active in assays with bud but not with root chromatin. Organ
specificity in respect of hormone-reactive protein was thus indicated. There
was evidence that the hormone-reactive protein bound to chromatin in the
presence of auxin.

A protein factor that stimulated RNA synthesis by E. coli RNA polymerase in the presence of freshly prepared plant chromatin or purified DNA
was isolated from extracts of shoots of pea and corn by affinity chromatography [562] . The protein adsorbed to 2,4-D-lysine coupled to agarose,
whereas no protein adsorbed to the unmodified agarose. Stimulation of RNA
synthesis was not dependent on the addition of 2,4-D to the reaction mixtures,
perhaps because of transformation of the protein by the 2,4-D to an active
configuration during passage through the column, after which the continued
presence of auxin is unnecessary. Alternatively, the auxin might be involved
only in transport of the factor from cytoplasm to nucleus. Preliminary evidence suggested that the factor influenced initiation of RNA chains.

The evidence from these in vitro experiments suggests that the primary
action of auxins in the stimulation of cell elongation is a reaction with
hormone-reactive protein that subsequently in some way stimulates RNA

polymerase activity. Such a reaction is entirely consistent with the auxin-receptor reaction deduced from structure-activity relationships.

b. Induction of other specific enzyme activities. Auxins induce specific enzyme activities other than that of RNA polymerase [563]. Well-established examples, where stimulation of the enzyme activity occurred to a greater extent than the general synthesis of protein and was inhibited by inhibitors of RNA or protein synthesis, include the IAA-induced synthesis of cellulase in pea epicotyls [521-524, 604, 605] and isocitrate lyase in potato [564], and the auxin-induced synthesis in pea epicotyls of the enzyme(s) responsible for conjugating IAA, naphthalene-1-acetic acid, or benzoic acid to aspartic acid [565-569]. These conjugate-forming activities were induced by IAA or other auxins, including 2,4-D, but not by antiauxins. However, of the auxin inducers, only IAA and naphthalene-1-acetic acid served as substrates, whereas benzoic acid, which served as a substrate, was not an inducer [568]. A second enzyme system which conjugated benzoic acid with malic acid was induced by auxins and many nonauxinic aromatic acids [569]. The mechanism of induction of specific plant enzymes by auxins is not known, but if it resembles enzyme induction in bacteria [570] it would involve reaction with a protein (repressor protein) which controls transcription [568]. The possibility that enzyme induction might be a response to ethylene rather than auxin was disproved for cellulase induction in the pea [571].

High levels of cellulase developed in the roots of 2,4-D-treated honey-vine milkweed, possibly as a result of induction by the auxin [572]. The cellulase activity may be responsible for root dysfunction and destruction which are features of the herbicidal action of 2,4-D on this plant.

c. Allosteric effects. The mechanism suggested in the preceding section for induction of the IAA- or naphthalene-1-acetic acid-aspartate conjugating enzyme(s) implies an allosteric interaction between auxin and protein [568], but this reaction remains to be demonstrated. However, the effects of IAA on the activity of citrate synthase (condensing enzyme) from maize mitochondria indicate such an interaction between IAA and the enzyme [573]. Sulfhydryl groups seemed to be involved in the allosteric interaction, which is consistent with concepts of the reaction of auxins with receptor molecules. Stimulation of the citrate synthase activity occurred at a very low auxin level; hence it seems unlikely to play a role in the action of phenoxy herbicides, but it does indicate an additional potential of auxins to affect cell metabolism through allosteric activation of preformed enzymes. The effects of 2,4-D on the activity of crystalline glyceraldehyde-3-phosphate dehydrogenase, glucose-6-phosphate dehydrogenase, isocitric dehydrogenase, and horseradish peroxidase [574] may also have involved allosteric auxin-enzyme interactions.

6. Herbicidal Action of Phenoxyalkanoic Acids

From the foregoing discussion it is evident that the phenoxyalkanoic acids used as herbicides are those with the highest auxin activity or, in the case of the phenoxybutyric acid herbicides, those that can be converted in plants to highly active phenoxyacetic acids. The action of these compounds on plant tissues and at the molecular level resembles the action of the natural plant hormone, IAA. However, there is an important difference between endogenous IAA controlling the growth of plants and phenoxyalkanoic acids exerting herbicidal effects, namely, concentrations of the former in different plant tissues are carefully controlled and regulated by biosynthetic and catabolic reactions of the plant, whereas concentrations of the latter are not [575]. The uptake and translocation of a phenoxy herbicide applied to the surface of a plant will result in increased auxin concentrations in many of the tissues, including those where the concentration is normally low. Furthermore, the synthetic auxin may persist in tissues for much longer than the natural auxin [575, 576]. The logical consequence is a disturbed and abnormal development of the plant, as some tissues are stimulated to regrowth and others, in which auxin concentration becomes supraoptimal and high levels of ethylene are perhaps produced, are inhibited in their development.

The present review of the action of phenoxyalkanoic acid herbicides has concentrated on their role in cell expansion, as this is the role that is best understood from the biochemical point of view. Of even greater significance, but not yet understood, are the effects on cell division and differentiation [577]. Stimulated cell division is associated with stimulated DNA and RNA synthesis [578], but the mechanisms that trigger division are unknown. Stimulated cell division and massive proliferation of tissue are characteristic responses of the stem-tap root axis, leading to swelling, phloem disruption and destruction, and excessive initiation of lateral root formation [470, 479-481, 577, 579]. By contrast, division in apical meristems is inhibited [479, 481, 577, 578], albeit only temporarily [479, 578], with consequent reduction and disturbance of normal root and leaf production [481, 577]. Leaves also fail to differentiate normally [579]. In the opinion of Hanson, Slife, and colleagues [470, 481] and Van Overbeek [575], death of the plant probably results from functional failure of the absorptive organs, i.e., leaves and shoots. Besides being directly inhibited in their development or developing abnormally, these organs rapidly lose essential materials to the proliferating tissues of the axis. Disruption of the vascular system in the axis presumably also plays a role. As indicated earlier, ethylene production is apparently not responsible for the death of plants treated with phenoxy herbicides, but it seems to have a significant influence in determining symptoms associated with aberrant growth.

The phenoxyalkanoic acid herbicides, because of their effects on such fundamental processes as the synthesis of nucleic acids and proteins, the activity of enzymes, and the stimulation or inhibition of cell division, affect a great many processes in plants (see Wort [580] and Penner and Ashton [581]). Some of the effects may be direct inhibitory or toxic effects unrelated to auxin action, for example, the inhibition of mitochondrial oxidative phosphorylation, which is caused by antiauxinic phenoxyalkanoic acids [582] as well as by high concentrations of the herbicides [582-585].

C. Selectivity

The effectiveness of phenoxyalkanoic acids as herbicides against dicotyledonous, but not monocotyledonous, plants seems to be determined primarily by differences in plant structure and rate of herbicide translocation. Thus, destruction of the phloem of dicotyledons as a result of abnormal herbicide-induced tissue proliferation [480] seems to be avoided in the resistant monocotyledons where the phloem is scattered in bundles, each surrounded by protective schlerenchyma tissue [4, 577]. Also, the absence of auxin-sensitive cambium and pericycle from the vascular bundles of monocotyledons is possibly an important resistance factor [470]. Comparisons of phenoxy herbicide movement in plants of the two classes indicate that translocation of foliar-applied herbicide from the site of uptake is much more restricted in monocotyledons than in susceptible dicotyledons [103, 139, 142, 586-588]. The intercalary meristem of stems and the leaves of young plants have been suggested as a barrier to herbicide translocation in monocotyledons [139, 586]. Rapid metabolism of phenoxy herbicides in at least some monocotyledons may help to ensure resistance to these compounds [104, 112, 139, 142]. Sensitivity to phenoxy herbicides is occasionally encountered among the monocotyledons, for example, 2,4-D-sensitive corn varieties [103], yellow nutsedge (Cyperus esculentus L.) [589], and purple nutsedge (Cyperus rotundus L.) [579, 590]. Relatively high 2,4-D translocation rates were observed in the sensitive corn varieties [103] and yellow nutsedge [589], and histological effects fundamentally similar to those in the bean, in the purple nutsedge [579, 590].

Among dicotyledons, resistance or susceptibility is determined by the rate or extent of herbicide uptake, translocation, metabolism, and, in a few cases, excretion. The review of Holly [300] gives valuable comparative information on the rates of uptake and factors that influence uptake by individual plant species. Ineffective or poor herbicidal action was associated with poor translocation in 2,4-D-treated mature Bermuda buttercup [111], honeyvine milkweed when root growth was not active [357], bur cucumber [588, but see 109], and 2,4-DB-treated soybean [311]. Translocation of herbicide to the roots, and hence effective control, is especially a problem in the case of woody species [106, 171, 180, 312, 354, 363]. Excretion of herbicide from the roots is apparently an important resistance mechanism in 2,4-D-treated jimsonweed [152] , and may be important in 2,4-D-treated nightflowering catchfly [155] and honeyvine milkweed [357, 588] .

The rapid degradation of the 2,4-D side chain appears to be responsible for the lack of toxicity of this compound toward red currants [99] , Cox's Orange Pippin [100] and McIntosh apples [101] , several strawberry varieties, and the garden lilac [100] . Replacement of the para-chlorine in 2,4-D with fluorine makes the molecule resistant to degradation by McIntosh apples; hence, 2-chloro-4-fluorophenoxyacetic acid is an effective growth regulator for this apple variety [101] .

The red currant is relatively resistant to MCPA and 4-chlorophenoxyacetic acid, which it decarboxylates at a rapid rate, but it is highly susceptible to 2,4,5-T, although this compound is decarboxylated as rapidly as 2,4-D, MCPA, and 4-chlorophenoxyacetic acid [99] . Luckwill and Lloyd-Jones [100] suggested that 2,4,5-trichlorophenol was mainly responsible for the damage to 2,4,5-T-treated red currants, which showed quite different symptoms to 2,4,5-T-treated black currants that could not degrade the 2,4,5-T side chain. Formation of the chlorophenols might also be responsible for the toxic effects of MCPA and 4-chlorophenoxyacetic acid on red currants and for the high toxicity of these compounds and 2,4,5-T toward strawberries [100] . The resistance of bedstraw to MCPA was attributed to the ability of the plant to remove the MCPA side chain; resistance and side-chain degradation were both overcome by the introduction of a methyl group into the side chain to give mecoprop [102] .

The resistance of red and black currant to 2-chlorophenoxyacetic acid [99] and of bur cucumber to 2,4-D [109] was correlated with the rapid formation of unidentified metabolites of these phenoxyacetic acids. By contrast, the bur cucumber was susceptible to 2,4,5-T, which was converted to unidentified metabolites only very slowly [109] . Cultivated cucumber also metabolized 2,4-D rapidly and 2,4,5-T slowly, but was susceptible to the action of both herbicides [109] . The susceptibility of the cultivated cucumber to 2,4-D coincided with higher 2,4-D concentrations in the plant, possibly the result of greater uptake, than in the 2,4-D-resistant wild cucumber. The high resistance of Bermuda buttercup to 2,4-D was attributed in part to the rapid metabolism of the herbicide to volatile or unstable metabolites [111] .

The rapid degradation of 2,4-D in the roots of big leaf maple compared with the slow degradation in the stems and leaves [106] , combined with poor translocation [172, 311] , explains the resistance of this plant to the herbicidal action of foliar-applied 2,4-D. Applications of 2,4-D that kill the tops of the plants often fail to kill the roots, which later give rise to regrowth. The big leaf maple roots were less effective in degrading 2,4,5-T [106] , but as this herbicide was also poorly translocated [172, 311] , it was not particularly effective [106]. The resistance of nightflowering catchfly to 2,4-D was associated with herbicide metabolism in the roots, but not the tops, as well as root excretion [155].

The use of MCPB, 2,4-DB, and 2,4,5-TB as weedkillers depends on their β oxidation to MCPA, 2,4-D, or 2,4,5-T by the target weeds [158, 159]. The crop plant, on the other hand, should be unable to carry out this activation reaction or should be resistant to the herbicide for another reason, for example, poor penetration or translocation to the site of action, or metabolism to nonherbicidal degradation products. Wain [158, 159] and Shaw and Gentner [591] cite examples of susceptible weeds and resistant crops on which the phenoxybutyric herbicides may be used. These herbicides are important for use with legumes, but inability to β-oxidize the phenoxybutyric acid to an active phenoxyacetic acid is probably not the main resistance mechanism in these plants, except perhaps in the case of 2,4,5-TB (see Sec. II). Degradation by lengthening of the side chain is probably an important resistance mechanism [119, 121, 169].

Limited β oxidation of 2,4-DB in the tops of big leaf maple and a high level of β oxidation in the roots provide an explanation for the effectiveness of this herbicide compared with other phenoxy acids [171]. Translocation of the nonherbicidal 2,4-DB from tops to roots was considerably improved over that of herbicidal phenoxyacetic and 2-phenoxypropionic acids, with the result that higher levels of herbicide were attained in the roots following translocation and β oxidation of 2,4-DB than when 2,4-D, 2,4,5-T, dichlorprop, or silvex was used.

The ability of a plant to degrade an active phenoxy herbicide does not automatically make the plant resistant. Thus, low rates of side-chain degradation are characteristic of susceptible species, as are many of the other reactions described in Sec. II. The decisive factor seems to be the rate of degradation. Also, certain metabolites of phenoxy herbicides may retain growth-regulating activity, for example, N-(2,4-dichlorophenoxyacetyl)aspartic acid [58], N-(2,4-dichlorophenoxyacetyl)glutamic acid, and 2,4-dichloro-3-hydroxyphenoxyacetic acid [131].

IV. CONCLUSIONS

The success of the phenoxyalkanoic acids as agents for the control of dicotyledonous weed species depends on both their mode of action and biodegradability. As active analogs of the plant growth hormone, IAA, they profoundly affect growth and cause death of susceptible plants at low application rates. Having achieved their task of weed control, they are biodegraded in the vegetation or soil; hence, there is no lasting hazard to susceptible crops or persisting contamination of the environment. There seems to be no general danger to livestock health from the agricultural use of these herbicides; recent work suggests that effects on gamebirds are unlikely. The much-publicized teratogenicity of 2,4,5-T for mice was observed at dose rates that are unlikely to be encountered under normal conditions of herbicide usage.

Recent evidence for the primary action of auxins has vindicated the conclusions from earlier studies of structure-activity relationships, namely, that stereospecific interactions with cell proteins mediate auxin action. The resulting stimulation of RNA and DNA polymerase and other enzyme activities can explain the effects of auxins on plant growth. A common mode of action for auxin herbicides of different chemical classes, suggested by much experimental data, is consistent with this mechanism.

REFERENCES

1. W. G. Templeman, Ann. Appl. Biol., 42, 162 (1955).

2. G. E. Peterson, Agr. History, 41, 243 (1967).

3. G. C. Klingman, Weed Control: As a Science, Wiley, New York, 1961.

4. A. S. Crafts, The Chemistry and Mode of Action of Herbicides, Wiley (Interscience), New York, 1961.

5. J. D. Fryer and S. A. Evans (eds.), Weed Control Handbook, 5th ed., Vols. 1 and 2, Blackwell, Oxford, 1968.

6. Herbicide Handbook of the Weed Science Society of America, 2nd ed., W. F. Humphrey, Geneva and New York, 1970.

7. U.S. Department of Agriculture, The Pesticide Review 1970, U.S.D.A., Stabil. and Conserv. Serv., Washington, D.C., 1970.

8. F. H. Tschirley, Science, 163, 779 (1969).

9. E. W. Pfeiffer, Sci. J., 5, 34 (1969).

10. G. H. Orians and E. W. Pfeiffer, Science, 168, 544 (1970).

11. P. M. Boffey, Science, 171, 43 (1971).

12. L. J. King, J. A. Lambrech, and T. P. Finn, Contrib. Boyce Thompson Inst., 16, 191 (1951).

13. B. Nelson, Science, 166, 977 (1969).

14. Chem. Eng. News, 48, 60 (1970).

15. K. D. Courtney, D. W. Gaylor, M. D. Hogan, H. L. Falk, R. R. Bates, and I. Mitchell, Science, 168, 864 (1970).

16. G. R. Higginbotham, A. Huang, D. Firestone, J. Verrett, J. Ress, and A. D. Campbell, Nature, 220, 702 (1968).

17. G. L. Sparschu, F. L. Dunn, and V. K. Rowe, Food Cosmet. Toxicol., 9, 405 (1971).

18. D. G. Crosby, A. S. Wong, J. R. Plimmer, and E. A. Woolson, Science, 173, 748 (1971).

19. E. A. Woolson, R. F. Thomas, and P. D. J. Ensor, J. Agr. Food Chem., 20, 351 (1972).

20. A. R. Isensee and G. E. Jones, J. Agr. Food Chem., 19, 1210 (1971).

21. D. R. Roll, Food Cosmet. Toxicol., 9, 671 (1971).

22. J. L. Emerson, D. J. Thompson, R. J. Strebing, C. G. Gerbig, and V. B. Robinson, Food Cosmet. Toxicol., 9, 395 (1971).

23. Y. Lutz-Ostertag and H. Lutz, C. R. Acad. Sci. Paris, Ser. D, 271, 2418 (1970).

24. G. W. Bailey and J. L. White, Residue Rev., 10, 7 (1965).

25. F. A. Gunther, W. E. Westlake, and P. S. Jaglan, Residue Rev., 20, 1 (1968).

26. N. N. Melnikov, Residue Rev., 36, 157 (1971).

27. U.S. Department of Agriculture, Suggested Guide for Weed Control, Uses of Herbicides, Agriculture Handbook No. 332, U.S. Govt. Printing Office, Washington, D.C., 1969.

28. V. H. Freed, in The Physiology and Biochemistry of Herbicides (L. J. Audus, ed.), Academic Press, New York, 1964, Chap. 2.

29. V. H. Freed, in Pesticides and Their Effects on Soils and Water, A.S.A. Spec. Publ. No. 8, Soil Sci. Soc. Am., Madison, Wisconsin, 1966, pp. 25-43.

30. Encyclopedia of Chemical Technology, 2nd ed., Vol. 22, Wiley, New York, 1970, pp. 179-183.

31. H. Veldstra and H. L. Booij, Biochim. Biophys. Acta, 3, 278 (1949).

32. R. C. Weast (ed.), Handbook of Chemistry and Physics, 49th ed., Chemical Rubber Co., Cleveland, Ohio, 1968.

33. M. Alexander and M. I. H. Aleem, J. Agr. Food Chem., 9, 44 (1961).

34. R. S. Bandurski, Bot. Gaz., 108, 446 (1947).

35. B. Warskowsky and E. J. Schantz, Anal. Chem., 22, 460 (1950).

36. N. Gordon and M. Beroza, Anal. Chem., 24, 1968 (1952).

37. K. P. Dorschner and K. P. Buchholtz, Weeds, 5, 102 (1957).

38. E. Behm and I. Schröder, Weed Res., 9, 43 (1969).

39. H. A. Glastonbury and M. D. Stevenson, J. Sci. Food Agr., 10, 379 (1959).

40. A. Fredga and B. Åberg, Ann. Rev. Plant Physiol., 16, 53 (1965).

41. R. H. F. Manske (to U.S. Rubber Co.), U.S. Pat. 2,471,575 (1949).

42. M. J. Skeeters (to Diamond Alkali Co.), U.S. Pat. 2,740,810 (1956).

43. J. A. Lambrech (to Union Carbide and Carbon Corp.), U.S. Pat. 2,573,769 (1951).

44. W. D. Harris and A. W. Feldman (to U.S. Rubber Co.), U.S. Pat. 2,828,198 (1958).

45. L. C. Erickson and H. Z. Hield, J. Agr. Food Chem., 10, 204 (1962).

46. G. Yip, J.A.O.A.C., 45, 367 (1962).

47. A. Bevenue, G. Zweig, and N. L. Nash, J.A.O.A.C., 45, 990 (1962).

48. R. P. Marquardt, H. P. Burchfield, E. E. Storrs, and A. Bevenue, in Analytical Methods for Pesticides, Plant Regulators and Food Additives, Vol. 4 (G. Zweig, ed.), Academic Press, New York, 1964, pp. 95-116.

49. R. D. Hagin and D. L. Linscott, J. Agr. Food Chem., 13, 123 (1965).

50. W. H. Gutenmann and D. J. Lisk, J. Agr. Food Chem., 11, 304 (1963).

51. W. H. Gutenmann, D. D. Hardee, R. F. Holland, and D. J. Lisk, J. Dairy Sci., 46, 991 (1963).

52. D. E. Clark, J. Agr. Food Chem., 17, 1168 (1969).

53. J. E. Scoggins and C. H. Fitzgerald, J. Agr. Food Chem., 17, 156 (1969).

54. W. R. Meagher, J. Agr. Food Chem., 14, 374 (1966).

55. W. H. Gutenmann and D. J. Lisk, J.A.O.A.C., 47, 353 (1964).

56. D. W. Woodham, W. G. Mitchell, C. D. Loftis, and C. W. Collier, J. Agr. Food Chem., 19, 186 (1971).

57. P. L. Pursley and E. D. Schall, J.A.O.A.C., 48, 327 (1965).

58. J. W. Wood and T. D. Fontaine, J. Org. Chem., 17, 891 (1952).

59. C. F. Krewson, C. H. H. Neufeld, T. F. Drake, T. D. Fontaine, J. W. Mitchell, and W. H. Preston, Weeds, 3, 28 (1954).

60. C. F. Krewson, T. F. Drake, C. H. H. Neufeld, T. D. Fontaine, J. W. Mitchell, and W. H. Preston, J. Agr. Food Chem., 4, 140 (1956).

61. J. F. Carmichael, E. J. Sagesse, J. S. Ard, C. F. Krewson, and E. M. Shantz, J. Agr. Food Chem., 12, 434 (1964).

62. C. F. Krewson, T. F. Drake, J. W. Mitchell, and W. H. Preston, J. Agr. Food Chem., 4, 690 (1956).

63. C. F. Krewson, J. F. Carmichael, T. F. Drake, J. W. Mitchell, and B. C. Smale, J. Agr. Food Chem., 8, 104 (1960).

64. F. Feigl and R. Moscovici, Analyst, 80, 803 (1955).

65. V. H. Freed, Science, 107, 98 (1948).

66. D. Le Tourneau and N. Krog, Plant Physiol., 27, 822 (1952).

67. R. P. Marquardt and E. N. Luce, Anal. Chem., 23, 1484 (1951).

68. R. P. Marquardt and E. N. Luce, J. Agr. Food Chem., 3, 51 (1955).

69. E. Sawicki, R. H. Thomas, and M. Sylvester, Anal. Chem., 34, 1460 (1962).

70. O. M. Aly and S. D. Faust, Anal. Chem., 36, 2200 (1964).

71. R. P. Marquardt and E. N. Luce, J. Agr. Food Chem., 9, 266 (1961).

72. C. M. Asmundson, J. M. Lyons, and F. H. Takatori, J. Agr. Food Chem., 14, 627 (1966).

73. D. E. Clark, F. C. Wright, and L. M. Hunt, J. Agr. Food Chem., 15, 171 (1967).

74. C. A. Bache, D. J. Lisk, and M. A. Loos, J.A.O.A.C., 47, 348 (1964).

75. A. Guardigli, W. Chow, and M. S. Lefar, J. Agr. Food Chem., 19, 1181 (1971).

76. L. C. Mitchell, J.A.O.A.C., 39, 891 (1956).

77. L. C. Mitchell, J.A.O.A.C., 40, 294 (1957).

78. L. C. Mitchell, J.A.O.A.C., 41, 781 (1958).

79. L. C. Mitchell, J.A.O.A.C., 44, 643, 720 (1961).

80. D. C. Abbott, H. Egan, E. W. Hammond, and J. Thomson, Analyst, 89, 480 (1964).

81. D. G. Crosby and M.-Y. Li, in Degradation of Herbicides (P. C. Kearney and D. D. Kaufman, eds.), Dekker, New York, 1969, Chap. 12.

82. R. P. Wayne, Photochemistry, Butterworths, London, 1970, pp. 242–249.

83. D. J. Jensen and E. D. Schall, J. Agr. Food Chem., 14, 123 (1966).

84. G. W. Flint, J. J. Alexander, and O. P. Funderburk, Weed Sci., 16, 541 (1968).

85. A. D. Baskin and E. A. Walker, Weeds, 2, 280 (1953).

86. R. W. Holley, F. P. Boyle, and D. B. Hand, Arch. Biochem. Biophys., 27, 143 (1950).

87. R. L. Weintraub, J. W. Brown, M. Fields, and J. Rohan, Am. J. Bot., 37, 682 (1950).

88. W. C. Shaw, J. L. Hilton, D. E. Moreland, and L. L. Jansen, U.S. Dept. Agr., ARS 20-9, 119 (1960).

89. L. J. Audus, in Encyclopedia of Plant Physiology, Vol. 14 (W. Ruhland, ed.), Springer, Berlin, 1961, pp. 1061-1067.

90. J. L. Hilton, L. L. Jansen, and H. M. Hull, Ann. Rev. Plant Physiol., 14, 353 (1963).

91. V. H. Freed and M. L. Montgomery, Residue Rev., 3, 1 (1963).

92. R. C. Brian, Weed Res., 4, 105 (1964).

93. A. S. Crafts, in The Physiology and Biochemistry of Herbicides (L. J. Audus, ed.), Academic Press, New York, 1964, Chap. 3.

94. M. A. Loos, in Degradation of Herbicides (P. C. Kearney and D. D. Kaufman, eds.), Dekker, New York, 1969, Chap. 1.

95. J. E. Casida and L. Lykken, Ann. Rev. Plant Physiol., 20, 607 (1969).

96. M. M. Robertson and R. C. Kirkwood, Weed Res., 10, 94 (1970).

97. M. A. Loos, in Pesticide Terminal Residues (A. S. Tahori, ed.), Butterworths, London, 1971, pp. 291-304.

98. R. L. Weintraub, J. W. Brown, M. Fields, and J. Rohan, Plant Physiol., 27, 293 (1952).

99. L. C. Luckwill and C. P. Lloyd-Jones, Ann. Appl. Biol., 48, 613 (1960).

100. L. C. Luckwill and C. P. Lloyd-Jones, Ann. Appl. Biol., 48, 626 (1960).

101. L. J. Edgerton and M. B. Hoffman, Science, 134, 341 (1961).

102. E. L. Leafe, Nature, 193, 485 (1962).

103. R. L. Weintraub, J. H. Reinhart, and R. A. Scherff, in A Conference on Radioactive Isotopes in Agriculture, A.E.C. Rep. No. TID-7512, 1956, pp. 203-208.

104. P. W. Morgan and W. C. Hall, Weeds, 11, 130 (1963).

105. E. Basler, Weeds, 12, 14 (1964).

106. L. A. Norris and V. H. Freed, Weed Res., 6, 212 (1966).

107. M. J. Canny and K. Markus, Aust. J. Biol. Sci., 13, 486 (1960).

108. M. C. Williams, F. W. Slife, and J. B. Hanson, Weeds, 8, 244 (1960).

109. F. W. Slife, J. L. Key, S. Yamaguchi, and A. S. Crafts, Weeds, 10, 29 (1962).

110. P. N. Chow, O. C. Burnside, T. L. Lavy, and H. W. Knoche, Weeds, 14, 38 (1966).

111. N. G. Marinos, F. H. Chapman, and L. H. May, Aust. J. Biol. Sci., 17, 631 (1964).

112. M. L. Montgomery, Y. L. Chang, and V. H. Freed, J. Agr. Food Chem., 19, 1219 (1971).

113. R. L. Weintraub, J. N. Yeatman, J. A. Lockhart, J. H. Reinhart, and M. Fields, Arch. Biochem. Biophys., 40, 277 (1952).

114. S. C. Fang, E. G. Jaworski, A. V. Logan, V. H. Freed, and J. S. Butts, Arch. Biochem. Biophys., 32, 249 (1951).

115. D. I. Chkanikov, N. N. Pavlova, and D. F. Gertsuskii, Khim. v Sel'skom Khoz., 3, 56 (1965) (in Russian); through C.A., 63, 8250c (1965).

116. C. H. Fitzgerald, C. L. Brown, and E. G. Beck, Plant Physiol., 42, 459 (1967).

117. H. Beevers, Respiratory Metabolism in Plants, Harper, New York, 1961, pp. 54-57.

118. M. K. Bach, Plant Physiol., 36, 558 (1961).

119. D. L. Linscott and R. D. Hagin, Weed Sci., 18, 197 (1970).

120. R. D. Hagin, D. L. Linscott, and J. E. Dawson, J. Agr. Food Chem., 18, 848 (1970).

121. D. L. Linscott, R. D. Hagin, and J. E. Dawson, J. Agr. Food Chem., 16, 844 (1968).

122. R. W. Holley, Arch. Biochem. Biophys., 35, 171 (1952).

123. C. H. Fawcett, R. M. Pascal, M. B. Pybus, H. F. Taylor, R. L. Wain, and F. Wightman, Proc. Roy. Soc. (London), B150, 95 (1959).

124. M. Wilcox, D. E. Moreland, and G. C. Klingman, Physiol. Plant., 16, 565 (1963).

125. E. W. Thomas, B. C. Loughman, and R. G. Powell, Nature, 199, 73 (1963).

126. E. W. Thomas, B. C. Loughman, and R. G. Powell, Nature, 204, 286 (1964).

127. D. G. Crosby, J. Agr. Food Chem., 12, 3 (1964).

128. E. W. Thomas, B. C. Loughman, and R. G. Powell, Nature, 204, 884 (1964).

129. R. H. Hamilton, J. Hurter, J. K. Hall, and C. D. Ercegovich, J. Agr. Food Chem., 19, 480 (1971).

130. G. Guroff, J. W. Daly, D. M. Jerina, J. Renson, B. Witkop, and S. Udenfriend, Science, 157, 1524 (1967).

131. C.-s. Feung, R. H. Hamilton, and F. H. Witham, J. Agr. Food Chem., 19, 475 (1971).

132. J. Fleeker and R. Steen, Weed Sci., 19, 507 (1971).

133. D. J. Collins and J. K. Gaunt, Biochem. J., 118, 54P (1970).

134. D. J. Collins and J. K. Gaunt, Biochem. J., 124, 9P (1971).

135. H. D. Klämbt, Planta, 57, 339 (1961).

136. K. Ojima and L. Gamborg, in Biochemistry and Physiology of Plant Growth Substances (F. Wightman and G. Setterfield, eds.), The Runge Press, Ottawa, 1968, pp. 857-865.

137. W. A. Andreae and N. E. Good, Plant Physiol., 32, 566 (1957).

138. E. G. Jaworski and J. S. Butts, Arch. Biochem. Biophys., 38, 207 (1952).

139. S. C. Fang and J. S. Butts, Plant Physiol., 29, 56 (1954).

140. E. G. Jaworski, S. C. Fang, and V. H. Freed, Plant Physiol., 30, 272 (1955).

141. S. C. Fang, Weeds, 6, 179 (1958).

142. J. S. Butts and S. C. Fang, in A Conference on Radioactive Isotopes in Agriculture, A.E.C. Rep. No. TID-7512, 1956, pp. 209-214.

143. M. K. Bach and J. Fellig, in Plant Growth Regulation (R. M. Klein et al., eds.), Iowa State University Press, Ames, 1961, pp. 273-287.

144. M. K. Bach and J. Fellig, Plant Physiol., 36, 89 (1961).

145. M. K. Bach and J. Fellig, Nature, 189, 763 (1961).

146. J. R. Corbett and C. S. Miller, Weeds, 14, 34 (1966).

147. W. R. Meagher, J. Agr. Food Chem., 14, 599 (1966).

148. H. O. Tutass, Ph.D. Thesis, University of California, Davis, 1967.

149. E. A. Erickson, B. L. Brannaman, and C. W. Coggins, J. Agr. Food Chem., 11, 437 (1963).

150. R. L. Weintraub, J. H. Reinhart, R. A. Scherff, and L. C. Schisler, Plant Physiol., 29, 303 (1954).

151. D. L. Linscott and M. K. McCarty, Weeds, 10, 65 (1962).

152. R. C. Fites, F. N. Slife, and J. B. Hanson, Weeds, 12, 180 (1964).

153. E. Basler, C. C. King, A. A. Badiei, and P. W. Santelman, Proc. South. Weed Conf., 17, 351 (1964).

154. W. M. Blacklow and D. L. Linscott, Weed Sci., 16, 516 (1968).

155. R. W. Neidermyer and J. D. Nalewaja, Weed Sci., 17, 528 (1969).

156. H. D. Coble, F. W. Slife, and H. S. Butler, Weed Sci., 18, 653 (1970).

157. H. L. Morton, Weeds, 14, 136 (1966).

158. R. L. Wain, Ann. Appl. Biol., 42, 151 (1955).

159. R. L. Wain, J. Agr. Food Chem., 3, 128 (1955).

160. R. L. Wain, Adv. Pest Control Res., 2, 263 (1958).

161. R. L. Wain, in The Physiology and Biochemistry of Herbicides (L. J. Audus, ed.), Academic Press, New York, 1964, Chap. 16.

162. C. H. Fawcett, H. F. Taylor, R. L. Wain, and F. Wightman, in The Chemistry and Mode of Action of Plant Growth Substances (R. L. Wain and F. Wightman, eds.), Butterworths, London, 1956, pp. 187-194.

163. D. L. Linscott, J. Agr. Food Chem., 12, 7 (1964).

164. C. H. Fawcett, J. M. A. Ingram, and R. L. Wain, Proc. Roy. Soc. (London), B142, 60 (1954).

165. R. L. Wain and F. Wightman, Proc. Roy. Soc. (London), B142, 525 (1954).

166. C. H. Fawcett, H. F. Taylor, R. L. Wain, and F. Wightman, Proc. Roy. Soc. (London), B148, 543 (1958).

167. M. E. Synerholm and P. W. Zimmerman, Contrib. Boyce Thompson Inst., 14, 369 (1947).

168. R. L. Wain and H. F. Taylor, Nature, 207, 167 (1965).

169. S. Wathana and F. T. Corbin, J. Agr. Food Chem., 20, 23 (1972).

170. P. G. Balayannis, M. S. Smith, and R. L. Wain, Ann. Appl. Biol., 55, 261 (1965).

171. L. A. Norris and V. H. Freed, Weed Res., 6, 283 (1966).

172. S. N. Fertig, M. A. Loos, W. H. Gutenmann, and D. J. Lisk, Weeds, 12, 147 (1964).

173. D. L. Linscott and R. D. Hagin, Proc. Northeast. Weed Control Conf., 17, 260 (1963).

174. D. L. Linscott, R. D. Hagin, and M. J. Wright, Crop. Sci., 5, 455 (1965).

175. A. S. Crafts, Weeds, 8, 19 (1960).

176. S. S. Szabo, Weeds, 11, 292 (1963).

177. C. E. Hagen, C. O. Clagett, and E. A. Helgesen, Science, 110, 116 (1949).

178. D. J. Morré and B. J. Rogers, Weeds, 8, 436 (1960).

179. J. v. d. W. Jooste and D. E. Moreland, Phytochemistry, 2, 263 (1963).

180. O. A. Leonard, D. E. Bayer, and R. K. Glenn, Bot. Gaz., 127, 193 (1966).

181. J. R. Baur, R. W. Bovey, and J. D. Smith, Weed Sci., 17, 567 (1969).

182. G. Yip and R. E. Ney, Weeds, 14, 167 (1966).

183. C. H. Fawcett, R. C. Seeley, H. F. Taylor, R. L. Wain, and F. Wightman, Nature, 176, 1026 (1955).

184. P. S. Nutman, H. G. Thornton, and J. H. Quastel, Nature, 155, 498 (1945).

185. H. R. De Rose, Bot. Gaz., 107, 583 (1946).

186. J. W. Mitchell and P. C. Marth, Bot. Gaz., 107, 408 (1946).

187. O. H. Kries, Bot. Gaz., 108, 510 (1947).

188. H. R. De Rose and A. S. Newman, Proc. Soil Sci. Soc. Am., 12, 222 (1948).

189. J. W. Brown and J. W. Mitchell, Bot. Gaz., 109, 314 (1948).

190. T. P. Hernandez and G. F. Warren, Proc. Am. Soc. Hort. Sci., 56, 287 (1950).

191. A. S. Newman and J. R. Thomas, Proc. Soil Sci. Soc. Am., 14, 160 (1950).

192. A. S. Newman, J. R. Thomas, and R. L. Walker, Proc. Soil Sci. Soc. Am., 16, 21 (1952).

193. L. J. Audus, in Herbicides and the Soil (E. K. Woodford and G. R. Sagar, eds.), Blackwell, Oxford, 1960, pp. 1-19.

194. L. J. Audus, in The Physiology and Biochemistry of Herbicides (L. J. Audus, ed.), Academic Press, New York, 1964, Chap. 5.

195. J. D. Fryer and K. Kirkland, Weed Res., 10, 133 (1970).

196. E. K. Akamine, Bot. Gaz., 112, 312 (1951).

197. F. T. Corbin and R. P. Upchurch, Weeds, 15, 370 (1967).

198. L. J. Audus, Plant Soil, 2, 31 (1949).

199. L. J. Audus, Nature, 166, 356 (1950).

200. L. J. Audus, Plant Soil, 3, 170 (1951).

201. L. J. Audus, J. Sci. Food Agr., 3, 268 (1952).

202. W. C. Evans and B. S. W. Smith, Biochem. J., 57, xxx (1954).

203. W. C. Evans and P. Moss, Biochem. J., 65, 8P (1957).

204. W. C. Evans, B. S. W. Smith, P. Moss, and H. N. Fernley, Biochem. J., 122, 509 (1971).

205. H. N. Fernley and W. C. Evans, Biochem. J., 73, 22P (1959).

206. W. C. Evans, B. S. W. Smith, H. N. Fernley, and J. I. Davies, Biochem. J., 122, 543 (1971).

207. J. K. Gaunt and W. C. Evans, Biochem. J., 79, 25P (1961).

208. J. K. Gaunt and W. C. Evans, Biochem. J., 122, 519 (1971).

209. Y. Gamar and J. K. Gaunt, Biochem. J., 122, 527 (1971).

210. J. K. Gaunt and W. C. Evans, Biochem. J., 122, 533 (1971).

211. R. L. Walker and A. S. Newman, Appl. Microbiol., 4, 201 (1956).

212. T. I. Steenson and N. Walker, Plant Soil, 8, 17 (1956).

213. T. I. Steenson and N. Walker, J. Gen. Microbiol., 16, 146 (1957).

214. G. R. Bell, Can. J. Microbiol., 3, 821 (1957).

215. G. R. Bell, Can. J. Microbiol., 6, 325 (1960).

216. C. Stapp and G. Spicher, Zentr. Bakteriol. Parasitenk. Abt. II, 108, 113 (1954).

217. G. Spicher, Zentr. Bakteriol. Parasitenk. Abt. II, **108**, 225 (1954).

218. J.-M. Bollag, C. S. Helling, and M. Alexander, Appl. Microbiol. ,
      15, 1393 (1967).

219. R. S. Horvath, Bull. Environ. Contam. Toxicol., **5**, 537 (1970).

220. M. H. Rogoff and J. J. Reid, J. Bacteriol., **71**, 303 (1956).

221. H. L. Jensen and H. I. Petersen, Nature, **170**, 39 (1952).

222. H. L. Jensen and H. I. Petersen, Acta Agr. Scand., **2**, 215 (1952).

223. M. A. Loos, R. N. Roberts, and M. Alexander, Can. J. Microbiol.,
      13, 679 (1967).

224. M. A. Loos, R. N. Roberts, and M. Alexander, Can. J. Microbiol.,
      13, 691 (1967).

225. M. A. Loos, J.-M. Bollag, and M. Alexander, J. Agr. Food Chem.,
      15, 858 (1967).

226. C. S. Helling, J.-M. Bollag, and J. E. Dawson, J. Agr. Food Chem.,
      16, 538 (1968).

227. J.-M. Bollag, C. S. Helling, and M. Alexander, J. Agr. Food Chem.,
      16, 826 (1968).

228. J.-M. Bollag, G. G. Briggs, J. E. Dawson, and M. Alexander, J.
      Agr. Food Chem., **16**, 829 (1968).

229. J. M. Tiedje, J. M. Duxbury, M. Alexander, and J. E. Dawson,
      J. Agr. Food Chem., **17**, 1021 (1969).

230. J. M. Duxbury, J. M. Tiedje, M. Alexander, and J. E. Dawson,
      J. Agr. Food Chem., **18**, 199 (1970).

231. O. B. Weeks, J. Bacteriol., **69**, 649 (1955).

232. H. C. Bounds and A. R. Colmer, Weeds, **13**, 249 (1965).

233. R. J. W. Byrde and D. Woodcock, Biochem. J., **65**, 682 (1957).

234. J. K. Faulkner and D. Woodcock, J. Chem. Soc., **1961**, 5397.

235. S. M. Bocks, J. R. Lindsay-Smith, and R. O. C. Norman, Nature,
      201, 398 (1964).

236. D. R. Clifford and D. Woodcock, Nature, **203**, 763 (1964).

237. J. K. Faulkner and D. Woodcock, Nature, **203**, 865 (1964).

238. J. K. Faulkner and D. Woodcock, J. Chem. Soc., **1965**, 1187.

239. K. Hurle and B. Rademacher, Weed Res., **10**, 159 (1970).

240. M. A. Loos, manuscript in preparation, 1975.

241. M. A. Loos, unpublished data, 1975.

242. T. I. Steenson and N. Walker, J. Gen. Microbiol., 18, 692 (1958).

243. A. B. Pardee, in The Bacteria, Vol. 3 (I. C. Gunsalus and R. Y. Stanier, eds.), Academic Press, New York, 1962, Chap. 12.

244. L. N. Ornston, Bacteriol. Rev., 35, 87 (1971).

245. J. M. Tiedje and M. Alexander, J. Agr. Food Chem., 17, 1080 (1969).

246. K. Burger, I. C. MacRae, and M. Alexander, Proc. Soil Sci. Soc. Am., 26, 243 (1962).

247. M. Alexander, Ann. Rev. Microbiol., 18, 217 (1964).

248. M. Alexander, Adv. Appl. Microbiol., 7, 35 (1965).

249. M. Alexander, Proc. Soil Sci. Soc. Am., 29, 1 (1965).

250. D. M. Webley, R. B. Duff, and V. C. Farmer, Nature, 179, 1130 (1957).

251. D. M. Webley, R. B. Duff, and V. C. Farmer, J. Gen. Microbiol., 18, 733 (1958).

252. D. M. Webley, R. B. Duff, and V. C. Farmer, Nature, 183, 748 (1959).

253. H. F. Taylor and R. L. Wain, Proc. Roy. Soc. (London), B156, 172 (1962).

254. W. H. Gutenmann, M. A. Loos, M. Alexander, and D. J. Lisk, Proc. Soil Sci. Soc. Am., 28, 205 (1964).

255. J. S. Whiteside and M. Alexander, Weeds, 8, 204 (1960).

256. W. H. Gutenmann and D. J. Lisk, J. Agr. Food Chem., 12, 322 (1964).

257. I. C. MacRae, M. Alexander, and A. D. Rovira, J. Gen. Microbiol., 32, 69 (1963).

258. I. C. MacRae and M. Alexander, J. Bacteriol., 86, 1231 (1963).

259. I. C. MacRae and M. Alexander, Agron. J., 56, 91 (1964).

260. I. C. MacRae and M. Alexander, J. Agr. Food Chem., 13, 72 (1965).

261. K. Kirkland and J. D. Fryer, Weed Res., 12, 90 (1972).

262. O. M. Aly and S. D. Faust, J. Agr. Food Chem., 12, 541 (1964).

263. G. W. Bailey, A. D. Thruston, J. D. Pope, and D. R. Cochrane, Weed Sci., 18, 413 (1970).

264. A. J. Vlitos, Contrib. Boyce Thompson Inst., 17, 127 (1953).

265. A. J. Vlitos, Contrib. Boyce Thompson Inst., 16, 435 (1952).

266. L. J. Audus, Nature, 170, 886 (1952).

267. A. J. Vlitos and L. J. King, Nature, 171, 523 (1953).

268. L. J. Audus, Nature, 171, 523 (1953).

269. R. B. Carroll, Contrib. Boyce Thompson Inst., 16, 409 (1952).

270. S. Levey and H. B. Lewis, J. Biol. Chem., 168, 213 (1947).

271. S. Khanna and S. C. Fang, J. Agr. Food Chem., 14, 500 (1966).

272. D. E. Clark, J. E. Young, R. L. Younger, L. M. Hunt, and J. K. McLaren, J. Agr. Food Chem., 12, 43 (1964).

273. D. J. Lisk, W. H. Gutenmann, C. A. Bache, R. G. Warner, and D. G. Wagner, J. Dairy Sci., 46, 1435 (1963).

274. C. A. Bache, D. J. Lisk, D. G. Wagner, and R. G. Warner, J. Dairy Sci., 47, 93 (1964).

275. L. E. St. John, D. C. Wagner, and D. J. Lisk, J. Dairy Sci., 47, 1267 (1964).

276. K. Erne, Acta Vet. Scand., 7, 240 (1966).

277. D. E. Clark and J. S. Palmer, J. Agr. Food Chem., 19, 761 (1971).

278. M. L. Leng, Down to Earth, 28, 12 (1972).

279. W. L. Zielinski and L. Fishbein, J. Agr. Food Chem., 15, 841 (1967).

280. W. H. Gutenmann, D. D. Hardee, R. F. Holland, and D. J. Lisk, J. Dairy Sci., 46, 1287 (1963).

281. J. W. Mitchell, R. E. Hodgson, and C. F. Gaetjens, J. Anim. Sci., 5, 226 (1946).

282. C. A. Bache, D. D. Hardee, R. F. Holland, and D. J. Lisk, J. Dairy Sci., 47, 298 (1964).

283. D. L. Klingman, C. H. Gordon, G. Yip, and H. P. Burchfield, Weeds, 14, 164 (1966).

284. K. Erne, Acta Vet. Scand., 7, 264 (1966).

285. W. Grunow, C. Böhme, and B. Budczies, Food Cosmet. Toxicol., 9, 667 (1971).

286. D. E. Clark, J. Agr. Food Chem., 17, 1168 (1969).

287. J. E. Coakley, J. E. Campbell, and E. F. McFarren, J. Agr. Food Chem., 12, 262 (1964).

288. C. A. Rodgers and D. L. Stalling, Weed Sci., 20, 101 (1972).

289. F. C. Wright, J. C. Riner, and B. N. Gilbert, J. Agr. Food Chem., 17, 1171 (1969).

290. F. C. Wright, J. C. Riner, J. S. Palmer, and J. C. Schlinke, J. Agr. Food Chem., 18, 845 (1970).

291. A. S. Crafts, J. Agr. Food Chem., 1, 51 (1953).

292. A. S. Crafts, Ann. Rev. Plant Physiol., 4, 253 (1953).

293. J. Van Overbeek, Ann. Rev. Plant Physiol., 7, 355 (1956).

294. A. S. Crafts, Adv. Pest Control Res., 1, 39 (1957).

295. E. K. Woodford, K. Holly, and C. C. McCready, Ann. Rev. Plant Physiol., 9, 311 (1958).

296. H. B. Currier and C. D. Dybing, Weeds, 7, 195 (1959).

297. I. W. Mitchell, B. C. Smale, and R. L. Metcalf, Adv. Pest Control Res., 3, 359 (1960).

298. A. S. Crafts, in Encyclopedia of Plant Physiology, Vol. 14 (W. Ruhland, ed.), Springer, Berlin, 1961, pp. 1044-1054.

299. A. S. Crafts and C. L. Foy, Residue Rev., 1, 112 (1962).

300. K. Holly, in The Physiology and Biochemistry of Herbicides (L. J. Audus, ed.), Academic Press, New York, 1964, Chap. 15.

301. C. L. Foy, J. Agr. Food Chem., 12, 473 (1964).

302. J. A. Sargent, Ann. Rev. Plant Physiol., 16, 1 (1965).

303. M. M. Robertson and R. C. Kirkwood, Weed Res., 9, 224 (1969).

304. W. Franke, Ann. Rev. Plant Physiol., 18, 281 (1967).

305. E. W. Hauser, Agron. J., 47, 32 (1955).

306. W. H. Orgell and R. L. Weintraub, Bot. Gaz., 119, 88 (1957).

307. J. A. Sargent and G. E. Blackman, J. Exp. Bot., 13, 348 (1962).

308. C. G. Greenham, Weed Res., 8, 272 (1968).

309. A. S. Crafts, Science, 108, 85 (1948).

310. A. S. Crafts, Hilgardia, 26, 335 (1956).

311. S. Wathana, F. T. Corbin, and T. W. Waldrep, Weed Sci., 20, 120 (1972).

312. L. A. Norris and V. H. Freed, Weed Res., 6, 203 (1966).

313. A. S. Crafts, Hilgardia, 26, 287 (1956).

314. J. E. Pallas, Forest Sci., 9, 485 (1963).

315. O. A. Leonard and R. K. Glenn, Weed Sci., 16, 352 (1968).

316. K. Holly, Ann. Appl. Biol., 44, 195 (1956).

317. J. A. Sargent, Meded. Landbhogesch. Opzoekingstat. Staat Gent, 1964, 656.

318. D. W. Staniforth and W. E. Loomis, Science, 109, 628 (1949).

319. J. W. Mitchell and P. J. Lindner, Science, 112, 54 (1950).

320. M. E. Gertsch, Weeds, 2, 33 (1953).

321. O. A. Leonard, Hilgardia, 28, 115 (1958).

322. L. L. Jansen, W. A. Gentner, and W. C. Shaw, Weeds, 9, 381 (1961).

323. L. L. Jansen, J. Agr. Food Chem., 12, 223 (1964).

324. L. L. Jansen, Weeds, 12, 251 (1964).

325. G. B. Wortmann, Z. Pflkrankh. Pflpath. Pflschutz, 72, 641 (1965).

326. C. G. L. Furmidge, J. Sci. Food Agr., 10, 274 (1959).

327. R. P. Upchurch, J. A. Keaton, and H. D. Coble, Weed Sci., 17, 505 (1969).

328. R. E. Temple and H. W. Hilton, Weeds, 11, 297 (1963).

329. C. G. L. Furmidge, J. Sci. Food. Agr., 10, 419 (1959).

330. H. M. Hull, Weeds, 4, 22 (1956).

331. S. S. Szabo and K. P. Buckholtz, Weeds, 9, 177 (1961).

332. H. A. Brady, Weed Sci., 18, 204 (1970).

333. E. R. Burns, G. A. Buchanan, and A. E. Hiltbold, Weed Sci., 17, 401 (1969).

334. E. L. Rice, Bot. Gaz., 109, 301 (1948).

335. G. E. Barrier and W. E. Loomis, Plant Physiol., 32, 225 (1957).

336. J. E. Pallas, Plant Physiol., 35, 575 (1960).

337. J. A. Sargent and G. E. Blackman, J. Exp. Bot., 16, 24 (1965).

338. E. Basler, G. W. Todd, and R. E. Meyer, Plant Physiol., 36, 573 (1961).

339. J. E. Pallas and G. G. Williams, Bot. Gaz., 123, 175 (1962).

340. M. G. Merkle and F. S. Davis, Weeds, 15, 10 (1967).

341. H. A. Brady, Weed Sci., 17, 320 (1969).

342. H. B. Currier, E. R. Pickering, and C. L. Foy, Weeds, 12, 301 (1964).

343. O. A. Leonard and A. S. Crafts, Hilgardia, 26, 366 (1956).

344. A. S. Crafts and S. Yamaguchi, Hilgardia, 27, 421 (1958).

345. O. A. Leonard and R. J. Weaver, Hilgardia, 31, 327 (1961).

346. M. A. Radwan, C. R. Stocking, and H. B. Currier, Weeds, 8, 657 (1960).

347. J. W. Mitchell and J. W. Brown, Bot. Gaz., 107, 393 (1946).

348. R. J. Weaver and H. R. de Rose, Bot. Gaz., 107, 509 (1946).

349. P. J. Linder, J. W. Brown, and J. W. Mitchell, Bot. Gaz., 110, 628 (1949).

350. L. M. Rohrbaugh and E. L. Rice, Bot. Gaz., 111, 85 (1949).

351. R. L. Weintraub and J. W. Brown, Plant Physiol., 25, 140 (1950).

352. L. Eliasson, Physiol. Plant., 18, 506 (1965).

353. B. E. Day, Plant Physiol., 27, 143 (1952).

354. B. O. Blair and W. H. Fuller, Bot. Gaz., 113, 368 (1952).

355. J. R. Hay, Weeds, 4, 349 (1956).

356. A. S. Crafts, Plant Physiol., 34, 613 (1959).

357. H. D. Coble, F. W. Slife, and H. S. Butler, Weed Sci., 18, 653 (1970).

358. E. Basler, F. W. Slife, and J. W. Long, Weed Sci., 18, 396 (1970).

359. A. A. Badiei, E. Basler, and P. W. Santelmann, Weeds, 14, 302 (1966).

360. M. A. Chlor, A. S. Crafts, and S. Yamaguchi, Plant Physiol., 37, 609 (1962).

361. M. A. Chlor, A. S. Crafts, and S. Yamaguchi, Plant Physiol., 38, 501 (1963).

362. M. A. Chlor, A. S. Crafts, and S. Yamaguchi, Weeds, 12, 194 (1964).

363. J. R. Hay, Weeds, 4, 218 (1956).

364. A. S. Crafts and S. Yamaguchi, Am. J. Bot., 47, 248 (1960).

365. S. Yamaguchi, Hilgardia, 36, 349 (1965).

366. O. A. Leonard, L. A. Lider, and R. K. Glenn, Weed Res., 6, 37 (1966).

367. Y. Eshel and G. N. Prendeville, Weed Res., 7, 242 (1967).

368. G. N. Prendeville, Y. Eshel, M. M. Schreiber, and G. F. Warren, Weed Res., 7, 316 (1967).

369. P. W. Perry and R. P. Upchurch, Weed Sci., 16, 32 (1968).

370. D. L. Sutton and S. W. Bingham, Weed Sci., 18, 193 (1970).

371. P. C. Scott and R. O. Morris, Plant Physiol., 46, 680 (1970).

372. K. V. Thimann, in The Physiology of Plant Growth and Development (M. B. Wilkins, ed.), McGraw-Hill, London, 1969, Chap. 1.

373. J. A. Bentley, in Encyclopedia of Plant Physiology, Vol. 14 (W. Ruhland, ed.), Springer, Berlin, 1961, pp. 485-500.

374. R. L. Wain and C. H. Fawcett, in Plant Physiology, Vol. 5A (F. C. Steward, ed.), Academic Press, New York, 1969, Chap. 4.

375. P. Larsen, in Encyclopedia of Plant Physiology, Vol. 14 (W. Ruhland, ed.), Springer, Berlin, 1961, pp. 521-582.

376. D. J. Osborne and R. L. Wain, Science, 114, 92 (1951).

377. C. H. Fawcett, D. J. Osborne, R. L. Wain, and R. D. Walker, Ann. Appl. Biol., 40, 231 (1953).

378. L. J. Audus (ed.), The Physiology and Biochemistry of Herbicides, Academic Press, New York, 1964.

379. H. Veldstra, in The Chemistry and Mode of Action of Plant Growth Substances (R. L. Wain and F. Wightman, eds.), Butterworths, London, 1956, pp. 117-133.

380. H. Linser, in The Chemistry and Mode of Action of Plant Growth Substances (R. L. Wain and F. Wightman, eds.), Butterworths, London, 1956, pp. 141-158.

381. A. Jönsson, in Encyclopedia of Plant Physiology, Vol. 14 (W. Ruhland, ed.), Springer, Berlin, 1961, pp. 958-1006.

382. N. P. Kefford and O. H. Caso, Bot. Gaz., 127, 159 (1966).

383. C. S. James and R. L. Wain, Ann. Appl. Biol., 61, 295 (1968).

384. H. Veldstra, Ann. Rev. Plant Physiol., 4, 151 (1953).

385. R. M. Muir and C. Hansch, Ann. Rev. Plant Physiol., 6, 157 (1955).

386. K. V. Thimann and A. C. Leopold, in The Hormones: Physiology, Chemistry and Applications, Vol. 3 (G. Pincus and K. V. Thimann, eds.), Academic Press, New York, 1955, pp. 1-56.

387. C. Hansch and R. M. Muir, in Plant Growth Regulation (R. M. Klein et al., eds.), Iowa State University Press, Ames, 1961, pp. 431-448.

388. H. Veldstra, in Comprehensive Biochemistry, Vol. 11 (M. Florkin and E. H. Stotz, eds.), Elsevier, Amsterdam, 1963, Chap. 9.

389. J. B. Koepfli, K. V. Thimann, and F. W. Went, J. Biol. Chem., 122, 763 (1938).

390. G. J. M. van der Kerk, M. H. van Raalte, A. Kaars Sijpesteijn, and R. van der Veen, Nature, 176, 308 (1955).

391. K. Rothwell and R. L. Wain, Ann. Appl. Biol., 51, 161 (1963).

392. R. L. Wain and F. Wightman, Ann. Appl. Biol., 45, 140 (1957).

393. B. Åberg, Ann. Rev. Plant Physiol., 8, 153 (1957).

394. B. Hansen, Bot. Notiser, 1954, 230.

395. B. Åberg, Kungl. Lantbrukshögsk. Ann., 20, 241 (1953).

396. B. Åberg, Kungl. Lantbrukshögsk. Ann., 23, 375 (1957).

397. B. Åberg, Lantbrukshögsk. Ann., 33, 613 (1967).

398. B. Åberg, Lantbrukshögsk. Ann., 33, 625 (1967).

399. B. Åberg, Lantbrukshögsk. Ann., 36, 235 (1970).

400. B. Åberg, Lantbrukshögsk. Ann., 36, 323 (1970).

401. B. Åberg, Swed. J. Agr. Res., 1, 23 (1971).

402. B. Åberg, Swed. J. Agr. Res., 1, 33 (1971).

403. M. S. Smith, R. L. Wain, and F. Wightman, Ann. Appl. Biol., 39, 295 (1952).

404. R. L. Wain and F. Wightman, Ann. Appl. Biol., 40, 244 (1953).

405. C. H. Fawcett, R. L. Wain, and F. Wightman, Ann. Appl. Biol., 43, 342 (1955).

406. J. Toothill, R. L. Wain, and F. Wightman, Ann. Appl. Biol., 44, 547 (1956).

407. M. B. Pybus, M. S. Smith, R. L. Wain, and F. Wightman, Ann. Appl. Biol., 47, 173 (1959).

408. H. Vitou and R. L. Wain, Ann. Appl. Biol., 47, 752 (1959).

409. G. G. Clarke and R. L. Wain, Ann. Appl. Biol., 51, 453 (1963).

410. B. Åberg, in The Chemistry and Mode of Action of Plant Growth Substances (R. L. Wain and F. Wightman, eds.), Butterworths, London, 1956, pp. 93-116.

411. B. Åberg, in Plant Growth Regulation (R. M. Klein et al., eds.), Iowa State University Press, Ames, 1961, pp. 219-232.

412. R. D. Firn and R. L. Wain, Ann. Appl. Biol., 69, 73 (1971).

413. K. V. Thimann, Plant Physiol., 27, 392 (1952).

414. R. M. Muir and C. Hansch, Plant Physiol., 28, 218 (1953).

415. M. Matell, Kungl. Lantbrukshögsk. Ann., 20, 206 (1953).

416. M. S. Smith, J. Sci. Food Agr., 13, 48 (1962).

417. C. G. Greenham, Aust. J. Sci., 20, 212 (1958).

418. H. Burström, Physiol. Plant., 3, 277 (1950).

419. H. Burström, Physiol. Plant., 4, 470 (1951).

420. D. H. McRae and J. Bonner, Physiol. Plant., 6, 485 (1953).

421. C. Wilske and H. Burström, Physiol. Plant., 3, 58 (1950).

422. A. Jönsson, G. Nilsson, and H. Burström, Acta Chem. Scand., 6, 993 (1952).

423. A. Fredga, Arkiv Kemi, 11, 23 (1957).

424. A. Takeda, J. Org. Chem., 22, 1096 (1957).

425. A. Takeda and J. Senda, Ber. Ohara Inst. Landwirtsch. Biol., Okayama Univ., 11, 1 (1957); through C.A., 52, 10879g (1958).

426. M. Matell, Acta Chem. Scand., 7, 228 (1953).

427. A. Takeda, S. Wada, and M. Fugimoto, Rep. Ohara Inst. Agr. Biol., 10, 98 (1956); through C.A., 51, 8026e (1957).

428. H. Burström, B. Sjöberg, and B. A. M. Hansen, Acta Agr. Scand., 6, 155 (1956).

429. C. G. Greenham, Aust. J. Biol. Sci., 10, 180 (1957).

430. C. G. Greenham, Aust. J. Sci., 16, 66 (1953).

431. H. Veldstra, W. Kryt, E. J. van der Steen, and B. Åberg, Rec. Trav. Chim., 73, 23 (1954).

432. R. L. Wain and H. F. Taylor, Nature, 207, 167 (1965).

433. D. B. Harper and R. L. Wain, Nature, 213, 1155 (1967).

434. J. Chatt, L. A. Duncanson, and L. M. Venanzi, Nature, 177, 1042 (1956).

435. B. Åberg, Kungl. Lantbrukshögsk. Ann., 26, 239 (1960).

436. J. L. Garraway and R. L. Wain, Ann. Appl. Biol., 50, 11 (1962).

437. R. M. Muir, C. Hansch, and J. Gally, Plant Physiol., 36, 222 (1961).

438. B. Sjöberg, Arkiv Kemi, 12, 251 (1958).

439. B. Sjöberg, Arkiv Kemi, 13, 1 (1958).

440. E. N. Ugochukwu and R. L. Wain, Ann. Appl. Biol., 61, 121 (1968).

441. R. A. Heacock, R. L. Wain, and F. Wightman, Ann. Appl. Biol., 46, 352 (1958).

442. P. W. Zimmerman, W. R. Smith, E. A. Prill, and A. E. Hitchcock, Contrib. Boyce Thompson Inst., 18, 453 (1957).

443. A. Jönsson, Svensk. Kem. Tidskr., 67, 166 (1955).

444. V. K. Chamberlain and R. L. Wain, Ann. Appl. Biol., 69, 65 (1971).

445. B. Åberg, Physiol. Plant., 7, 241 (1954).

446. R. M. Muir and C. Hansch, Plant Physiol., 26, 369 (1951).

447. K. V. Thimann, in Plant Growth Regulation (R. M. Klein et al., eds.), Iowa State University Press, Ames, 1961, pp. 444-447.

448. W. L. Porter and K. V. Thimann, Phytochemistry, 4, 229 (1965).

449. C. Hansch, R. M. Muir, and R. L. Metzenberg, Plant Physiol., 26, 812 (1951).

450. J. Grundy, Chem. Ind., 1954, 1071.

451. R. M. Muir, T. Fujita, and C. Hansch, Plant Physiol., 42, 1519 (1967).

452. K. Fukui, C. Nagata, and T. Yonezawa, J. Am. Chem. Soc., 80, 2267 (1958).

453. R. M. Muir and C. Hansch, in Plant Growth Regulation (R. M. Klein et al., eds.), Iowa State University Press, Ames, 1961, pp. 249-258.

454. K. Koshimuzu, T. Fujita, and T. Mitsui, J. Am. Chem. Soc., 82, 4041 (1960).

455. M. Cocordano and J. Ricard, Physiol. Veg., 1, 129 (1963).

456. C. Hansch, P. P. Maloney, T. Fujita, and R. M. Muir, Nature, 194, 178 (1962).

457. C. Hansch, R. M. Muir, T. Fujita, P. P. Maloney, F. Geiger, and M. Streich, J. Am. Chem. Soc., 85, 2817 (1963).

458. D. H. McRae and J. Bonner, Plant Physiol., 27, 834 (1952).

459. O. L. Hoffmann, Plant Physiol., 28, 622 (1953).

460. T. Ingestad, Physiol. Plant., 6, 796 (1953).

461. R. J. Foster, D. H. McRae, and J. Bonner, Proc. Natl. Acad. Sci. (U.S.), 38, 1014 (1952).

462. D. H. McRae, R. J. Foster, and J. Bonner, Plant Physiol., 28, 343 (1953).

463. R. J. Foster, D. H. McRae, and J. Bonner, Plant Physiol., 30, 323 (1955).

464. J. Bonner and R. J. Foster, in The Chemistry and Mode of Action of Plant Growth Substances (R. L. Wain and F. Wightman, eds.), Butterworths, London, 1956, pp. 295-309.

465. S. Housley, in Encyclopedia of Plant Physiology, Vol. 14 (W. Ruhland, ed.), Springer, Berlin, 1961, pp. 1007-1043.

466. A. W. Galston and W. K. Purves, Ann. Rev. Plant Physiol., 11, 239 (1960).

467. B. Åberg, Physiol. Plant., 5, 567 (1952).

468. L. J. Audus and M. E. Shipton, Physiol. Plant., 5, 430 (1952).

469. D. E. Moreland, Ann. Rev. Plant Physiol., 18, 365 (1967).

470. J. B. Hanson and F. W. Slife, Residue Rev., 25, 59 (1969).

471. A. W. Galston and P. J. Davies, Science, 163, 1288 (1969).

472. J. L. Key, Ann. Rev. Plant Physiol., 20, 449 (1969).

473. J. Silberger and F. Skoog, Science, 118, 443 (1953).

474. H. M. Sell, R. W. Luecke, B. M. Taylor, and C. L. Hamner, Plant Physiol., 24, 295 (1949).

475. T. L. Rebstock, C. L. Hamner, and H. M. Sell, Plant Physiol., 29, 490 (1954).

476. S. H. West, J. B. Hanson, and J. L. Key, Weeds, 8, 333 (1960).

477. J. L. Key and J. B. Hanson, Plant Physiol., 36, 145 (1961).

478. M. J. Chrispeels and J. B. Hanson, Weeds, 10, 123 (1962).

479. J. L. Key, C. Y. Lin, E. M. Gifford, and R. Dengler, Bot. Gaz., 127, 87 (1966).

480. A. J. Eames, Am. J. Bot., 37, 840 (1950).

481. J. Cardenas, F. W. Slife, J. B. Hanson, and H. Butler, Weed Sci., 16, 96 (1968).

482. J. C. Shannon, J. B. Hanson, and C. M. Wilson, Plant Physiol., 39, 804 (1964).

483. J. L. Key and J. C. Shannon, Plant Physiol., 39, 360 (1964).

484. J. Ingle, J. L. Key, and R. E. Holm, J. Mol. Biol., 11, 730 (1965).

485. J. Ingle and J. L. Key, Plant Physiol., 40, 1212 (1965).

486. C. Y. Lin, J. L. Key, and C. E. Bracker, Plant Physiol., 41, 976 (1966).

487. J. L. Key and J. Ingle, Proc. Natl. Acad. Sci. (U.S.), 52, 1382 (1964).

488. J. L. Key, Plant Physiol., 41, 1257 (1966).

489. J. L. Key, N. M. Barnett, and C. Y. Lin, Ann. N.Y. Acad. Sci., 144, 49 (1967).

490. J. L. Key and J. Ingle, in Biochemistry and Physiology of Plant Growth Substances (F. Wightman and G. Setterfield, eds.), The Runge Press, Ottawa, 1968, pp. 711-722.

491. Y. Masuda and E. Tanimoto, Plant Cell Physiol., 8, 459 (1967).

492. Y. Masuda, in Biochemistry and Physiology of Plant Growth Substances (F. Wightman and G. Setterfield, eds.), The Runge Press, Ottawa, 1968, pp. 699-710.

493. R. Roychoudhury and S. P. Sen, Physiol. Plant., 17, 352 (1964).

494. R. Roychoudhury, A. Datta, and S. P. Sen, Biochim. Biophys. Acta, 107, 346 (1965).

495. S. C. Maheshwari, S. Guka, and S. Gupta, Biochim. Biophys. Acta, 117, 470 (1966).

496. J. H. Cherry, Ann. N.Y. Acad. Sci., 144, 154 (1967).

497. T. J. O'Brien, B. C. Jarvis, J. H. Cherry, and J. B. Hanson, in Biochemistry and Physiology of Plant Growth Substances (F. Wightman and G. Setterfield, eds.), The Runge Press, Ottawa, 1968, pp. 747-759.

498. T. J. O'Brien, B. C. Jarvis, J. H. Cherry, and J. B. Hanson, Biochim. Biophys. Acta, 169, 35 (1968).

499. A. G. Matthysse and C. Phillips, Proc. Natl. Acad. Sci. (U.S.), 63, 897 (1969).

500. H. R. Leffler, T. J. O'Brien, D. V. Glover, and J. H. Cherry, Plant Physiol., 48, 43 (1971).

501. J. Bonner, in Plant Growth Regulation (R. M. Klein et al., eds.), Iowa State University Press, Ames, 1961, pp. 307-328.

502. R. Cleland, Ann. N.Y. Acad. Sci., 144, 3 (1967).

503. J. A. Lockhart, C. Bretz, and R. Kenner, Ann. N.Y. Acad. Sci., 144, 19 (1967).

504. R. Cleland, in Biochemistry and Physiology of Plant Growth Substances (F. Wightman and G. Setterfield, eds.), The Runge Press, Ottawa, 1968, pp. 613-624.

505. D. J. Morré and W. R. Eisinger, in Biochemistry and Physiology of Plant Growth Substances (F. Wightman and G. Setterfield, eds.), The Runge Press, Ottawa, 1968, pp. 625-645.

506. L. D. Nooden and K. V. Thimann, Proc. Natl. Acad. Sci. (U.S.), 50, 194 (1963).

507. J. L. Key, Plant Physiol., 39, 365 (1964).

508. A. A. De Hertogh, D. M. McCune, J. Brown, and D. Antoine, Contrib. Boyce Thompson Inst., 23, 23 (1965).

509. D. J. Morré, Plant Physiol., 40, 615 (1965).

510. L. D. Nooden and K. V. Thimann, Plant Physiol., 40, 193 (1965).

511. L. D. Nooden and K. V. Thimann, Plant Physiol., 41, 157 (1966).

512. P. Penny and A. W. Galston, Am. J. Bot., 53, 1 (1966).

513. Y. Masuda, Plant Cell Physiol., 7, 75 (1966).

514. Y. Masuda, Plant Cell Physiol., 7, 573 (1966).

515. Y. Masuda, Ann. N.Y. Acad. Sci., 144, 68 (1967).

516. M. Black, C. Bullock, E. N. Chantler, R. A. Clarke, A. D. Hanson, and G. M. Jolley, Nature, 215, 1289 (1967).

517. J. S. Coartney, D. J. Morré, and J. L. Key, Plant Physiol., 42, 434 (1967).

518. L. D. Nooden, Plant Physiol., 43, 140 (1968).

519. E. Tanimoto and Y. Masuda, Physiol. Plant., 21, 820 (1968).

520. A. N. J. Heyn, Science, 167, 874 (1970).

521. D.-F. Fan and G. A. MacLachlan, Can. J. Bot., 44, 1025 (1966).

522. D.-F. Fan and G. A. MacLachlan, Plant Physiol., 42, 1114 (1967).

523. G. A. MacLachlan, E. Davies, and D.-F. Fan, in Biochemistry and Physiology of Plant Growth Substances (F. Wightman and G. Setterfield, eds.), The Runge Press, Ottawa, 1968, pp. 443-453.

524. A. H. Datko and G. A. MacLachlan, Plant Physiol., 43, 735 (1968).

525. S. Wada, E. Tanimoto, and Y. Masuda, Plant Cell Physiol., 9, 369 (1968).

526. R. Cleland, Science, 160, 192 (1968).

527. A. W. Ruesink, Planta, 89, 95 (1969).

528. A. C. Olson, J. Bonner, and D. J. Morré, Planta, 66, 126 (1965).

529. P. M. Ray and D. B. Baker, Plant Physiol., 40, 353 (1965).

530. R. Cleland, Plant Physiol., 38, 12 (1963).

531. M. A. Hall and L. Ordin, in Biochemistry and Physiology of Plant Growth Substances (F. Wightman and G. Setterfield, eds.), The Runge Press, Ottawa, 1968, pp. 659-671.

532. A. A. Abdul-Baki and P. M. Ray, Plant Physiol., 47, 537 (1971).

533. P. M. Ray and A. A. Abdul-Baki, in Biochemistry and Physiology of Plant Growth Substances (F. Wightman and G. Setterfield, eds.), The Runge Press, Ottawa, 1968, pp. 647-648.

534. M. L. Evans and P. M. Ray, J. Gen. Physiol., 53, 1 (1969).

535. G. M. Barkley and M. L. Evans, Plant Physiol., 45, 143 (1970).

536. M. L. Evans and D. L. Rayle, Plant Physiol., 45, 240 (1970).

537. D. L. Rayle, M. L. Evans, and R. Hertel, Proc. Natl. Acad. Sci. (U.S.), 65, 184 (1970).

538. D. Nissl and M. H. Zenk, Planta, 89, 323 (1969).

539. P. Penny, Plant Physiol., 48, 720 (1971).

540. R. Cleland, Planta, 99, 1 (1971).

541. P. W. Zimmerman and F. Wilcoxon, Contrib. Boyce Thompson Inst., 7, 209 (1935).

542. H. K. Pratt and J. D. Goeschl, Ann. Rev. Plant Physiol., 20, 541 (1969).

543. F. B. Abeles, Plant Physiol., 41, 585 (1966).

544. B. G. Kang, W. Newcomb, and S. P. Burg, Plant Physiol., 47, 504 (1971).

545. S. P. Burg and E. A. Burg, Proc. Natl. Acad. Sci. (U.S.), 55, 262 (1966).

546. S. P. Burg and E. A. Burg, in Biochemistry and Physiology of Plant Growth Substances (F. Wightman and G. Setterfield, eds.), The Runge Press, Ottawa, 1968, pp. 1275-1294.

547. S. P. Burg and E. A. Burg, Plant Physiol., 42, 144 (1967).

548. A. V. Chadwick and S. P. Burg, Plant Physiol., 42, 415 (1967).

549. W. A. Andreae, M. A. Venis, F. Jursic, and T. Dumas, Plant Physiol., 43, 1375 (1968).

550. A. V. Chadwick and S. P. Burg, Plant Physiol., 45, 192 (1970).

551. R. E. Holm and F. B. Abeles, Planta, 78, 293 (1968).

552. F. B. Abeles, Weed Sci., 16, 498 (1968).

553. P. W. Morgan and W. C. Hall, Physiol. Plant., 15, 420 (1962).

554. W. Crocker, A. E. Hitchcock, and P. W. Zimmerman, Contrib. Boyce Thompson Inst., 7, 231 (1935).

555. F. B. Abeles and R. E. Holm, Plant Physiol., 41, 1337 (1966).

556. R. E. Holm and F. B. Abeles, Plant Physiol., 42, 1094 (1967).

557. R. E. Holm, T. J. O'Brien, J. L. Key, and J. H. Cherry, Plant Physiol., 45, 41 (1970).

558. A. C. Hulme, M. J. C. Rhodes, and L. S. Wooltorton, Phytochemistry, 10, 749 (1971).

559. N. Marei and R. Romani, Plant Physiol., 48, 806 (1971).

560. J. W. Hardin, T. J. O'Brien, and J. H. Cherry, Biochim. Biophys. Acta, 224, 667 (1970).

561. A. G. Matthysse, Biochim. Biophys. Acta, 199, 519 (1970).

562. M. A. Venis, Proc. Natl. Acad. Sci. (U.S.), 68, 1824 (1971).

563. P. Filner, J. L. Wray, and J. E. Varner, Science, 165, 358 (1969).

564. A. Datta, S. P. Sen, and A. G. Datta, Biochim. Biophys. Acta, 108, 147 (1965).

565. M. H. Zenk, Planta, 58, 75 (1962).

566. J. Südi, Nature, 201, 1009 (1964).

567. M. A. Venis, Nature, 202, 900 (1964).

568. J. Südi, New Phytol., 65, 9 (1966).

569. M. A. Venis, Plant Physiol., 49, 24 (1972).

570. W. Hayes, The Genetics of Bacteria and Their Viruses, 2nd ed., Blackwell, Oxford, 1968, pp. 732-734.

571. I. Ridge and D. J. Osborne, Nature, 223, 318 (1969).

572. H. D. Coble and F. W. Slife, Weed Sci., 19, 1 (1971).

573. I. V. Sarkissian, in Biochemistry and Physiology of Plant Growth Substances (F. Wightman and G. Setterfield, eds.), The Runge Press, Ottawa, 1968, pp. 473-485.

574. V. H. Freed, F. J. Reithel, and L. F. Remmert, in Plant Growth Regulation (R. M. Klein et al., eds.), Iowa State University Press, Ames, 1961, pp. 289-306.

575. J. Van Overbeek, in The Physiology and Biochemistry of Herbicides (L. J. Audus, ed.), Academic Press, New York, 1964, Chap. 13.

576. W. A. Andreae, in Regulateurs Naturels de la Croissance Végétale (J. P. Nitsch, ed.), Centre National de la Recherche Scientifique, Paris, 1963, pp. 559-573.

577. O. Kiermeyer, in The Physiology and Biochemistry of Herbicides (L. J. Audus, ed.), Academic Press, New York, 1964, Chap. 6.

578. R. C. Fites, J. B. Hanson, and F. W. Slife, Bot. Gaz., 130, 118 (1969).

579. A. J. Eames, Science, 110, 235 (1949).

580. D. J. Wort, in The Physiology and Biochemistry of Herbicides (L. J. Audus, ed.), Academic Press, New York, 1964, Chaps. 9 and 10.

581. D. Penner and F. M. Ashton, Residue Rev., 14, 39 (1966).

582. G. Stenlid and K. Saddik, Physiol. Plant., 15, 369 (1962).

583. C. M. Switzer, Plant Physiol., 32, 42 (1957).

584. R. T. Wedding and M. K. Black, Plant Physiol., 37, 364 (1962).

585. P. B. Lotlikar, L. F. Remmert, and V. H. Freed, Weed Sci., 16, 161 (1968).

586. A. H. Gallup and F. G. Gustafson, Plant Physiol., 27, 603 (1952).

587. F. M. Ashton, Weeds, 6, 257 (1958).

588. A. G. Dexter, F. W. Slife, and H. S. Butler, Weed Sci., 19, 721 (1971).

589. V. M. Bhan, E. W. Stoller, and F. W. Slife, Weed Sci., 18, 733 (1970).

590. A. J. Eames, Am. J. Bot., 36, 571 (1949).

591. W. C. Shaw and W. A. Gentner, Weeds, 5, 75 (1957).

592. J. R. Fleeker, Phytochemistry, 12, 757 (1973).

593. R. C. Steen, I. R. Schultz, D. C. Zimmerman, and J. R. Fleeker, Weed Res., 14, 23 (1974).

594. C.-s. Feung, R. H. Hamilton, F. H. Witham, and R. O. Mumma, Plant Physiol., 50, 80 (1972).

595. C.-s. Feung, R. H. Hamilton, and R. O. Mumma, J. Agr. Food Chem., 21, 637 (1973).

596. J. D. Altom and J. F. Stritzke, Weed Sci., 21, 556 (1973).

597. A. E. Smith, Weed Res., 12, 364 (1972).

598. T. W. Donaldson, D. E. Bayer, and O. A. Leonard, Plant Physiol., 52, 638 (1973).

599. L. G. Chen, C. M. Switzer, and R. A. Fletcher, Weed Sci., 20, 53 (1972).

600. M. Q. Arens and E. R. Stout, Plant Physiol., 50, 640 (1972).

601. T. J. Guilfoyle and J. B. Hanson, Plant Physiol., 53, 110 (1974).

602. K. D. Johnson and W. K. Purves, Plant Physiol., 46, 581 (1970).

603. L. G. Chen, A. Ali, R. A. Fletcher, C. M. Switzer, and G. R. Stephenson, Weed Sci., 21, 181 (1973).

604. E. Davies and G. A. Maclachlan, Arch. Biochem. Biophys., 128, 595 (1968).

605. E. Davies and G. A. Maclachlan, Arch. Biochem. Biophys., 129, 581 (1969).

606. P. M. Ray, Plant Physiol., 51, 609 (1973).

607. P. M. Ray, Plant Physiol., 51, 601 (1973).

608. F. S. Spencer, G. Shore, B. Ziola, and G. A. Maclachlan, Arch. Biochem. Biophys., 152, 311 (1972).

609. D. Pope and M. Black, Planta, 102, 26 (1972).

610. A. Hager, H. Menzel, and A. Krauss, Planta, 100, 47 (1971).

611. D. L. Rayle and R. Cleland, Plant Physiol., 46, 250 (1970).

612. D. L. Rayle, P. M. Haughton, and R. Cleland, Proc. Natl. Acad. Sci. (U.S.), 67, 1814 (1970).

613. D. L. Rayle and R. Cleland, Planta, 104, 282 (1972).

614. G. M. Barkley and A. C. Leopold, Plant Physiol., 52, 76 (1973).

615. A. Apelbaum and S. P. Burg, Plant Physiol., 50, 117 (1972).

616. A. Apelbaum and S. P. Burg, Plant Physiol., 50, 125 (1972).

617. J. E. Sherwin and S. A. Gordon, Planta, 116, 65 (1974).

618. E. D. Kopischke, J. Wildl. Manage., 36, 1353 (1972).

619. J. D. Somers, E. T. Moran, B. S. Reinhart, and G. R. Stephenson, Bull. Environ. Contam. Toxicol., 11, 33 (1974).

620. J. D. Somers, E. T. Moran, and B. S. Reinhart, Bull. Environ. Contam. Toxicol., 11, 339 (1974).

621. J. D. Somers, E. T. Moran, and B. S. Reinhart, Bull. Environ. Contam. Toxicol., 11, 511 (1974).

Chapter 2

s-TRIAZINES

HERBERT O. ESSER,
GERARD DUPUIS, EDITH EBERT,
and CHRISTIAN VOGEL

Agrochemicals Division
Ciba-Geigy Ltd.
Basel, Switzerland

GINO J. MARCO
Agricultural Division
Ciba-Geigy Corporation
Greensboro, North Carolina

I. INTRODUCTION: PHYSICAL AND CHEMICAL PROPERTIES

The herbicidal properties of the s-triazines were discovered in 1952 by a research group of J. R. Geigy Ltd., in Basel, Switzerland. The first patent applications were made in 1954 covering 2-chloro-4, 6-bis(alkyl-amino)-s-triazines [1], 2-methoxy- and 2-methylthio-4, 6-bis(alkylamino)-s-triazines [2], and their influence on the growth of plants. The selective action of these compounds and their unique herbicidal properties compared with existing classes of herbicides were first reported in 1955 [3, 4].

Major research efforts and worldwide field tests following their original discovery established the outstanding selective herbicidal properties of the s-triazines. Atrazine, simazine, prometryn, and ametryn were the compounds that gained major recognition in agriculture. Among the numerous s-triazine derivatives investigated thereafter, only a limited number reached the marketing stage. In Table 1 are listed the more important of these commercial products, their corresponding structures, code numbers, the chemical and common names, and physical data of practical relevance.

Attempts to substitute the chlorine, methylthio, or methoxy groups with other substituents such as other halogen atoms or halogenated alkyl groups have resulted in compounds of no practical importance. Only the azido derivatives, first synthesized by Degussa [5], are of any particular interest, especially the compounds of Ciba Ltd. An example of this group is 2-methylthio-4-azido-6-isopropylamino-s-triazine, i.e., aziprotryn [6], in which both the methylthio and the azido group were introduced as substituents.

Variations of the N-alkylamino side chains enlarged the spectrum of available compounds with respect to biological activity and degradability. Alkoxyalkylamino groups were introduced in 2-alkylthio-s-triazines by Du Pont [7] and by Geigy [8]. Examples of these compounds were developed by Monsanto with 2-methylthio-4, 6-bis(3-methoxypropylamino)-s-triazine [9] and by Geigy with methoprotryn. In recent developments, the N-alkyl groups were further varied as, for example, in cyprozine of Gulf [10], which contains a cyclopropyl group, or in cyanazine of Shell [11], where a cyano group is attached to an isopropyl group.

Among the voluminous literature on s-triazines published to date, several papers review the relationships between structure, herbicidal

TABLE 1

Physical Properties of s-Triazine Herbicides

Structure and chemical denomination	Code number	Common name	Melting point (°C)	Solubility in water at 20–25°C (ppm)	Vapor pressure at 20°C (mm Hg)	pK$_a$, 21°C	Density (g/cm^3)
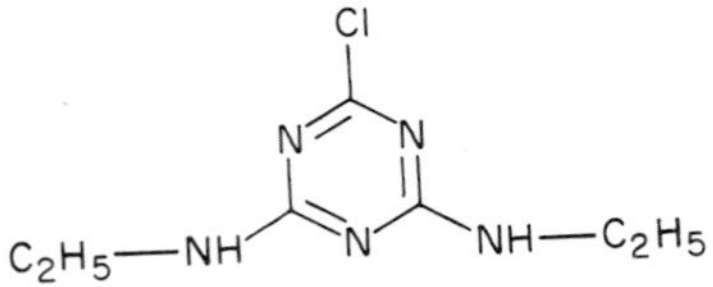 Simazine structure; 2-Chloro-4,6-bis(ethylamino)-s-triazine	G-27692	Simazine	225–227	5	6.1×10^{-9}	1.7	1.302
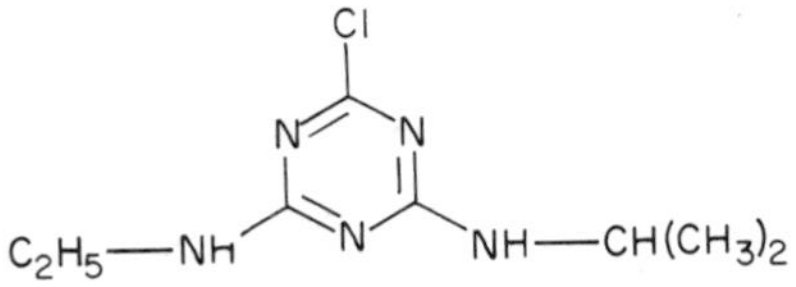 Atrazine structure; 2-Chloro-4-ethylamino-6-isopropylamino-s-triazine	G-30027	Atrazine	175–177	33	3.0×10^{-7}	1.7	1.187
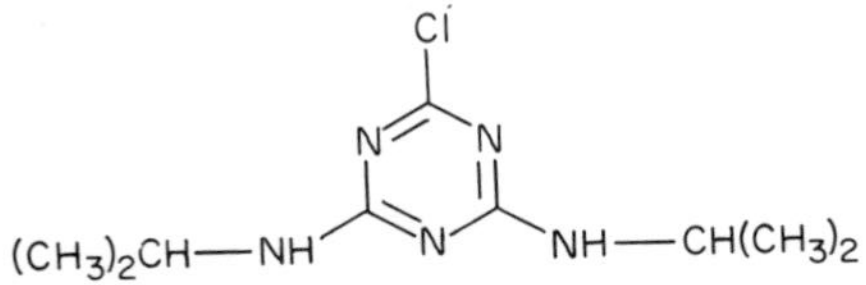 Propazine structure; 2-Chloro-4,6-bis(isopropylamino)-s-triazine	G-30028	Propazine	212–214	8.6	2.9×10^{-8}	1.7	1.162

(continued)

TABLE 1 (continued)

Structure and chemical denomination	Code number	Common name	Melting point (°C)	Solubility in water at 20-25°C (ppm)	Vapor pressure at 20°C (mm Hg)	pK$_a$, 21°C	Density (g/cm^3)
 2-Chloro-4-isopropylamino-6-cyclopropylamino-s-triazine	S-6115	Cyprozine[a]	167-168[a]	6.9[a]	3.0×10^{-7}	—	1.24
 2-Chloro-4-ethylamino-6-tert-butylamino-s-triazine	GS-13529	Terbuthyl-azine	177-179	8.5	1.12×10^{-6}	2.0	1.188
 2-Chloro-4-ethylamino-6-(1-cyano-1-methylethylamino)-s-triazine	SD-15418 WL-19805	Cyanazine[b]	166.5-167[b]	171[b]	1.6×10^{-9}	1.0	—

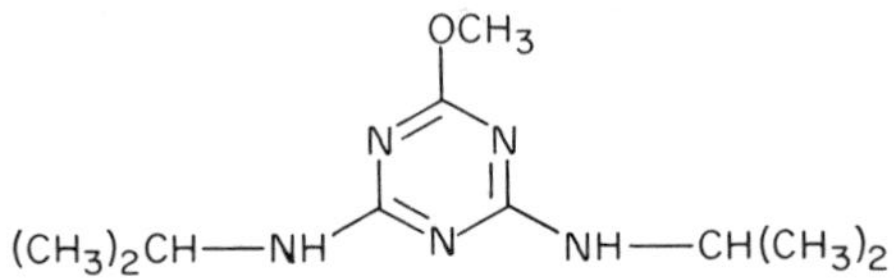 G-31435	Prometone	91–92	750	2.3×10^{-6}	4.3	1.088
GS-14254	Seebumetone[c]	86–88	620	7.3×10^{-6}	4.4	1.105
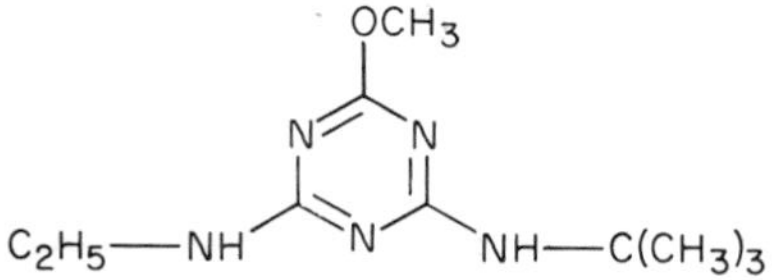 GS-14259	Terbumetone[c]	123–124	130	2.0×10^{-6}	4.6	1.081
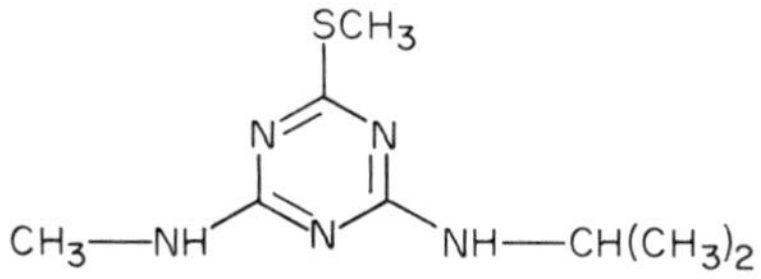 G-34360	Desmetryn	84–86	580	1.0×10^{-6}	4.0	1.172

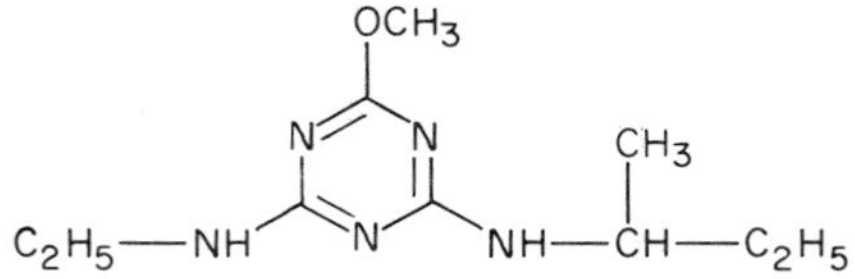

2-Methoxy-4,6-bis(isopropyl-amino)-s-triazine

OCH_3

$(CH_3)_2CH—NH$ … $NH—CH(CH_3)_2$

2-Methoxy-4-ethylamino-6-sec-butylamino-s-triazine

OCH_3

$C_2H_5—NH$ … $NH—CH—C_2H_5$ (CH$_3$)

2-Methoxy-4-ethylamino-6-tert-butylamino-s-triazine

OCH_3

$C_2H_5—NH$ … $NH—C(CH_3)_3$

2-Methylthio-4-methylamino-6-isopropylamino-s-triazine

SCH_3

$CH_3—NH$ … $NH—CH(CH_3)_2$

(continued)

TABLE 1 (continued)

Structure and chemical denomination	Code number	Common name	Melting point (°C)	Solubility in water at 20–25°C (ppm)	Vapor pressure at 20°C (mm Hg)	pKa, 21°C	Density (g/cm³)
SCH_3 ... $C_2H_5—NH$ ... $NH—CH(CH_3)_2$ 2–Methylthio–4–ethylamino–6–isopropylamino–s–triazine	G–34162	Ametryn	84–86	185	8.4×10^{-7}	4.1	1.190
SCH_3 ... $(CH_3)_2CH—NH$ ... $NH—CH(CH_3)_2$ 2–Methylthio–4,6–bis(isopropyl-amino)–s–triazine	G–34161	Prometryn	118–120	48	1.0×10^{-6}	4.1	1.157
SCH_3 ... $(CH_3)_2CH—NH$ ... $NH—(CH_2)_3—OCH_3$ 2–Methylthio–4–isopropylamino–6–(3–methoxypropylamino)–s–triazine	G–36393	Metoprotryn	68–70	320	2.85×10^{-7}	4.0	1.186

Structure	Code	Common name	mp				
SCH₃ / C₂H₅—NH / NH—C(CH₃)₃ triazine 2-Methylthio-4-ethylamino-6-tert-butylamino-s-triazine	GS-14260	Terbutryn	104–105	58	9.6×10^{-7}	4.3	1.115
SCH₃ / NH—CH(CH₃)₂ / N₃ triazine 2-Methylthio-4-azido-6-iso-propylamino-s-triazine	C-7019	Aziprotryn	95	50	2.0×10^{-6}	—	1.40
Cl / cyclopropyl—CH—NH / NH—C(CH₃)₂—CN triazine 2-Chloro-4-cyclopropylamino-6-(1-cyano-1-methylethyl-amino)-s-triazine	CGA-18762	—	168	300	1.0×10^{-8}	0.8	1.279

(continued)

TABLE 1 (continued)

Structure and chemical denomination	Code number	Common name	Melting point (°C)	Solubility in water at 20-25°C (ppm)	Vapor pressure at 20°C (mm Hg)	pK_a, 21°C	Density (g/cm³)

2-Methylthio-4-ethylamino-6-(1,2-dimethylpropylamino)-s-triazine

Dimeth-ametryn[c] — C-18898 — 65 — 50 — 1.4×10^{-6} — 4.0 — 1.098

2-Ethylthio-4,6-bis(isopropyl-amino)-s-triazine

Dipropetryn[c] — GS-16068 — 104-106 — 16 — 7.3×10^{-7} — 4.3 — 1.116

[a]Gulf Research and Development Company, Merriam, Kansas.

[b]Shell Development Company, Modesto, California.
The remaining physical data are from the Analytical Laboratories, Ciba-Geigy Ltd., Basel, Switzerland.

[c]ISO draft recommendation.

activity, and mode of action [12-19], the metabolic fate in different biological systems [20], and the methodology developed for the analysis of these
compounds [21-23]. Other papers reviewing selected aspects of the behavior
and properties of s-triazines are mentioned in the appropriate sections of
this chapter.

The present chapter reviews the published information on the reactivity
of s-triazines under purely chemical conditions and in different biological
systems. However, the selection of chemical reactions was restricted to
those reactions affecting the substituents and the heterocyclic ring of
s-triazine molecules which also participate in biotransformation reactions.

With few exceptions, the s-triazines currently used as selective or
general herbicides are substituted diamino-s-triazines which have a chlorine,
methoxy, methylthio, or azido group attached to the third ring carbon atom.
In order to facilitate the recognition and comparison of parent compounds
and reaction products, the following nomenclature is used throughout this
chapter (1):

(1)

The alkyl, alkoxyalkyl, or otherwise substituted amino groups are assigned
to C-4 and C-6; the halogen, alkoxy, or alkylthio groups are coordinated to
C-2. Unlike the nomenclature system of Chemical Abstracts, the substituents of the reaction products will maintain the numbers given to the original
ligands of the parent compound regardless of the alphabetical order of these
substituents. For example, the degradation product of 2-chloro-4-ethyl-
amino-6-isopropylamino-s-triazine, after hydrolysis and N-deethylation,
is designated as 2-hydroxy-4-amino-6-isopropylamino-s-triazine and not
as 2-amino-4-hydroxy-6-isopropylamino-s-triazine as would be required
when the Chemical Abstracts system is applied.

Comprehensive reviews on the preparation and chemical reactivity of
s-triazines have been published [24, 25]. Corresponding information on
herbicidally active s-triazines has also been compiled [26]. Certain physical properties of a number of herbicidal s-triazines have been compared
with those of representatives of other classes of herbicides [27].

The remarkable stability of s-triazine derivatives can be explained by
the electronic configuration of the heterocyclic ring which resembles that
of benzene to a certain extent. Both ring systems are stabilized by delocalization of their π electrons which are spread over all six ring atoms. However, essential differences exist in electronic configuration between

s-triazine and benzene as a consequence of the greater electronegativity of
the nitrogen atoms as compared to that of the carbon atoms. Thus the π
electrons in the s-triazine ring are localized in the vicinity of the nitrogen
atoms rather than being evenly distributed over the whole ring. A polar
mesomeric form (2) that bears an additional pair of unshared electrons on
the nitrogen atoms and a positive charge on the carbon atoms will therefore
contribute, to a certain degree, to the actual structure of the s-triazine
molecule. As a result, the aromatic character of the s-triazine is less
pronounced than that of benzene [Eq. (1)].

(1)

(2)

The same delocalization effect in combination with inductive and mesomeric effects exerted by the substituents at C-2, C-4, and C-6 greatly influences the chemical behavior and physical properties of s-triazine derivatives.
The relative electron deficiency of the ring carbon atoms makes them susceptible to nucleophilic attack. This attack is facilitated when electron-
withdrawing substituents such as chlorine are attached to the carbon atoms,
and is impeded when the electron density of the aromatic system is increased
by electron-supplying substituents such as amino groups. These facts are
clearly reflected by the high reactivity of cyanuric chloride and the possibility of a stepwise substitution of its three chlorine atoms, each subsequent
step requiring more drastic reaction conditions than the preceding one.

Herbicidally active dialkylamino-s-triazines behave as weak bases in
aqueous solution. Protonation occurs preferably on the ring nitrogen atoms
as shown by ultraviolet [28] (for triazine metabolites see also Boitsov et
al. [29]), infrared, and nuclear magnetic resonance studies [30].

It is evident from the pK_a data listed in Table 1 [31] that substituents
at the 2 position of 4,6-bis(alkylamino)-s-triazines most significantly affect
the basicity which increases in the order chloro-, methylthio-, methoxy-
s-triazines. The size and the degree of branching of N-alkyl groups at the
4 and 6 positions also have a distinct, but less pronounced effect: tert-
butylamino derivatives are the most basic representatives of an s-triazine
series with a common C-2 substituent. In both cases, increasing basicity
parallels increasing the electron-donating power of the substituents. The
2-hydroxy derivatives, which represent one of the main degradation products
of the herbicides in soil, are more basic than the parent compounds. They
are amphoteric in nature and, as a consequence, show two ionization constants. For example, the two pK_a values of the atrazine hydrolysis product,

2-hydroxy-4-ethylamino-6-isopropylamino-s-triazine, are 5.2 and 10.4, respectively [32].

Dialkylamino-s-triazines have low solubilities in water, the 2-chloro-s-triazines being less soluble than the 2-methylthio and 2-methoxy analogs. Aqueous solubility is, over a wide range, practically independent of the pH of the solution [33]. However, a pronounced increase in solubility is observed at pH values where strong protonation occurs, e.g., between pH 5.0 and 3.0 or lower for 2-methoxy- and 2-methylthio-s-triazines, and at pH 2.0 or lower for 2-chloro-s-triazines.

Structural modifications of the substituents at either the 2 or the 4 and 6 positions of dialkylamino-s-triazines significantly affect solubility at all pH levels. Generally, increasing solubility is associated with increasing electron-donating capacity of the substituents at C-2 and decreasing size and branching of the N-alkyl groups in the 4 and 6 positions. Differences in molecular symmetry, and therefore in molecular polarity, can account for the higher solubilities of 4,6-asymmetrically substituted 2-chloro-s-triazines such as atrazine as compared to that of symmetrically substituted compounds such as simazine and propazine.

A.   Physical Properties

Certain physical data for the herbicides discussed in this chapter are listed in Table 1.

B.   Chemical Properties

1.   Chemical Reactions

a.   Reactions affecting the substituents at C-2.   The chlorine atom in 2-chloro-4,6-bis(alkylamino)-s-triazines is readily displaced by a variety of nucleophiles. These reactions are reviewed in Table 2.

Although the products resulting from hydrolysis of 2-chloro-4,6-bis-(alkylamino)-s-triazines [and also of the 2-methoxy- and 2-methylthio-4,6-bis(alkylamino)-s-triazines] are usually designated as 2-hydroxy-4,6-bis-(alkylamino)-s-triazines, it should be noted that they are present in the form of their tautomeric 2-oxo-4,6-bis(alkylamino)-1,2-dihydro-s-triazines [41]; for a different interpretation of the dependency of the keto form on pH see Skipper et al. [42]. This latter structure, evident from infrared spectra, also explains various aspects of their chemical behavior such as the failure to form methoxy derivatives by usual methylating methods. In the corresponding S analogs, however, the thiol form prevails and therefore they are readily methylated to the respective methylthio derivatives.

TABLE 2

Nucleophilic Reactions Affecting the Chlorine Atom of
2-Chloro-4, 6-bis(alkylamino)-s-triazines

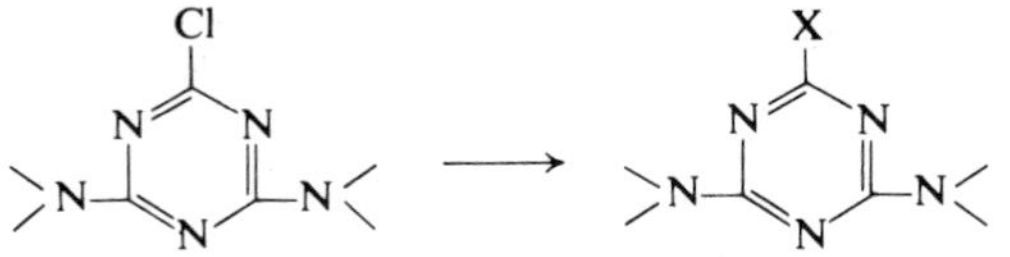

(2)

Attacking agent	Resultant substituent	Reference
H^+ or OH^-	$-OH$	24a
SH^-	$-SH$	24b
$NH_2CSNH_2 \cdot HCl/NaOH$	$-SH$	34
Alkyl-OH, OH^-	$-O$-alkyl	24c
Alkyl-SH, OH^-	$-S$-alkyl	35
NH_3, alkyl-NH_2, (alkyl)$_2$NH	$-NH_2$, NH-alkyl, N(alkyl)$_2$	24d
$N(CH_3)_3$	$-[N(CH_3)_3]^+ \cdot Cl^-$	36
NH_2NH_2, NH_2NH-alkyl	$-NHNH_2$, $-NHNH$-alkyl	37
KF	$-F$	38
KCN	$-CN$	39
NaN_3	$-N_3$	40

Treatment of 2-chloro-s-triazine herbicides with polysulfide ions has been proposed as a method for the destruction of unwanted residues in soil [43]. However, reports on the effectiveness of this procedure under practical conditions are controversial [44-46].

Special attention was drawn to the quaternary nitrogen compounds obtained by reaction with trimethylamine. The quaternary nitrogen group is extremely reactive. Using these intermediates for further reaction with some of the above-mentioned nucleophilic agents such as alkylthiols, KCN, and NaN_3, the respective end products are obtained more readily than by

direct reaction with the 2-chloro-4,6-bis(alkylamino)-s-triazines [40, 47].
Triethylamine does not form on analogous quaternary compound under
identical conditions [48].

Reference is also made to a reaction [49] that was considered to be a
model for the catalytic mechanism used by corn for the detoxication of
chlorotriazines by means of the DIMBOA system [Eq. (3); for details see
Sec. II, A, 2, a].

Acid or alkaline hydrolysis of 2-methoxy-4,6-bis(alkylamino)-s-
triazines leads to the corresponding 2-hydroxy-4,6-bis(alkylamino)-s-
triazines [48]. Heat causes isomerization into 2-oxo-1-methyl-1,2-dihydro-
s-triazines [50]. After heating prometone at 260°C, the following isomerization and dismutation products could be identified [Eq. (4)]:

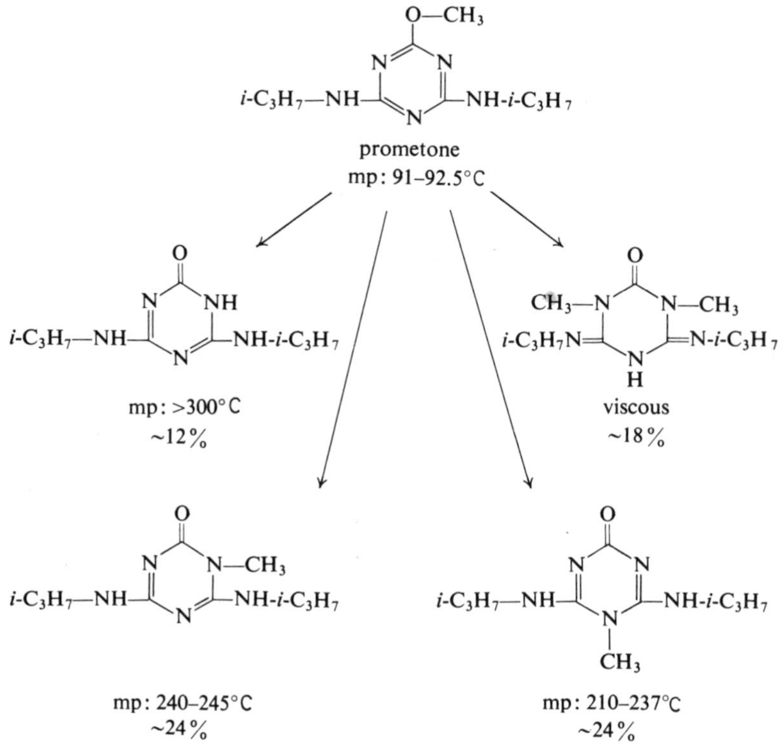

(4)

Acid or alkaline hydrolysis of 2-methylthio-4,6-bis(alkylamino)-s-triazines does not lead to the respective 2-mercapto-4,6-bis(alkylamino)-s-triazines, but again to the 2-hydroxy-4,6-bis(alkylamino)-s-triazines [51]. The 2-methyl-4,6-bis(alkylamino)-s-triazines can be oxidized by phthalomonoper acid whereby the corresponding sulfoxy and sulfono analogs are obtained [50]. Of special interest is the hydrolytic behavior of these two compounds when compared with the original methylthio analog. The data in Table 3

TABLE 3

Hydrolytic Behavior of Prometryn and Its Derivatives

Hydrolytic agent, 25°C	Time required for hydrolysis of		
	CH_3S-	CH_3SO-	CH_3SO_2-
0.1 N HCl	22 days	96 min	16 min
H_2O	> 5 years	150 days	3 days
0.01 N NaOH	> 5 years	20 min	2 min

were determined for prometryn and its derivative [52]. It is evident from Table 3 that under acid, neutral, and alkaline conditions the sulfone is more rapidly hydrolyzed than the sulfoxide. More important, from a practical point of view, is the increased rate of hydrolysis of both oxidation products when compared with parent prometryn.

b. Reactions affecting the substituents at C-4 and C-6. The secondary amino groups of 2-chloro-4,6-bis(alkylamino)-s-triazines do not readily undergo reaction. Acylation with the usual acylating agents normally leads to substitution of the chlorine atom by a hydroxy group. However, under special conditions, acylation yielding 2-chloro-4-acylalkylamino-6-alkyl-amino-s-triazines was possible [53]. Another method for the preparation of monoacylated derivatives of chlorotriazines made use of ketene [53].

The basicity of the amino groups of 2-methoxy- and 2-methylthio-4,6-bis(alkylamino)-s-triazines is greater than that of the corresponding 2-chloro analogs. Consequently, substitution at the lateral nitrogens occurs more easily and under conditions which do not significantly interfere with the methoxy or methylthio groups. By acylation with the usual acylation agents or ketene, N-mono- or N,N'-bis(acylalkylamino)-s-triazines were obtained [54].

Patents of BASF [55] and Degussa [56] describe N-nitroso derivatives of 2-chloro-4,6-bis(alkylamino)-s-triazines obtained by treatment with sodium nitrite. Other patents of Degussa [57-59] report the synthesis of N-trichloromethylthio, N-carbamido, N-sulfonamido, and N-phosphoryl-amido derivatives.

Treatment of simazine with concentrated nitric acid resulted in the substitution of the chlorine atom by a hydroxy group and simultaneous nitration of one of the ethylamino groups [49]. The N-nitroethylamino group readily undergoes nucleophilic replacement reactions.

N-Dealkylation of s-triazine herbicides, known to be an important mechanism of biological degradation, can be achieved chemically as shown with various model systems. The reaction products of 2-chloro-4,6-bis-(alkylamino)-s-triazines with free-radical generating systems such as Fenton's reagent (ferrous sulfate-hydrogen peroxide) were identified as mono- and didealkylated 2-chloro-4,6-diamino-s-triazines identical to those isolated from biological systems [60]. The evolution of olefins from various 4,6-bis(alkylamino)-s-triazines having different substituents in position 2 (including chloro, methoxy, methylthio, and hydroxy groups) by heating at 240-250°C is good evidence for a thermal monomolecular N-dealkylation by way of a cyclic transition state (Chugaev reaction) as outlined in Eq. (5) [61]. Ultrasonication of atrazine and propazine, suspended in water, produced the corresponding olefins with ratios similar to those obtained in thermal N-dealkylation [61].

$$(5)$$

c. Cleavage of the triazine ring. Direct chemical cleavage of the triazine ring of 2-chloro-, 2-methoxy-, or 2-methylthio-4,6-bis(alkyl-amino)-s-triazines has not been reported. Reports describing ring cleavage of other s-triazine derivatives are relatively few if one excludes such dras-tic conditions as heating with concentrated sulfuric acid, melting with alkali, and heating at very high temperatures [24e]. The splitting of cya-nuric chloride through reaction with dimethylformamide [62, 63], and the splitting of s-triazine itself [64-67], cannot be considered as reactions of general applicability because the special chemical character of these two compounds differs significantly from that of 4,6-bis(alkylamino)-s-triazines. Potential mechanisms for ring degradation have been suggested in reactions where unexpected end products were encountered [68-70].

2. Photochemical Reactions

Early studies showed that s-triazines irradiated with uv and sunlight responded with changes in their uv absorption spectrum and with the altera-tion of their white color into a light tan. Although different changes were observed with the two light sources [71, 72], the greatest change occurred after far-uv irradiation (with a peak at 253.7 nm).

Recent investigations substantiated the susceptibility of bis(alkylamino)-s-triazines to photodecomposition by uv irradiation. Different reaction products were isolated depending on the experimental conditions used. The 2 position of the s-triazines was primarily affected. In aqueous solution, the 2-chloro-s-triazines degraded to their 2-hydroxy analogs, whereas in alcoholic solution the respective 2-alkoxy derivatives were formed according to Eq. (6). Under these conditions, 2-methoxy- and 2-hydroxy-s-triazines do not undergo photochemical reactions. The methylthio group of 2-methyl-

thio-s-triazines was replaced by hydrogen regardless of the solvent used
[73], and 2-H analogs were obtained in good yield [Eq. (7)]. No oxidation
of the methylthio group occurred when solid prometryn was irradiated in air
with a sunlamp. Again, only 2,4-bis(isopropylamino)-s-triazine was formed,
but in low yield [74].

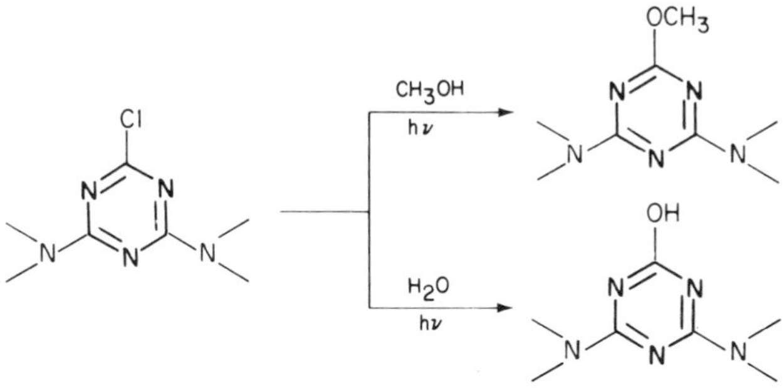

(6)

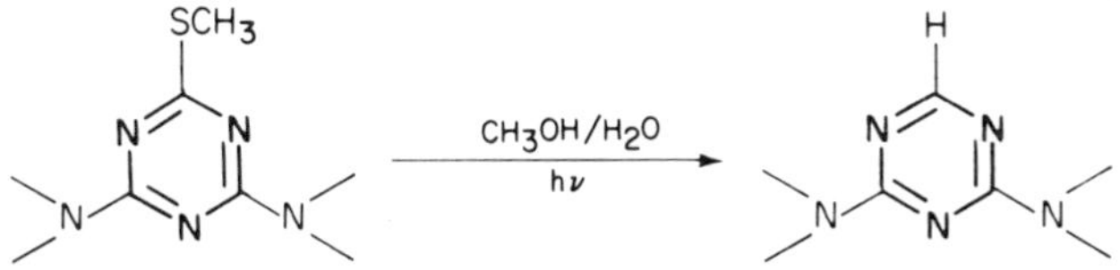

(7)

The major product of the photodecomposition of 2-methylsulfinyl-4,6-
bis(isopropylamino)-s-triazine with light of wavelength > 260 nm was the
2-hydroxy derivative. Loss of the methylsulfinyl group by reductive desulfuration [Eq. (8)] was also observed [75].

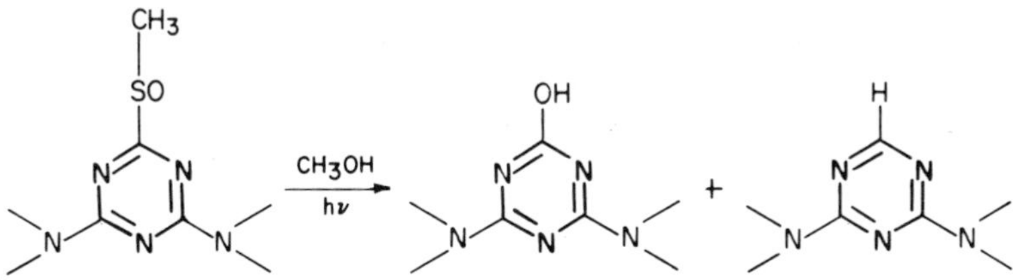

(8)

Replacement of an alkylamino side chain by hydrogen was observed
when simazine was photolyzed in methanolic solution at 220 nm. In addition
to the 2-methoxy and 2-H analogs, 2-methoxy-4-ethylamino-s-triazine
[Eq. (9)] was formed [76].

$$(9)$$

C. Synthesis

Cyanuric chloride is the starting material for the synthesis of all s-triazines described in the present chapter. It is obtained by trimerization of cyanogen chloride in organic solvents such as benzene, chloroform, or carbon tetrachloride in the presence of acidic catalysts, e.g., hydrogen chloride [77-84]. According to a process introduced in 1949, the trimerization is carried out in the gaseous phase at temperatures ranging from 200° to 500°C using activated charcoal as a catalyst [85-87]. Cyanogen chloride is produced by reacting hydrocyanic acid with chlorine [88], both chemicals being raw materials of easy accessibility in the chemical industry.

The reactivity of the three chlorine atoms of cyanuric chloride toward nucleophilic reagents such as amines, alkoxides, mercaptides, and azides decreases as substitution proceeds so that a stepwise displacement with different nucleophilic reactants is possible. Use is made of this property for the synthesis of all s-triazine herbicides [Eq. (10)].

Because of the high reactivity of two chlorine atoms of cyanuric chloride, special care has to be taken if only one chlorine atom is to be replaced by an amino group. In such a case, the temperature should be maintained between $-15°$ and $+5°C$ to obtain optimum results. It is necessary to use an acid acceptor that may be either a second mole of amine or an equivalent quantity of sodium hydroxide, sodium carbonate, or sodium bicarbonate.

(10)

To replace two chlorine atoms by amines, the temperature for the second step of the substitution must generally be raised to 20°-60°C and the quantities of the amine and the acid acceptor have to be doubled [77, 89-92]. Even when using two different amines for the substitution of two chlorine atoms in cyanuric chloride, it is possible to obtain asymmetrically alkylated 2-chloro-4,6-diamino-s-triazines of excellent quality in high yield. This applies also to the synthesis under technical conditions. Generally, the sequence in which the different amines are brought to reaction is of minor importance. However, some sterically hindered amines of low basicity such as α-aminoisobutyronitrile, an intermediate in the synthesis of cyanazine, react only with one chlorine atom of cyanuric chloride under normal conditions.

A wide variety of reaction media suitable for the preparation of 2,4-dichloro-6-alkylamino-s-triazines and 2-chloro-4,6-bis(alkylamino)-s-triazines have been reported: water [93], mixtures of water-acetone [91, 92, 94], or water-dioxane [95, 96]; two-phase systems such as water-benzene [95], water-chlorobenzene [95], or water-2-butanone [97]; and absolute solvents such as ether [77], benzene [95], dioxane [92], 2-butanone, or trichloroethane [98]. The remaining chlorine atom of 2-chloro-4,6-bis-(alkylamino)-s-triazines can easily be substituted by the methoxy or methyl-thio group under appropriate conditions. Thus, 2-methoxy-4,6-dialkyl-s-triazines are prepared by refluxing a methanolic solution of the respective chloro derivative in the presence of sodium hydroxide or sodium methoxide [99-101].

A similar procedure leads to 2-methylthio-4,6-bis(alkylamino)-s-triazines which are obtained either by heating the chlorine derivatives and methanethiol in aqueous sodium hydroxide under pressure [102] or by reacting methanethiol in methanol with trimethyltriazinylammonium salts. The latter are obtained by reaction of the chlorine derivatives with trimethyl-amine [103] according to Eq. (11):

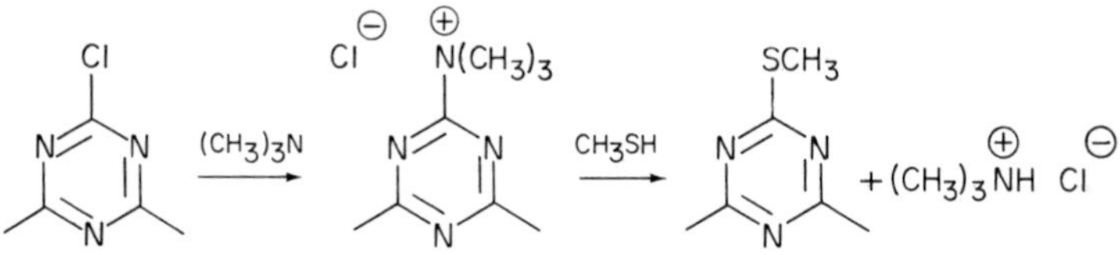

More convenient methods for the synthesis of 2-methylthio-4,6-bis(alkyl-amino)-s-triazines have been reported. Conversion of 2-chloro-4,6-bis-(alkylamino)-s-triazines with thiourea to 2-mercapto compounds [34] and subsequent alkylation of the mercapto group with methylating agents [Eq. (10)] give the respective methylthio-s-triazines in high yield [104-106]. An additional route to the same compounds, reversing the sequence of nucleophilic substitution of the three chlorine atoms in cyanuric chloride, is outlined in Eq. 10. Cyanuric chloride is reacted with methanethiol in the presence of a sterically hindered pyridine base such as collidine [107, 108] or, preferably, in a two-phase system in the presence of aqueous alkali [109] to give 2-methylthio-4,6-dichloro-s-triazine. The reaction of this intermediate with 2 moles of alkylamines in the presence of the appropriate amount of an acid acceptor leads to 2-methylthio-4,6-bis(alkylamino)-s-triazines [95]. The intermediate, 2-methylthio-4,6-dichloro-s-triazine, is also used in the preparation of 2-methylthio-4-azido-6-alkylamino-s-triazines, for example, aziprotryn. The two steps of the synthesis are carried out in mixtures such as water-dioxane, the first step with alkyl-amines and sodium hydroxide at room temperature, the second step with sodium azide at 75°C [110].

D. Formulations

The development of formulations for s-triazines was determined by two main factors: the mode of application and the physical-chemical properties of this class of herbicides. Owing to their high herbicidal activity after root uptake, formulations for the preemergent application of the triazines, especially of the chlorotriazines, were of primary interest.

The relatively low solubility of the compounds in water and common organic solvents did not permit the preparation of aqueous solutions or emulsifiable concentrates with high contents of active ingredient. Consequently, wettable powders were developed which meet the requirements for facile transport and application of the compounds.

Wettable powder formulations were also found suitable for postemergent applications. Apart from some exceptions within the chlorotriazine series, this type of application is mainly reserved for the methylthiotriazines and their combinations because this group can also exert their phytotoxic action via foliage.

Commercial wettable powder formulations may contain up to 80% of a triazine or mixtures of several triazines. Certain physical-chemical properties of particular triazines, however, limit the content in active ingredient to 25-50% or may require, as for example with triazines of relatively low melting points, the addition of specific carriers such as calcium silicates, sodium aluminum silicates, or precipitated silicic acid. By these auxiliary means, the formulation of such triazines can be greatly facilitated and the stability of the product during storage enhanced. The addition of agents preventing foaming may also be required.

Other important ingredients of such formulations are wetting agents, such as alkylarylsulfonates, and dispersing agents, such as lignin sulfonates, which may each account for up to 5% of the formulation. Frequently, chalk or clay is used as a carrier to make up the balance of the formulation.

Wettable powder formulations of s-triazines can generally be mixed in the calculated amount of water without prior pasting. Their suspensibility, when determined at concentrations of 0.1-2.0% of active ingredient, amounts to 65-75% or more. The wet-sieve residue, as measured on a sieve of 74 μ mesh size, is less than 2%, and the average bulk density normally ranges between 0.25 and 0.35 kg/liter. When stored in the original package under normal conditions, these triazine formulations are stable for at least three years without significant alteration of their physical and chemical properties.

Granules represent newer developments of highly concentrated nonliquid triazine formulations. They serve to meet specific needs arising, for example, with large-scale ground or aerial application where transport of large volumes of water must be avoided. These granular formulations, however, require the use of special application devices.

Liquid formulations in the form of emulsifiable concentrates represent exceptions to the previously discussed limitations. Recent developments have demonstrated that the triazines can also be formulated as suspension concentrates on a water or oil basis. These flowable formulations are resistant to frost and can be measured volumetrically, thus significantly facilitating their handling by the applicator. Since suspension concentrates, by nature, will sediment after prolonged periods of storage, they have to be well shaken prior to dilution for final use.

II. DEGRADATION PATHWAYS

In contrast to the fairly well defined conditions of chemical degradation experiments like those described in Sec. I, B, 1, the living organisms metabolizing the herbicides represent complex, differentiated structures. Furthermore, in practice they are closely correlated with their respective environment. Therefore, various concurrent effects of a physicochemical, as well as a biochemical and chemical nature, finally decide the extent and rate of fixation, translocation, and degradation of an applied herbicide.

Most of the experiments on the fate of s-triazine herbicides in biological systems to be discussed in the following sections were performed in the laboratory. No special attempt was made to relate the results of these studies directly with field data obtained by residue analysis. Since it is difficult to distinguish between the biochemical and the chemical origin of an observed structural alteration, the degradation products characterized were related to one of these specific degradative mechanisms only in selected cases.

The designation of degradation reactions and reaction products used in the following sections is based on the nomenclature system outlined in Sec. I.

A. Degradation in Plants

Numerous investigations have been and are still being performed to search for the basis of the outstanding herbicidal properties of the s-triazines. Factors determining uptake and translocation and the metabolic reactions detoxifying the compounds were the primary targets of scientific interest. Sensitive and resistant plants or varieties of a given plant were included in these studies to find an explanation for the remarkable selectivity of this class of herbicides.

1. Uptake, Translocation, and General Metabolism

Many studies have been concerned with the behavior of s-triazines after their uptake from nutrient solutions. Generally, no consideration has been given to the nature of the compounds actually incorporated by the plants under field conditions.

When plants are grown in nutrient solution containing radiolabeled s-triazines, root uptake readily occurs in all plants studied regardless of whether they are resistant or susceptible to the herbicides. Increased uptake was observed with increasing concentrations and time of exposure to the compounds. Higher temperatures and low relative humidity also accelerate uptake. After labeled s-triazines were absorbed by the roots, autoradiography and direct measurement showed the radioactivity to be evenly distributed by way of the xylem into all aerial parts of the plants. Generally, some accumulation is apparent in the marginal zones of leaves, especially of susceptible plants. Increasing the transpiration stream by hyperventilation or by decreasing the relative humidity also increased translocation.

When plants were transferred from nutrient solution containing labeled s-triazines into a herbicide-free medium, translocation of radioactivity from labeled leaves into newly formed leaves was negligible. Basipetal transport of foliar radioactivity was also negligible; however, some release of absorbed radioactivity from the roots into the herbicide-free medium was observed. After foliar application of labeled s-triazines, a relatively low rate of penetration and translocation of radioactivity is generally found. This can be enhanced by the addition of adjuvants like mineral oil and depends to some degree on the s-triazine used. Penetration of atrazine, for example, is more pronounced than that of simazine. Wet foliage before and after the application of atrazine plus adjuvant further increases foliar penetration in corn.

Radioactive studies on uptake and translocation were frequently extended by determining the concentration of unchanged parent compound in addition to the total radioactivity. After extraction, soluble radioactivity was generally fractionated into a lipophilic and a hydrophilic fraction by partitioning between chloroform and water. Since all s-triazine herbicides partition exclusively into chloroform, the presence of radioactivity in the aqueous phase and its increase with increasing time after application were used as a measure of degradation. These early investigations clearly established that a wide variety of plants are able to degrade the s-triazines. The extent of the metabolic activity of a given plant generally reflected its degree of resistance toward these compounds (see Sec. III, B for details). Chromatographic analysis of the extractable radioactivity from experiments with different s-triazines showed that both the chloroform and the aqueous phases contained several labeled compounds. Both the number and the concentration of compounds present in the lipophilic phase, in addition to unchanged parent, were small; however, the aqueous phase contained a pattern of metabolites the complexity of which varied with the type of plant and s-triazine used.

In order to facilitate a detailed pursuit in the literature of the topics briefly discussed in this section, only a limited number of the numerous publications was selected. The references are compiled in groups referring to uptake and translocation and to general metabolism and are subdivided according to the groups of plants in which the triazines were investigated: uptake and translocation in corn and cereals (chlorotriazines [111-133]; methoxy and methylthiotriazines [112, 113, 115]; hydroxytriazines [112]); in other plants (chlorotriazines [111, 112, 114-116, 119, 124, 127, 129-131, 134-146, 426]; methoxy and methylthiotriazines [134, 135, 140, 141, 147-151]; hydroxytriazines [134]); in weeds (chlorotriazines [124, 130, 131, 152-161, 427, 428, 438]; methylthiotriazines [151, 153]). General metabolism in corn and cereals (chlorotriazines [162-169]); in other plants (chlorotriazines [165, 167, 168, 170-175]; methylthiotriazines [173, 176]).

2. Degradation Reactions

a. Reactions of the C-2 substituents. In this chapter, the reaction products of chlorotriazine hydrolysis or of methoxy or methylthio group cleavage are designated as 2-hydroxy derivatives. Such terminology has often been used in the literature. For the actual configuration of these compounds see Sec. I, B, 1, a.

$$\text{(12)}$$

First evidence of the hydrolysis of chlorotriazines [Eq. (12)] was obtained with in vitro systems. Cell sap, homogenates, and even particulate free extracts from corn rapidly formed the corresponding hydroxy derivatives when incubated with simazine and atrazine [131, 162, 177-180] . The inability of the cell sap of oats to hydrolyze these compounds corresponded with the sensitivity of this plant to the s-triazines [178, 180] .

Further work was stimulated by the observation that resistant corn contains an active compound capable of rapidly inactivating simazine and atrazine. This compound had the same properties as a cyclic hydroxamate isolated from different varieties of corn and wheat having antifungal and insecticidal properties [181-185] . Its structure was shown to be 2, 4-dihydroxy-7-methoxy-1, 4(2H)-benzoxazin-3(4H)-one (DIMBOA), which was originally present in the form of its β-D-glucopyranoside (3) [186-191].

(3)

The same compound, but without the 7-methoxy group (DIBOA), was found to be the active principle in rye plants [192, 193] . The existence of these and similar compounds was confirmed in a variety of cereals and other plants [189, 190, 194] . Both the glucosides and their aglucones, which are easily released by enzymatic hydrolysis, show a similar degree of efficacy with respect to the chemical hydrolysis of chlorotriazines. The aglucones of DIMBOA and related benzoxazinones were found to decompose when stored in solution or when heated to yield 7-methoxy-1, 4(2H)-benzoxazol-3(4H)-one, which has no catalytic activity. The presence of DIMBOA in resistant and sensitive lines of different cereals and its correlation to the selectivity of s-triazines is discussed in Sec. III, B.

A large number of studies of the in vivo degradation of chlorotriazines demonstrated the formation of hydroxytriazines as primary degradation products. A wide variety of plants was found to have this hydrolytic capacity to a varying degree when grown in the presence of different chlorotriazines. Corn showed an outstanding capacity in this respect, when compared to other plants. Its hydrolytic activity was located in both roots and shoots. It must be stressed, however, that in early studies no discrimination was possible between hydroxy derivatives in contrast to the hydroxy dealkylated derivatives and polar conjugates identified at a later time. Thus, the hydroxy derivatives reported in these studies probably represented only a fraction of the aqueous part of the extractable radioactivity.

The following references give representative information about the in vivo formation of hydroxy derivatives: in corn and cereals from atrazine, simazine, and propazine [119, 126, 131, 166-168, 183, 194-201, 429]; from cyprozine [202]; from GS-13529 [203]; in other plants from atrazine and simazine [119, 140, 142, 167, 168, 204]; and in weeds from simazine and atrazine [161, 194, 205-208].

When the metabolism of atrazine after short-term root uptake in resistant and susceptible plant species was compared, chromatographic analysis of the water-soluble radioactivity indicated that no hydroxyatrazine was formed in sorghum, pea, soybean, and cotton plants. The radioactivity consisted of several compounds even more polar than hydroxyatrazine [198, 200, 209, 210]. An interesting variation with respect to hydrolysis was found in corn with cyanazine [i.e., 2-chloro-4-ethylamino-6-(1-cyano-1-methylethylamino)-s-triazine]. No parent hydroxy compound was detected. The hydroxy derivatives found either had lost their side-chain ethyl group or had their original cyano group converted to a carboxyl group [211-214].

(13)

In vitro studies on the cleavage of methoxy- and methylthiotriazine compounds [Eq. (13)] in corn extracts showed that the DIMBOA system catalyzed less than 20% of the hydrolysis of methoxy compounds as compared to 100% hydrolysis of the chloro compounds. Methylthio derivatives were unresponsive in this system [196, 215, 216]. Substitution of methoxy substituents by the hydroxy group was observed in alfalfa and sugarcane. After spray application of ^{14}C-GS-14254, alfalfa roots and shoots contained the parent hydroxy compound in addition to the hydroxy desethyl derivative [217]. Young sugarcane plants grown in nutrient solution with ^{14}C-GS-14259 contained only the corresponding hydroxy desethyl derivative [218].

Despite the ineffectiveness of the DIMBOA system with the methylthio compounds, the corresponding hydroxypropazine and hydroxyatrazine were detected when cotton, soybeans, peanuts, and carrots were treated with ^{14}C-ring-labeled prometryn [148, 168, 176, 196] and when sugarcane and banana plants were grown in nutrient solution with ^{14}C-ring-labeled ametryn [140, 173]. Supporting evidence for methylthio group cleavage came from studies with $S^{14}CH_3$-labeled prometryn in peas where moderate amounts of $^{14}CO_2$ were liberated [216].

Other recently developed methylthio compounds are methoprotryn [219] and terbutryn [220], of which the corresponding hydroxy parent compounds were found in wheat and sorghum after root uptake of the labeled compounds from nutrient solution. Sulfoxide and sulfone derivatives were claimed to

be hypothetical intermediates during the removal of the methylthio group
[221] . Experimental evidence was obtained with broad beans stem-injected
with ^{14}C-ring-labeled prometryn and analyzed by thin-layer chromatography
(tlc) and special hydrolytic procedures [216] . The same intermediates were
tentatively characterized in different tree species after root uptake of
prometryn [150] . ^{35}S-Labeled SO_4^{2-} was detected as one of the end products
of methylthio group oxidation of prometryn in mustard, pea, and carrot
plants [222] . Difficulties in detecting sulfoxide or sulfone metabolites in
greater amounts became understandable when the greatly enhanced hydrolytic lability of these intermediates is considered. Sulfone derivatives
seem to be more labile than sulfoxides as found with prometryn [223,
Table 3] .

The behavior of hydroxytriazines in plants as primary metabolites was
studied after root uptake. In short-term experiments with corn and sorghum,
less hydroxyatrazine was taken up from the nutrient solution than atrazine
[199] . Practically no difference in the translocation of both compounds was
observed in corn plants [197, 199] , whereas decreased translocation of the
hydroxy derivative was observed in sorghum [199] and soybeans [197] .
While intact s-triazines were observed to concentrate in the lysigenous
glands of cotton, hydroxysimazine and hydroxypropazine did not [112] .

$$(14)$$

Alteration of the azido group [Eq. (14)] of WL-9385 (i.e., 2-azido-4-ethyl-
amino-6-tert-butylamino-s-triazine) to yield the corresponding melamine
derivative, was observed in the foliage of wheat plants foliarly treated at the
three-leaf stage with the ^{14}C-labeled compound [224] .

b. Conjugation

$$(15)$$

New transformation products were detected when the unknown radioactivity in the aqueous phases of [^{14}C]atrazine root-treated sorghum and

leaf-treated corn was studied in more detail. Large amounts of these unknown metabolites were formed when sorghum discs or excised leaves were incubated with [^{14}C]atrazine or when young sorghum and corn plants were foliarly treated with the herbicide. Two of these products, representing the main transformation products of early atrazine metabolism, were isolated and found, after hydrolysis, to consist of equimolar amounts of hydroxyatrazine, glutamic acid, cysteine, and glycine. By means of amino acid sequence and spectral analysis, the primary metabolite was identified as a glutathione conjugate of atrazine; thus, R_1 of Eq. (15) is HOOC-CH(NH$_2$)-CH$_2$-CH$_2$-CO- and R_2 is -NH-CH$_2$-COOH. Of next importance was a conjugate which had lost its glycine moiety by enzymatic cleavage [225-227].

Conjugate formation was found to be the predominant mechanism for atrazine detoxication in corn and sorghum when the compound was introduced directly into the leaves. After 24 hr incubation, about two-thirds of the radioactivity in the leaf blades of a resistant corn line was in the form of peptide conjugates, whereas the concentration of hydroxyatrazine was small. Conversely, after 48 hr root uptake, the conjugates present were below 20%, whereas hydroxyatrazine increased to about one-third of the total radioactivity taken up. Very little conjugate was found in the roots [201, 228]. This behavior was confirmed in corn foliarly treated with an oil emulsion of atrazine and grown in nutrient solution. Equal distribution of the radioactivity in the shoots between conjugates and hydroxyatrazine was found [126]. This conjugation mechanism was also found operative in field experiments with [^{14}C]atrazine and sorghum. After five weeks, 50% of the absorbed radioactivity was present as glutathione conjugates. In contrast, no indication for the presence of this type of derivative was found at maturity [229]. In corn, rapid conjugation of foliarly applied cyprozine (i.e., 2-chloro-4-cyclopropylamino-6-isopropylamino-s-triazine) was observed, but decreased drastically in the course of three weeks. At that time the increasing hydroxy derivatives reached the concentration of the conjugates [202].

Weed plants like wild cane, large crabgrass, fall panicum, and giant foxtail also seem capable of forming peptide conjugates when grown in nutrient solution with simazine, atrazine, and propazine. Shortly after uptake of the triazine, they contained, in addition to hydroxy derivatives, hydrophilic metabolites which showed the same chromatographic behavior as the peptide conjugates of corn. Whereas detoxication by hydrolysis was of the same order with atrazine and simazine, conjugation of atrazine occurred at a rate five times higher than that of simazine [161, 208, 427].

The conjugation mechanism which results in a loss of phytotoxicity of the parent compounds was demonstrated to be generally applicable to chloro-triazines. A good correlation between the efficiency of the conjugation and the degree of tolerance of a given plant was observed [230].

The suggested influence of air pollutants on the ability of plants to metabolize pesticides was investigated in corn. Plants were treated at the six to seven leaf stage with a concentration of ozone just producing first signs of injury. Excised leaves of these plants showed no change in their content of hydroxy and conjugate type of derivatives when exposed to atrazine solutions [231].

A search for the enzyme responsible for the conjugation of chlorotriazine in resistant plants resulted in the identification of a glutathione s-transferase in corn, sorghum, sugarcane, johnsongrass, and Sudan grass [225, 232]. The enzyme was located mainly in the leaves, whereas roots contained no detectable activity. In contrast, when leaf extracts of susceptible species like pea, oats, barley, and pigweed were analyzed for the enzyme, no activity was detected.

The enzyme from the soluble fraction of corn leaves was purified 7.6-fold and showed a pH optimum of 6.6-6.8 and a substrate specificity for reduced glutathione. Other SH reagents could not be substituted for gluta-thione. A series of triazine structure-enzyme activity relationships was also established: 2-Methylthio and 2-methoxy as well as 2-hydroxy deriva-tives were not accepted as substrates by the enzyme. Chlorotriazines reacted in the decreasing order of GS-13529, atrazine, cyprozine, propazine, and simazine with no correlation to their water solubility or partition coef-ficients. Apparently, intact alkylamino substituents at both the C-4 and C-6 positions are preferred structures for conjugation since most of the enzyme activity was lost when one of the side-chain alkyl groups was re-moved. Direct chemical reaction of chlorotriazines with reduced glutathione resulted in yields ranging from 10 to 20% of that obtained from the enzyme-catalyzed reaction [232, 434]. The fate of these peptide conjugates was followed for 30 days in young sorghum plants root-treated for 2 days with ^{14}C-ring-labeled atrazine. The glutathione (4) and the glutamylcysteinyl (5) conjugates of Eq. (16), rapidly formed in the first phase, were found to be further transformed in a sequence of reactions which were reported [233] to be those shown in Eq. (16). The fate of the lanthionine derivative (7) is still to be determined. The importance of the conjugation mechanism for the selective action of the chlorotriazines is discussed in Sec. III, B.

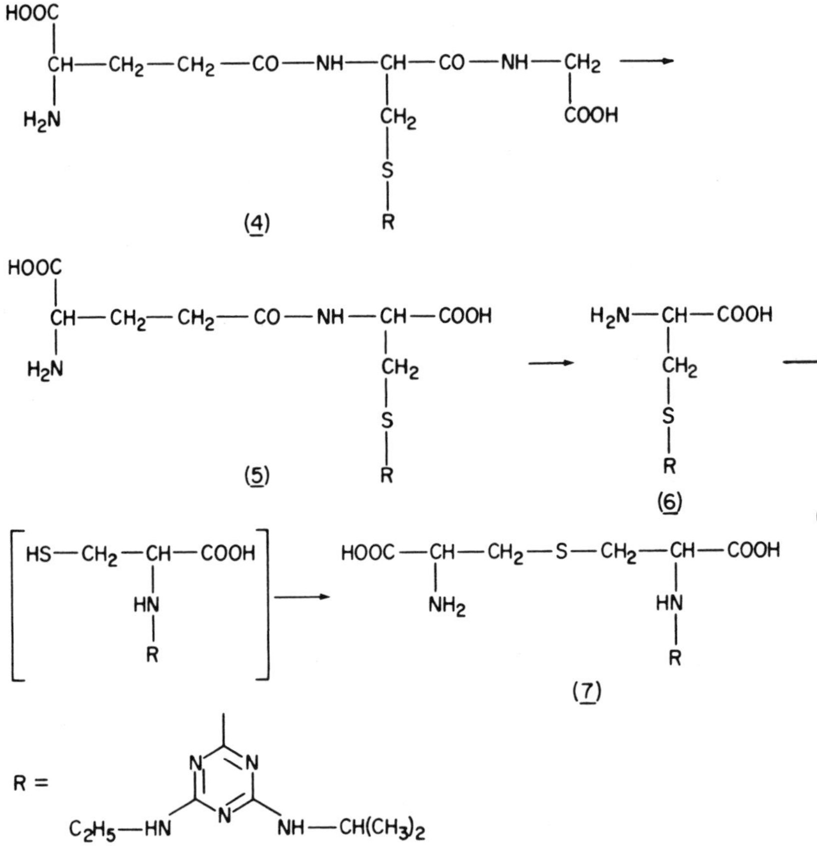

HOOC
|
CH—CH_2—CH_2—CO—NH—CH—CO—NH—CH_2 $\longrightarrow$
| | |
H_2N CH_2 $COOH$
                                     |
                                     S
                                     |
            (4)   R

HOOC
|
CH—CH_2—CH_2—CO—NH—CH—$COOH$ H_2N—CH—$COOH$
| | |
H_2N CH_2 $\longrightarrow$ CH_2 $\longrightarrow$
                                  | |
                                  S   S
                                  | |
            (5)   R   R
                                                            (6)
                                                                        (16)

$$\left[\begin{array}{c} HS—CH_2—CH—COOH \\ | \\ HN \\ | \\ R \end{array} \right] \longrightarrow$$ HOOC—CH—CH_2—S—CH_2—CH—COOH
                                              | |
                                             NH_2 HN
                                                                      |
                                                                      R

                                              (7)

R =

C_2H_5—HN   NH—$CH(CH_3)_2$

c. N-Dealkylation

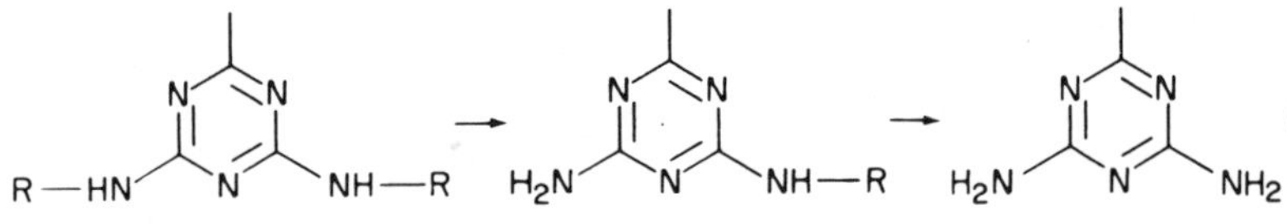

R—HN   NH—R H_2N NH—R H_2N NH_2 (17)

R = alkyl groups

The first evidence for the removal of the N-alkyl groups [Eq. (17)] came from experiments with corn, cotton, and soybeans grown in nutrient solution containing ^{14}C-side-chain-labeled simazine. Appreciable amounts of $^{14}CO_2$ were liberated under these conditions [195]. This was confirmed for corn when up to 70% of the side-chain radioactivity of atrazine taken up by the plants was evolved as $^{14}CO_2$ [216].

The C fragments of cleaved side chains were not only liberated as CO_2 but were also found to be used for endogenous amino acid synthesis as demonstrated with side-chain-labeled atrazine and corn. Roots and shoots of resistant corn incorporated about 2.2 and 0.9%, respectively, of the assimilated atrazine into different amino acids [234]. N-Dealkylated metabolites remaining after the cleavage of side chains were identified in different plants cultivated in the presence of ^{14}C-ring-labeled s-triazines. In peas, a metabolite of prometryn that had lost one of its isopropyl groups was isolated [216]. One of the two possible dealkylation products of atrazine was identified as the major metabolite in pea plants. Its structure, elucidated by means of thin-layer and gas chromatography and by ir spectroscopy, was shown to be 2-chloro-4-amino-6-isopropylamino-s-triazine. Later, the alternative desisopropyl derivative was found, but always in smaller concentrations than the desethyl derivative [116, 198, 235].

This type of reaction was established as a principal degradation mechanism which is followed in resistant as well as in susceptible plants. A selection of the available information on the cleavage of ethyl and isopropyl groups to form monodealkylated parent compounds is given in the following references: in corn and sorghum with atrazine [131, 198-200], with cyprozine [202], with cyanazine [211-214]; in wheat with WL-9385 [224], with GS-14260 [220]; in cotton with atrazine [210]; in soybeans with atrazine [198, 209]; in sugarcane with GS-14259 [218], with ametryn [236]; and in weeds with simazine [205-207].

Relatively easy cleavage of the N-methoxypropyl group of methoprotryn was observed in wheat [219]. In alfalfa plants sprayed with GS-14254, cleavage of the more bulky secondary butyl group, in addition to that of the ethyl group, was found to yield 2-methoxy-4,6-diamino-s-triazine among other metabolites [217]. Dealkylated derivatives with a 2-hydroxy substituent were found as products of a more advanced metabolism in different plants treated with various triazine herbicides in the laboratory. The smaller N-alkyl group was predominantly cleaved in all cases studied, for example, in corn with atrazine and GS-13529 [203]; in alfalfa with GS-14254 [217]; and in wheat with terbutryn [220] and methoprotryn [219].

Sugarcane is apparently able to dealkylate ametryn very efficiently. Following a 30-day growth period in nutrient solution containing ^{14}C-ring-labeled herbicide, roots were reported to contain 34.4% ammeline (absolute identification still tentative) and 12.5% 2-methylthio-4,6-diamino-s-triazine [236]. Both degradative mechanisms were also found in weeds where ammeline and 2-hydroxy-4-amino-6-ethylamino-s-triazine were detected after simazine treatment [205, 206]. Ammeline is the metabolite potentially common to all types of s-triazines.

The dealkylation reaction is not restricted to parent compounds with an intact C-2 position. This was demonstrated with ^{14}C-ring-labeled hydroxyatrazine in corn and sorghum and with ^{14}C-ring-labeled hydroxysimazine in

corn and, possibly, in cotton. In the case of hydroxyatrazine, both dealkylation products, i.e., 2-hydroxy-4-amino-6-isopropylamino- and 6-ethyl-
amino-s-triazine, were found. Again, cleavage of the ethyl group occurred
at a much higher rate than that of the isopropyl group as found with the
dealkylation of parent atrazine [199]. With hydroxysimazine, 2-hydroxy-4-
amino-6-ethylamino-s-triazine was found as the major metabolite in corn
plants. In addition, chromatographic evidence was obtained for the presence
of ammeline [237]. Potent dealkylating activity was also observed in sugarcane which evolved about 23% of the ^{14}C-ethyl label of hydroxyatrazine as
$^{14}CO_2$ within 30 days [236].

In conclusion, monodealkylated parent compounds generally represent
transient metabolites that are further degraded to the corresponding hydroxy
or, possibly, didealkylated derivatives. Among the plants studied, intermediately susceptible peas and cotton seem to be an exception, as larger
amounts of monodealkylated metabolites were found. Whereas the hydroxylation reaction results in a complete detoxication of herbicidally active s-
triazines, monodealkylation of parent compounds liberates metabolites still
showing intermediate phytotoxicity [116]. In contrast, didealkylated metabolites with an intact C-2 substituent no longer possess phytotoxic properties.
Monodealkylated parent compounds lose their phytotoxic properties when the
C-2 substituent is substituted by a hydroxy group, thus behaving identically
to the parent hydroxy derivatives.

d.   Side-chain modifications

$$(18)$$

Oxidation of an N-alkyl side chain was another degradative mechanism
found to be operable in alfalfa plants sprayed with GS-14254 [Eq. (18)]. In
addition to 2-methoxy-4,6-diamino-s-triazine, 2-methoxy-4-amino-6-(1-
methyl-2-hydroxypropylamino)-s-triazine was detected, thus demonstrating
oxidation of the secondary butyl group to a primary alcohol [217].

$$(19)$$

An example of the dealkylation of a methoxy group placed in a side chain to yield a primary alcohol [Eq. (19)] was observed in young wheat plants grown in nutrient solution with methoprotryn. This cleavage, occurring before hydrolysis or dealkylation started, seemed to be the primary reaction yielding 2-methylthio-4-isopropylamino-6-(3-hydroxypropylamino)-s-triazine, most of which was in the form of a glucoside. However, its concentration decreased from 29% of the absorbed radioactivity after one week to 5% after seven weeks [219].

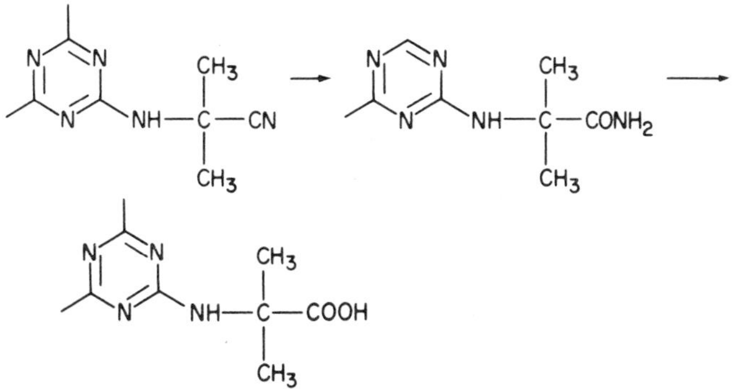

(20)

Another mechanism producing polar metabolites was detected in the case of cyanazine. Following root uptake of the herbicide, corn plants contained metabolites in which the cyano group had been transformed into an amide group and the side-chain ethyl group partially removed. Further transformation of the amide group into a carboxyl group was accompanied by the removal of the chlorine substituent, thus yielding the corresponding 2-hydroxy derivatives [211-214] [Eq. (20)].

e. Deamination

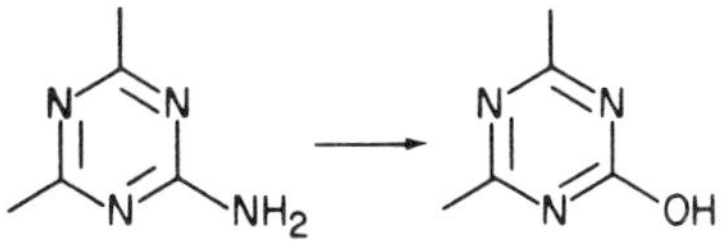

(21)

Replacement of unsubstituted amino groups by hydroxy substituents represents one of the final steps in s-triazine degradation [Eq. (21)]. Metabolites like ammelide (2,4-dihydroxy-6-amino-s-triazine) and cyanuric acid (2,4,6-trihydroxy-s-triazine) are to be expected from such reactions.

Ammelide was detected in corn exposed to a nutrient solution containing hydroxysimazine [237]. The metabolic fate of ^{14}C-labeled 2-hydroxy-4-amino-6-isopropylamino-s-triazine (GS-17794), a known intermediate of atrazine and propazine degradation, was followed in corn plants exposed to GS-17794 for 4 days and then grown for six weeks without the metabolite [238].

A surprisingly high rate of deamination was observed instead of the expected cleavage of the isopropyl group. GS-11957 (i.e., 2,4-dihydroxy-6-isopropyl-amino-s-triazine) represented the major metabolite, although traces of ammeline and larger amounts of unchanged GS-17794 were also detected.

2-Hydroxy-4-amino-6-ethylamino-s-triazine (GS-17792) was found to behave similarly under these conditions; however, dealkylation was also operative. Thus, ammelide and greater amounts of cyanuric acid were observed in addition to 2,4-dihydroxy-6-ethylamino-s-triazine [238]. In similar experiments, corn and alfalfa plants deaminated ^{14}C-labeled ammeline to yield ammelide and cyanuric acid as the principal metabolites in addition to $^{14}CO_2$ [238, 239].

In these studies with triazine metabolites, virtually all plant residues were extractable. This leads to the conclusion that the nonextractable metabolites found in varying amounts in plants after triazine herbicide treatment are formed before the stage of ammeline is reached.

f. Triazine ring cleavage

$$\text{triazine ring} \longrightarrow \left[\text{Unknowns?}\right] \longrightarrow CO_2 \qquad (22)$$

Several of the plant metabolism studies described above examined the ultimate fate of the triazine ring [Eq. (22)]. Using ^{14}C-ring-labeled triazines, liberation of the conveniently detectable $^{14}CO_2$ was taken as conclusive proof of the final oxidation of the heterocycle. However, it is apparent from the different studies published that the amount of $^{14}CO_2$ liberated depended on several experimental factors such as plant species, length of experimental period, growth of plants in light or darkness, ventilation of the plants with normal or carbon dioxide-free air, concentration of the herbicide in the nutrient solution, etc.

Generally, traces or small amounts of $^{14}CO_2$, in the order of a few percent of the radioactivity absorbed by the plants, were found in experiments lasting for not longer than one week. This was observed with atrazine in corn, alfalfa, and cucumber [196, 216]; with propazine in sorghum [180]; with simazine in corn, cotton, and cucumber [163, 240, 241]; with prometryn in peas and broad beans [216]; and with ametryn in sugarcane during 30 days [236]. It is of interest to note that in the propazine sorghum experiment removal of the atmospheric carbon dioxide from the flush gas resulted in the complete interruption of the $^{14}CO_2$ evolution. Possibly, this fact was the reason for the absence of an evolution of $^{14}CO_2$ in young cotton plants grown in the presence of 0.25 ppm labeled prometryn and ventilated with carbon dioxide-free air [176]. Only in one case was there reported a high rate of $^{14}CO_2$ liberation from ^{14}C-ring-labeled simazine and atrazine in corn plants. The liberation was increased when the plants were kept in the dark [164].

Cleavage of the triazine ring followed by $^{14}CO_2$ evolution was found to be of minor significance during atrazine metabolism in corn, cotton, and soybeans for 4 days [114]. The same result was obtained in corn and sorghum cultivated for 7 days in nutrient solution containing approximately 0.35 ppm labeled atrazine [198] and in corn and carrots treated for 3 days with labeled prometryn [196]. Failure to detect $^{14}CO_2$ in cotton for three nights following root treatment with labeled ipazine (2-chloro-4-diethyl-amino-6-isopropylamino-s-triazine) was also reported [170]. It is apparent from these short-term experiments that degradation of the triazine molecule to a stage favorable for ring cleavage is a relatively slow process that may become important only when long periods of time are considered. However, it has been shown that the triazine heterocycle can, in fact, be cleaved, resulting in the liberation of $^{14}CO_2$. Sugarcane plants, efficient in ametryn dealkylation, metabolized ^{14}C-ring-labeled hydroxyatrazine to $^{14}CO_2$, which accounted for about 8% of the applied radioactivity after 30 days [236].

Corn and alfalfa plants were able to cleave the triazine ring when culti-vated in the presence of ^{14}C-labeled ammeline. After an uptake period lasting from 4 days to four weeks followed by a depletion period of up to six weeks, corn and alfalfa plants evolved up to 27.3 and 5.9%, respectively, of the absorbed radioactivity as $^{14}CO_2$ [239]. ^{14}C-Labeled cyanuric acid, under the same conditions, liberated 13.2 and 7.9% of the absorbed radio-activity as $^{14}CO_2$ from corn and alfalfa, respectively [239]. These results confirm earlier studies performed with this compound in corn [216].

In other studies, triazine derivatives still having one alkylamino group, e.g., 2-hydroxy-4-amino-6-isopropylamino-s-triazine (GS-17794), and the corresponding 6-ethylamino derivative (GS-17792) were included. It was observed that removal of the second N-alkyl group apparently represents a rate-determining step in the ultimate degradation of the triazine ring. In corn 0.9 and 4.0% of the absorbed radioactivity were evolved as $^{14}CO_2$ fol-lowing treatment with GS-17794 and GS-17792, respectively. As mentioned in Sec. II,A,2,e, replacement of the free amino group by a hydroxy group predominates over dealkylation and final ring cleavage [238, 239].

Biguanides were proposed as intermediates in final triazine ring cleavage. They would be formed after hydrolytic removal of the C-2 atom as CO_2 and would be readily susceptible to further hydrolysis [13]. Such compounds have not yet been positively identified. However, some indication was reported for their existence in corn after propazine treatment [196] and in sugarcane following ametryn application. In the latter case, small amounts of a volatile basic metabolite were observed.

B.   Degradation in Soil

The heterogeneous system of soil is capable of executing diversified reactions of a physical, chemical, and biochemical nature. While the physical properties of the soil responsible for adsorption and translocation of a herbicide can be measured relatively easily, the separate analysis of its chemical and biochemical degradative capacity is problematic. This is related to the difficulty of efficiently sterilizing soil without altering its chemical and physical properties.

It is outside the scope of this chapter to discuss the influence of soil factors on the behavior of s-triazines; therefore, the interested reader is referred to a review on the behavior and fate of s-triazines [242] and to the monograph on triazines published in the Residue Reviews series [243]. The latter source presents comprehensive information concerning adsorption, movement, volatilization, and persistence of this class of herbicides as well as effects on soil microorganisms. Also reviewed in that volume are microbial and nonbiological degradation of s-triazines. A comparison of the degradation reactions of s-triazines in soil with those of other classes of pesticides is also available [244].

1. General Evidence for s-Triazine Degradation

The first evidence of the contribution of biochemical factors, in addition to chemical processes, in s-triazine dissipation was obtained from studies which followed the bioavailability of these herbicides for different plants under various environmental conditions. Greatest inactivation was generally correlated to conditions optimum for microbial growth [245-249]. Laboratory experiments comparing sterile versus nonsterile conditions supported this observation [241, 250]. Acceleration of s-triazine degradation by application of microbial nutrient broth to soil also falls in line with these reports [251]. More direct evidence for the participation of microorganisms in degradation came from experiments with isolated soil fungi and bacteria. Several species were capable of growing in nutrient media containing s-triazines as the sole source of carbon, nitrogen, and, in the case of methyl-thiotriazines, of sulfur (see for simazine and atrazine [245, 252-261]; for ametryn and prometryn [245, 260, 262, 263]; for atratone and simetone [245, 259]). A list of the microorganisms effective in degrading s-triazines has also been published [242].

2. Degradation Reactions

a. Reactions of the C-2 substituents

$$R = Cl, OCH_3, or SCH_3$$

(23)

The occurrence of the corresponding hydroxy derivatives of simazine, atrazine, and propazine as major degradation products in soil sterilized by sodium azide or heat is clear evidence for the participation of nonbiological mechanisms in s-triazine degradation [264-266]. Several factors were reported to affect this detoxication reaction. Increasing temperature and moisture, low pH, and high organic matter content favored the hydrolysis of chlorotriazines [264, 267, 268]. However, with the exception of low pH, these are conditions that are also favorable for microbial growth.

From available information it is apparent that the role of the different adsorption sites in the hydrolysis of s-triazines depends, in practice, on the ratio of the clay and organic matter content of the soil under study. High adsorption of atrazine and simazine to clay or montmorillonite was found in all studies reported. In some studies only a slight effect of this adsorption on hydrolysis was noted [264, 265]. In others, infrared spectral evidence was obtained for the formation of protonated hydroxy derivatives of adsorbed atrazine and propazine [269, 270] and of other triazines [271]. These spectra also proved the products of hydrolysis to be predominantly in the keto form [269]. From studies with ^{35}Cl- and ^{14}C-ring-labeled simazine, it was concluded that organic matter in the soil, because of its catalytic properties, is responsible for an increase in degradation which yields the corresponding hydroxy derivative. Similar results were also found with atrazine and propazine [265, 272, 432].

The absence of a lag phase and the first-order kinetics observed in the degradation of chlorotriazines in soil also support the idea of the chemical nature of the hydrolytic process [273, 274]. Furthermore, the occurrence of the corresponding hydroxy derivatives after treatment with simazine, atrazine, propazine [275-279], GS-13529 [203], and ametryn [236] was reported. The same result was obtained with cyprozine [202] and cyanazine [211-214]. However, the hydroxy derivative of the latter compound showed additional cyano group conversion to a carboxyl group.

Although following a reaction mechanism apparently different from that found with chlorotriazines, hydroxypropazine was also observed in prometryn soil incubation studies [280, 281]. In these studies, chromatographic evidence for the occurrence of trace amounts of transient sulfoxide and sulfone intermediates was obtained.

Limited information is available on the conversion of triazines to their hydroxy compounds by isolated microorganisms. Incubation of Fusarium roseum (Lk; Snyder and Hansen) with ^{14}C-ring-labeled atrazine resulted in hydroxyatrazine as the predominant metabolite when analyzed by paper chromatography [282]. Whether direct hydroxylation followed by dealkylation or the reverse sequence occurred with simazine incubated with Aspergillus fumigatus Fres. cannot be decided on the basis of the ammelide identified chromatographically in this study [283].

b. N-Dealkylation

R—HN⟶triazine⟶NH—R   H$_2$N⟶triazine⟶NH—R   H$_2$N⟶triazine⟶NH$_2$ (17)

R = alkyl groups

The first indication of the cleavage of alkyl groups [Eq. (17)] in soil came from experiments designed to study the reason for the pronounced lag phase in the phytotoxic action of chlorazine [2-chloro-4,6-bis(diethylamino)-s-triazine] in oats [284] and of ipazine (2-chloro-4-diethylamino-6-iso-propylamino-s-triazine) in pine seedlings [285]. The sharp increase of phytotoxicity several months after application of both compounds may be explained by successive dealkylations transforming chlorazine into tri-etazine and simazine, and ipazine into atrazine.

Further evidence for the oxidative removal of N-alkyl groups was ob-tained in experiments with isolated microorganisms cultivated with ^{14}C-side-chain-labeled simazine or atrazine. Variable amounts of ^{14}CO$_2$ ranging from 1 to 39% of the radioactivity applied to the nutrient medium were liberated by a variety of fungi and bacteria [255, 259, 266, 282, 286, 287]. Dilute soil suspensions [288], soil perfusate [276], and intact soil [279] also gave rise to small amounts of ^{14}CO$_2$ when incubated with ^{14}C-side-chain-labeled atrazine.

Dealkylated metabolites still having intact C-2 substituents were observed in both isolated microorganisms and soil. With simazine and Aspergillus fumigatus Fres., 2-chloro-4-amino-6-ethylamino-s-triazine was identified by tlc, ir, nmr, and mass spectroscopy [283]. Among the degradation products formed by a variety of soil fungi cultivated in the presence of ^{14}C-ring-labeled atrazine, both 2-chloro-4-amino-6-isopropyl-amino- and -6-ethylamino-s-triazine were chromatographically character-ized [287]. The importance of N-dealkylation in microbial s-triazine degradation was also shown with several representatives of the chloro-, methoxy-, and methylthio-s-triazine groups using Aspergillus fumigatus Fres. [433].

The importance of the dealkylation reaction observed in the studies with isolated microorganisms was confirmed with intact soil under both laboratory and field conditions. Parent monodealkylated derivatives were detected after the application of simazine and atrazine [229, 277, 278, 289, 443], GS-13529 [203], GS-14254 [217], and terbutryn [220]. As observed in plants, the ease of alkyl group cleavage by microorganisms and soil decreases in the sequence ethyl, isopropyl, and larger or more branched alkyl groups.

Both dealkylated and hydroxylated parent compounds are subject to further degradation. Both types of products may give rise to common metabolites, namely 2-hydroxy-4-amino-6-alkylamino-s-triazines. The existence of such compounds was reported in the case of simazine and atrazine [229, 278], GS-13529 [203], ametryn [236, 290], and terbutryn [220].

Although information on the enzymic basis of the N-dealkylation mechanism is not yet available, speculation on the participation of free radicals in the dealkylation process has been made. In model experiments (Sec. I, B, 1, b), dealkylated compounds were found as a result of s-triazine free-radical reactions [60].

c. Side-chain modification

$$(20)$$

As found in plants, the cyano group of cyanazine was also converted in soil to the carboxyl group by the amide group [Eq. (20)]. The metabolites characterized still had an intact chlorine atom in the case of the amide group or were hydrolyzed to the corresponding hydroxy derivatives in the case of the carboxyl group [211-214].

d. Deamination

$$(21)$$

Examples of deamination of s-triazines [Eq. (21)] by microorganisms and soil are known. Ammelide was characterized by tlc as one of the degradation products formed by Aspergillus fumigatus Fres. cultivated in the presence of simazine [283]. Incubation of soil with the known simazine and atrazine metabolites, 2-hydroxy-4-amino-6-ethylamino- or -6-isopropyl-amino-s-triazine, for up to three months yielded the corresponding

2,4-dihydroxy derivatives as primary metabolites. This demonstrates that, as in plants, deamination proceeds more rapidly than cleavage of the final N-ethyl or isopropyl groups [238]. In contrast, ammeline was degraded within two months to CO_2, cyanuric acid, and traces of ammelide [239]. Cyanuric acid was also detected as a natural soil constituent [291-293].

e.   Triazine ring cleavage

$$\text{[triazine ring structure]} \longrightarrow \left[\text{Unknowns ?} \right] \longrightarrow CO_2 \qquad (22)$$

In contrast to the high rate of $^{14}CO_2$ evolution from ^{14}C-side-chain-labeled s-triazines, only small amounts of the radioactivity of ring-labeled s-triazines were converted to $^{14}CO_2$ by the soil [Eq. (22)]. Generally, $^{14}CO_2$ was evolved in the order of a few percent of the applied radioactivity within one to four months from atrazine [229, 247, 266, 267, 277, 294-296], simazine [241, 294, 296], and ipazine [170, 294]. An increase in $^{14}CO_2$ liberation was observed when the soil was treated with glucose [247, 295]. Sterilizing the soil for 20 min at 120°C on three consecutive days [241] or the presence of anaerobic conditions [267, 277] prevented $^{14}CO_2$ liberation, pointing to the microbial origin of this reaction.

Among the reaction conditions determining the degradative capacity of a given soil, the length of the observation period seems to be of special significance. Approximately 2.5% of the ^{14}C-ring label of simazine was evolved from different soils within five weeks. This period was followed by a lag phase of up to 15 weeks, after which a second period of $^{14}CO_2$ evolution started. This period came to an end after 30 weeks. Depending on the soil type, the cumulative amount of $^{14}CO_2$ liberated varied from 6 to 17% [297]. An interesting result was obtained with simazine adsorbed to peat. When the adsorbed simazine was incorporated in soil a remarkable increase in the liberation of $^{14}CO_2$ was observed in contrast to the control in which simazine was added directly [298].

Degradation of triazine metabolites was examined and compared with that of atrazine. About 10% decomposition of hydroxyatrazine to CO_2 was observed under both aerobic and anaerobic conditions in soil and lake sediments high in organic matter content [299]. Although higher $^{14}CO_2$ liberation occurred with hydroxyatrazine than with atrazine in aerobically incubated soils of different organic matter content, the absolute rate remained in the order of 2%. The amount of liberated $^{14}CO_2$ was doubled when the soil was treated with a supplemental carbon and nitrogen source [266, 277, 279].

As indicated in Sec. II, B, 2, d, 2-hydroxy-4-amino-6-ethylamino- and -6-isopropylamino-s-triazine were preferentially degraded to the corresponding 2,4-dihydroxy derivatives. Nevertheless, 21.7 and 4.3% of the

[14]C-ring-labeled ethylamino and isopropylamino derivatives, respectively, were oxidized and liberated as $^{14}CO_2$ during a three-month incubation period [238, 239].

Ammeline degradation in soil seems to occur most rapidly. Approximately 50-80% of the ^{14}C-labeled ammeline was converted into $^4CO_2$ within two or three months, respectively. The remainder of the ^{14}C was present primarily as cyanuric acid. The same high rate of ring cleavage was observed with cyanuric acid, although no intermediates were observed in this case [238, 239]. These results confirm earlier findings on the ready availability of the triazine nitrogen of ammeline, ammelide, and cyanuric acid for plant [300, 301] and microbial [302] growth.

C. Degradation in Animals

Animals represent the third biological system of importance that is engaged in the degradation of s-triazines and their metabolites. Since s-triazine-treated plants may serve as an animal feed, results from studies with ruminants and monogastric animals provide necessary information on the residues possibly occurring in products such as meat, milk, and eggs which are used for human nutrition. Furthermore, animal model studies contribute essentially to an understanding of the behavior of these compounds in the human.

1. Absorption, Distribution, and Excretion

Despite a remarkably low toxicity to mammals, additional information on the metabolic fate of the parent s-triazines and their important plant metabolites is required to establish the safety of these compounds. For this purpose, balance studies were performed with laboratory and farm animals in which the routes and the rate of excretion of these compounds were determined. Ready absorption of parent compounds from the digestive tract was generally observed when radiolabeled s-triazines in amounts of up to 70 mg/kg of body weight were orally administered to laboratory animals. The same observation was made in goats and cows which were given different s-triazines in the order of single milligram-per-kilogram doses either in gelatine capsules or with their feed.

Rapid and complete excretion of radioactivity after dosing laboratory and farm animals with different types of ^{14}C-ring-labeled s-triazines was found. Generally, the urine was the favored route of excretion, although the ratio of excretion in urine and feces varied from compound to compound. In addition, factors such as particle size and form of the applied material, i.e., in solution or suspension, may influence the rate of absorption of compounds of such low solubility and, consequently, their excretion. An apparent exception was found in the case of terbutryn in which less than 30%

was excreted in the urine. This behavior was not caused by incomplete absorption but by the efficient excretion of the absorbed radioactivity into the gut via the bile [220].

When the rate of excretion of various s-triazines was determined, the time required for the excretion of 50% of the applied dose in laboratory animals was 12-26 hr, whereas the corresponding figures for ruminants ranged from 1 to 2 days. The 2-methoxy-s-triazines seem to be the most rapidly eliminated derivatives as shown with prometone, GS-14254, and GS-14259.

In contrast to the feces, where small amounts of parent s-triazines were observed in some cases, there was no indication that intact parent compounds were eliminated in the urine when ^{14}C-labeled materials had been used. Small amounts of parent simazine and atrazine were reported to be excreted in the urine of cows fed the unlabeled herbicides for 4 days. However, the analytical procedure used did not allow discrimination between parent compounds and metabolites still containing the chlorine substituent [303, 304].

Residual radioactivity in the essential organs of rats and ruminants was determined at the end of most of the balance studies or was followed in a time-course study after single- or multiple-dose applications. These studies demonstrated that, even after dosages as high as 50 mg/kg or more, residues in the tissue were continuously and progressively eliminated from the body. The results obtained depended on the dose applied and ranged from 3 days for complete elimination to 12 days with still detectable, but constantly decreasing, radioactive residues. No organ-specific retention or accumulation of radioactive residues was observed. Among the organs investigated, those with abundant blood supply such as liver and kidney showed the highest transient radioactivity which rapidly disappeared when the excretion came to an end. Generally, fat and muscle contained the lowest concentrations of radioactivity.

Feeding goats and cows with single doses of radiolabeled s-triazines in the order of 1 mg/kg resulted in transient milk residues which showed a maximum concentration between 8 and 24 hr after application. The residues declined rapidly, reaching the limit of detection after an additional 2-4 days.

Details of the available information on absorption, excretion, and organ residues can be obtained in the following references: simazine in rats [305-307], in ruminants [305, 308]; atrazine in rats [229, 307, 309, 310], in ruminants [229, 308]; propazine in rats [311-313], in ruminants [314]; GS-13529 in rats and ruminants [203]; cyanazine in rats [315]; prometone in rats [313]; GS-14254 in rats [217, 316, 317], in ruminants [217, 318]; GS-14259 in rats [218]; ametryn in rats [319], in ruminants [290]; prometryn in rats [306, 320], in ruminants [320]; terbutryn in rats [220]; G-36393 in rats [219].

2. Degradation Reactions

a.   Reactions of the C-2 substituents

$$R = Cl,\ OCH_3, or\ SCH_3 \tag{23}$$

The detection of parent hydroxy derivatives in the course of chlorotria-zine degradation by animals has been reported. In rats dosed with ^{14}C-labeled atrazine [168, 229, 310] , simazine [305] , and propazine [311] , only small amounts, on the order of a few percent, were observed in urine and feces. Excretion of parent hydroxy derivatives of the methoxytriazines seems even more limited than with the chlorotriazines, as was found, for example, in rats fed GS-14259 [218] . Nevertheless, replacement of the methoxy substituent by a hydroxy group takes place since greater amounts of hydroxy derivatives with modified side chains were detected in the urine of rats and cows fed GS-14254 [316-318] .

Parent hydroxy derivatives were also excreted to a limited extent by rats after methylthiotriazine dosing. Less than 4% of applied [^{14}C]prometryn occurred in urine and feces as the corresponding hydroxy derivative [320] . About 6 and 11% of the parent hydroxy derivative were excreted in the urine and feces of rats dosed with 3 and 70 mg/kg of terbutryn, respectively. However, there was an indication that this metabolite, like other polar metabolites found, was originally present in the form of a weakly bound conjugate which might be cleaved during ion-exchange chromatography [220] .

Several studies investigated the fate of parent hydroxy derivatives applied directly to rats. Using ^{14}C-ring-labeled hydroxyatrazine [229, 310, 321] , hydroxysimazine [321] , hydroxyterbutryn [238] , and hydroxy-GS-14254 [316, 317] , excretion of radioactivity was rapid and complete; the urine was the main route of excretion in most cases. At the end of the excretion phase, no detectable amounts of radioactive residues in tissue were found.

An unusual degradation sequence reportedly occurred in rats and rabbits dosed with large amounts of nonlabeled prometryn. Two urinary metabolites were characterized by tlc as the free mercapto derivative and its correspond-ing disulfide. Each was concomitantly N-depropylated at C-4 [322] [Eq. (24)].

$$\tag{24}$$

b. Conjugation

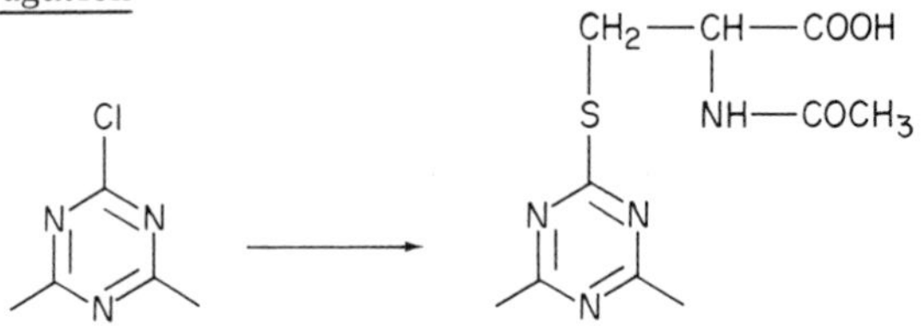

(25)

The first example of the conjugation of the triazine ring system with an
amino acid in animals was observed with cyanazine. Mercapturic acid formation, a mechanism well known from the degradation of organic halides,
and N-deethylation were found to yield N-acetyl-S-[4-amino-6-(1-methyl-
1-cyanoethylamino)-s-triazinyl-2]-L-cysteine as a major metabolite in the
urine [315] [Eq. (25)].

c. N-Dealkylation

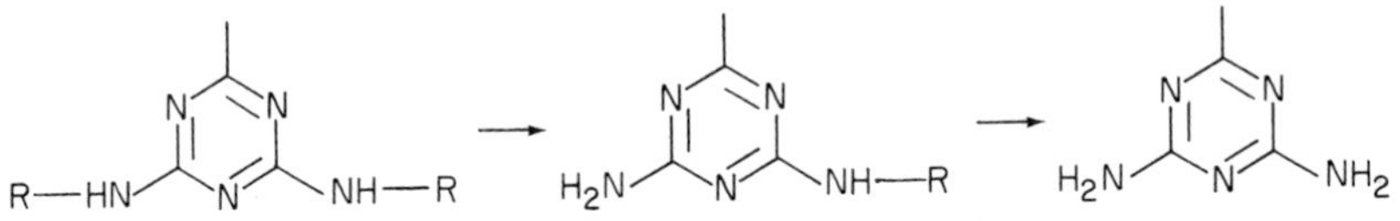

R = alkyl groups   (17)

Indirect proof of the removal of the alkyl groups [Eq. (17)] came from
experiments with ^{14}C-side-chain-labeled s-triazines. Efficient and rapid
oxidative cleavage of ethyl and isopropyl groups to yield $^{14}CO_2$ in the range
of 20-75% of the applied dose was detected with simazine and atrazine in the
rat [307] and sheep [308], propazine in the rat [313] and sheep [314], and
ametryn in the rat [319]. Mono- and didealkylated parent compounds resulting from this degradation mechanism were identified in the urine of rats
and rabbits after the application of several hundred milligrams per kilogram
of nonlabeled simazine, atrazine, propazine, prometone, and prometryn
[322, 323].

Formation of dealkylated derivatives was confirmed in several studies
where ^{14}C-ring-labeled chloro-, methoxy-, and methylthio-s-triazines had
been applied to rats in doses ranging from 1 to 15 mg/kg. Generally, only
traces or a few percent of the applied dose were recovered from the excreta
as parent monodealkylated derivatives, as shown, for example, in studies
with cyanazine [315], GS-13529 [203], and GS-14260 [220]. The corresponding 2-hydroxy derivatives of the respective monodealkylated parent
compounds were isolated from the urine of rats treated with ^{14}C-ring-labeled
atrazine. It is possible, however, that these compounds were originally
excreted as monodealkylated parent compounds which lost their chlorine

atom during ion-exchange chromatography used for the separation of metabolites [310]. Monodealkylated parent compounds also seem to be minor metabolites in cows, as shown in a study with GS-14254. However, removal of both alkyl groups yielded 2-methoxy-4,6-diamino-s-triazine, which was of more importance [217, 318].

Completely dealkylated derivatives with intact C-2 substituents were also detected as urinary metabolites in the rat following GS-14254 [316, 317] and atrazine [229, 310] application. Hydroxy derivatives corresponding to each type of dealkylated metabolite were also detected in these studies. Ammeline was found in high concentration in rat and cow urine after application of atrazine [310] and GS-14254 [317, 318]. However, metabolites still containing intact C-2 substituents were not separated from the group of hydroxy metabolites prior to ion-exchange chromatography. Thus, possible replacement of the C-2 substituents by hydroxy groups does not permit determination of the actual excreted ratio of 2-chloro- or 2-methoxy-4,6-diamino-s-triazines and ammeline.

When methylthio-containing metabolites had been previously removed, almost 20% of the radioactivity of a 2 or 70 mg/kg dose of terbutryn fed to rats was excreted in the form of 2-hydroxy-4-amino-6-tert-butylamino-s-trazine in contrast to only 2.4% of parent monodesethyl derivative. Ammeline excretion did not exceed 1% [220]. Characterization of urinary radioactivity of rats dosed with [^{14}C]simazine also demonstrated the excretion of substantial amounts of metabolites simultaneously hydroxylated and dealkylated. The ratio of 2-hydroxy-4-amino-6-ethylamino-s-triazine to ammeline was found to be 6-14% in the 24-hr sample. The ammeline content was raised to more than 31% after acid hydrolysis, indicating the presence of completely dealkylated C-2 conjugates [305].

A number of studies followed the fate of significant triazine plant metabolites in rats. The most striking observation in contrast to parent compounds was the simple pattern of metabolites found after the application of hydroxy parent and hydroxy monodealkylated derivatives. Generally, parent hydroxy compounds were excreted partly unchanged or as 2-hydroxy-4-amino-6-alkylamino derivatives. In contrast, when the latter metabolites are given directly to rats, they leave the organism practically unchanged. These results have been obtained with hydroxyatrazine [238, 310], hydroxy-GS-14254 [317], hydroxyterbutryn [238], 2-hydroxy-4-amino-6-tert-butylamino-s-triazine [238], 2-hydroxy-4-amino-6-ethyl- or -6-sec-butylamino-s-triazine [317], and 2-hydroxy-4-amino-6-isopropylamino-s-triazine [239]. The fate of radioactive residues left in [^{14}C]atrazine-treated sorghum plants after routine extraction procedures was followed in the rat and sheep. After feeding of the methanol-extracted plant shoots, almost exclusive excretion of the radioactivity in the feces indicated that no absorption of the nonextractable radioactivity from the digestive tract took place in either species [324].

d. Side-chain modification

R = alkyl groups   (26)

Thus far, only one example for the oxidation of branched-side-chain groups yielding alcohols has been reported [Eq. (26)]. Metabolites that contained a hydroxy group at C-1 or C-3 of the sec-butyl substituent were identified in the urine of both the rat and cow fed ^{14}C-ring-labeled GS-14254. These metabolites were also deethylated and had their C-2 substituent intact or replaced by a hydroxy group [217, 316-318].

Omega oxidation of alkyl side chains resulting in the formation of carboxy acids [Eq. (26)] occurred with different types of triazines. Small amounts of monocarboxy acids of simazine, atrazine, propazine, and prometone were excreted in the urine of rats and rabbits dosed with large amounts of the nonlabeled herbicides. All of the metabolites were mono-dealkylated [322, 323]. Following the application of ^{14}C-labeled atrazine to the rat [310], GS-14254 to the cow [318], and GS-13529 to the rat [203], ω oxidation of the more bulky isopropyl and sec-butyl groups yielded small amounts of the respective carboxy acids of the otherwise intact parent compounds.

e. Deamination and ring cleavage

Ammeline (8) was a urinary metabolite in cows fed GS-14254 [318] and in rats treated with atrazine [310], but no ammelide (9), cyanuric acid (10), or CO_2 was observed. When ammeline was directly fed to rats, however, CO_2 and the intermediate metabolites, ammelide and cyanuric acid, were found [239]. Cyanuric acid, when given to rats, was rapidly excreted in an unchanged form [239]. Ring cleavage with final liberation of CO_2 was found to be of no importance in animals since only traces of $^{14}CO_2$, if evolved at all, were detected after the application of atrazine [307, 310], ametryn [319], and GS-14254 [316] to rats. Similar observations were made when propazine and prometone were fed to rats and goats [313, 314].

The results discussed in this section (II, C) show that, when parent s-triazines are fed to animals, degradation essentially ends at the stage of 2-hydroxy-4-amino-6-alkylamino derivatives or at 4,6-diamino compounds with an intact C-2 substituent. Both types can be directly cleared by the kidneys. Therefore, stages of further degradation occur in animals in significant amounts only when these compounds are fed directly in the form of plant metabolites.

III. MODE OF ACTION

A. Biochemical Processes Affected

Intensive research was initiated in many areas of plant biochemistry to relate the striking physiological effects of the s-triazines with basic biochemical and biophysical processes. Four of the most important of these processes have been selected for discussion in this section. A summary of the available information has already been published [325] and a comprehensive review will be published elsewhere [326].

1. Photosynthesis

The triazine herbicides, like the substituted phenylureas, acylanilides, and carbamates, are classified as powerful inhibitors of photosynthesis in plants by interrupting the light-driven flow of electrons from water to nicotinamide adenine dinucleotide phosphate (NADP) (for general reviews on this topic see Refs. 327-331). It is well documented that the chlorotriazines [332-339], methylthiotriazines [339], and methoxytriazines [336, 337, 339] drastically inhibit the Hill reaction, i.e., the evolution of oxygen from water in the presence of chloroplasts and a suitable electron acceptor. This effect was observed in isolated chloroplasts [332, 334-337, 339], algae [338], and submerged Elodea sp. [333]. It was suggested that the inhibition of photochemical activity in chloroplasts results from interference of the herbicide with the mechanism for oxygen production. The exact site of action of the inhibitor molecule seems to be at the water-splitting site of the photosystem since simazine does not affect photoreduction in algae such as Scenedesmus sp. which are capable of photoreduction in a hydrogen atmosphere through hydrogenase enzymes [340].

The levels of Hill reaction inhibition in chloroplasts [13, 332, 339] show that the phytotoxic triazines inhibit this reaction in the same range of concentrations as CMU (monuron), i.e., 10^{-6}-10^{-7} M. Inhibition of energy transfer in chloroplasts is apparently essential for the herbicidal action, as nonphytotoxic metabolites like hydroxysimazine or cyanuric acid do not reveal any

activity in this specific test. However, the degree of activity in the chloroplast system is not correlated with the selective herbicidal action of the
s-triazines on plants, since chloroplast breakdown was shown to occur to
the same extent in both susceptible and resistant plant species [342, 343].

Most of the speculations about the mechanism of action of the herbicidal triazines during energy transfer of the photosynthetic systems are
based on the general idea that the triazines form hydrogen bonds. This is
possible through the free amino hydrogens and the ring nitrogens of the
triazines binding to the carbonyl oxygen of a protein molecule and reacting
in this way with active centers in the chloroplasts. A comparison of the
chemical structures of herbicides that exhibit a similar mode of action
reveals that most of them have the common chemical feature of a NH group
attached to an electron-deficient carbon atom such as C=O groups in the
case of the phenylureas, anilides, uracils, and carbamates, or C=N arrangements in the case of the triazines. Planar arrangement of the system also
seems to be a prerequisite for herbicidal activity [17, 329, 344, 345].
This stresses the importance of charge transfer and dipole interchanges in
addition to hydrogen bonding. Chlorazine [2-chloro-4, 6-bis(diethylamino)-
s-triazine] inhibits the Hill reaction less than other chlorotriazines [346]
($I_{50} = 1.2 \times 10^{-4}$ M compared to 1.3×10^{-6} M for atrazine [341]). This property of chlorazine is surprising in the light of the distinct herbicidal activity
of the compound. Probably, prior metabolic conversion is required for the
development of phytotoxic action of the compound. Studies correlating pI_{50}
values and structural elements of phenylurea and acylanilide herbicides
resulted in a similar conclusion [347].

Chlorophyll is thought to be the principal pigment involved in triazine
phytotoxicity in plants [348]. The relationship between light quantity as well
as light quality and herbicidal action was investigated thoroughly. No toxicity occurred in the dark. The degree of phytotoxicity was an increasing
function of light intensity. Also, light quality greatly influenced toxicity. A
maximum of toxicity occurred at 428 and 658 nm. The spectrum of injury
closely followed the adsorption spectra of chlorophylls a and b. The
adsorption maxima of carotenoids did not coincide with these values [348].

In an attempt to explain the s-triazine phytotoxicity, the formation of
phytotoxic substances involving chlorophylls a and b upon illumination has
been proposed [342, 348]. This hypothesis was suggested during an investigation on the mode of action of the phenylureas [349]. However, the substance involved has not been isolated so far. It seems likely that the configuration of the total chlorophyll-protein complex undergoes changes in the
secondary or tertiary structure of the protein part as a result of hydrogen
bonding with the triazines, thereby impairing its function. The buildup of
the triazine-inhibitor complex was investigated by means of partition analysis,
inhibition kinetics, and adsorption and distribution pattern analysis of atrazine in spinach chloroplasts [350]. The sites of inhibition were calculated

to be one for every 2500 chlorophyll molecules or even less [351], which would correspond to the estimated size of one photosynthetic unit [352].

Determination of fluorescence by chlorophyll has been studied as an indicator of energy conversion during photosynthesis. In a Chlorella sp. suspension a straight line relationship was found between the percent increase in fluorescence and the logarithm of the triazine concentrations. The concentration at which fluorescence increased by 50% corresponded very well to that which inhibits oxygen evolution in Chlorella sp. by 50% [338]. During illumination the electrons leave their site in the chlorophyll-protein complex and are taken up by appropriate electron acceptors. As the triazines block the further flow of electrons from water to chlorophyll, oxidation of the chlorophyll probably takes place [329] and gradual destruction of the lamellar structure of the chloroplast occurs [342]. According to another theory, the triazine molecule is operating by binding the proton which is released during photolysis and hindering its transfer to NADP [353].

Interference of the triazine herbicides with energy transfer during oxygen evolution in the Hill reaction consequently affects the reduction of NADP [332, 335] and the production of ATP. The stoichiometrically coupled noncyclic photophosphorylation is inhibited, whereas cyclic photophosphorylation remains intact. The concentrations inhibiting the cyclic photophosphorylation are 50-100 times those required for stopping the Hill reaction [328, 354]. The generation of reduced nicotinamide adenine dinucleotide phosphate (NADPH$_2$) and the formation of adenosine triphosphate (ATP) by the photosynthetic systems are prerequisites for subsequent assimilation of CO_2. The CO_2 fixation was shown to be drastically inhibited by the triazines in the light [355, 356] as was the formation of starch in leaves of Coleus sp. [357] and Chlorella vulgaris [358]. The blockage preceded the synthesis of sucrose since the addition of sucrose overcame this block [357, 358]. Simazine injury to barley plants could be overcome by supplying glucose through the leaf tips [337].

As a result of the direct effect of triazines on photosynthesis, some indirect effects, such as the reduction of transpiration, are seen [359, 360]. Microscopic examinations revealed that the closing of the stomata is the reason for this and not an impairment of the water uptake by the roots. A probable explanation is that the blockage of photosynthesis increases the CO_2 concentration in the cell and inhibits the glycolic acid synthesis which has its special role in the synthesis of carbohydrates in the guard cells [361].

2. Plant Growth Regulation

After the introduction of the triazine herbicides into agriculture, it became evident that some plants from simazine- or atrazine-treated plots grew more vigorous and healthier than plants from control plots where the weeds were removed manually. This stimulation of plant growth by triazines

not only affects shoot length [362-372]; the leaf blade [368, 373] and the
stem diameter are increased as well [374]. By the application of higher
amounts of the triazine herbicides, inhibition can be obtained [367, 371, 372].
Like the "protein effect" of triazines (see Sec. III,A,3), growth stimulations
or inhibitions are dependent to a large extent on the concentration applied
and the plant species under investigation. They include triazine-resistant
[364, 370] and triazine-sensitive plants [365] as well as annuals [364, 370]
and perennials [363, 366, 368, 369, 371, 372].

The conclusion was drawn that plant hormone metabolism may be influenced by the triazine herbicides. Experiments have been performed to
determine the influence of atrazine on auxins, cytokinins, and gibberellins
[374-376]. Wheat coleoptiles and cress roots react positively in simple
plant physiological tests with low levels of atrazine [374]. The indole
acetic acid-peroxidase system was used as a biochemical test to detect
changes in auxin metabolism by measuring the rate of decarboxylation of
labeled indole acetic acid (IAA). A strong influence of atrazine on this
particular enzyme could be demonstrated in wheat and oat coleoptiles but
not in systems in vitro [374].

A "greening effect" quite often becomes obvious during the growth of
triazine-treated plants and is reminiscent of fertilizer effects or a possible
kinetin effect which delays the catabolism of the chlorophyll [362, 363, 366-
369] and senescence of plants [363]. Senescence biotests with leaf discs of
Rumex acetosa or Taraxacum officinale gave no effect from atrazine [375].
Atrazine could not replace the kinetin in the agar medium for growing soybean tissue [374]. However, 50% growth inhibition of tobacco callus tissue
takes place after 10^{-6} M atrazine is added to the nutrient medium containing
kinetin [377]. Promotions of soybean tissue growth occurred in the same
nutrient medium containing 0.05 ppm kinetin and 10^{-12} M atrazine [374].
Interactions between atrazine and kinetin were demonstrated in chlorophyll
retention studies in corn leaf discs [376]. Ametryn exerted a somewhat
less, but nevertheless significant effect on chlorophyll loss from corn leaf
discs. It is clear from these studies that atrazine does not act as a cytokinin
per se in the plant tissue, but that it influences this metabolism indirectly.
The influence of atrazine on gibberellic acid metabolism has been investigated with the barley endosperm biotest, but there was no effect of atrazine
[375, 378].

Attempts have been made to explain the effects of the triazines on plant
growth by effects at the RNA level [379]. Owing to insufficient basic information at the molecular level, the visible effects of triazines on plant growth
cannot yet be related to definite biochemical processes.

3. Nitrogen Metabolism

It has frequently been reported that triazines have an influence on the
dry matter content of plants and seeds, especially in the protein fraction

[362, 366, 367, 369, 380-392]. This so-called protein effect of the triazines has been given considerable attention since it might be of economic value in agriculture. Most triazine derivatives that are herbicidal can provoke this effect, whereas the nonherbicidal hydroxytriazines produce no additional protein or yield increase in plants [339, 382, 393]. Of particular interest is that occasionally seeds from various plant species treated with simazine and grown to maturity contain more dry matter content and more protein [386-390]. Resistance or sensitivity of the individual plant species to the triazine herbicides is not of importance here. Dry matter and/or protein increases are just as likely in triazine-resistant plants as in triazine-susceptible plants.

It was shown in field work and laboratory tests that the uptake of mineral nutrients can be appreciably higher in triazine-treated plants than in control plants [367, 370, 394]. The nitrogen uptake by plants was followed quantitatively in several studies [367, 370, 382, 395] and increases occurred only when nitrate was supplied as the nitrogen source but not when ammonium was used [395]. The most spectacular increases occurred when nitrate was very low in the nutrient medium and the triazines were at a sublethal concentration [369, 382, 395].

The site of action at which the triazines facilitate the uptake of mineral nutrients has not been investigated. It might be located in the area of the plasmalemma membranes. Suction pressure and water permeability were changed in protoplasts of epidermal cells of Solanum tuberosum petiols after treatment with simazine [396]. The observations indicate membrane changes which could facilitate greater ion uptake. Another possibility is that root growth is promoted by sublethal levels of triazine herbicides [374, 397, 398], and the enlarged root surface contributes to the increased ion absorption.

Nitrate absorbed by triazine-treated plants is readily transformed to organic nitrogen and found in the amino acid and protein fractions, as in normal nitrogen metabolism. By monitoring the conversion of the nitrate from triazine-treated plants into plant constituents, it was demonstrated that nitrate reductase was affected [395, 399]. Enhancement of nitrate reduction by simazine occurred mainly at low night temperatures and at low nitrate level in the nutrient solution. Nitrate reductase activity was not stimulated by atrazine in experiments in vitro with plant material [399] or with the pure enzyme system [400]. Thus, activation of enzyme activity seems unlikely. Increases in the nitrate level of plant tissue and increases in potassium ion concentration in the cells are responsible for the inductive formation of nitrate reductase; this may be the reason for the de novo synthesis of this enzyme.

An increase in the quantity of free amino acids has been reported in experiments with both higher plants and microorganisms under the influence of triazines [401-403]. This increase reportedly arises from the utilization of carbon and nitrogen fragments from triazine degradation [404] as well

as from the extra nitrogen taken up. However, owing to the small amounts
of triazines actually available to the plants after regular field application,
the first effect can only be of minor importance. There is also evidence
from CO_2-fixation studies, as well as from metabolism studies with $[^{14}C]$-
sucrose and $[^{14}C]$serine in plants grown under light and dark conditions, that
atrazine changed the relative distribution of the ^{14}C-labeled compounds
especially favoring the amino acid fraction [356, 405].

Available information explaining the effects of the triazine herbicides
on protein synthesis is still quite fragmentary and of hypothetical character
[386, 402, 406]. The protein pattern of triazine-treated plants was the
same as in the control plants according to amino acid analysis [375, 386].
The same types of proteins are apparently synthesized. From protein fractionation studies, it is quite clear that in oat leaves the chloroplast proteins
were affected by simazine whereas the proteins of the mitochondrial, microsomal, and cytoplasmic fractions remained nearly the same after 8 days
[407]. Drastic reductions of the chloroplast proteins are normally accompanied by the destruction of chloroplasts [342] and by chlorosis. Conversely,
protein increases in leaves are often parallel to chlorophyll increases. The
leaves appear dark green and remind one of a fertilizer effect. This "greening effect" from triazines is not caused by providing more nutrients to the
plant through interactions of the herbicide with the microflora of the soil
or the plant rhizosphere since it could be demonstrated in plants growing in
a completely sterile medium [375]. The diameter of chloroplasts from
atrazine-treated corn plants which showed the greening effect was increased
and these chloroplasts were heavier as well. They sedimented quicker during
centrifugation than those of the control plants, and they contained more
structural proteins. It was confirmed by these studies that the chloroplast
protein fraction has a unique role in protein increases in plants.

There is evidence in the voluminous literature available about protein
formation in plants following treatment with triazines that a specific triazine
concentration is necessary to produce stimulations. This specific concentration of triazine varies largely among the different plant species and even
within development stages of a single plant species. By increasing this
optimal and mostly subherbicidal level of the triazines, reductions of dry
matter with or without increases of proteins can occur. Yield and protein
increases in plants after triazine treatment do not occur consistently,
especially under practical conditions. Unfortunately, the environmental
and agricultural conditions necessary for the triazines to enhance protein
production in plants cannot be sufficiently controlled. Therefore, this worldwide interesting "protein effect" of s-triazines remains mostly of academic
interest.

4. Nucleic Acid Metabolism

Chemical effects on growth regulation and protein or energy metabolism may be correlated with effects on nucleic acid metabolism. Therefore, it became of interest to investigate whether s-triazine treatments result in any influence on the nucleic acid metabolism of plants and microorganisms. Although the s-triazines and the pyrimidines are of different chemical reactivity, their similar physical and chemical properties led some researchers to speculate that s-triazine derivatives may possibly be substituted for pyrimidine bases, thus interfering with nucleic acid metabolism [408-410].

Traces of radioactivity were found in the DNA and RNA fractions when the triple auxotroph Escherichia coli 15 arg⁻, t⁻, u⁻, which requires arginine, thymine, and uracil, was incubated with ^{14}C-labeled prometryn and cyanuric acid instead of thymine or uracil [408]. Since no labeled nucleotides were isolated and no positive control was used in these studies, the significance of the results is open to question. It was clearly demonstrated that this organism did not have the capacity to incorporate atrazine or prometryn as base analogs into the DNA and RNA fractions. In contrast, 5-bromo[2-^{14}C]uracil, known to substitute for thymine, was strongly incorporated in the DNA fraction by these organisms [411]. No incorporation of [^{14}C]prometryn into the total nucleic acid fraction during the growing phase was observed in studies with E. coli s-26, nor did the incorporation of ^{32}PO$_4$ differ from the controls [409].

Information concerning the influence of s-triazines on the nucleic acid metabolism of higher plants is very limited at the present time. No incorporation of the ring moiety from ^{14}C-ring-labeled atrazine was found in the nucleic acid fractions of resistant and susceptible selections of Zea mays L., GT-112-R, and GT-112-S [412]. In a study with etiolated corn seedlings cultivated in the presence of ^{14}C-ring-labeled simazine, some radioactivity was found associated with different RNA fractions. However, the nature of the radioactivity was not characterized [410].

Such physical effects as intercalations with the bases in DNA or allosteric binding to proteins must also be considered. These processes are of importance in nucleotide or nucleic acid metabolism and have to be taken into consideration as possibly causing biological effects. Several types of possible physical interactions of prometryn with nucleic acids were investigated. Equilibrium dialysis of a single- and double-stranded phage DNA against [^{14}C]prometryn and denaturation studies which should demonstrate intercalating or nonenzymatic binding showed no effect of [^{14}C]prometryn. Atrazine, prometryn, and simazine had no effect on a Bacillus subtilis transformation system nor on the metabolism of viruses in bacteria which

were grown with unlabeled triazines [409] . These data support the results
of cytological and genetic studies. In Vicia faba stipular cells and in
Tradescantia sp. hair cells treated with low concentrations of atrazine,
mitosis proceeded and finished normally from early prophase [413] . Cell
division in root tips of V. faba, Hordeum vulgare, and Allium cepa, as well
as in pollen mother cells of Sorghum vulgare, was normal under the influence of the triazines. The rate of cell division was not changed and there
was no deviation in the structure of the chromosomes [339, 414] . No mutagenic effects occurred following several generations of Arabidopsis thaliana
[339] .

Kiermayer's chapter in the book edited by L. J. Audus, The Physiology
and Biochemistry of Herbicides [415] , reports that triazine herbicides cause
nuclear aberrations and refers to an earlier paper [416] . The study reported
on chromosome breakage induced in V. faba root tips by 2, 4, 6-triethylen-
imino-s-triazine (TEM), a compound that has been used in cancer chemo-
therapy because of its alkylating properties. However, this is misleading
since the triazine herbicides are not alkylating agents like TEM. The triazine ring itself, as well as melamine (2, 4, 6-triamino-s-triazine), are
genetically ineffective. In comparison, trimethylolmelamine (2, 4, 6-tri-
hydroxymethylamino-s-triazine) and TEM have a strong mutagenic effect
which is based on the alkylating action of the three methylol and ethylene
substituents, respectively [417] .

Possible influences of the triazines on RNA and protein synthesis are
proposed on the basis of biochemical studies in higher plants [379, 407] .
The effect of atrazine and a series of other herbicides on the synthesis of
RNA and proteins was investigated in excised tissue assays [378, 418] .
Incorporation of [^{14}C]leucine into the proteins of plant material as a measure
of the de novo synthesis of proteins was not affected by atrazine [378, 418] ,
nor was the induction and activity of gibberellin-induced α-amylase [378] .
The latter process is taken as a measure of effects on RNA synthesis necessary for the induction of α-amylase by gibberellin. Incorporation of [^{14}C]-
orotic acid as a precursor for nucleic acid pyrimidines was stimulated by
atrazine at concentrations of 2.10^{-4} M [378] .

Since nucleic acid and protein syntheses are energy-requiring processes,
any compound that interferes with ATP production by uncoupling energy-
transfer mechanisms will affect these syntheses indirectly [419] . Other
effects, such as (a) the induction of nitrate reductase, which requires prior
nucleic acid synthesis [395, 399] , or (b) the influence on plant hormones
known to change RNA contents of ribosomes, or (c) their influence on the
translation system [420] , may be considered as factors explaining the effects
of s-triazines on nucleic acid metabolism. However, since the information
available on this topic is rather limited, the results from experiments of
photosynthetically active herbicides on nucleic acid metabolism of higher
plants should be interpreted carefully.

B. Selectivity

When it was discovered that s-triazine herbicides are potent inhibitors of photosynthesis, it was logical that an attempt be made to correlate their mode of action with their striking selectivity. However, it soon became apparent that the inhibition of this fundamental process does not explain the selectivity of various s-triazine structures, because they inhibit the Hill reaction of isolated chloroplasts of both susceptible and resistant plants [337] (see also Sec. III,A,1 for details). Consequently, factors determining uptake, translocation, and detoxication moved into the focus of scientific interest.

The bioavailability of soil-applied herbicides for plants is essentially governed by the dynamic equilibrium existing between the adsorbability of the compounds to the soil colloids and their solubility in the soil-water phase. Consequently, an appreciable influence of these parameters on the selective action of s-triazines must be expected. Because of the relatively strong adsorption of the s-triazines, selective phytocidal behavior should primarily show up in cases where deep-rooted crops are mixed with subsurface-rooted weeds. Chlorotriazines, in particular, do find a widespread use in weed control of perennial woody crops like grapevines, fruit trees, forest nurseries, etc.

It is difficult to establish generally valid rules for the correlation between adsorption (usually observed to decrease in the order methylthio, methoxy, hydroxy, and chloro derivatives) and selectivity of s-triazines. At best, it can be deduced that the more soluble and more basic compounds show the higher degree of adsorption and generally the smaller margin in selective phytotoxicity. Thus, methoxytriazines, possessing by far the highest water solubility among the triazines, generally show the lowest degree of selectivity. With few exceptions, this class of s-triazines is therefore especially useful for industrial weed control. As a result of additional properties, such as higher vapor pressure and higher phytocidal activity via foliar application, methylthiotriazines show the highest variability in selectivity among the s-triazines. They are used as both pre- and postemergent herbicides in an extreme variety of crops either alone or in combination with other triazines.

Special application techniques may improve the selective action of s-triazines. Banana plants in Hawaiian soils are more sensitive to the more mobile atrazine than to the more tightly adsorbed ametryn [140] . Selective weed control in new alfalfa can be more effectively accomplished by subsurface band application of the less mobile simazine, rather than by atrazine or methoprotryn [421] .

Results obtained with plants grown in nutrient solution containing radiolabeled s-triazines clearly demonstrate that absorption itself does not contribute appreciably to the selectivity of these herbicides. Root uptake occurred in both resistant and susceptible species of crops and weeds. Triazine concentration variations found in the roots were not analogous to variations found in shoots, which would be necessary to explain the wide differences in selectivity observed in practice [111, 121, 167, 422].

These studies indicate that translocation is also of no practical relevance to explaining the selectivity of s-triazines. Rather, they are thus in agreement with the studies discussed in Sec. II, A, 1, which demonstrated the nonspecific distribution of the triazines within plants through the transpiration stream. One exception to this was observed in cotton plants where a preferential accumulation of ^{14}C-ring-labeled simazine in the lysigenous glands occurred. This compartmentalization was interpreted as a protective mechanism which prevented the triazine from reaching the sensitive sites [111, 134]. However, when glanded and nonglanded isogenic lines of cotton were examined for their atrazine storing and degrading capacity, the lysigenous glands neither increased the tolerance to triazines nor decreased the inhibition of photosynthesis by atrazine. Instead, the degradative capacity of both lines was found to be the real basis for the intermediate susceptibility of the cotton plant [210].

The ability of plants to detoxify s-triazines by specific metabolic reactions is an additional factor necessary to fully explain their selectivity as susceptibility to s-triazines can be correlated with the content of unchanged triazine available for action at the sensitive sites [167, 200, 228, 423].

Soon after its discovery in plants, hydrolysis of chlorotriazines was related to cyclic hydroxamates like DIMBOA which catalyze hydrolysis nonenzymatically. However, the degree of tolerance to chlorotriazines did not coincide with the content of these compounds. Cyclic hydroxamates were found not only in highly resistant and susceptible corn lines [424] but also in susceptible wheat [195] and rye [192, 193], but not in resistant sorghum or susceptible oats and barley [194]. A corn mutant with a tenfold decreased DIMBOA content showed an insignificantly reduced tolerance although only a little hydroxysimazine was formed when homogenates were incubated with simazine [425]. Thus, primary hydrolysis of chlorotriazines represents an essential degradation mechanism but contributes to selectivity apparently only in those plants that contain cyclic hydroxamates catalyzing this reaction. This reaction is quite efficient when a high DIMBOA content is present in roots and shoots in conjunction with a preferential uptake of chlorotriazines by the roots. Corn preemergently treated with atrazine serves as an example. Insufficient information is available at present on the mechanisms involved in the hydrolysis of methoxy- and methylthio-triazines. Compounds similar to DIMBOA show only a very restricted activity on these compounds.

Intact chlorotriazine that escapes hydrolysis in the roots and is translocated to the leaves, or is directly absorbed by the leaves, can be detoxified by means of conjugation with glutathione [227, 230] , a reaction catalyzed by the enzyme glutathione-S-transferase. Unlike the benzoxazinone system, this enzyme is active only in resistant plants like sorghum, corn, sugarcane, and some grasses. In contrast, susceptible plants like oats, wheat, and pea are devoid of the enzyme [232] or, like barley, have only slight enzyme activity [233] . Since the first conjugation step with glutathione results in the complete detoxication of chlorotriazines, the further conversion of the conjugate is of no relevance to selectivity. Furthermore, since the enzymatic process is efficiently operative only with intact chlorotriazines, plants confronted with methoxy-, methylthio-, and monodealkylated parent s-triazines must rely on the hydrolytic and oxidative degradation reactions already mentioned, to detoxify these compounds effectively and to maintain sufficient tolerance.

Such oxidative degradation reactions as N-dealkylation play a significant role in s-triazine selectivity, especially in intermediately susceptible plants like pea and cotton [116, 210, 235] . This reaction also contributes appreciably to selectivity in resistant plants such as sorghum, corn [199] , and sugarcane [218, 236] by providing transient metabolites that are less phytotoxic than the parent compounds and are subject to further degradation. Insufficiently rapid degradation of these monodealkylated parent compounds in a given plant results in their temporary accumulation, which allows that plant to have only an intermediate tolerance.

Selective action of s-triazines toward plants is thus primarily determined by the pathway and the rate of detoxication of these compounds in a given plant. Hydrolytic and conjugation processes allow resistant plants to detoxify the s-triazines rapidly, whereas plants of moderate susceptibility degrade these compounds more slowly to intermediates which still possess partial phytocidal activity. Compared to these metabolic factors, soil adsorption and translocation phenomena are of secondary importance for selectivity.

IV.   CONCLUSIONS

In this chapter the physical and chemical properties of the s-triazines have been presented. The biochemical mechanisms in different living organisms that degrade and detoxify this important class of herbicides have also been discussed. Analysis shows that both chemical and biochemical reactions occur at similar sites of the triazine molecule. Only a small number of biochemical reactions is utilized by nature to achieve detoxication and/or to

render these compounds sufficiently water soluble for excretion. Three of
these reactions govern the metabolic behavior of s-triazines to a large
extent: (a) replacement of the substituent of C-2 by a hydroxy group (hydrolysis), (b) replacement by peptides and amino acids (conjugation), and
(c) removal of the side chains (N-dealkylation). A number of supplementary
reactions, such as side-chain oxidation and glucoside formation, aid in preventing these compounds from reaching sensitive sites in plants. In animals,
for example, formation of hydroxy or carboxy groups on branched side
chains aids in the excretion of the applied products.

In most cases, degradation by mammals stops at the stage of 2-hydroxy-
4-amino-6-alkylamino derivatives which are completely excreted even when
they are directly applied to animals. This was confirmed with hydroxy
parent compounds that partially undergo monodealkylation and are excreted
either unchanged or in the form of 2-hydroxy-4-amino-6-alkylamino derivatives. As a consequence of efficient excretion, no organ-specific retention
or accumulation of s-triazines or their metabolites has been observed in
mammals. Final cleavage of the ring during s-triazine degradation in soil
and plants is a slow process which presumably proceeds through ammeline.
Once this stage is reached the triazine ring can be cleaved by the biological
systems investigated to date.

Except for the conjugation mechanism which has not yet been identified
in soil or isolated microorganisms, the three major degradation pathways
are utilized to various degrees by all biological organisms so far investigated. Hydrolysis and conjugation result in a complete loss of phytocidal
properties. The ability of plants to achieve these two reactions is the basis
of their resistance to the s-triazines. Plants that use N-dealkylation instead
of the other two mechanisms are of intermediate tolerance, since this reaction results only in a partial loss of phytotoxicity. Removal of both alkyl
side chains with the C-2 position still intact produces intermediates that no
longer possess phytocidal activity.

The three major degradation reactions account for further decreases
in the already low mammalian toxicity of the s-triazines. The vast majority
of parent compounds have LD_{50} values in the range of 2000 to more than
5000 mg/kg body wt (rat, oral application). Introduction of a hydroxy group
in the C-2 position or complete dealkylation of chlorotriazines yields 2-
hydroxy-4-amino derivatives as well as 2-chloro-4,6-diamino-s-triazine
which have LD_{50} values of more than 5000 mg/kg. Monodealkylation of
parent compounds results in a small increase in the LD_{50} values but the
compounds formed are transient metabolites, which are subject to further
degradation and, consequently, to further detoxication.

Our understanding of the enzymatic mechanisms responsible for s-
triazine degradation and detoxication increased with the successful isolation
and characterization of the plant enzyme catalyzing the conjugation of

chlorotriazines with glutathione. This reaction was found to be important to detoxication and plant selectivity. Hopefully, further investigations on such basic enzymatic processes in plants, soil, and animals will be undertaken. Future findings of this nature will increase our understanding of the biochemical processes and pathways as they relate to the numerous facts reported on the fate and behavior of s-triazines in approximately 2000 publications.

V. ADDENDUM

Several papers which substantiate or extend the information on certain topics presented in Sec. II have been published during the period between preparation and printing of the book.

A. Degradation in Plants

Final identification of several intermediates of the glutathione conjugation pathway used by tolerant crops to detoxify s-triazines was achieved with atrazine in sorghum. The S-cysteine conjugate [see Eq. (16)] was identified by derivatization techniques and mass spectrometry. The structure of the N-cysteine conjugate, its chemical transformation product, which was isolated in the form of the disulfide dimer, was also established. The structure of the lanthionine derivative, the last member of this metabolic sequence currently known, was determined by mass spectrometry after this metabolite had been transformed by Raney nickel reduction or esterification. The lanthionine metabolite apparently is not a direct precursor of the insoluble residue since, after its injection into sorghum seedlings, water-soluble transformation products were mainly obtained [434, 435].

The properties and behavior of glutathione S-transferase, the enzyme starting the glutathione detoxication pathway in plants and animals, have been reviewed [436]. N-Dealkylation of atrazine, though not a major pathway, was also operative in sorghum since the 2-chloro-4,6-diamino-, 2-hydroxy-4,6-diamino-, and 2-hydroxy-4-isopropylamino-6-amino-s-triazine derivatives were identified in this plant [434, 435]. The root microflora was shown to be responsible for efficient N-dealkylation of simazine in citrus hydroponic cultures yielding 2-chloro-4-ethylamino-6-amino- and 2-chloro-4,6-diamino-s-triazine derivatives. No degradation was observed in the absence of the roots [437].

B. Degradation in Soil

Liberation of substantial amounts of $^{14}CO_2$ from ^{14}C-ring-labeled s-triazines in long-term studies was confirmed with atrazine in a sandy loam.

After a lag phase of about one month, about 18% of the incubated label was liberated as $^{14}CO_2$ within 18 months. Under saturated conditions liberation of $^{14}CO_2$ was apparently suppressed since not more than 1.5% $^{14}CO_2$ was measured over the same period. Very efficient cleavage of the s-triazine ring was observed in these studies when the known metabolites 2-chloro-4,6-diamino-s-triazine and cyanuric acid were incubated. Up to 40% $^{14}CO_2$ within six months and more than 90% within one month were liberated, respectively [439].

C. Metabolism in Animals

N-Dealkylation was found as the predominant degradation reaction of prometone in the rat. 2-Methoxy-4,6-diamino-s-triazine and ammeline represented about 11 and 31%, respectively, of the 92% of radioactivity excreted with the urine. Replacement of the methoxy group by a hydroxy group and oxidation of the isopropyl groups to yield alcohol and carboxy derivatives gave rise to various minor metabolites [440].

Use of all known major degradation reactions resulted in a complex pattern of metabolites in rats fed ^{14}C-ring-labeled cyanazine. N-Deethylation and conjugation to yield 2-mercapturic acid derivatives were the favored mechanisms used to form urinary metabolites. In contrast, in feces 2-hydroxy-6-carboxylic acid derivatives were found as the major excretion products, indicating intensive hydrolysis of the cyano group [441].

N-Dealkylation and glutathione conjugation of atrazine and some of its metabolites were studied in detail in vitro with different rat liver fractions. N-Dealkylation was found to reside in the microsomal fraction and to require the presence of NADPH and oxygen. In contrast to plants and soil, the isopropyl group was more susceptible to this reaction than the ethyl group. 2-Hydroxy derivatives reacted only to a small extent. Generally, the first N-dealkylation step proceeded at a higher rate than the second one.

Conjugation of atrazine and its 2-chloro derivatives was shown to occur in the cytosol (100,000 × g supernatant). Both the monodealkylated derivatives and the 2-chloro-4,6-diamino derivatives were conjugated to a similar order of magnitude as was atrazine. When the atrazine glutathione conjugate was injected intravenously into rats, neither unchanged conjugate nor 2-hydroxyatrazine was found in the urine. The excreted radioactivity behaved electrophoretically like the corresponding mercapturic acid, thus indicating that the glutathione conjugates are easily accessible to further metabolic attack [442].

REFERENCES

1. H. Gysin and E. Knüsli (to J. R. Geigy S.A.), Swiss Pats. 329,277;
   342,784; 342,785 (1954); U.S. Pat. 2,891,855 (1955).

2. H. Gysin and E. Knüsli (to J. R. Geigy S.A.), Swiss Pats. 337,019;
   340,372; 340,373 (1955); U.S. Pat. 2,909,420 (1955).

3. A. Gast, E. Knüsli, and H. Gysin, Experientia, 11, 107 (1955).

4. A. Gast, E. Knüsli, and H. Gysin, Experientia, 12, 146 (1956).

5. W. Schwarze (to Degussa), Ger. Pat. 1,187,241 (1958).

6. E. Nikles and L. Ebner (to Ciba), Swiss Pat. 480,793; BE 656,233;
   NE 6,413,689 (1963).

7. D. S. Acker (to du Pont de Nemours), U.S. Pats. 3,185,561 (1957);
   3,267,099 (1965).

8. H. Gysin and E. Knüsli (to J. R. Geigy S.A.), Swiss Pat. 409,516;
   Fr. Pat. 1,239,782 (1958).

9. Anon., Chem. Eng. News, 41, 33 (1963).

10. R. P. Neighbors and L. V. Phillips (to Gulf), U.S. Pat. 3,451,802
    (1966).

11. W. Schwarze (to Degussa), U.S. Pat. 3,505,325 (1967).

12. E. Knüsli, Phytiat.-Phytopharm., 7, 81 (1958).

13. H. Gysin and E. Knüsli, Adv. Pest Control Res., 3, 289 (1960).

14. H. Gysin, Weeds, 8, 541 (1960).

15. A. S. Crafts, The Chemistry and Mode of Action of Herbicides, Wiley
    (Interscience), New York, 1961.

16. H. Gysin and E. Knüsli, Proc. Brit. Weed Control Conf., 1960, 1.

17. H. Gysin, Chem. Ind. (London), 1962, 1392.

18. L. J. Audus (ed.), The Physiology and Biochemistry of Herbicides,
    Academic Press, New York, 1964.

19. E. Knüsli, Proc. 7th Brit. Weed Control Conf., 1964, 287.

20. R. H. Shimabukuro, G. L. Lamoureux, D. S. Frear, and J. E. Bakke,
    in Pesticide Terminal Residues, IUPAC Int. Symp. Tel Aviv, 1971
    (A. S. Tahori, ed.), Butterworths, London, 1971, p. 323.

21. E. Knüsli, Analytical Methods for Pesticides, Plant Growth Regulators
    and Food Additives, Vol. 4 (G. Zweig, ed.), Academic Press, New
    York, 1964, pp. 13, 27, 33, 171, 179, 187, 213.

22. R. Delley, K. Friedrich, B. Karlhuber, G. Székely, and K. Stammbach, Z. Anal. Chem., 228, 23 (1967).

23. A. M. Mattson, R. A. Kahrs, and R. T. Murphy, Residue Rev., 32, 371 (1970).

24. E. M. Smolin and L. Rapoport, Chemistry of Heterocyclic Compounds, Vol. 13, Wiley (Interscience), New York, 1959, (a) p. 293; (b) p. 301; (c) p. 285; (d) p. 351; (e) pp. 44, 45, 53, 77, 78, 102, 162, 163, 206, 221, 328, 399, 401, 407.

25. H. R. Allcook, Heteroatom Ring Systems and Polymers, Academic Press, New York, 1967.

26. H. Gysin and E. Knüsli, Adv. Pest Control Res., 3, 289 (1960).

27. G. W. Bailey and J. L. White, Residue Rev., 10, 97 (1965).

28. T. M. Ward and J. B. Weber, Spectrochim. Acta, 25A, 1167 (1969).

29. E. N. Boitsov, A. I. Finkel'Shtein, and V. A. Petukhov, Optics Spectrosc., 7, 151 (1962).

30. G. Morimoto, J. Chem. Soc. Japan, 87, 790 (1966).

31. J. B. Weber, Spectrochim. Acta, 23A, 458 (1967).

32. Ciba-Geigy, unpublished work, 1971.

33. T. M. Ward and J. B. Weber, J. Agr. Food Chem., 16, 959 (1968).

34. E. Knüsli, W. Schäppi, and D. Berrer (to J. R. Geigy S.A.), Swiss Pat. 393,344; U.S. Pat. 3,145,208 (1961).

35. H. Schrameck and S. Stäubli (to J. R. Geigy S.A.), Swiss Pat. 396,021 (1961).

36. W. Kloetzer, Monatsh., 93, 1055 (1962).

37. A. Staehelin and A. Hüni (to Ciba), Swiss Pat. 395,116 (1960).

38. E. Knüsli, W. Stammbach, and H. Gysin (to J. R. Geigy S.A.), Swiss Pat. 376,926; U.S. Pat. 3,122,541 (1959).

39. E. Knüsli (to J. R. Geigy S.A.), Swiss Pat. 389,633 (1960).

40. H. Schulz and W. Schwarze (to Degussa), Ger. Pat. 1,172,684 (1962).

41. Ciba-Geigy, unpublished work, 1957.

42. H. D. Skipper, R. Frech, and V. V. Volk, J. Agr. Food Chem., in press; Ph.D. thesis, Oregon State University, Corvallis, 1970.

43. P. Castelfranco and B. Deutsch, Weeds, 10, 244 (1962).

44. Ciba-Geigy, unpublished work, 1962.

45. I. J. Zemanek, Rostlinna Vyroba, 10, 959 (1964).

46. C. I. Harris, J. Agr. Food Chem., 15, 157 (1967).

47. E. Knüsli and W. Stammbach (to J. R. Geigy S.A.), Swiss Pat. 407,138 (1962).

48. Ciba-Geigy, unpublished work, 1956-1961.

49. Ciba-Geigy, unpublished work, 1959-1961.

50. Ciba-Geigy, unpublished work, 1962-1964.

51. Ciba-Geigy, unpublished work, 1959-1963.

52. Ciba-Geigy, unpublished work, 1963-1964.

53. E. Knüsli, I. Rumpf, and G. A. Klein (to J. R. Geigy S.A.), Swiss Pats. 380,733; 393,829; U.S. Pats. 3,195,998; 3,224,712 (1960).

54. E. Knüsli, I. Rumpf, and H. Gysin (to J. R. Geigy S.A.), Swiss Pats. 394,703; 381,699; U.S. Pats. 3,218,148; 3,236,846 (1958).

55. H. Cordes, A. Fischer, and H. Stummeyer (to BASF), DAS 1,184,148 (1963).

56. Degussa, BE 663,328 (1964).

57. H. Huemer (to Degussa), DAS 1,138,978 (1960).

58. Degussa, BE 665,127 (1964).

59. W. Schwarze (to Degussa), Ger. Pat. 1,445,663 (1963).

60. J. R. Plimmer, P. C. Kearney, and U. I. Klingebiel, J. Agr. Food Chem., 19, 572 (1971).

61. Z. D. Tadic and S. K. Ries, J. Agr. Food Chem., 19, 46 (1971).

62. H. Gold (to Bayer), Ger. Pat. 1,108,209 (1958).

63. H. Gold, Angew. Chem., 72, 956 (1960).

64. C. Grundmann, Angew. Chem., 75, 39 (1963).

65. A. Kreutzberger, Angew. Chem., 77, 1086 (1965).

66. A. Kreutzberger, Arch. Pharm., 299, 897a, 984b (1966).

67. D. Wöhrle, Tetrahedron Lett., 46, 1969 (1971).

68. A. Hantsch, Berichte, 61B, 1776 (1928).

69. B. S. Joshi, R. Srinivasan, R. V. Talavdekar, and K. Venkataraman, Tetrahedron, 11, 133 (1960).

70. D. R. Osborne and R. Levine, J. Org. Chem., 28, 2933 (1939).

71. L. S. Jordan, B. E. Day, and W. A. Clerx, Weeds, 12, 5 (1964).

72. L. S. Jordan, J. D. Mann, and B. E. Day, Weeds, 13, 43 (1965).

73. B. E. Pape and M. J. Zabik, J. Agr. Food Chem., 18, 202 (1970).

74. J. R. Plimmer, P. C. Kearney, and U. I. Klingebiel, Tetrahedron Lett., 44, 3891 (1969).

75. J. R. Plimmer, P. C. Kearney, and H. Chisaka, Abstr. Meeting Weed Sci. Soc. Am., 1970, 87.

76. J. R. Plimmer, Residue Rev., 33, 64 (1970).

77. O. Diels, Berichte, 32, 691 (1899).

78. W. N. Oldham (to American Cyanamid), U.S. Pat. 2,417,659 (1944).

79. T. P. Metcalfe (to I.C.I.), U.S. Pat. 2,414,655 (1944).

80. A. E. W. Smith, R. H. Stanley, and C. W. Scaife (to American Cyanamid), U.S. Pat. 2,416,656 (1944).

81. I. G. Farben-Industrie, Fiat Fin. Rep. 1313, 1, 249 (1948).

82. A. J. Cofrancesco (to General Aniline & Film Corp.), U.S. Pat. 2,692,880 (1951).

83. A.-J. Courtier and H. Jean (to Comp. de Saint-Gobain), Fr. Pat. 1,251,359 (1959).

84. M. Teysseire (to Lonza), Swiss Pats. 337,539; 342,231 (1956).

85. J. T. Thurston (to American Cyanamid), U.S. Pat. 2,491,459 (1945).

86. H. Huemer (to Degussa), Ger. Pat. 833,490 (1949).

87. H. Huemer and H. Schulz (to Degussa), Ger. Pat. 842,067 (1950).

88. H. Schulz and H. Huenes, Review in Degussa Festschrift "ACS Forschung und Produktion," Frankfurt am Main, 1953.

89. A. W. Hofmann, Berichte, 18, 2753 (1885).

90. I. H. Witt and C. S. Hamilton, J. Am. Chem. Soc., 67, 1078 (1945).

91. W. M. Pearlman and C. K. Banks, J. Am. Chem. Soc., 70, 3726 (1948).

92. J. T. Thurston, J. R. Dudley, D. W. Kaiser, J. Heckenbleikner, F. C. Schaefer, and D. Holm-Hansen, J. Am. Chem. Soc., 73, 2981 (1951).

93. B. S. Corton and B. L. Williams (to Monsanto), U.S. Pat. 2,770,622 (1954).

94. W. Schwarze (to Degussa), DOS 1,670,528 (1966); U.S. Pat. 3,505,325 (1967).

95. Ciba-Geigy, unpublished work, 1954.

96. D. Holm-Hansen (to American Cyanamid), U.S. Pat. 2,513,264 (1946).

97. Th. Gauer (to J. R. Geigy S. A.), Swiss Pat. 457,469 (1965); U.S. Pat. 3,376,302 (1966).

98. G. A. Saul (to J. R. Geigy S.A.), Swiss Pat. 508,640 (1967); U.S. Pat. 3,436,394 (1968).

99. J. Controulis and C. K. Banks, J. Am. Chem. Soc., 67, 1946 (1945).

100. W. M. Pearlman, J. D. Mitulski, and C. K. Banks, J. Am. Chem. Soc., 71, 3248 (1949).

101. J. R. Dudley, J. T. Thurston, F. C. Schaefer, D. Holm-Hansen, C. J. Hull, and P. Adams, J. Am. Chem. Soc., 73, 2986 (1951).

102. W. Rufener, J. Riethmann, and R. Berger (to J. R. Geigy S.A.), U.S. Pat. 3,558,622 (1968); Swiss Pat. 515,253 (1969).

103. E. Knüsli and W. Stammbach (to J. R. Geigy S.A.), Swiss Pat. 407,138 (1962).

104. C. Grundmann and A. Kreutzberger, J. Am. Chem. Soc., 76, 632 (1954).

105. C. Grundmann, H. Ulrich, and A. Kreutzberger, Berichte, 86, 181 (1953).

106. A. Hüni and A. Staehelin (to Ciba), Belg. Pat. 552,496 (1956).

107. J. H. Uhlenbrock, H. H. Haeck, I. Daams, and M. I. Kopmans, Rec. Trav. Chim. Pays-Bas, 78, 967 (1959).

108. H. Koopman and I. Daams, Rec. Trav. Chim. Pays-Bas, 79, 83 (1960).

109. W. Schwarze and W. Weigert (to Degussa), DOS 1,670,585 (1967); U.S. Pat. 3,544,569 (1968).

110. E. Nikles and L. Ebner (to Ciba), Swiss Pat. 480,793 (1963); U.S. Pat. 3,415,827 (1964).

111. D. E. Davis, H. H. Funderburk, and N. G. Sansing, Weeds, 7, 300 (1959).

112. C. L. Foy, Weeds, 12, 103 (1964).

113. H. W. Hilton and N. Nomura, Weed Res., 4, 216 (1964).

114. D. E. Davis, J. V. Gramlich, and H. H. Funderburk, Weeds, 13, 252 (1965).

115. A. G. Dexter, O. C. Burnside, and T. L. Lavy, Weeds, 14, 222 (1966).

116. R. H. Shimabukuro, J. Agr. Food Chem., 15, 557 (1967).

117. R. H. Shimabukuro and A. J. Linck, Weeds, 15, 175 (1967).

118. M. Y. Berezovskii and R. P. Radzhaeva, Weed Abstr., 16, 398 (1967).

119. D. E. Roberts, Diss. Abstr., 27, 3002-B (1967).

120. T. L. Lavy, Soil Sci. Soc. Am. Proc., 32, 377 (1968).

121. H. L. Wheeler and R. H. Hamilton, Weed Sci., 16, 7 (1968).

122. E. F. Eastin, Abstr. Meeting Weed Sci. Soc. Am., 1969, 184.

123. T. J. Muzik, Abstr. Meeting Weed Sci. Soc. Am., 1969, 16.

124. J. F. Peacock and C. D. Dybing, Proc. North Central Weed Control Conf., 24, 80 (1969).

125. J. W. Schrader, Thesis, Michigan State University, East Lansing, 1970.

126. L. Thompson, F. W. Slife, and H. S. Butler, Weed Sci., 18, 509 (1970).

127. A. Wallace, R. T. Mueller, and A. M. El Gazzar, Agron. J., 62, 373 (1970).

128. W. D. Reynolds and J. A. Tweedy, Weed Sci., 18, 270 (1970).

129. D. Penner, Proc. North Central Weed Control Conf., 25, 72 (1970).

130. F. W. Roeth and T. L. Lavy, Weed Sci., 19, 93 (1971).

131. F. W. Roeth and T. L. Lavy, Weed Sci., 19, 98 (1971).

132. M. G. T. Shone and A. V. Wood, J. Exp. Bot., 23, 1 (1972).

133. A. Walker, Pestic. Sci., 3, 139 (1972).

134. C. L. Foy, Plant Physiol., 37, 25 Suppl. (1962).

135. P. K. Biswas, Weeds, 12, 31 (1964).

136. F. W. Freeman, D. P. White, and M. J. Bukovac, Forest Sci., 10, 330 (1964).

137. W. H. Minshall, Abstr. Meeting Weed Sci. Soc. Am., 1964, 80.

138. O. A. Leonard, L. A. Lider, and R. K. Glenn, Weed Res., 6, 37 (1966).

139. S. D. Hocombe, Weed Res., 8, 68 (1968).

140. R. C. Barba and R. R. Romanovski, Weed Res., 9, 114 (1969).

141. H. W. Hilton, Q. H. Yuen, and N. S. Nomura, J. Agr. Food Chem., 18, 217 (1970).

142. L. S. Jordan, Abstr. Meeting Weed Sci. Soc. Am., 1970, 62.

143. K. Moody, C. A. Kust, and K. P. Bucholtz, Weed Sci., 18, 214 (1970).

144. K. Moody, C. A. Kust, and K. P. Bucholtz, Weed Sci., 18, 642 (1970).

145. G. Rieder, K. P. Buchholtz, and C. A. Kust, Weed Sci., 18, 101 (1970).

146. H. J. Vostral, K. P. Buchholtz, and C. A. Kust, Weed Sci., 18, 115 (1970).

147. C. L. Foy and T. Bisalputra, Plant Physiol., 39, 68 Suppl. (1964).

148. H. C. Sikka and D. E. Davis, Weed Sci., 16, 474 (1968).

149. Y. Eshel, Weed Sci., 17, 492 (1969).

150. L. Taimr, H. Cervinkova, and E. Bergmannova, Commun. Inst. For. Cech., 6, 143 (1969).

151. J. N. Sing, E. Basler, and P. W. Santelmann, Pestic. Biochem. Physiol., 2, 143 (1972).

152. H. H. Funderburk and J. M. Lawrence, Weeds, 11, 217 (1963).

153. H. H. Funderburk and J. M. Lawrence, Weed Res., 3, 304 (1963).

154. L. M. Wax and R. Behrens, Weeds, 13, 107 (1965).

155. O. E. Strand, Thesis, University of Minnesota, Minneapolis, 1969.

156. D. L. Sutton and S. W. Bingham, Weed Sci., 17, 431 (1969).

157. L. Thompson and F. W. Slife, Weed Sci., 17, 251 (1969).

158. D. G. Swan, Res. Prog. Rep. West. Weed Control Conf., Feb., 104 (1969).

159. L. Thompson and F. W. Slife, Weed Sci., 18, 349 (1970).

160. A. B. Rogerson and S. W. Bingham, Weed Sci., 19, 325 (1971).

161. L. Thompson, J. M. Houghton, F. W. Slife, and H. S. Butler, Weed Sci., 19, 409 (1971).

162. W. Roth, C. R. Séances Acad. Sci., 245, 942 (1957).

163. C. L. Foy and P. Castelfranco, Plant Physiol., 35, 28 Suppl. (1960).

164. M. Montgomery and V. H. Freed, Weeds, 9, 231 (1961).

165. T. J. Sheets, Weeds, 9, 1 (1961).

166. P. H. Plaisted and D. P. Ryskiewich, Plant Physiol., 37, 25 Suppl. (1962).

167. N. S. Negi, H. H. Funderburk, and D. E. Davis, Weeds, 12, 53 (1964).

168. P. H. Plaisted and M. L. Thornton, Contrib. Boyce Thompson Inst., 22, 399 (1964).

169. P. H. Plaisted, D. P. Ryskiewich, and C. D. Ercegovich, Plant Physiol., 39, 68 Suppl. (1964).

170. R. H. Hamilton and D. E. Moreland, Weeds, 11, 213 (1963).

171. S. K. Uhlig, Phytopathol. Z., 50, 289 (1964).

172. P. K. Biswas and D. D. Hemphill, Nature, 207, 215 (1965).

173. H. W. Hilton, Residue Rev., 15, 1 (1966).

174. P. S. Dhillon, W. R. Byrnes, and C. Merritt, Weed Sci., 16, 374 (1968).

175. J. R. Wichman and W. R. Byrnes, Abstr. Meeting Weed Sci. Soc. Am., 1970, 33.

176. D. C. Whitenberg, Weeds, 13, 68 (1965).

177. M. Montgomery and V. H. Freed, Res. Prog. Rep. West. Weed Control Conf., 1960, 71.

178. P. Castelfranco, C. L. Foy, and D. B. Deutsch, Weeds, 9, 580 (1961).

179. V. H. Freed, M. Montgomery, and M. Kief, Proc. Northeast. Weed Control Conf., 15, 6 (1961).

180. C. L. Foy, Weed Abstr., 12, 44 (1963).

181. W. Roth and E. Knüsli, Experientia, 17, 312 (1961).

182. P. Castelfranco and M. S. Brown, Weeds, 10, 131 (1962).

183. R. H. Hamilton and D. E. Moreland, Science, 135, 373 (1962).

184. R. N. Andersen, Weeds, 12, 60 (1964).

185. C. L. Foy, Plant Physiol., 42, 51 Suppl. (1967).

186. R. S. Loomis, S. E. Beck, and J. F. Stauffer, Plant Physiol., 32, 379 (1957).

187. A. I. Virtanen, P. K. Hietala, and Oe. Wahlroos, Arch. Biochem. Biophys., 69, 486 (1957).

188. Oe. Wahlroos and A. I. Virtanen, Acta Chem. Scand., 13, 1906 (1959).

189. C. L. Tipton, J. A. Klun, R. R. Husted, and M. D. Pierson, Biochemistry, 6, 2866 (1967).

190. J. Hofman and O. Hofmanova, Eur. J. Biochem., 8, 109 (1969).

191. C. L. Tipton, R. R. Husted, and F. H. C. Tsao, J. Agr. Food Chem., 19, 484 (1971).

192. A. I. Virtanen and P. K. Hietala, Acta Chem. Scand., 14, 499 (1960).

193. P. K. Hietala and A. I. Virtanen, Acta Chem. Scand., 14, 502 (1960).

194. R. H. Hamilton, J. Agr. Food Chem., 12, 14 (1964).

195. H. H. Funderburk, Jr., and D. E. Davis, Weeds, 11, 101 (1963).

196. M. Montgomery and V. H. Freed, J. Agr. Food Chem., 12, 11 (1964).

197. D. R. Roberts, D. E. Davis, and H. H. Funderburk, Abstr. Meeting Weed Sci. Soc. Am., 1964, 71.

198. R. H. Shimabukuro, Plant Physiol., 42, 1269 (1967).

199. R. H. Shimabukuro, Plant Physiol., 43, 1925 (1968).

200. R. H. Shimabukuro and H. R. Swanson, J. Agr. Food Chem., 17, 199 (1969).

201. R. H. Shimabukuro, H. R. Swanson, and W. C. Walsh, Plant Physiol., 46, 103 (1970).

202. R. S. Schroeder, Abstr. 159th Meeting Am. Chem. Soc., 1970.

203. Ciba-Geigy, unpublished work, 1970, 1971.

204. K. Lund-Höie, Weed Res., 9, 142 (1969).

205. J. Hurter, Experientia, 22, 741 (1966).

206. J. Hurter, Thesis Nr. 4038, ETH Zürich, Switzerland, 1967.

207. J. Hurter, Abstr. 6th Int. Congress of Plant Protection, 1967, 398.

208. L. Thompson, Jr., Weed Sci., 20, 153 (1972).

209. R. H. Shimabukuro and H. R. Swanson, Abstr. 155th Meeting Am. Chem. Soc., 1968, A-24.

210. R. H. Shimabukuro and H. R. Swanson, Weed Sci., 18, 231 (1970).

211. T. Chapman, D. Jordan, D. H. Payne, W. J. Hughes, and R. H. Schieferstein, Proc. 9th Brit. Weed Control Conf., 2, 1018 (1968).

212. K. I. Beynon, G. Stoydin, and A. N. Wright, Pestic. Sci., 3, 293 (1972).

213. K. I. Beynon, G. Stoydin, and A. N. Wright, Pestic. Sci., 3, 379 (1972).

214. K. I. Beynon, G. Stoydin, and A. N. Wright, Pestic. Biochem. Physiol., 2, 153 (1972).

215. P. W. Müller, Biochem. J., 101, 1 P (1966).

216. P. W. Müller and P. H. Payot, Isotopes in Weed Research, Int. Atomic Energy Agency, Vienna, 1966, p. 61.

217. Ciba-Geigy, unpublished work, 1968-1970.

218. Ciba-Geigy, unpublished work, 1971, 1972.

219. Ciba-Geigy, unpublished work, 1971.

220. Ciba-Geigy, unpublished work, 1968-1970.

221. H. Gysin, Chem. Ind. (London), 1962, 1393.

222. E. Bergmannova, L. Taimr, M. Kutacek, and A. Kudelova, III Konferenz für Pflanzenschutz, Prague, 1970, p. 246.

223. E. Knüsli, Proc. 7th Brit. Weed Control Conf., 1, 287 (1964).

224. K. I. Beynon and A. N. Wright, J. Sci. Food Agr., 20, 21 (1969).

225. G. L. Lamoureux, R. H. Shimabukuro, H. R. Swanson, and D. S. Frear, Abstr. 157th Meeting Am. Chem. Soc., 1969, AGFD-16.

226. R. H. Shimabukuro and H. R. Swanson, Abstr. Meeting Weed Sci. Soc. Am., 1969, 197.

227. G. L. Lamoureux, R. H. Shimabukuro, H. R. Swanson, and D. S. Frear, J. Agr. Food Chem., 18, 81 (1970).

228. R. H. Shimabukuro, D. S. Frear, H. R. Swanson, and W. C. Walsh, Plant Physiol., 47, 10 (1971).

229. Ciba-Geigy, unpublished work, 1964-1971.

230. G. L. Lamoureux, L. E. Stafford, R. H. Shimabukuro, and R. G. Zaylskie, J. Agr. Food Chem., 21, 1020 (1973).

231. R. H. Hodgson, Abstr. Meeting Weed Sci. Soc. Am., 1970, 15.

232. D. S. Frear and H. R. Swanson, Phytochemistry, 9, 2123 (1970).

233. G. L. Lamoureux, L. E. Stafford, and R. H. Shimabukuro, Abstr. 163rd Meeting Am. Chem. Soc., 1972, Pest 36.

234. D. P. Schultz and B. G. Tweedy, Weed Sci., 19, 133 (1971).

235. R. H. Shimabukuro, R. E. Kadunce, and D. S. Frear, J. Agr. Food Chem., 14, 392 (1966).

236. K. P. Goshwami, Thesis, University of Hawaii, Honolulu, May 1972.

237. M. L. Montgomery, D. L. Botsford, and V. H. Freed, J. Agr. Food Chem., 17, 1241 (1969).

238. Ciba-Geigy, unpublished work, 1964, 1969-1972.

239. G. Dupuis, T. Laanio, P. Marbach, and H. O. Esser, Abstr. 7th Int. Congress of Plant Protection, 1970, 711.

240. D. E. Davis, H. H. Funderburk, and N. G. Sansing, Proc. South. Weed Conf., 12, 172 (1959).

241. M. T. H. Ragab and J. P. McCollum, Weeds, 9, 72 (1961).

242. C. I. Harris, D. D. Kaufman, T. J. Sheets, R. G. Nash, and P. C. Kearney, Adv. Pest Control Res., 8, 1 (1968).

243. The Triazine Herbicides, Residue Rev. (F. Gunther, ed.), 32, 93, 267 (1970).

244. H. O. Esser, Meded. Rijks Fakult. Landbouwetensch. Gent, 35, 753 (1970).

245. W. B. Duke, Thesis, Oregon State University, Corvallis, 1964.

246. R. E. Talbert and O. H. Fletchall, Weeds, 12, 33 (1964).

247. L. L. McCormick and A. E. Hiltbold, Weeds, 14, 77 (1966).

248. L. C. Liu, H. R. Cibes-Viadé, and J. Gonzalez-Ibanez, J. Agr. Univ. Puerto Rico, 54, 631 (1970).

249. J. P. Martin and J. O. Ervin, Proc. Calif. Weed Conf., 22, 83 (1970).

250. O. C. Burnside, E. L. Schmidt, and R. Behrens, Weeds, 9, 477 (1961).

251. G. W. McClure, Contrib. Boyce Thompson Inst., 24, 235 (1970).

252. J. Guillemat, Compt. Rend., 250, 1343 (1960).

253. J. Guillemat, M. Charpentier, P. Tardieux, and J. Pochon, Ann. Epiphyties, 11, 261 (1960).

254. M. Charpentier and J. Pochon, Ann. Inst. Pasteur, 102, 501 (1962).

255. D. D. Kaufman and P. C. Kearney, Abstr. Meeting Weed Sci. Soc. Am., 1964, 12.

256. Gy. Pantos, P. Gyurko, T. Takacs, and L. Varga, Acta Agron. Acad. Sci. Hung., 13, 21 (1964).

257. F. H. Farmer, R. E. Benoit, and W. E. Chappell, Proc. Northeast. Weed Control Conf., 19, 350 (1965).

258. H. Bortels, E. Fricke, and R. Schneider, Nachrichtenbl. Dtsch. Pflanzenschutzd. (Braunschweig), 19, 101 (1967).

259. D. D. Kaufman, P. C. Kearney, and T. J. Sheets, J. Agr. Food Chem., 13, 238 (1965).

260. A. V. Manorik, V. F. Vasil'cenko, N. M. Mandrovskaja, and S. M. Malicenko, Agrochim. Moskva, 4, 123 (1968).

261. G. Voinova and D. Bakalivanov, Meded. Rijks Fakult. Landbouwetensch. Gent, 35, 839 (1970).

262. W. L. Rieck, 60th Ann. Meeting Am. Soc. Agron., Nov. 10-15, 1968.

263. D. S. Murray, W. L. Rieck, and J. Q. Lynd, Appl. Microbiol., 19, 11 (1970).

264. D. E. Armstrong, G. Chesters, and R. F. Harris, Soil Sci. Soc. Am. Proc., 31, 61 (1967).

265. C. I. Harris, J. Agr. Food Chem., 15, 157 (1967).

266. H. D. Skipper, C. M. Gilmour, and W. R. Furtick, Soil Sci. Soc. Am. Proc., 31, 653 (1967).

267. O. Agundis, Thesis, University of Minnesota, Minneapolis, 1966.

268. S. R. Obien and R. E. Green, Weed Sci., 17, 509 (1969).

269. J. D. Russell, M. Cruz, J. L. White, G. W. Bailey, W. R. Payne, J. D. Pope, and J. I. Teasley, Science, 160, 1340 (1968).

270. M. Cruz, J. L. White, and J. D. Russell, Israel J. Chem., 6, 315 (1968).

271. C. B. Brown and J. L. White, Agron. Abstr., Nov. 1968, 89.

272. N. P. Agnihotri, Diss. Abstr. Int., 31, No. 12, Pt. 1, 7042-B (1971).

273. R. J. Hance, J. Sci. Food Agr., 20, 144 (1969).

274. R. L. Zimdahl, V. H. Freed, M. L. Montgomery, and W. R. Furtick, Weed Res., 10, 18 (1970).

275. C. I. Harris, Weed Res., 5, 275 (1965).

276. V. E. Leiniger, Diss. Abstr. Int., 30, 3943-3944B (1970).

277. K. P. Goswami and R. E. Green, Environ. Sci. Technol., 5, 426 (1971).

278. K. A. Ramsteiner, W. Hörmann, and D. Eberle, Z. Pflanzenkrankheiten, Sonderheft, 6, 43 (1972).

279. H. D. Skipper and V. V. Volk, Weed Sci., 20, 344 (1972).

280. P. C. Kearney and J. R. Plimmer, Abstr. 158th Meeting Am. Chem. Soc., 1970, Pest 30.

281. J. R. Plimmer, P. C. Kearney, and H. Chisaka, Abstr. Meeting Weed Sci. Soc. Am., 1970, 87.

282. R. W. Couch, J. V. Gramlich, D. E. Davis, and H. H. Funderburk, Proc. South. Weed Conf., 18, 623 (1965).

283. P. C. Kearney, D. D. Kaufman, and T. J. Sheets, J. Agr. Food Chem., 13, 369 (1965).

284. T. J. Sheets, A. S. Crafts, and H. R. Drever, J. Agr. Food Chem., 10, 458 (1962).

285. T. T. Kozlowski, Nature, 205, 104 (1965).

286. D. D. Kaufman, P. C. Kearney, and T. J. Sheets, Science, 142, 405 (1963).

287. D. D. Kaufman and J. Blake, Abstr. Meeting Weed Sci. Soc. Am., 1969, 230.

288. F. W. Roeth, T. L. Lavy, and O. C. Burnside, Weed Sci., 17, 202 (1969).

289. K. A. Ramsteiner, W. D. Hörmann, and D. O. Eberle, Meded. Rijks Fakult. Landbouwetensch. Gent, 36, 1119 (1971).

290. Ciba-Geigy, unpublished work, 1969, 1971.

291. L. E. Wise and E. H. Walters, J. Agr. Res., 10, 85 (1917).

292. E. H. Walters and L. E. Wise, J. Am. Chem. Soc., 39, 2472 (1917).

293. Ichikawa Chikabumi, J. Agr. Chem. Soc. Japan, 12, 898 (1936).

294. I. C. MacRae and M. Alexander, J. Agr. Food Chem., 13, 72 (1965).

295. G. H. Wagner and K. S. Chahal, Soil Sci. Soc. Am. Proc., 30, 752 (1966).

296. S. R. Obien, Thesis, University of Hawaii, Honolulu, 1970.

297. A. Süss, private communication, Bayer. Landesanstalt für Boden-kultur, Pflanzenbau und -schutz, Munich, 1967.

298. A. Süss, C. Eben, and H. Siegmund, Z. Pflanzenkrankheiten, Sonder-heft, 6, 43 (1972).

299. R. J. Hance and G. Chesters, Soil Biol. Biochem., 1, 309 (1969).

300. R. D. Hauck and H. F. Stephenson, J. Agr. Food Chem., 12, 147 (1964).

301. G. L. Terman, J. D. DeMent, C. M. Hunt, J. T. Cope, and L. E. Ensminger, J. Agr. Food Chem., 12, 151 (1964).

302. H. L. Jensen and A. S. Abdel-Ghaffar, Arch. Mikrobiol., 67, 1 (1969).

303. L. E. St. John, D. E. Wagner, and D. J. Lisk, J. Dairy Sci., 47, 1267 (1964).

304. L. E. St. John, J. W. Ammering, D. G. Wagner, R. G. Warner, and D. J. Lisk, J. Dairy Sci., 48, 502 (1965).

305. Ciba-Geigy, unpublished work, 1964-1965, 1971.

306. J. S. Bowman, Hazleton Laboratories, Palo Alto, California, Report, April 29, 1960.

307. J. E. Bakke and J. D. Robbins, Abstr. 153rd Meeting Am. Chem. Soc., 1967, A 49.

308. J. E. Bakke and J. D. Robbins, Abstr. 155th Meeting Am. Chem. Soc., 1968, A 43.

309. J. S. Bowman, Hazleton Laboratories, Palo Alto, California, Report, July 15, 1960.

310. J. E. Bakke, J. D. Larson, and C. E. Price, J. Agr. Food Chem., 20, 602 (1972).

311. Ciba-Geigy, unpublished work, 1965.

312. B. Q. Richman, J. H. Kay, and J. C. Calandra, Ind. Bio-Test Lab., Northbrook, Illinois, Report, Sept. 30, 1964.

313. J. E. Bakke, J. D. Robbins, and V. J. Feil, J. Agr. Food Chem., 15, 628 (1967).

314. J. D. Robbins, J. E. Bakke, and V. J. Feil, J. Agr. Food Chem., 16, 698 (1968).

315. D. H. Hutson, E. C. Hoadley, N. H. Griffiths, and C. Donninger, J. Agr. Food Chem., 18, 507 (1970).

316. J. D. Larson, Thesis, North Dakota State University, Fargo, 1970.

317. J. D. Larson and J. E. Bakke, North Dakota Academy of Science Annual Meeting, May 1970.

318. J. E. Bakke, J. D. Robbins, and V. J. Feil, J. Agr. Food Chem., 19, 462 (1971).

319. W. H. Oliver, G. S. Born, and P. L. Ziemer, J. Agr. Food Chem., 17, 1207 (1969).

320. Ciba-Geigy, unpublished work, 1963-1965.

321. J. Donoso and J. J. Peterson, Woodard Research Corp., Herndon, Virginia, Report, Oct. 11, 1963.

322. C. Boehme and F. Baer, Food Cosmet. Toxicol., 5, 23 (1967).

323. C. Boehme, Dissertation, Universitaet Berlin, 1965.

324. J. E. Bakke, R. H. Shimabukuro, K. L. Davison, and G. L. Lamoureux, Chemosphere, 1, 21 (1972).

325. E. Ebert and P. W. Mueller, Experientia, 24, 1 (1968).

326. E. Ebert and S. W. Dumford, Residue Rev., to be published.

327. J. L. Hilton, L. L. Jansen, and H. M. Hull, Ann. Rev. Plant Physiol., 14, 353 (1963).

328. D. E. Moreland, Ann. Rev. Plant Physiol., 18, 365 (1967).

329. A. Barth and H.-J. Michel, Pharmazie, 24, 11 (1969).

330. G. M. Cheniae, Ann. Rev. Plant Physiol., 1970, 467.

331. A. Trebst, Ber. Dtsch. Bot. Ges., 83, 373 (1970).

332. B. Exer, Experientia, 14, 135 (1958).

333. W. Roth, Experientia, 14, 137 (1958).

334. D. E. Moreland, W. A. Gentner, J. J. Hilton, and K. L. Hill, Plant Physiol., 34, 432 (1959).

335. B. Exer, Weed Res., 1, 233 (1961).

336. N. E. Good, Plant Physiol., 36, 788 (1961).

337. D. E. Moreland and K. L. Hill, Weeds, 10, 229 (1962).

338. G. Zweig, I. Tamas, and E. Greenberg, Biochim. Biophys. Acta, 66, 196 (1963).

339. Ciba-Geigy, unpublished work, 1972.

340. N. J. Bishop, Biochim. Biophys. Acta, 57, 186 (1962).

341. Ciba-Geigy, unpublished work, 1974.

342. F. M. Ashton, E. M. Gifford, and T. Bisalputra, Bot. Gaz., 124, 336 (1963).

343. E. R. Hill, E. C. Putala, and J. Vengris, Weed Sci., 16, 377 (1968).

344. K. H. Buechel, Pestic. Sci., 3, 89 (1972).

345. H. Gysin and E. Knuesli, Proc. 4th Brit. Weed Control Conf., 1, 34 (1960).

346. R. C. Brian, Chem. Ind. (London), 1965, 1955.

347. C. Hansch, Prog. Photosynth. Res., 3, 1685 (1969).

348. F. M. Ashton, Weeds, 13, 164 (1965).

349. P. B. Sweetser, C. W. Todd, and R. T. Hersh, Biochim. Biophys. Acta, 51, 509 (1961).

350. S. Izawa and N. E. Good, Biochim. Biophys. Acta, 102, 20 (1965).

351. T. J. Monaco, Ph.D. Thesis, North Carolina State University, Raleigh, 1968.

352. R. B. Park and J. Bigging, Science, 144, 1009 (1964).

353. J. A. Melnikova, Ju. A. Baskakov, et al., Report on the Third Symposium of RGW on Metabolism of Herbicides, Moscow 1967 (cited in Ref. 329).

354. C. C. Black and L. Myers, Weeds, 14, 331 (1966).

355. F. M. Ashton, G. Zweig, and G. W. Mason, Weeds, 8, 448 (1960).

356. G. Zweig and F. M. Ashton, J. Exp. Bot., 13, 5 (1962).

357. A. Gast, Experientia, 14, 134 (1958).

358. F. M. Ashton, T. Bisalputra, and E. B. Risley, Am. J. Bot., 53, 217 (1966).

359. D. Smith and K. P. Buchholtz, Science, 136, 263 (1962).

360. G. D. Wilis, D. E. Davis, and H. H. Funderburk, Jr., Weeds, 11, 253 (1963).

361. P. E. Waggoner and I. Zelitch, Science, 150, 1413 (1965).

362. C. Bartley, Agr. Chem., 12, 34, 113 (1957).

363. A. Gast, Schweiz. Z. Obst. Weinbau, 69, 203 (1960).

364. G. G. Lorenzoni, Maydica, 7, 115 (1962).

365. B. Swietochowski and S. Miklaszewski, Zesz. Nauk Wyzs. Szkoly. Poln. Wroclawiu, 15, 91 (1962).

366. S. K. Ries, R. P. Larsen, and A. L. Kenworthy, Weeds, 11, 270 (1963).

367. M. L. de Vries, Weeds, 11, 220 (1963).

368. A. Gast and H. Grob, VIIth Brit. Weed Control Conf., 1964, 217.

369. H. Karnatz, Z. Pflanzenkrh.-Pflanzenschutz, 2, 175 (1964).

370. J. R. Freney, Aust. J. Agr. Res., 16, 257 (1965).

371. R. Goren and S. P. Monselise, Weeds, 14, 141 (1966).

372. J. Mueller, Arch. Forstwes., 15, 85 (1966).

373. K. Hafner, Z. Pflanzenkrh.-Pflanzenschutz, 2, 172 (1964).

374. E. Ebert and Ch. J. Van Assche, Experientia, 25, 758 (1969).

375. E. Ebert and Ch. J. Van Assche, Abstr. 7th Int. Congr. Plant Protection, 1970, 286.

376. L. G. Copping and D. E. Davis, Weed Sci., 20, 86 (1972).

377. L. S. Jordan, T. Murashige, J. D. Mann, and B. E. Day, Weeds, 14, 134 (1966).

378. D. E. Moreland, S. S. Malhotra, R. D. Gruenhagen, and E. H. Shokrah, Weed Sci., 17, 556 (1969).

379. H. Graeser, Wiss. Z. Univ. Rostock, Mathem.-Naturwiss. Reihe, 16, 565 (1967).

380. A. Gast and H. Grob, Pestic. Technol., 3, 68 (1960).

381. H. Domanska, Combaterea Buruienelor cu Ajutorul Erbicidelor, Consfatuirea Internationala, Bucharest 1962, Weed Abstr., 14, 59 (1965).

382. S. K. Ries and A. Gast, Weeds, 13, 272 (1965).

383. V. A. Alabushev, Nauchn. Dokl. Vyssh. Shkoly, Biol. Nauk., 1965, 163, in C.A., 65, 1319 (1966).

384. P. V. Saburowa and A. A. Petunova, Dokl. Akad. Nauk SSSR, 160, 1215 (1965), in Weed Abstr., 15, 111 (1966).

385. J. V. Gramlich and D. E. Davis, Weeds, 15, 157 (1967).

386. S. K. Ries, H. Chmiel, D. R. Dilley, and P. Filner, Proc. Natl. Acad. Sci. (U.S.), 58, 256 (1967).

387. S. K. Ries, C. J. Schweizer, and H. Chmiel, Biol. Sci., 18, 205 (1968).

388. C. J. Schweizer and S. K. Ries, Science, 165, 73 (1969).

389. D. W. Allison and R. A. Peters, Agron. J., 62, 246 (1970).

390. S. K. Ries, O. Moreno, W. F. Meggitt, C. J. Schweizer, and S. K. Ashkar, Agron. J., 62, 746 (1970).

391. W. G. Monson, G. W. Burton, W. S. Wilkinson, and S. W. Dumford, Agron. J., 63, 928 (1971).

392. R. J. Fink and O. H. Fletchall, Weeds, 15, 272 (1967).

393. E. F. Eastin and D. E. Davis, Weeds, 15, 306 (1967).

394. V. B. Bagaev and L. J. Kobazeva, Dokl. TSKHA, 109, 207 (1965).

395. J. A. Tweedy and S. K. Ries, Plant Physiol., 42, 280 (1967).

396. J. Krihning, Phytopathol. Z., 53, 65, 241, 372 (1965).

397. R. J. C. Holloway, Ann. Rep. East Mall. Res. Station, 1965, 94 (1966).

398. S. J. Wiemann and A. P. Appleby, Weed Res., 12, 65 (1972).

399. J. V. Gramlich, D. E. Davis, and H. H. Funderburk, Proc. South. Weed Control Conf., 18, 611 (1965).

400. W. Zumpft, University of Erlangen, Germany, personal communication, 1969.

401. N. Balicka and H. Bilodub-Pantera, Acta Microbiol. Pol., 13, 149 (1964), in Weed Abstr., 15, 59 (1966).

402. H. Graeser, Flora, 158A, 493 (1967).

403. B. Swietochowski, M. Ploszynsky, and H. Zurawski, Pamietnik Pulawski, 21, 211 (1966), in Landw. Zbl. Abt. Pflanz. Prod., 12, 1362 (1967).

404. J. Krakkai, Abstr. Herbicid-Kongr., Sofia, 1965, cited in H. Graeser and A. Gzik, Int. Symp. Biol. Zentralanstalt Berlin, 1968, Tagungsbericht No. 109, 69 (1970).

405. F. M. Ashton and E. G. Uribe, Weeds, 10, 295 (1962).

406. B. Singh and D. K. Salunkhe, Can. J. Bot., 48, 2213 (1970).

407. R. P. Singh and S. H. West, Weeds, 15, 31 (1967).

408. A. Temperli, H. Tuerler, and C. D. Ercegovich, Z. Naturforsch., 21b, 903 (1966).

409. N. G. Sausing and Y. P. Cho, Proc. 5th Weed Control Conf., 23, 320 (1970).

410. H. Graeser, Tag.-Ber. Dtsch. Akad. Landw.-Wiss., Berlin, Nr., 109, 97 (1970).

411. W. Muecke, P. W. Mueller, and H. O. Esser, Experientia, 25, 353 (1969).

412. D. P. Schultz and B. G. Tweedy, Plant Physiol., 43, Suppl. 3 (1968).

413. S. Sawamura, Cytologia, 30, 325 (1965).

414. A. Mueller, E. Ebert, and A. Gast, Experientia, 28, 704 (1972).

415. O. Kiermayer, in The Physiology and Biochemistry of Herbicides (L. J. Audus, ed.), Academic Press, New York, 1964, p. 207.

416. R. Wakonig and T. J. Arnason, Can. J. Bot., 37, 403 (1959).

417. G. Roehrborn, Z. Vererbungslehre, 93, 1 (1962).

418. D. J. Mann, L. S. Jordan, and B. E. Day, Plant Physiol., 40, 840 (1965).

419. R. D. Gruenhagen and D. E. Moreland, Weed Sci., 19, 319 (1971).

420. J. L. Key, Ann. Rev. Plant Physiol., 20, 449 (1969).

421. D. L. Lindscott and R. D. Hagin, Weed Sci., 17, 46 (1969).

422. H. L. Wheeler and R. H. Hamilton, Proc. Northeast. Weed Control Conf., 20, 618 (1966).

423. I. E. Fong and R. F. Norris, Abstr. Meeting Weed Sci. Soc. Am., 1972, 181.

424. R. D. Palmer and C. D. Grogan, Weeds, 13, 219 (1965).

425. R. H. Hamilton, Weeds, 12, 27 (1964).

426. A. Walker and R. M. Featherstone, J. Expt. Bot., 24, 450 (1973).

427. L. Thompson, Weed Sci., 20, 584 (1972).

428. R. P. Thompson and F. W. Slife, Abstr. Meeting Weed Sci. Soc. Am., 1973, 64-65.

429. R. H. Shimabukuro and V. J. Masteller, Abstr. Meeting Weed Sci. Soc. Am., Feb. 1971, 43.

430. D. F. Paris and D. L. Lewis, Residue Rev., 45, 95 (1973).

431. R. J. Hanes, Pestic. Sci., 4, 817 (1973).

432. D. C. Nearpass, Proc. Soil Sci. Soc. Am., 36, 606 (1972).

433. D. D. Kaufman and J. R. Plimmer, Abstr. Meeting Weed Sci. Soc. Am., 1971, 18.

434. G. L. Lamoureux, L. E. Stafford, R. H. Shimabukuro, and R. G. Zaylskie, J. Agr. Food Chem., 21, 1020 (1973).

435. R. H. Shimabukuro, W. C. Walsh, G. L. Lamoureux, and L. E. Stafford, J. Agr. Food Chem., 21, 1031 (1973).

436. D. S. Frear, H. R. Swanson, and F. S. Tanaka, in Structural and Functional Aspects of Phytochemistry, Academic Press, New York, 1972, p. 225.

437. L. S. Jordan and V. A. Jolliffe, Pestic. Sci., 4, 467 (1973).

438. G. W. Burt, Weed Sci., 22, 116 (1974).

439. D. C. Wolf, Dissertation, University of California, Riverside, August 1973.

440. J. F. Bakke and C. E. Price, J. Agr. Food Chem., 21, 640 (1973).

441. J. V. Crayford and D. H. Hutson, Pestic. Biochem. Physiol., 2, 295 (1972).

442. W. C. Dauterman and W. Muecke, Pestic. Biochem. Physiol., 4, 212 (1974).

443. D. O. Eberle and J. A. Guth, Proc. 2nd Congr. Chemistry in Agriculture, Bratislava, Czechoslovakia, June 27-30, 1972, Section C 35, p. 1.

Chapter 3

THE SUBSTITUTED UREAS

HANS GEISSBÜHLER, HENRY MARTIN,
and GÜNTHER VOSS

Agrochemicals Division
Ciba-Geigy Ltd.
Basel, Switzerland

I. INTRODUCTION: DEVELOPMENT AND CHEMISTRY OF UREA HERBICIDES

A. History

The discovery and development of the herbicidal properties of substituted ureas began shortly after the end of the second world war. In 1946, Thompson et al. [1] related the biological effects of a large number of different compounds, among them 82 urea derivatives, to those of 2,4-D. Included were 1-n-butyl-3-(2-chlorophenyl)urea and 1-n-butyl-3-(3-chlorophenyl)urea. In their conclusions they suggested further examination of the herbicidal potential of the various urea compounds.

In 1951, Bucha and Todd [2] of E. I. du Pont de Nemours & Company described the herbicidal properties of 3-(p-chlorophenyl)-1,1-dimethylurea (CMU, monuron), whereby they stressed the toxic effect of this compound on annual and perennial grasses. Todd [3] was granted a series of U.S. patents in 1953 (priority date February 14, 1952) which covered the use of 1,1-dialkyl-3-(halophenyl)ureas as herbicides. These patents were part of a more general one that had been submitted in 1949 and which was subsequently abandoned. In a corresponding German patent [4] about 1.3 billion alkylated arylureas were revealed and claimed to be herbicidal.

As a consequence of these scientific efforts, Du Pont developed the following substituted ureas for commercial applications: fenuron, monuron, diuron, and neburon (Table 1). The technical foundation for the practical use of ureas had thus been established. Although initially these compounds were primarily suggested as industrial weedkillers, in subsequent years their potential for selective application in agriculture was recognized. This

TABLE 1

Designations, Physical and Chemical Properties of Commercial Urea Herbicides

Structural formula	Chemical designation	Common and trade names[a]	Mol wt; sol. H_2O (ppm)	Melting point (mp, °C); vapor pressure (vp, mm Hg)
	1,1-Dimethyl-3-phenylurea	Fenuron (Dybar)	164.2 3850 (25°)	vp: 1.6×10^{-4} (60°)
	3-(p-Chlorophenyl)-1,1-dimethylurea	Monuron (Telvar)	198.7 230 (25°)	mp: 176–177 vp: 5×10^{-7} (25°)
	3-(3,4-Dichlorophenyl)-1,1-dimethylurea	Diuron (Karmex)	233.1 42 (25°)	mp: 158–159 vp: 0.31×10^{-5} (50°)
	3-(3-Chloro-4-methylphenyl)-1,1-dimethylurea	Chlortoluron (Dicuran)	212.7 70 (20°)	mp: 147–148 vp: 3.6×10^{-8} (20°)

(continued)

TABLE 1 (continued)

Structural formula	Chemical designation	Common and trade names [a]	Mol. wt; sol. H_2O (ppm)	Melting point (mp, °C); vapor pressure (vp, mm Hg)
	1,1-Dimethyl-3-(α,α,α-trifluoro-m-tolyl)urea	Fluometuron (Cotoran)	232.1 90 (25°)	mp: 163–164.5 vp: 5×10^{-7} (20°)
	3-(3-Chloro-4-methoxyphenyl)-1,1-dimethylurea	Metoxuron (Dosanex)	228.7 678 (24°)	mp: 126–127 vp: 3.5×10^{-8} (20°)
	1,1-Dimethyl-3-[4-(4-methoxyphenoxy)phenyl]urea	Difenoxuron (Lironion)	286.2 20 (20°)	mp: 138–139 vp: 10^{-11} (20°)
	3-[p-(p-Chlorophenoxy)-phenyl]-1,1-dimethylurea	Chloroxuron (Tenoran)	290.7 3.7 (20°)	mp: 151–152 vp: 1.8×10^{-9} (20°)

3-Benzoyl-3-(3,4-dichlorophenyl)-1,1-dimethylurea	Fenobenzuron (Benzomarc)	337 16 (22°)	mp: 119
3-[3-(N-t-Butylcarbamoyloxy)phenyl]-1,1-dimethylurea	Karbutilate (Tandex)	279.4 325 (20°)	mp: 176–176.5
3-Cyclooctyl-1,1-dimethylurea	Cycluron (ingredient of Alipur)	198.2 1200 (20°)	mp: 138 vp: 10^{-7} (20°)
3-(Hexahydro-4,7-methano-indan-5-yl)-1,1-dimethyl-urea	Norea (Herban)	222.3 150	mp: 168–169

(continued)

TABLE 1 (continued)

Structural formula	Chemical designation	Common and trade names[a]	Mol wt; sol. H_2O (ppm)	Melting point (mp, °C); vapor pressure (vp, mm Hg)
	1-Butyl-3-(3,4-dichloro-phenyl)-1-methylurea	Neburon (Kloben)	275.2 4.8 (24°)	mp: 102–103
	3-(p-Chlorophenyl)-1-methyl-1-(1-methyl-2-propynyl)urea	Buturon (Eptapur)	236.7 30 (20°)	mp: 145–146 vp: $< 10^{-7}$ (20°)
	1-(2-Methylcyclohexyl)-3-phenylurea	Siduron (Tupersan)	232.3 18 (25°)	mp: 133–138
	3-(p-Chlorophenyl)-1-methoxy-1-methylurea	Monolinuron (Aresin)	214.6 580 (20°)	mp: 76–78 vp: 1.5×10^{-4} (22°)
	3-(p-Bromophenyl)-1-methoxy-1-methylurea	Metobromuron (Patoran)	259.1 330 (20°)	mp: 95–96 vp: 3×10^{-6} (20°)

Structure	Name	Trade name		mp / vp
	3-(3,4-Dichlorophenyl)-1-methoxy-1-methylurea	Linuron (Afalon, Lorox)	249.0 75 (25°)	mp: 93-94 vp: 1.5×10^{-5} (24°)
	3-(4-Bromo-3-chloro-phenyl)-1-methoxy-1-methylurea	Chlorbromuron (Maloran)	293.6 50 (20°)	mp: 94-96 vp: 4×10^{-7} (20°)
	1-Methyl-3-(2-benzo-thiazolyl)urea	Benzthiazuron (Gatnon)	207.3 12 (20°)	mp: 287 vp: $< 1 \times 10^{-6}$ (20°)
	3-(2-Benzothiazolyl)-1,3-dimethylurea	Methabenz-thiazuron (Tribunil)	221.3 59 (20°)	mp: 119 vp: 10^{-5} (20°)

[a]Trade names are in parentheses.

mainly applied to diuron, which was recommended for use in cotton, sugar-cane, pineapple, grapes, apples, citrus, and alfalfa [5].

The initial discoveries stirred the imagination of numerous industrial chemists to embark on further exploration of the ureas for herbicidal activity. From statements made in the literature, it seems certain that literally thousands of urea compounds have found their way from chemical laboratories to greenhouses to be scrutinized for selective properties by biologists [6-8]. Aliphatic as well as cycloaliphatic, aromatic and heterocyclic urea derivatives have been prepared. As the urea molecule is readily accessible even to multiple substitutions with the same or different moieties, the possibilities of structural variation are practically unlimited. For example, two published French patents reveal 20 billion and 17 trillion different urea structures, respectively [9, 10].

Today about 20-25 different urea herbicide preparations with various use patterns are available commercially (Table 1). Additional compounds are under development. Continued scientific efforts are required to cope with such needs as lower rates of application, increased selectivity, ease of degradation in different types of soil, and changing weed populations.

In addition to Du Pont, the following companies have been instrumental in the development of urea herbicides [8]: Badische Anilin- und Sodafabrik AG (BASF), Bayer, Ciba-Geigy (formerly Ciba Ltd.), Farbwerke Hoechst, Monsanto, and Péchiney-Progil. After more than 20 years of urea herbicide research, it is an interesting task to trace the different approaches taken and to classify the ideas and chemical patterns which have emerged to lead to practical success.

1. Phenylureas

Although Thompson et al. [1] alluded to the herbicidal properties of 1-alkyl-3-chlorophenylureas, it is C. W. Todd of Du Pont who deserves the credit for having developed the first technically useful herbicides with structure (1), wherein R represents a nonhalogenated or a halogenated aromatic hydrocarbon moiety.

$$R-NH-CO-N\begin{array}{c} CH_3 \\ \\ CH_3 \end{array}$$

(1)

Commercial representatives of this class of herbicides are fenuron, monuron, diuron, fluometuron, and chlortoluron.

The pertinent patent priority dates demonstrate that the lead originally taken by Du Pont was followed only hesitatingly and did not produce tangible results for some time. The well-documented observation showing that the selectivities of fenuron, monuron, and diuron were somewhat marginal prevented chemists from further probing the area of dimethylureas substituted with a chlorinated aromatic hydrocarbon moiety. In this respect the recent development of chlortoluron by Ciba [11] for selective use in cereals came somewhat as a surprise.

Substitution of a 1-methyl by a 1-butyl group in the 1,1-dimethyl moiety of ureas, as exemplified by neburon (2), was early recognized as reducing herbicidal activity but increasing selectivity.

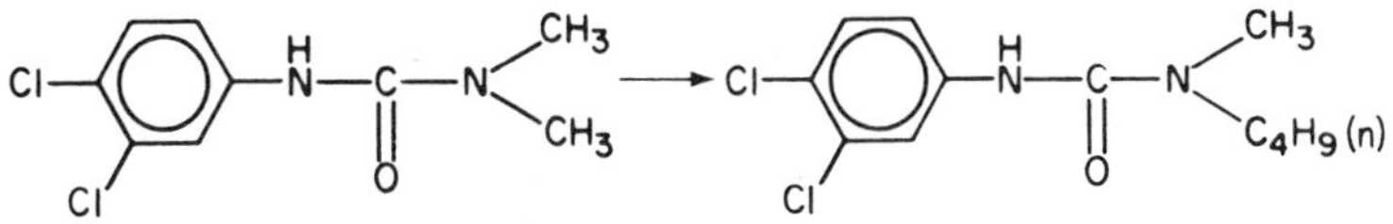

| Diuron | Neburon (2) |

Thus, whereas diuron is active as a total herbicide, neburon lacks this property; however, the latter compound is more selective, for example in cereals.

Additional compounds which resulted from this approach of increasing the 1-alkyl chain are buturon (Table 1) and 3-(p-chlorophenyl)-1-methyl-1-propylurea (3).

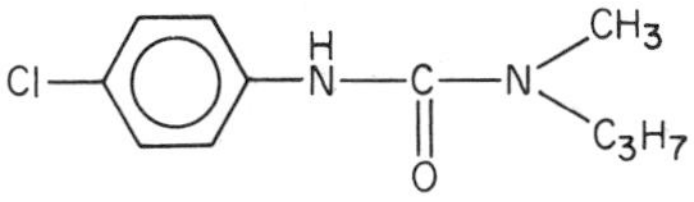

(3)

A similar effect is obtained when substituting a 1-methoxy for a 1-methyl group as shown in linuron (4).

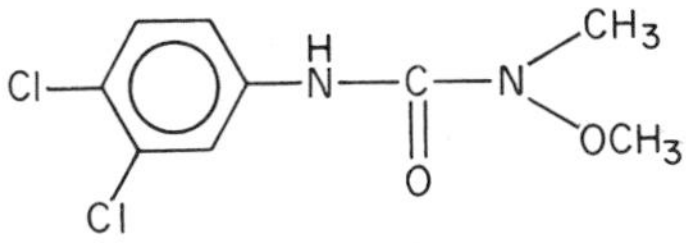

Linuron (4)

Although the 1-methyl-1-methoxy configurations of monolinuron and linuron represent only a slight modification of the structure of monuron and diuron, respectively, the differences in general and specific biological properties between the two types are considerable. Scherer et al. [12] have described these differences as follows: "Monuron and diuron exhibit little selectivity and are relatively persistent in soil. On the other hand, the 1-methyl-1-methoxy analogs are quite selective in some important food crops and their persistence in soil is low." Other representatives of this type of compound are metobromuron and chlorbromuron (Table 1).

Substitution of one chlorine atom in diuron by a methoxy group, as demonstrated by metoxuron (5), has the same basic effects, i.e., decreasing herbicidal activity, improving selectivity, and reducing persistence [13].

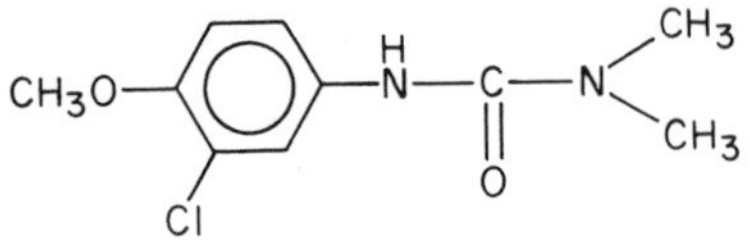

Metoxuron (5)

Replacement of a chlorine by a methoxy moiety is a well-recognized mechanism of modifying the biological activity of pesticides as shown with the two earlier ureas chloroxuron and difenoxuron, with triazene herbicides (see Chapter 2), and also with insecticides such as DDT and methoxychlor.

2. Hydroaromatic Ureas

In the late fifties and early sixties efforts were made to replace the aromatic nucleus of the urea herbicides by a saturated hydrocarbon moiety [14, 15]. Compounds of this type which have reached the commercial stage are 3-[1- or 2-(4,5,6,7,8,9-hexahydro-4,7-methanoindanoyl)]-1,1-dimethyl urea, cycluron, and noruron (6).

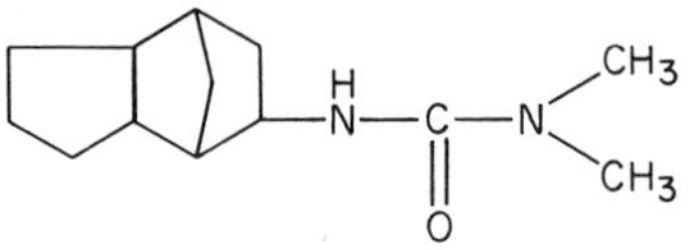

Noruron (6)

3. Heterocyclic Ureas

A third variation in structural configuration was introduced by replacing the aromatic moiety with a heterocyclic ring system. The pioneering effort was again provided by Du Pont, which patented 2-benzothiazolylureas as herbicides [16]. Whereas Du Pont recognized the selective action of 3-(2-benzothiazolyl)-1,3-dimethylurea in cotton as early as 1955, Bayer [17] in 1963 introduced 1-methyl-3-(2-benzothiazolyl)urea (benzthiazuron) as a selective herbicide in carrots and in 1965 3-(2-benzothiazolyl)-1,3-dimethylurea (methabenzthiazuron) (7) for weed control in cereals [18].

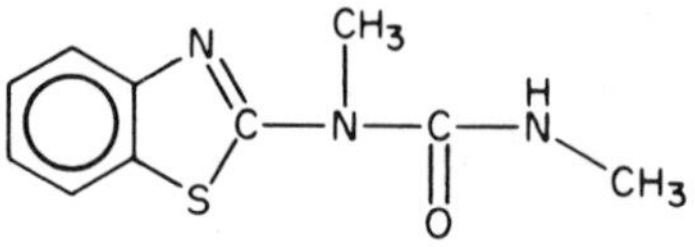

Methabenzthiazuron (7)

A more extensive description of the various approaches taken in the development of urea herbicides than is possible in this brief review is provided by Wegler [8] in his recent monograph.

B. Physical and Chemical Properties

The main physical properties of the present family of commercial herbicides are listed in Table 1. The table demonstrates that in general substituted ureas exhibit intermediate melting ranges, low water solubilities, and low vapor pressures. These properties not only determine their uses in the field but also affect their behavior in various biological environments. Chemical features that are relevant to their reaction mechanisms, their degradation, and their mode of action are discussed in the corresponding sections.

C. Synthesis

The following summary deals with the principal routes of synthesis applied to the herbicidal, asymmetrically substituted ureas of the general formula (8):

$$R_1-NH-CO-N\begin{array}{c} \diagup R_2 \\ \diagdown R_3 \end{array}$$

(8)

1. Reaction of Isocyanates with Amines (or Adducts Thereof)

Isocyanates are known to be very reactive with ammonia and amines containing an active hydrogen atom [19] . The synthesis takes advantage of this reactivity and represents the most common procedure for the preparation of asymmetrically substituted ureas [Eq. (1)] .

$$R_1-N=C=O + HN\begin{array}{c} \diagup R_2 \\ \diagdown R_3 \end{array} \rightarrow R_1-NH-CO-N\begin{array}{c} \diagup R_2 \\ \diagdown R_3 \end{array} \qquad (1)$$

A simple method is provided by combining the solutions of isocyanate and amine in an inert organic solvent such as acetonitrile. The ureas produced in this exothermic reaction precipitate as highly insoluble products [20] .

2. Reaction of Carbamic Acid Chlorides with Amines

As an alternative to the isocyanate procedure, the substituted ureas can be prepared by reacting the aliphatic carbamic acid chloride with the appropriate aromatic amine according to reaction (2) [21] .

$$\qquad \qquad \qquad \qquad \qquad \qquad \qquad \qquad \qquad \qquad (2)$$

3. Reaction of Carbamic Acid Phenylesters with Amines

In an early patent granted to Rohm [22] it was demonstrated that carbamic acid phenylesters easily react with amines at room temperature to yield the corresponding phenylurea and phenol [Eq. (3)] .

$$\qquad \qquad \qquad \qquad \qquad \qquad \qquad \qquad \qquad \qquad (3)$$

4. Transamidation of Ureas with Amines

Whereas the reactions described in the preceding sections required phosgene for the preparation of the appropriate intermediates, the following procedures are carried out under phosgene-free conditions.

Transamidation of ureas with amines proceeds according to reaction (4) [23] .

$$R\text{-}NH\text{-}CO\text{-}NH_2 + NH\!\!\begin{array}{c} CH_3 \\ \\ CH_3 \end{array} \rightarrow R\text{-}NH\text{-}CO\text{-}N\!\!\begin{array}{c} CH_3 \\ \\ CH_3 \end{array} + NH_3 \qquad (4)$$

This method offers certain technical advantages when compared with the isocyanate procedure. The starting material, i.e., the aromatic urea, is prepared by reacting potassium cyanate with an amine salt [Eq. (5)] .

$$R\text{-}NH_3 \cdot HX + KCNO \rightarrow R\text{-}NH\text{-}CO\text{-}NH_2 + KX \qquad (5)$$

The transamidation is facilitated by carrying out the reaction in organic solvents such as dichlorobenzene, anisole, or dimethylformamide at temperatures between 50° and 120°C.

5. Alkylation of Hydroxamic Acid Derivatives

Alkylalkoxyphenylureas are prepared by alkylation of the corresponding hydroxamic acid derivatives [24, 25], as shown for linuron [Eq. (6)]. This reaction is of commercial significance.

However, 1-alkyl-1-alkoxyphenylureas may also be synthesized by condensation of the phenylisocyanates with alkylalkoxyamines as described in Sec. I,C,1.

6. Further Syntheses

A number of additional routes of synthesis for asymmetrically substituted ureas have been described in the literature. So far these reactions have been of limited significance or commercial importance in urea herbicide synthesis. However, they are briefly summarized to allow access to pertinent publications.

To synthesize mono- and disubstituted ureas, the amine can be reacted with an excess of nitrourea [26]. The latter intermediate is prepared from urea nitrate in cold sulfuric acid. For synthesizing ureas with higher alkyl chains, nitrosourea is used instead of nitrourea.

Several amines react with cyanogen bromide in aqueous solution to form ureas [27]. Frequently the cyanamides which appear as intermediates in this reaction may be hydrolyzed under alkaline or acid conditions to yield ureas substituted in only one of their N atoms.

A number of methods have been described which permit desulfuration of thioureas to form the corresponding ureas [20]. Oxidants used in these reactions are potassium cyanide, ferric chloride, potassium permanganate, potassium chlorate, halogens, and others. As an alternative, desulfuration could be accomplished in the presence of heavy metal salts, such as mercuric oxide or lead oxide. Thioureas used as starting materials in these procedures can be prepared by reacting formanilides with sulfuryl chloride/thionyl chloride to yield isocyanide chlorides [28] , which are then transformed to isothiocyanates in the presence of sulfide salts. The isothiocyanates react with secondary amines to give thioureas as described in Sec. I,C,1.

N-Phenylphosphoroazoanilides prepared from aromatic amines and phosphorus trichloride react with primary or secondary aliphatic amines and carbon dioxide to yield urea derivatives [29, 30] . Preparation of mon-uron and diuron by this procedure has been described by Sedlmaier [31] .

Tertiary N-phosphines, synthesized according to Michaelis [32] , also react with carbamic acids to yield substituted ureas [31] .

A novel synthesis for the preparation of 1,1-dialkyl-3-arylureas has been described by Franz et al. [33, 34] . Carbon monoxide and sulfur were used to prepare carboxy sulfide [Eq. (7)] , which was then reacted with an aromatic amine to form a thiolcarbamate salt [Eq. (8)] . Under the experimental conditions described, the latter compound decomposed into the isocyanate, which was condensed with a secondary amine to yield the substituted urea [Eq. (9)] .

$$S + CO \rightleftarrows COS \tag{7}$$

$$R_1\text{-}NH_2 + COS \rightleftarrows R_1\text{-}NH\text{-}COSH \cdot R_1\text{-}NH_2 \rightleftarrows$$

$$R_1\text{-}N{=}C{=}O + R_1\text{-}NH_2 \cdot H_2S \tag{8}$$

$$R_1\text{-}N{=}C{=}O + \begin{matrix} R_2 \\ \diagdown \\ NH \\ \diagup \\ R_3 \end{matrix} \rightarrow R_1\text{-}NH\text{-}CO\text{-}N\begin{matrix} R_2 \\ \diagup \\ \diagdown \\ R_3 \end{matrix} \tag{9}$$

If the amines and/or isocyanates are not easily obtained, the corresponding carbonic acid amides, carbonic acid hydrazides or hydroxamic acids, respectively, can be used as starting materials. Thus, in the Hoffmann reaction [35] a carbonic acid amide is treated with sodium hypochlorite or hypobromite, whereby the carbonyl moiety of the amide is evolved as carbon dioxide. The corresponding isocyanate, which is formed as an intermediate in this reaction, can be trapped with a secondary aliphatic amine to yield the asymmetrical urea derivative.

In the Curtius reaction [36] an acyl hydrazide is used as the starting material. Upon heating in organic solvents, this compound evolved nitrogen and rearranged to the isocyanate. By condensation with a secondary amine the desired substituted urea was formed. Brunnett and McCarthy [37] have described the synthesis of 1,1-diethyl-3-(3-thienyl)urea (9) by this reaction.

(9)

Synthesis of asymmetrical isoureas [38], such as trimeturon [1-(p-chlorophenyl)-2,3,3-trimethylpseudourea] has been accomplished by either treating aryliminohalogenocarbonic acid amides with methanol [Eq. (10)] or by reacting aryliminohalogenocarbonic acid alkyl esters with primary and secondary amines [Eq. (11)].

$$
Cl\text{—}C_6H_4\text{—}N{=}C(Cl)\text{—}N(CH_3)_2 \xrightarrow{CH_3OH} Cl\text{—}C_6H_4\text{—}N{=}C(OCH_3)\text{—}N(CH_3)_2 \quad (10)
$$

$$
Cl\text{—}C_6H_4\text{—}N{=}C(OCH_3)\text{—}Cl \xrightarrow{HN(CH_3)_2} Cl\text{—}C_6H_4\text{—}N{=}C(OCH_3)\text{—}N(CH_3)_2 \quad (11)
$$

D. Chemical Reactions

Substituted ureas may be subjected to a variety of chemical reactions. From present data it appears that these purely chemical reactions are not a significant factor in causing structural transformations of the urea herbicides upon their application under field conditions. However, some importance can be attributed to photochemical reactions of ureas as described in Volume 2, Chapter 17.

On the other hand, the reactivity of substituted ureas has been utilized for analytical purposes and also for the preparation of herbicidal products with modified biological properties. The following section summarizes such reactions briefly.

1. Salt Formation

The simple substituted ureas have basic properties and, like urea itself, are thus able to undergo salt formation. With increasing complexity of substitution, the basicity disappears.

Of practical importance are the trichloroacetic acid salts of fenuron and monuron [39] which are exploited commercially as total herbicides under the common names fenuron TCA (10) and monuron TCA, respectively.

$$
C_6H_5\text{—}NH\text{—}CO\text{—}N(CH_3)_2 \cdot CCl_3\text{—}COOH
$$

Fenuron TCA (10)

2. Hydrolysis

Substituted ureas may be hydrolyzed, whereby the end products are carbon dioxide and the corresponding amines [Eq. (12)]. Hydrolysis proceeds under acid as well as alkaline conditions.

$$\text{(structure)} -NH-CO-N\overset{CH_3}{\underset{CH_3}{\diagup}} + H_2O \longrightarrow \text{(structure)} -NH_2 + CO_2 + HN\overset{CH_3}{\underset{CH_3}{\diagup}} \qquad (12)$$

This reaction is used for the quantitative analysis of active ingredients and/or residues of urea herbicide preparations by determining either the aromatic or the aliphatic amine moiety [40, 41].

3. Halogenation

Phenylureas can easily be halogenated in the para position. Advantage is taken of such a halogenation procedure [42] for synthesizing the brominated phenylureas metobromuron and chlorbromuron (Table 1) according to reaction (13).

$$\text{(structure)} -NH-CO-N\overset{CH_3}{\underset{OCH_3}{\diagup}} \xrightarrow{Br_2} Br-\text{(structure)} -NH-CO-N\overset{CH_3}{\underset{OCH_3}{\diagup}} + HBr$$

Metobromuron   (13)

A similar procedure has been described for the iodination of phenylureas in the para position [43]. Iodination has also been utilized for the preparation of halogenated anilines of phenylureas which are sensitive enough for electron capture detection in gas chromatography [44].

4. Nitrosation

Substituted ureas with a free hydrogen on the 1-nitrogen atom can be subjected to nitrosation with nitrous acid. Nitrosation of phenylureas according to reaction (14) has been described by Fujita et al. [45]. Compounds of this type, in addition to being slightly herbicidal, have a sterilizing effect on insects.

$$Cl-\text{(structure)} -NH-CO-N\overset{CH_3}{\underset{H}{\diagup}} \longrightarrow Cl-\text{(structure)} -NH-CO-N\overset{CH_3}{\underset{N=O}{\diagup}} \qquad (14)$$

5. Chlorosulfonation

Chlorosulfonation of substituted ureas to form the corresponding sulfonamides has been described in several patents [46, 47]. As an example, the formation of the herbicidal 3-(3-chloro-4-dimethylaminosulfonylphenyl)-1,1-dimethylurea from 3-(3-chlorophenyl)-1,1-dimethylurea is shown in reaction scheme (15).

$$(15)$$

6. Acylation

A number of proposals have been made to N-acylate substituted phenylureas by reacting them with different compounds such as chloroacetyl chloride [48], unsubstituted and substituted benzoyl chloride [49, 50], and five- and six-membered heterocyclic carbonic acid chloride [50]. A commercial herbicidal benzoyl derivative prepared by this reaction is fenobenzuron [3-benzoyl-3-(3,4-dichlorophenyl)-1,1-dimethylurea] (11).

Fenobenzuron (11)

Ureas containing a phenolic hydroxyl group may be O-acylated with several reactants, including alkyl isocyanates [51], dialkylphosphorochloridates [52], and dialkylsulfonyl chlorides [53]. A typical example of this reaction is the synthesis of the commercial herbicide karbutilate, 3-[3-(N-tert-butylcarbamoyloxi)phenyl]-1,1-dimethylurea, as indicated in reaction (16).

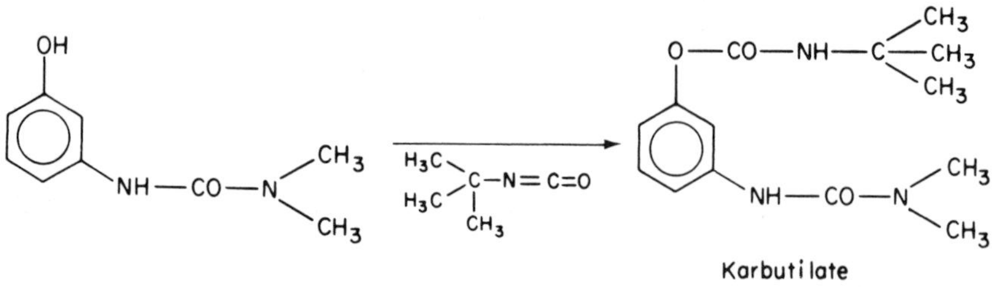

Karbutilate

(16)

7. Alkylation

N-Alkylation of trisubstituted ureas has been described by Gobeil and Luckenbaugh [54]. Diuron was treated with sodium methylate to form the sodium salt of the urea derivative. By reacting this intermediate with cyanogen chloride the desired N-substituted product was obtained [reaction (17)].

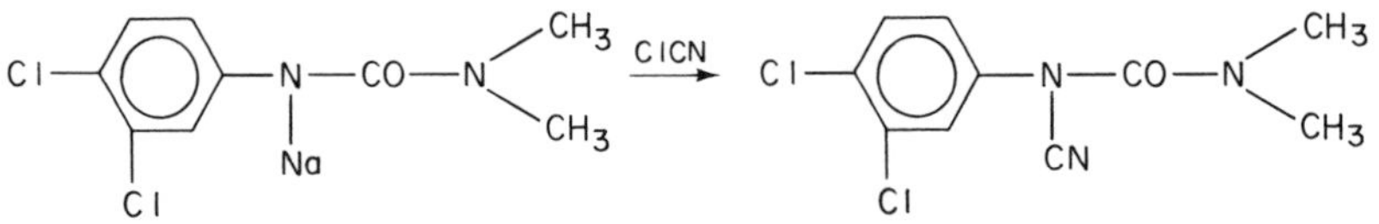

(17)

Ureas containing a phenolic hydroxyl group may be O-alkylated with several reactants, including propinyl bromide [55]. This reaction (18) yields the corresponding phenylether.

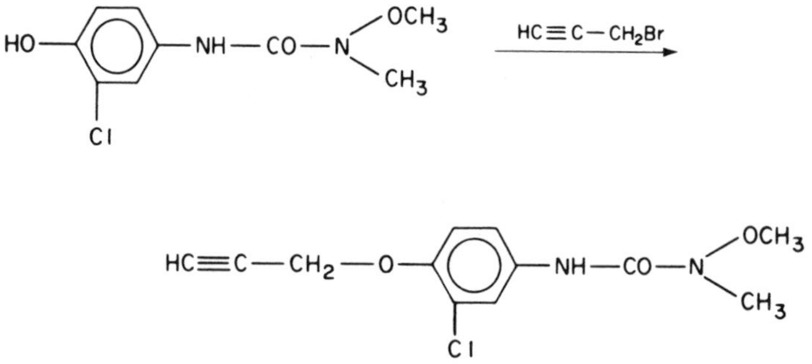

(18)

8. Cyclization

Numerous patents demonstrate that herbicidal ureas have been subjected to cyclization in order to obtain derivatives with different biological properties. The pertinent literature covers the formation of both five- and six-membered heterocyclic ring systems. However, none of these derivatives, at present, has become significant as a commercial herbicide. The main reactions are therefore listed by their products only.

The range of five-membered heterocycles covers 2-iminooxazolidines [56], 2-oxoimidazolidines [57, 58], hydantoins [59], 2,4,5-trioxoimidazolidines [60-63], oxyhydantoins [64], 3,5-dioxo-, 3-thiono-5-oxo-1,2,4-ozadiazolidines [65-68], and 3,5-dioxo-1,2,4-thiadiazolidines [62, 69].

The six-membered heteocyclic derivatives include 2-oxohexahydrotriazines [70], 3,5-dioxotetrahydro-1,2,4-oxadiazines [71], and 2,4-dioxodecahydroquinazolines [72].

E. Formulation

Certain physical properties of the urea herbicides (relatively high melting points, low water solubilities, limited solubilities in organic solvents) limit the possibilities of formulating them for field application.

The most common formulations are wettable powders. These are prepared according to known techniques that involve the mixing and grinding of the active ingredient with carriers, diluents, dispersing and wetting agents.

Another more recent type of formulation consists of suspended concentrates of active ingredients in water or oil. These suspensions are prepared by wet grinding in the presence of surfactants and stabilizers.

Both types of formulations can easily be dispersed in water and applied as sprays in desired concentrations by conventional equipment.

Granular formulations, especially microgranules, are a new development in the urea herbicide field. They offer the advantage of not requiring large volumes of water in large-scale ground or aerial applications. These granular formulations, however, need specialized equipment. The successful performance of this type of formulation is dependent on the relations between the particle size, the content of the active ingredient, and its rate of release.

II. BEHAVIOR IN SOILS, PLANTS, AND ANIMALS

The metabolism and the mode of action of substituted ureas not only depend on their chemical characteristics as described in Sec. I but also on the physical and chemical features of the biological environments to which they are exposed after application. This section describes some of the common basic parameters that affect the interactions between urea herbicides and soil, plant, and animal systems.

A. Soil-Herbicide Interactions

1. Effects of Soil Physical-Chemical Properties

In the last 20 years, numerous soil experiments have been undertaken and the following processes have emerged as being potentially involved in dissipating soil-applied urea herbicides [73, 74]:

1. Removal through leaching and/or runoff by rainfall and irrigation water; adsorption to soil particles; volatilization, and uptake by plants.

2. Degradation by chemical, photochemical, and biochemical mechanisms.

The significance and relative importance of dissipation by removal processes have been dealt with in recent reviews and articles [73, 75, 76] and are considered here only as far as relevant to degradation mechanisms.

Examination of the herbicidal activity of soil-applied urea compounds under a variety of field conditions immediately demonstrated that certain soil properties and soil environmental factors had a prominent effect not only on their performance but also on their persistence [77-79]. These initial observations have led to a large number of soil-behavior experiments, mostly of a bioassay type, which have been reviewed in detail by Sheets [73]. A number of studies suggest that adsorption of the urea herbicides to certain soil constituents (organic particles, various types of clay colloids) is an important factor not only in controlling their removal from soil layers by leaching but also in regulating their rates of degradation by microorganisms. Soil adsorption and desorption equilibria determine the concentration of each particular compound in the soil solution and thus its availability to microbial decomposition. Adsorption appears to be lowest in sandy soil, intermediate in clay loams, and highest in organic soils with a high organic matter content [80-102]. In addition, there are significant differences among the various urea herbicides with regard to their adsorption in a particular soil or to a particular soil constituent. This is demonstrated in Fig. 1, which compares the adsorption isotherms of seven substituted ureas as determined in a model system consisting of water and ethyl cellulose [103]. Although no extensive comparison of the adsorptive properties of all commercial ureas in different types of soil is available at this time, there are sufficient data to demonstrate that among the chlorinated 1,1-dialkylphenyl derivatives,

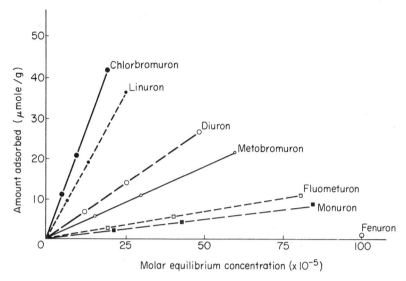

FIG. 1.   Isothermal equilibrium adsorption of seven phenylurea herbicides on ethyl cellulose. The molar quantities of herbicide fixed per gram
of adsorbent have been plotted versus the molar concentrations in the aqueous phase on linear scales. The phenylureas were readily desorbed from
ethyl cellulose with water, demonstrating that the adsorption mechanism
was physical in nature. (Redrawn with the permission of Weber [103].)

adsorption increases according to the following sequence: fenuron < monuron
< diuron < neburon < chloroxuron [84-87, 91, 103, 104]. Indications are that
1-alkyl-1-alkoxy derivatives, such as monolinuron and linuron, are subject
to a different type of bonding mechanism and are slightly more adsorbed
than their corresponding 1,1-dialkyl analogs [91, 103]. Replacement of
ring chlorine by bromine increases adsorption, as demonstrated by the comparison of linuron and chlorbromuron [103]. Apparently, no change in
adsorption is observed when the ring is substituted with trifluoromethyl
instead of a chlorine moiety [103, 105]. On the other hand, the replacement
of a halogen ring substituent by a methyl or methoxy moiety as exemplified
by chlortoluron and metoxuron, respectively, apparently decreases
adsorption [106].

Since the availability in the soil solution is inversely related to adsorption, it is to be expected that, in a particular soil and under the same environmental conditions, the more strongly adsorbed urea herbicides are less
rapidly decomposed by microorganisms. In support of this assumption,
fenuron was inactivated most rapidly and diuron least rapidly in a large
number of California soils, whereas the rate of monuron was intermediate
[83]. Similarly, when examined with the same soil, monuron was found to
be more persistent than fenuron [107], but less persistent than neburon

[108, 109]; linuron was degraded less rapidly than monolinuron [11 -112] and fluometron; and noruron disappeared at a faster rate than diuron [113, 114]. However, when extending the comparison to different soil types, it must be remembered that adsorption is highest in the organic soil which, at the same time, is most favorable for microbial growth and thus potentially most efficient for biochemical degradation. This phenomenon may mask or offset the rates of microbial degradation which would be predicted on the basis of adsorptive behavior only.

2. Chemical and Photochemical Degradation

Although published information is very limited at this time, the aqueous stability studies routinely carried out in industrial laboratories indicate that urea herbicides are sufficiently stable under practical temperature and soil conditions to resist hydrolytic breakdown, oxidation, or reduction by purely chemical (i.e., nonbiological) means [106, 115]. Following the suggestion forwarded by Hartley [76] that hydrolysis and oxidation reactions might be accelerated by the adsorption of herbicides to soil particles, Hance [116] submitted a number of herbicides, including the two ureas diuron and linuron, to high-temperature treatments in the presence of aqueous slurries of soils and clay colloids. Velocity constants of the decomposition reactions undergone by the compounds at 85°, 95°, and 107° were extrapolated to 20°C. The half-lives estimated for diuron and linuron at this temperature were of the order of 10-70 years. Although Hance regarded these results as rough approximations owing to several limitations of the experiment, he concluded that nonbiological or chemical processes are not important in the loss of diuron and linuron from soils.

In contrast to their chemical stability, urea herbicides seem to be more readily decomposed by photochemical means [117-124]. Such processes apparently are of some practical importance especially when the herbicides are applied in arid regions which are exposed to intense sunlight. The significance and the reaction mechanisms of photodecomposition of substituted ureas are described in detail in Volume 2, Chapter 17.

3. Microbial Degradation

A number of early investigations demonstrated that conditions favoring growth of microorganisms such as elevated temperatures, high moisture content, presence of organic matter, and soil cultivation hasten the inactivation of urea herbicides applied to soils [78, 125-127]. Further proof of microbial activity as an important factor in the degradation of urea compounds has been obtained by comparing the rates of disappearance of herbicidal activity in nonsterilized soil samples and in samples sterilized by autoclaving or by chemical treatment [84, 110, 115, 124, 128]. A typical example of such an experiment with diuron is shown in Fig. 2. Because of the physical and chemical changes induced in soils by sterilizing processes,

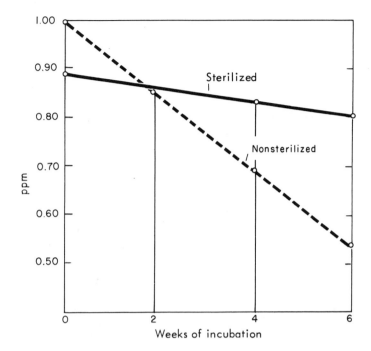

FIG. 2.   Rate of disappearance of diuron in chloropicrin-sterilized and
nonsterilized Cecil loamy sand. Compound applied at the rate of 1 ppm
based on weight of air-dry soil. Soil samples watered and then stored at
80° F and 60% relative humidity. Concentrations of herbicide determined by
bioassay using oat seedlings as indicator plants. (Redrawn with the permission of Hill et al. [115].)

the validity of such experiments may be questioned [74, 129]. However, the
acknowledged resistance of ureas to chemical breakdown, as noted above,
and the drastic decrease in degradation observed after soil sterilization,
are sufficient proof to indicate that microorganisms are a significant factor
in the transformation of urea compounds.

Hill et al. [115] were the first to search systematically for soil organisms capable of decomposing phenylureas and were able to isolate from a
Brookston silty clay loam a bacterium of the genus Pseudomonas which
utilized monuron as a sole source of carbon. When measuring its respiration by the Warburg technique, the authors demonstrated that, after the
compound was added to the buffered medium, total oxygen uptake increased
with increasing concentrations of herbicide. Another bacterium of the
Pseudomonas genus could rapidly oxidize monuron, provided the medium

was supplemented with an appropriate growth factor such as yeast extract. Extending the same approach to other groups of microorganisms, Hill and McGahen [130] observed that other bacteria, such as Xanthomonas, Sarcina, and Bacillus spp., all of which are common soil microorganisms, were able to utilize monuron as a carbon source in agar media.

This initial search for defined bacterial and fungal species that are able to metabolize substituted ureas was abandoned for more than 10 years in favor of degradation experiments with actual soils or with undefined mixed microbial populations. Attempts to isolate and biochemically characterize microorganisms have only recently been resumed. Another bacterium of the Pseudomonas sp., isolated from a Hawaiian soil, utilized diuron as a respiratory substrate [131]. Three different Aspergillus spp. exhibited varying capacities to degrade fenuron, monuron, diuron, noruron, and fluometuron [132]. Börner et al. [133] subjected 81 soil fungal and 13 bacterial species cultured in liquid media to degradation experiments with monuron, diuron, monolinuron, and linuron. Efficiency of herbicide degradation was determined by measuring the residual amounts of total aniline-containing compounds remaining in the cultures after exposure periods of 14-21 days. Most of the organisms examined appeared to be active to some extent in herbicide transformations. Although the analytical method used in these experiments permits no more than a cursory approach to the measurement of degradation efficiency, the data indicate that the capacity of microbes to metabolize urea herbicides is rather widespread. On the other hand, Wallnöfer [134] examined about 300 pure cultures of soil bacteria and detected only one strain of Bacillus sphaericus that was able to degrade monolinuron. On a more limited scale Tweedy et al. [135] observed that the two fungi Talaromyces wortmanii and Fusarium oxysporum metabolized ^{14}C-labeled metobromuron, whereas a Bacillus sp. and the alga Chlorella vulgaris were unable to degrade the same herbicide. Another fungus, Rhizoctonia solani, was shown by Weinberger [136, 309] to metabolize chlorbromuron and several other phenylureas.

Although the list of microbes that are able to transform ureas biochemically has increased, the present data do not allow the classification of the most common soil microorganisms according to their degradative capacity. In addition, several observations indicate that the rate and pattern of degradation depend on the growth rate of the organisms and on the type and amount of energy source available in the soil [129, 132, 137, 138, 307 and 308]. To further complicate matters, one particular species or strain may be able to transform one type of urea molecule, but then lack this property when exposed to derivatives with slightly altered chemical structures [139].

The kinetics of the disappearance of urea herbicides from soil have not yet been clearly defined. The degradation curves of monuron presented by Hill et al. [115] were interpreted by Audus [140] to follow the "enrichment type," i.e., showing a definite initial lag phase during which the bacteria adapt enzymatically to cope with the foreign substrate to which they are exposed. However, it is difficult to visualize how a mixed soil bacterial and fungal population would be enzymatically adapted or built up within a few days when exposed to such minute quantities of herbicide. Physical factors, such as slow dissolution from carrier particles or the gradual establishment of soil solution equilibria, would also be expected to cause a certain initial lag period. That the total rate of decomposition is likely the result of several processes is shown by a recent attempt to define the kinetics of degradation of linuron in two different soils and at four levels of application [141]. Measurements conducted over a three-month period indicated that neither zero-order, half-order, first-order, or Michaelis-Menten equations were sufficient to describe the disappearance of the herbicide adequately.

Enzymatic adaptation of soil bacteria or buildup of a soil microbial population with enhanced degradative capacity may affect the rates of urea herbicide decomposition over prolonged periods of time [129, 142]. On the other hand, it has to be borne in mind that fungal species, which represent the major portion of living soil protein, have only a limited ability to form adaptive enzymes [143].

B.   Uptake by and Movement in Plants

1.   Absorption and Translocation after Soil Application

The site and rate of degradation and the mode of action of urea herbicides in plants are intimately connected with the capacity of these plants to take up the compounds from the environment and to translocate them into various organs and tissues. The absorption and movement of substituted ureas have mainly been investigated with the aid of biological and gross autoradiographic methods [144-159]. In some instances these techniques have been supplemented by parallel quantitative measurements [147, 153, 155, 160-162]. In a recent study microautoradiography has been used to follow the movement of diuron in cotton at the cellular and subcellular levels [163].

There is general agreement that all ureas so far examined are easily taken up from nutrient and soil solutions by root systems and most are rapidly translocated into stems and leaves by the transpiration stream. In autoradiographic studies, radioactivity has consistently been observed to be concentrated in those organs and tissues which exhibit high rates of water flow, such as actively absorbing branch roots and mature, well-expanded leaves. It is therefore not surprising that factors which lower the rate of transpiration, including high external water vapor pressure, low temperature, or closure of foliar stomata, also decrease the amounts of herbicides absorbed and translocated [145, 153, 155].

The microautographic study mentioned above, in which [^{14}C]carbonyl-labeled diuron was applied to the roots of cotton seedlings, has shown that

CHLOROXURON   METOBROMURON   FLUOMETURON
2.5ppm herbicide (equal specific activity)

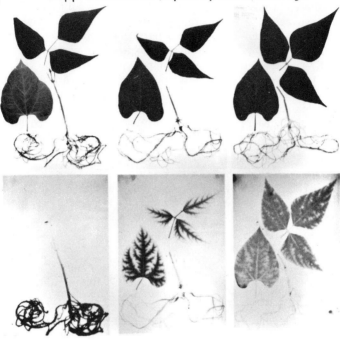

FIG. 3. Comparative uptake by roots and translocation to aerial parts of ^{14}C-labeled chloroxuron, metobromuron, and fluometuron in French dwarf bean seedlings, cultured in nutrient solution supplied with 2.5 ppm (and equal concentrations of radioactivity) of each herbicide. Plants were kept in herbicide solution for 24 hr and then cultured in regular nutrient solution for an additional 24 hr. Top: mounted plants; bottom: radioautographs of same plants. (Adapted from Voss and Geissbühler [157].)

the herbicide apparently moved from the root surface to the xylem through the walls of the cortical cells. The endodermis did not seem to be a significant barrier to the passage of radioactivity into the xylem vessels. Once translocated to the leaves, lateral movement of radioactivity into the cortical region and epidermal layers again was observed to be mainly associated with the cell walls. These data have substantiated the assumption originally made by Crafts [150] that soil-applied substituted ureas primarily move in the plant apoplast.

    In a number of investigations, in which a particular urea herbicide was applied to different plants under the same external conditions, it was observed that the rates of uptake and translocation varied from one plant species to another [153, 158-161, 164]. These different distribution patterns of the same compound, which apparently depend on some inherent physiological properties of the plants exposed to the herbicide, have been postulated to be involved in certain selectivity mechanisms (see Sec. IV, B).

Of particular importance in connection with the mode of action and the formation of terminal residues by degradation is the fact that different urea herbicides exhibit different mobilities in plant systems [157]. This is demonstrated in Fig. 3 which compares uptake and translocation of chloroxuron, fluometuron, and metobromuron by bean plants when supplied to nutrient cultures in equal concentrations (essentially equimolar) and specific radioactivities. Whereas the movement of chloroxuron to aerial parts was found to be restricted under the short-term conditions applied, both fluometuron and metobromuron were rapidly translocated into leaves. However, the two latter compounds differed strikingly with regard to their distribution pattern within the leaves. In contrast to metobromuron, which was mainly confined to tracheal veins, fluometuron had almost completely moved out into the mesophyll tissues. Although no extensive comparisons of the movement of different urea structures are available at this time, it would appear that their apoplastic distribution in plants is regulated in part by the same physical-chemical phenomena which control the mobility and availability of the compounds in soils.

2. Absorption and Movement after Foliar Application

When applied to leaf surfaces, the substituted ureas penetrate cuticular and epidermal layers to varying degrees. This entry apparently may be enhanced by the addition of suitable surfactants [165-167]. Subsequently a fraction of the compound not only reaches the photosynthesizing mesophyll cells but also the tracheal veins by which it is moved in peripheral and/or acropetal direction. However, there is little or no entry into the phloem system and therefore practically no translocation into stem, neighboring leaves, flowers, or fruit by the assimilate stream. Even in organs of high phloem and seemingly low xylem activity, such as soybean cotyledons, symplastic movement of diuron was observed to be practically nonexistent [156].

Suggestions that certain surfactants may induce phloem or "downward" movement [168] have not been substantiated by actual measurements.

3. Site of Urea Herbicide Degradation in Plants

From the results of a number of investigations, it may be concluded that leaves, roots, and even hypocotyls and stems are potentially capable of transforming urea herbicides by biochemical means. Swanson and Swanson [169] incubated leaf discs of cotton, plantain, soybean, and corn with aqueous solutions of ^{14}C-labeled monuron and diuron. Leaves from all species except corn were able to completely or partially demethylate the herbicides. Frear et al. [170, 171] isolated enzyme preparations that were active in the N-demethylation of substituted ureas from etiolated cotton seedling hypocotyls and also from leaves of cotton, plantain, buckwheat, and broadbean. After injection of ^{14}C-labeled metobromuron into the stem of Galium aparine, Fykse [159] observed extensive metabolism of the compound in the aerial plant tissues. In contrast, when isolated roots of Vicia sativa were cultured under sterile conditions in a nutrient solution supplied with [^{14}C]chloroxuron, abundant amounts of mono- and demethylated metabolites were recovered from the culture medium and the root tissues [172]. When sugarbeet plants were divided into root and aerial sections and separately exposed to [^{14}C]fenuron, Schütte et al. [173] observed that both sections were able to metabolize the herbicide.

Under actual growing conditions in the field, the site of urea herbicide degradation would thus appear to depend on the plant species involved and especially on the mobility of the particular herbicide within the plant. Compounds that are rapidly translocated to aerial parts, such as fenuron, fluometuron, noruron, monuron, and monolinuron, would be expected to be metabolized in the leaves, whereas for less mobile compounds, such as neburon and chloroxuron, degradation in root tissue may be more important.

Apparently active photosynthesis is not required for the metabolism of ureas in those plant tissues normally exposed to light. Parallel studies of samples of leaf discs maintained in the dark and in the light showed about the same rate of monuron degradation under the short-term conditions applied [169].

C. Absorption, Distribution, and Elimination in Animals

In contrast to the extensive data on the behavior of substituted ureas in soil and plant environments, published information on the absorption, distribution, and elimination of these herbicides and/or their metabolites in mammals and other animal species is somewhat limited at this time.

When a number of phenylureas labeled with ^{14}C in different positions were administered orally to rats, it was observed that total radioactivity was rapidly and efficiently eliminated in the urine and feces. For the following compounds 90% or more of the radioactivity applied was accounted for in the excreta within 72 hr after administration: [carbonyl-^{14}C]chloroxuron [174], [carbonyl-^{14}C]chlorbromuron [175], [ring-^{14}C]metobromuron [176], and [tolyl-^{14}C]chlortoluron [177]. With most compounds mentioned, the major portion of the label was eliminated in the urine, whereas the amount of fecal radioactivity was usually no more than 20% in terms of the dose applied. Within the range of herbicide concentrations examined

(2-10 mg/kg) the rate of elimination was observed to be neither dose nor sex dependent.

The apparent lack of accumulation or storage of phenylureas and/or their metabolites in mammalian organs and body fluids, as indicated by the elimination experiments just mentioned, has been confirmed by a series of chemical and radiochemical tissue analyses. In chronic toxicity studies, diuron and linuron were continuously fed to rats and dogs for a period of two years at dietary levels ranging from 25 to 2500 ppm [178, 179]. Upon termination of the experiments, the total aniline-containing residues in blood and various tissues (muscle, fat, liver, kidney, spleen) were a few to 100 ppm. These residues represented only a minute fraction of the large quantities of herbicide ingested by the animals. Similar residue data were obtained in the short-term elimination experiments described above. The amounts of radioactivity which persisted in blood and different tissues at the end of the 72-hr observation periods were no more and usually much less than 1% of the dose applied.

In a recent study [tolyl-^{14}C]chlortoluron was administered daily to rats for a period of two weeks [177]. In spite of continued feeding of labeled herbicide, concentrations of radioactivity in blood and various organs reached their highest level on the first or second day after application and did not increase for the remainder of the experiment. In addition, when feeding of label was abandoned, radioactivity in all tissues decreased below detectable levels within a period of 9 days.

Though the abovementioned investigations have been confined to a limited number of compounds, the present data seem to demonstrate that phenylureas and/or their transformation products are efficiently and rapidly eliminated from the mammalian body and do not accumulate in significant quantities in any of the different tissues analyzed.

III. METABOLIC PATHWAYS

A. Basic Patterns and Concepts

Present knowledge on the metabolism of urea herbicides is mainly confined to the group of phenyl derivatives. Only limited information is available for compounds substituted with a saturated hydrocarbon or a heterocyclic ring system. A survey of the literature demonstrates that pathways of transformation of phenylureas in plants, animals, and soil have sufficient common features to obviate separate discussion of degradation in the three substrates [180]. However, the significance of each type of reaction with regard to the three biological systems will be mentioned in the corresponding section and a tabulated summary (Table 3) is provided in the concluding section.

In presenting the main transformation mechanisms of ureas the suggestion of Williams and Parke [181-183] to distinguish between phase I and
phase II reactions is followed. Phase I reactions are defined as oxidative,
reductive, or hydrolytic transformations of phenylureas. They introduce or
lead to relatively polar and biochemically reactive groups (such as $-NH_2$,
$-OH$, and $-COOH$). With phase II reactions these groups then react with or
are bound to endogenous animal, plant, or microbial metabolites.

The following phase I reactions have so far been demonstrated for
phenylureas: N-demethylation, N-demethoxylation, ring hydroxylation, oxidation of ring substituents, and aniline formation. Phase II reactions include
acetylation of anilines, glucoside/glucuronide formation, and bonding to
nitrogen constituents of undetermined nature.

Techniques used for characterizing phenylurea metabolites have ranged
from simple thin-layer chromatographic separation and color reactions to
elaborate fractionation procedures and spectral analyses. It is evident that
depending on the methods applied, the reliability and accuracy of structural
assignments vary substantially. To account for such variation the present
review attempts to differentiate in terminology and to use expressions like
"identified" only when really justified. The authors feel that the significance
of ultraviolet, infrared, and mass-spectral analyses in identification of
metabolites is difficult to evaluate when the key spectra have not been published and properly interpreted.

B.   N-Dealkylation and N-Dealkoxylation

By the time the first attempts were made to elucidate the pathways of
urea herbicide metabolism in various living systems, it had already been
established with certain other pesticides and pharmaceuticals that alkylated
amines, amides, carbamates, or sulfonamides are subjected to oxidative
N-dealkylation in animals tissues [184, 185]. The initial indications that
such a mechanism might also apply to 1,1-dimethyl-substituted phenylureas
had been provided by Hill et al. [115]. These authors observed that upon
treatment of moist samples of clay loam with ^{14}C-methyl-labeled monuron,
radioactive carbon dioxide was evolved at a slight but steady rate for a period
of 70 days. At the end of the experiment about 10% of the radioactivity originally applied was accounted for by $^{14}CO_2$.

In a more detailed study, Geissbühler et al. [124] exposed [carbonyl-
^{14}C]chloroxuron to liquid cultures of mixed soil bacteria. The two major
radioactive metabolites, separated by column and thin-layer chromatography
and characterized by their infrared spectra, were identified as 1-methyl-3-
(p-chlorophenoxy)phenylurea and 3-(p-chlorophenoxy)phenylurea, respectively. The following stepwise N-demethylation was therefore proposed for
chloroxuron:

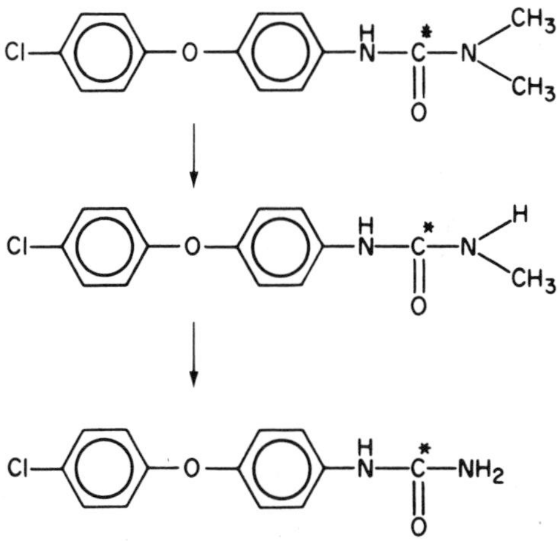

(19)

The same mechanism of sequential N-demethylation of chloroxuron was also observed in several crop and weed species.

An entirely analogous soil degradation pathway was demonstrated for diuron by Dalton et al. [186] after they had analyzed soil samples from herbicide-treated cotton fields and characterized the following metabolites: 1-methyl-3-(3,4-dichlorophenyl)urea and 3-(3,4-dichlorophenyl)urea [reaction (20)].

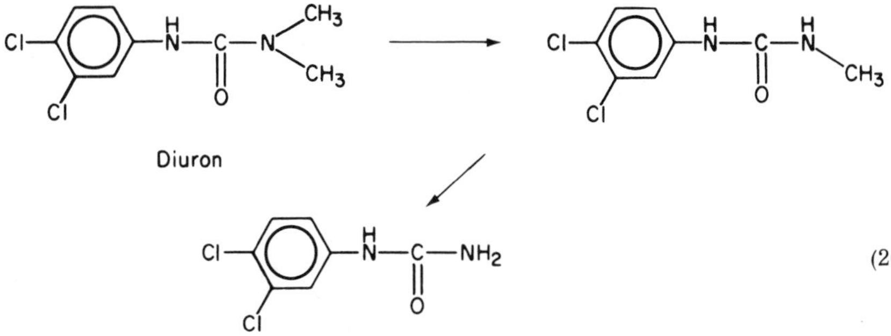

Diuron

(20)

N-Dealkylation of phenylureas by mammalian tissues was first reported by Ernst and Böhme [187, 188]. After oral administration of unlabeled monuron and diuron to male albino rats, a number of metabolites (see Sec. III, C), including the monomethylated and demethylated derivatives, were recovered from the urine of these animals. The same authors observed that the N-methyl moiety of the 1-methyl-1-methoxyphenylureas monolinuron and linuron was also subjected to oxidative removal.

N-Dealkylation, which represents the major initial pathway of urea herbicide metabolism, has now been verified for a number of dialkyl- and

TABLE 2

Recognized N-Demethylation and/or N-Demethoxylation Reactions of Phenylurea Herbicides in Biological Environments[a]

Herbicide	Soil/microbial media	Plants	Mammals
Fenuron		Sugar beet [173]	
Monuron		Cotton [161, 169], soybean [161, 169], plantain [169], bean [198], corn [198]	Rat [187]
Diuron	Cotton field soil [186]	Cotton [161, 169], soybean [161, 169], plantain [169], corn [161, 197], oat [161]	Rat [188], dog [178], humans [192]
Fluometuron	Sandy loam soil [138]	Cotton [157, 160, 171, 193], cucumber [160], plantain [171], wheat [157], weed sp. [193]	Rat [191]
Chlortoluron	Silty clay soil [106]	Wheat [106]	Rat [177]
Chloroxuron	Microbial suspension [124]	Bean [124], broadbean [124], weed sp. [124]	Rat [174], rumen [189]
Metoxuron	Wheat [190]		
Monolinuron	Talaromyces wortmanii[b] [135]	Corn[b] [133], weed sp.[b] [133]	Rat[b] [187]
Metobromuron	Fusarium oxyporum[b] [135]	Potato[b] [157], tobacco[b] [106]	Rat[b] [176]
Linuron	Aspergillus sp.[b] [133]	Corn [194], soybean [194], carrot[b] [196], weed sp. [194, 196]	Rat[b] [188]
Chlorbromuron	Rhizoctonia solani [136, 309]	Corn[b] [195], cucumber[b] [195]	Rat[b] [175]

[a]References are cited in brackets.

[b]N-Demethoxylation demonstrated.

alkoxyalkylphenylureas with a variety of soil/microbial media, plants, and mammals, including humans. Table 2 summarizes the more important of these experiments by listing them according to compounds and substrates.

The rate of formation of N-dealkylated metabolites depends on the type of biological system involved. In soil/microbial media and in plants the first dealkylation step of 1, 1-dimethyl compounds generally appears to proceed at a faster rate than the second step since the monomethylated derivatives have normally been observed to be more abundant than the unsubstituted ureas (see Fig. 4). On the other hand, the latter metabolites are present in larger quantities in animal excreta, indicating that complete dealkylation is accomplished more easily by mammalian tissues.

From the terminal residue point of view, depending on the plant species or the type of soil involved and the time elapsed after application, dealkylated metabolites may be more abundant than the parent compounds (see Fig. 4). This fact is accounted for by the residue analytical methods which are routinely applied to phenylureas and which determine the total quantity of aniline-containing transformation products [40, 41, 44, 199].

Factors that regulate the rate of N-dealkylation in various biological substrates are not yet defined. In one of the few comparative plant studies it was shown that cotton dealkylated monuron more rapidly than diuron, whereas with plantain, the rate of degradation of the two compounds was reversed [169]. This observation no more than indicates that inherent physiological features of the biological systems involved might be more important than the chemical properties of the herbicides in controlling their rates of N-dealkylation.

The N-monomethyl metabolites are usually less phytotoxic than their dialkylated parent compounds and phytotoxicity disappears completely with the second demethylation step [161-163]. The consequences of this phenomenon, which has been related to certain selectivity mechanisms in plants, are discussed in more detail in Sec. IV, B. With regard to mammalian toxicity, acute oral LD_{50} values of monomethylated and desmethylated metabolites have been shown to be of the same low order of magnitude as those of their parent herbicides [199, 201].

Biochemical removal of the N-alkoxy group of 1-methyl-1-methoxy phenylureas was first implicated in the animal experiments of Ernst and Böhme [187, 188]. These authors observed that both monolinuron and linuron were converted to the corresponding unsubstituted phenylureas by rats. Apparently the N-methyl and/or N-methoxy intermediates, which are expected to be involved in such a two-step reaction, were eliminated only in traces in the urine. N-Demethoxylation has meanwhile been confirmed to occur with a number of 1-methyl-1-methoxyphenylureas in plants and soil microbial species. The pertinent herbicides and biological substrates are listed in the lower part of Table 2. In some of these experiments the

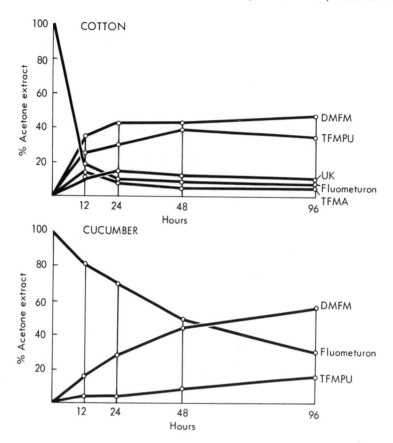

FIG. 4.   Amounts of [^{14}C]fluometuron and its radioactive metabolites as a function of time in cotton and cucumber leaves. Plants cultured in nutrient solution supplied with 1 ppm of [trifluoromethyl-^{14}C]fluometuron. Plant material extracted with acetone and fluometuron and metabolites separated by thin-layer chromatography. Key: DMFM, 1-methyl-3-(m-trifluoromethylphenyl)urea; TFMPU, m-trifluoromethylphenylurea; TFMA, m-trifluoromethylaniline; UK, unknown metabolite. (Redrawn with the permission of Rogers and Funderburk [160] .)

intermediates mentioned above were present in sufficient quantities to be adequately identified [106, 135] . Although the N-methyl moiety normally seemed to be more quickly and more easily removed than the N-methoxy group, the data obtained in different studies do not allow the assignment of a definite sequence of the two reactions. For the time being the combined N-methylation/N-demethoxylation pathway of 1-methyl-1-methoxyphenyl-ureas is therefore visualized as shown for metobromuron in scheme (21).

The exact mechanism of N-demethoxylation is still unknown, although in some experiments there were indications that the corresponding hydroxyl-amides are present as transient intermediates [106, 133, 157] . It might be difficult to detect such intermediates in sufficient quantities, since the related arylhydroxylamines are rapidly reduced to the corresponding amines in animal tissues [184] . The question of whether the mechanism of N-demethoxylation is in fact initiated by an O-demethylation reaction cannot be properly answered at this time.

C. Ring Hydroxylation

In 1949, Bray et al. [202] demonstrated that the ureido grouping of nonalkylated N-phenylureas and N-tolylureas was biologically rather stable and that ring hydroxylation, preferably in the para position, was a prominent pathway of transformation with these compounds in mammalian metabolism. It was therefore to be expected that herbicidal phenylureas would follow a similar pattern of animal metabolism. In 1965 Ernst and Böhme [187, 188] reported extensive degradation experiments in male albino rats with monuron, diuron, monolinuron, and linuron. The urinary metabolites were separated and characterized by several extraction and thin-layer chromatographic steps. The identity of the main transformation products is said to have been confirmed by infrared spectroscopy and melting-point determinations. Although none of the herbicides administered was eliminated unchanged in the urine, a major portion of the identified metabolites retained the intact ur grouping, -NH-CO-N<. In addition to the N-dealkylated and N-dealkoxy degradation products already mentioned in Sec. III, B, a number of ring-hydroxylated metabolites were described. A major portion of these pher metabolites was found to be eliminated as glucuronides or ethereal sulfa The orientation of the hydroxyl moieties apparently followed steric effec The monochloro compounds were hydroxylated in ortho and to a smaller tent in meta positions, whereas the dichloro compounds were converted mainly to the 6-hydroxy derivatives (see Fig. 5). From these experiments Ernst [203] suggested the composite N-dealkylation/N-dealkoxylation and ring-hydroxylation pathways shown in scheme (22) for monuron and mono-linuron, and in scheme (23) for diuron and linuron, respectively. Scheme

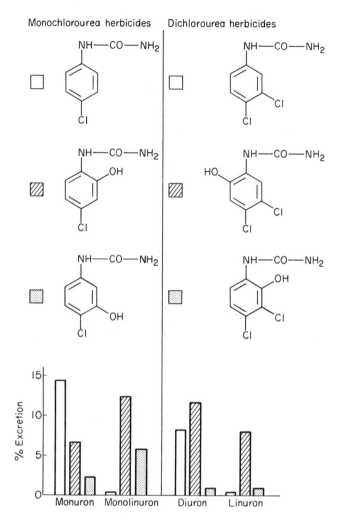

FIG. 5. Structure and relative amounts of metabolites eliminated in rat urine after oral administration of monuron, monolinuron, diuron, and linuron. Values which represent cumulative excretion within 8 days after administration are expressed in percent of the dose applied. (Redrawn with the permission of Ernst [203].)

(23) includes as a minor pathway the hydroxylation of dichloro compounds in the 1 position. Since the position is sterically hindered by the adjacent chlorine atom, the reaction appears to be doubtful and ought to be verified by additional experiments.

(22)

(23)

Quantitative analyses of the different metabolites eliminated in rat urine demonstrated that N-demethylation without parallel ring hydroxylation occurred only with the dimethyl derivatives monuron and diuron, whereas the methylmethoxy compounds monolinuron and linuron were mainly eliminated as hydroxy metabolites (Fig. 5).

Ring hydroxylation of phenylureas in animal tissues as described by Ernst and Böhme has been observed with fluometuron [191], metobromuron [176], and chlorbromuron [175]. However, in these experiments characterization of metabolites was confined to thin-layer chromatographic comparison with the corresponding reference compounds.

In a detailed and excellent study on the mammalian metabolism of siduron, Belasco and Reiser [204] demonstrated that both the phenyl and cyclohexyl rings are subjected to hydroxylation in the dog. Positive identification of the metabolites was achieved by infrared, mass-spectral, and nuclear magnetic resonance spectral analyses. The pathways proposed in scheme (24) show that siduron is either hydroxylated at the para position of the phenyl nucleus or at the 4 position of the 2-methylcyclohexyl moiety. Both metabolites are further hydroxylated to form 1-(4-hydroxy-2-methyl-cyclohexyl)-3-(p-hydroxyphenyl)urea. All three identified metabolites were

(24)

eliminated as highly water-soluble conjugates. When urine samples of both dogs and rats were subjected to hydrolysis and analyzed for aminophenol and aniline, it was observed that hydroxylation of the phenyl moiety was more prominent in the rat than in the dog. Apparently rats exhibit a particular efficiency in hydroxylating the aromatic nucleus of phenylureas.

In another study in which the ^{14}C-carbonyl-labeled compound was applied, Belasco and Langsdorf [205] showed that microbial degradation was the major route of siduron disappearance in the soil. Metabolites separated on thin-layer plates had the same characteristics as those identified in the urine, and the authors therefore concluded that ring hydroxylation was also operative in soil environments. As siduron is hydroxylated in both animals and soil, claims by Splittstoesser and Hopen [206] that this herbicide is not metabolized in plants would appear to require reexamination.

In view of the general mechanisms of microbial degradation of aromatic compounds it is surprising that ring hydroxylation of phenylureas has not been observed more frequently in soils. Since hydroxylation mechanisms are the accepted preliminary steps in enzymatic fission of aromatic nuclei [207], a more thorough study of ureas with regard to such mechanisms appears to be desirable.

Some preliminary evidence suggests that ring hydroxylation of phenyl-ureas does occur in plants. A metabolite of fluometron fulfilling a number of criteria for a compound hydroxylated in the 6 position was isolated from cotton tissue; however, positive identification has not yet been achieved [106]. In a recent study with excised bean and corn leaves, Lee and Fang [198] proposed ring hydroxylation of ^{14}C-carbonyl-labeled monuron in the ortho position. The following metabolites were separated as β-D-glucosides on thin-layer plates: 1,1-dimethyl-3-(2-hydroxy-4-chlorophenyl)urea, 1-methyl-3-(2-hydroxy-4-chlorophenyl)urea, and 3-(2-hydroxy-4-chloro-phenyl)urea. However, the identity of these transformation products needs to be validated by more rigid analytical procedures.

D.   Oxidation of Ring Substituents

At present there is no evidence to indicate that Cl or Br substituents on the aromatic nucleus of phenylureas are subjected to biochemical removal by either oxidative or reductive dehalogenation mechanisms. In photolysis experiments with metobromuron, monuron, and linuron it was observed that both chlorine and bromine, when attached in the para position, were replaced in part by hydroxy substituents [121, 122]. In addition, monuron was recently shown to be dechlorinated to fenuron when illuminated under anaerobic conditions [208]. In spite of the photochemical reactivity of the halogen substituents, such replacement reactions have not yet been demonstrated in biological systems.

The 3-trifluoromethyl moiety of fluometuron seems to be accessible to some extent to biochemical transformation reactions. A slight but measurable amount of radioactivity (< 5% of the dose applied) evolved as carbon dioxide during the 52-day observation period on incubation of the ^{14}C-tri-fluoromethyl-labeled herbicide in soils [138]. Oxidation of the trifluoromethyl carbon represents a minor pathway of degradation for fluometuron in the soil.

Alkyl or alkoxy ring substituents, as present in chlortoluron, metoxuron, and difenoxuron (Table 1) may be susceptible to biochemical oxidation reactions. So far verification of this assumption has been confined to chlortoluron. In a detailed investigation of the fate of the ^{14}C-tolyl-labeled herbicide in the rat. Mucke et al. [177] demonstrated that the methyl ring substituent was subjected to stepwise oxidation, yielding first the corresponding hydroxymethyl

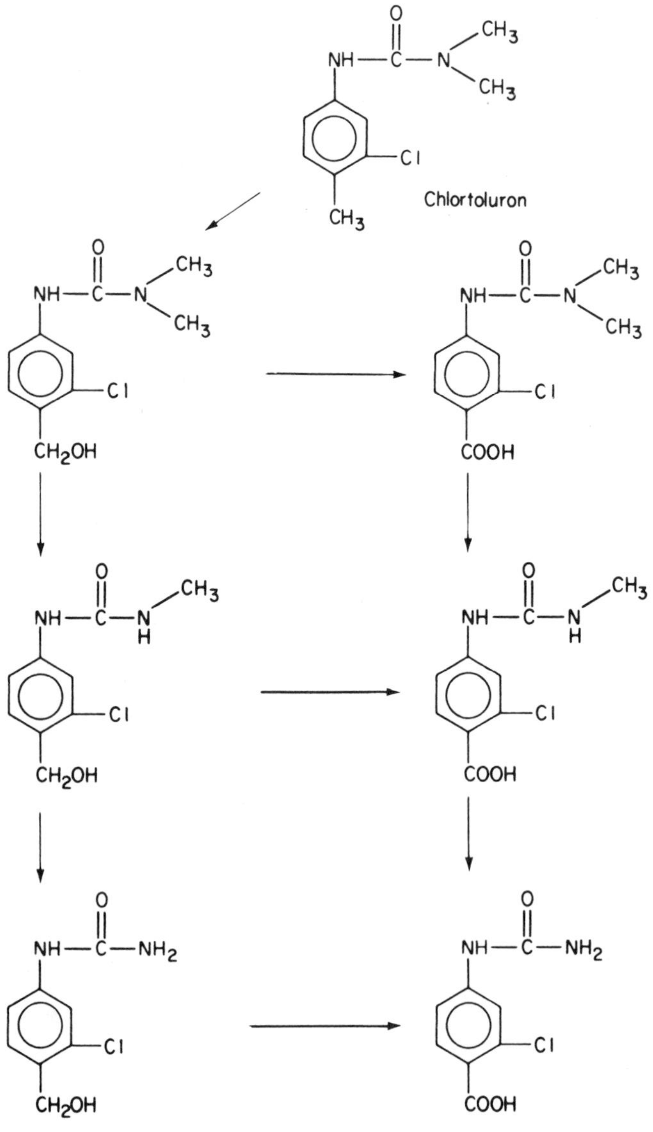

Chlortoluron

(25)

and then the carboxy derivative. The identity of the main metabolites recovered from urine by ether extraction was confirmed by infrared, mass-spectral, and nuclear magnetic resonance spectral analyses. The combined N-demethylation/side-chain oxidation pathway of chlortoluron in rats is depicted in scheme (25). The metabolism of chlortoluron in wheat exhibited essentially the same qualitative pattern of degradation. However, the major portion of the hydroxymethyl metabolite [3-(3-chloro-4-hydroxymethylphenyl)-1,1-dimethylurea] was first trapped as water-soluble conjugate(s). Further oxidation to the corresponding carboxy derivative occurred only in later stages of plant growth. In addition, N-demethylation did not proceed beyond the N-monomethyl stage [106].

E. Aniline Formation and Further Pathways

1. Extent of Aniline Formation

In many investigations on phenylurea metabolism it has been postulated that stepwise N-demethylation and/or N-demethoxylation are followed by hydrolysis of the dealkylated ureas to the corresponding anilines [Eq. (26)].

$$Cl-\langle\bigcirc\rangle-\underset{H}{N}-\underset{\underset{O}{\|}}{C}-NH_2 \longrightarrow Cl-\langle\bigcirc\rangle-NH_2 + CO_2 + NH_3 \quad (26)$$

Upon close examination of the various data available it is concluded that evidence supporting aniline formation and subsequent conversions is in part rather circumstantial and based on limited structural characterization. In some instances such evidence has even been deduced from results obtained with the related acylanilide and phenylcarbamate herbicides. Since there appear to be differences in the rate of aniline formation from phenylureas in various substrates, the following sections differentiate between aniline formation in plants, animals, and soils.

When whole plants or tissue sections were exposed to a variety of different phenylureas and subjected to chemical and/or radiochemical analyses, the amounts of free anilines detected at different time intervals were consistently small, i.e., less than 5% of the dose applied [124, 133, 160, 161, 194-197]. Characterization of these trace quantities was in most studies limited to chromatographic separation and normally no provisions were made to determine if such small amounts represented herbicide impurities or were formed by purely hydrolytic or photochemical reactions before and during extraction. In extensive plant experiments with several compounds labeled with ^{14}C in either the phenyl ring or in one of its substituents, free radioactive anilines were not observed or were present in such small quantities that positive identification was impossible [106].

The rate of aniline formation in plant tissues might presumably be masked by further rapid transformations of the aromatic amine. However, as is discussed below (see Sec. III, E, 2), no confirmed evidence for such conversion reactions in plants has been forthcoming.

In a recent study, Hoagland and Graf [209] subjected three herbicides, including fenuron, to hydrolysis experiments with crude enzyme preparations derived from roots and leaves of 19 different genera of weeds. Hydrolysis was measured by the rate of aniline formation. Although the unspecific assay procedure permitted no more than preliminary indications, it was observed that fenuron was hydroxylated by only one weed species, whereas propanil yielded the corresponding aniline with the majority of plant extracts examined.

From the data presented above and from additional experiments which demonstrate that N-dealkylated phenylurea metabolites are efficiently trapped by conjugation mechanisms (see Sec. III, F, 2), it is concluded that aniline formation by hydrolysis of either the original herbicides or their dealkylated phenylurea metabolites is a minor pathway in plants.

The same arguments with regard to the formation of aromatic amines apply to mammalian organisms. The quantities of free anilines recovered from the excreta of rats and dogs exposed to a number of different phenylureas were consistently low [174-178, 187, 188, 191]. When [carbonyl-^{14}C]chloroxuron and [^{14}C]chlorbromuron were administered to rats, no more than 0.1-2% of the radioactivity applied was accounted for by carbon dioxide in the expired air [174, 175]. Neither the unchanged herbicides nor their metabolites has been decarboxylated to any major extent in these animals. In addition, no significant metabolites of anilines have thus far been observed to occur in mammalian excreta or tissues, even in those experiments in which a ^{14}C balance with ring- or substituent-labeled phenylureas was established [176, 191].

Degradation of phenylureas and/or their metabolites to the corresponding anilines appears to proceed more easily in soil and microbial systems. However, present data do not provide information concerning the rate and significance of this reaction under practical conditions. The amounts of free anilines detected in soil samples varied from a few percent to almost 50% in terms of the total aniline-containing residues [110, 124, 133, 135, 138, 186]. Similar variation was observed when different investigators applied carbonyl-labeled phenylureas to soil and microbial systems [110, 124, 134, 137, 139]. The rate of formation of labeled carbon dioxide, which was usually followed for one to two months after application, ranged from a few percent to over 80% of the radioactivity applied. In those experiments in which high rates of $^{14}CO_2$ evolution were observed with 1-alkyl-1-alkoxyphenylureas, the corresponding free anilines did not accumulate in expected proportions. The anilines were proposed to disappear by different mechanisms [110, 139]. Variations in aniline formation are difficult to evaluate and, since no more consistent and extensive data are available, it can be assumed that this reaction varies from substrate to substrate or from compound to compound.

Most studies with soils and microbial systems have demonstrated that the pathway of aniline formation succeeds by N-demethylation/N-demethoxylation as shown in scheme (26). However, Wallnöfer and Bader [134, 139, 210] have recently presented evidence for direct hydrolysis of 1-methyl-1-methoxyphenylureas by a strain of <u>Bacillus sphaericus</u> isolated from soil treated with monolinuron. In nutrient cultures the corresponding anilines accumulated in sufficient quantity to be identified by infrared spectroscopy. In further experiments Engelhardt et al. [212-214] followed the fate of the alkylalkoxy moiety and proposed N,O-dimethylhydroxylamine as a metabolite, which was characterized by its dinitrophenyl derivative. The pathway. of direct hydrolysis of 1-alkyl-1-alkoxyphenylureas was suggested to proceed according to scheme (27). In addition to linuron, <u>Bacillus sphaericus</u> hydro-

lyzed monolinuron and metobromuron but failed to degrade the 1,1-dimethyl derivative fluometuron and the heterocyclic 1,3-dimethyl compound methabenzthiazuron. The significance of this hydrolytic pathway in the conversion of phenylureas under field conditions remains to be determined.

2. Further Conversions of Anilines

If phenylureas are degraded to the corresponding aromatic amines in certain biological systems, the anilines could be metabolized by a number of different reactions. Evidence for the occurrence of such aniline conversion products after application of phenylureas to various substrates is not extensive at this time.

One aniline transformation reaction has so far been described to occur in plant tissues. Evidence for this reaction was provided by Onley et al. [197] , who applied ^{14}C-ring-labeled diuron to corn seedlings in nutrient solution. In addition to small quantities of 3,4-dichloroaniline, they also isolated by gas chromatography and identified by ir and mass spectra

3,4-dichloronitrobenzene. Oxidation of the aniline was postulated to occur according to reaction (28). Unfortunately, the authors were not in a position

$$\text{Cl}-\underset{\text{Cl}}{\bigcirc}-\text{NH}_2 \longrightarrow \text{Cl}-\underset{\text{Cl}}{\bigcirc}-\text{NO}_2 \tag{28}$$

to decide if nitrobenzene formation represented a chemical or enzymatic oxidative process in plants, or a buildup of an artifact during extraction, or a combination of these phenomena. A deliberate search for the formation of nitrobenzenes after application of [trifluoromethyl-^{14}C]fluometuron and [ring-^{14}C]metobromuron to various crop species failed to substantiate this reaction [106]. Additional data are therefore required to verify the presence of this pathway in plants.

No confirmed aniline conversion products have yet been observed in tissues and excreta of mammals exposed to phenylureas. After two-year chronic feeding of high concentrations of unlabeled diuron to dogs and rats, trace quantities of both 3,4-dichloroaniline and 3,4-dichlorophenol were recovered from the urine of these animals [178]. However, the latter metabolite was not sufficiently characterized nor was the pathway of its formation verified. Reaction products of hydroxylation and acetylation of anilines have been detected in mammalian metabolism studies of phenylcarbamate herbicides [215-217]. However, the occurrence of such metabolites has not yet been described for phenylureas.

Anilines derived from phenylureas in soil systems and microbial substrates have been speculated to undergo the same reactions which apply to anilide and phenylcarbamate herbicides. However, only one confirmed conversion reaction has been described in the literature. Tweedy et al. [135, 218] supplied nutrient solutions containing ^{14}C-ring-labeled metobromuron to four microorganisms and recovered 4'-bromoacetanilide from the culture media of the two fungi <u>Talaromyces wortmanni</u> and <u>Fusarium oxysporum</u>. Since both organisms and also a <u>Bacillus</u> sp. and the alga <u>Chlorella vulgaris</u> efficiently acetylated 4-bromoaniline, formation of the latter was proposed to proceed according to scheme (29). The significance of this acetylation

$$\text{Br}-\bigcirc-\text{NH}_2 \longrightarrow \text{Br}-\bigcirc-\overset{\text{H}}{\underset{}{\text{N}}}-\underset{\underset{\text{O}}{\parallel}}{\text{C}}-\text{CH}_3 \tag{29}$$

reaction with regard to the phenylureas should be confirmed with additional herbicides and actual soil samples.

A further aniline-transformation pathway in soil [Eq. (30)] has recently been observed to occur with ^{14}C-ring-labeled 3,4-dichloroaniline [219]. Among other conversion products 3',4'-dichloroformylanilide was identified

$$Cl-\bigcirc(Cl)-NH_2 \longrightarrow Cl-\bigcirc(Cl)-NHCHO \qquad (30)$$

by mass-spectral analysis and by comparison with an authentic sample. The occurrence of this reaction with soil-applied phenylureas remains to be determined.

In a number of investigations phenylureas have been implicated to give rise to symmetrical and asymmetrical azobenzenes, azoxybenzenes, and other polymeric materials in soil and plant environments [139, 220, 221]. Such transformation products were observed when relatively high concentrations of anilide herbicides or free anilines were applied in a series of model and field experiments [220-231]. However, the authors are not aware of any experimental data that would demonstrate that phenylureas are converted into azobenzenes under either model or field conditions. To the contrary, extensive analyses by various laboratories failed to reveal detectable residues of azobenzene metabolites in a large number of soil samples collected from fields that had been treated with different urea herbicides [106, 232, 233]. When high concentrations of [trifluoromethyl-^{14}C]fluometuron and [ring-^{14}C]metobromuron were applied to strongly irradiated cotton and tobacco plants, no labeled azobenzene derivatives were observed [106]. In addition, Maier-Bode [234] analyzed numerous crop samples grown on phenylurea-treated soils but was unable to detect azobenzenes in plant material. Apparently the rate of formation of free anilines was too slow to permit the buildup of high concentrations required for condensation or coupling reactions. From data currently available, it is concluded that azobenzenes or similar materials do not represent metabolites or terminal residues of phenylureas in either soils or plants.

That the aniline moiety does disappear from soil environments has been demonstrated by many degradation experiments in which the total residue method was applied [41, 130, 186, 235, 236]. A typical example of such a degradation curve is shown in Fig. 6. The procedure used in these experiments determines the aromatic amine moiety after strong alkaline or combined acid-alkaline digestion of the entire substrate or sample [40, 41, 44, 235]. Hence, it not only accounts for the original herbicide, its N-dealkylated metabolites, and free aniline but also for conjugated or "bound" aromatic amines. Such bonding of halogenated anilines to certain soil constituents, including humic acid fractions, has recently been reported in the literature [237, 238]. Additional discussion of the metabolic fate of halogenated anilines in soils is presented in Volume 2, Chapter 12.

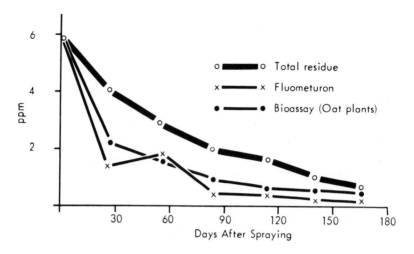

FIG. 6. Rate of disappearance of fluometuron in the 0- to 5-cm layer of a plant-free Swiss clay loam under field conditions. Soil composition: 18% clay, 12% silt, 67% sand, < 1% humus; pH 5.7. Herbicide applied at the rate of 4 kg active ingredient/ha. Total trifluoromethylaniline-containing residue determined chemically by subjecting the soil sample to alkaline hydrolysis, steam distillation extraction, and colorimetric measurement [40]. Structurally unchanged fluometuron measured after exhaustive soil extraction with acetone and separation of the herbicide from its potential metabolites on a Florisil column. Bioassay carried out with oat seedlings. (Reprinted with the permission of Guth et al. [235].)

F. Conjugation and Complex Formation

Enough experimental evidence has been accumulated to demonstrate that phenylureas, their N-dealkylated, and/or hydroxylated metabolites are conjugated with or bound to endogenous carbohydrate and peptide materials in plants and animals. However, whereas carbohydrate conjugates have been sufficiently identified to postulate definite pathways, the exact nature and metabolic significance of peptide or protein complexes remain to be determined.

No conjugation or complexing reactions have yet been described to occur with phenylureas or their degradation products in soil environments and microbial substrates. As already mentioned in Sec. III, E, 2, recent investigations indicate that free halogenated anilines, when applied to soils, are in part bound to such soil constituents as humic acids [237, 238]. The relevance of these sorption reactions to soil-applied phenylureas has not yet been established.

1. Protein-Peptide Complexes

In the first study on plant metabolism of urea herbicides, Fang et al. [152] observed that [carbonyl-^{14}C]monuron, when applied to bean leaves, was subjected to a complexing mechanism. Upon acid hydrolysis this complex yielded the unchanged herbicide. Within a period of 12 days about 20% of the original compound had been converted to the complexed form, which was later suggested to be of low molecular protein or peptide nature [239]. In a recent study Lee and Fang [198] applied the same labeled herbicide to excised bean and corn leaves and again detected formation of the described monuron complex. By gel filtration its molecular weight was estimated to be greater than 5000. An additional polypeptide conjugate was postulated to be formed by the N-monomethyl metabolite of monuron. Since characterization of both complexes was confined to acid hydrolyses and ninhydrin reactions, their structural properties need to be verified by more extensive biochemical and instrumental procedures before they can be included in any transformation scheme.

Sorption of phenylureas to plant protein fractions was also described by Nashed et al. [194, 195] who investigated the metabolism of the two alkylalkoxy compounds linuron and chlorbromuron in corn and cucumber. After the removal of acetone-extractable metabolites, additional quantities of "bound" aniline-containing materials were released from the extracted plants by alkaline digestion or hydrolysis with proteolytic enzymes. Further characterization of these proposed complexes is required to establish their role in phenylurea metabolism.

In model experiments, Camper and Moreland [240] demonstrated sorption of different phenylureas to bovine serum albumin. The sorption process was observed to be dependent on pH, ionic strength, and temperature, and was postulated to involve bond formation between either or both the amide hydrogen and the carbonyl oxygen of the herbicides with free amino groups of the protein. So far it has not been determined to what extent such sorption phenomena might account for the various protein and polypeptide complexes described above.

2. Carbohydrate Conjugates

The metabolism studies of Voss and Geissbühler [157] and Swanson and Swanson [169] with carbonyl and ^{14}C-ring-labeled phenylureas indicated that, in addition to N-dealkylated metabolites, significant quantities of much more polar transformation products were removed from plants by either methanol or acetonitrile extraction. Further work conducted independently by both the Ciba-Geigy and Fargo laboratories indicated that these polar metabolites represented glucosidic conjugates of hydroxymethyl intermediates formed during N-dealkylation [180, 241]. The formation and structure of these conjugates have now been verified in a detailed and excellent study by Frear and Swanson [242] . Two polar metabolites were recovered from excised cotton leaves in experiments with [trifluoromethyl-^{14}C]fluometuron and with [^{14}C]monuron labeled in different positions. After purification by thin-layer and column chromatography the two conjugates were separated by gel filtration. The aglycone and carbohydrate moieties were characterized by mild acid or glucosidase hydrolysis and revealed the structures of the two β-D-glucosides to be as shown for monuron in formulas (12) and (13):

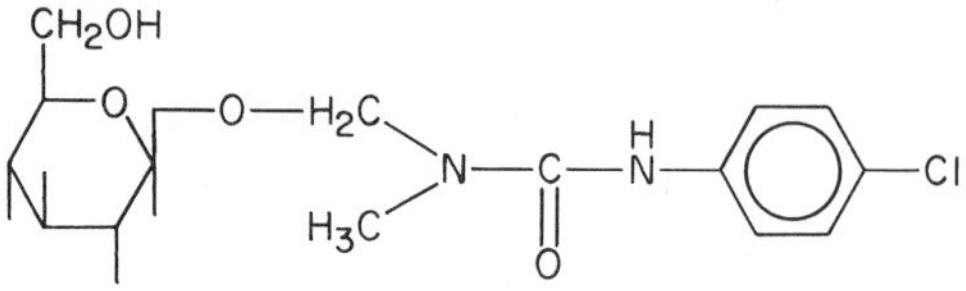

(12)

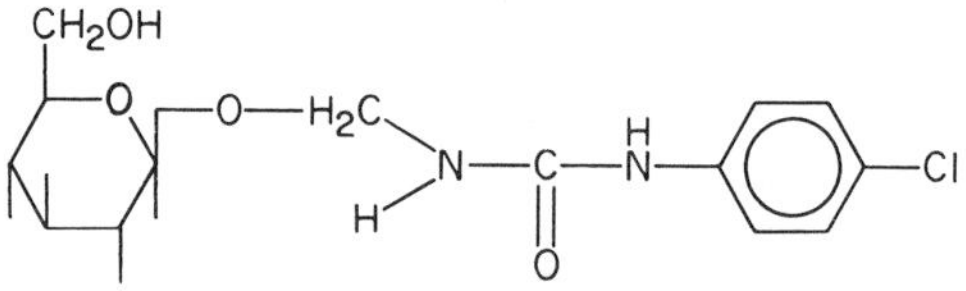

(13)

Formation of the two carbohydrate conjugates was suggested to be catalyzed by a glucosyl transferase reaction with the unstable 1-methyl-1-hydroxy-methyl and the 1-hydroxymethyl intermediates, respectively (for formation and significance of hydroxymethyl metabolites see Sec. III, G). In these short-term experiments the two conjugates represented about 25% of the methanol-soluble radioactivity 24 hr after application. Additional methanol-soluble and -insoluble conjugates were observed which presumably were formed from 1-hydroxymethyl-3-(p-chlorophenyl)urea and which were described to involve higher molecular weight carbohydrate fractions.

In most mammalian metabolism experiments with phenylureas the urine of the animals was subjected to routine enzymatic hydrolysis with β-glucu-ronidase/aryl sulfatase preparations [174-176, 187, 188, 204] . By this treatment significant quantities of ring-hydroxylated and in some investi-gations also N-dealkylated metabolites were released from the aqueous phase into organic fractions. This observation demonstrated that the men-tioned metabolites were wholly or in part eliminated as β-D-glucuronides or ethereal sulfates. However, neither the defined structures nor the proportions of such conjugates were established in the various animal exper-iments. In view of the identified 1-hydroxymethyl conjugates as described above for plant tissues, a search for such conjugates in mammalian excreta seems to be warranted.

Except for the ring-hydroxylated compounds, all of the protein-polypeptide complexes and carbohydrate conjugates formed by phenylureas and their metabolites still contain the unchanged aniline moiety. Provided they do occur as rather persistent metabolites in certain biological media, they would be accounted for by the chemical residue method as described in Sec. I, D, 2.

G.   Biochemical and Enzymatic Mechanisms

The structure of 1, 1-dialkyl- and 1-alkyl-1-alkoxyphenylureas might suggest that the classical urease type of reaction would cleave these compounds directly to their corresponding anilines, carbon dioxide, and dialkyl- or alkylalkoxyamines, respectively. In spite of the widespread distribution of urease in microorganisms and plants, the enzyme apparently exhibits an absolute specificity for urea. Sumner and Somers [243] examined many potential substrates, including substituted ureas and related compounds, but reported no hydrolysis of these substances by urease.

Although urease itself is inactive, direct hydrolysis of 1-alkyl-1-alkoxyphenylureas was demonstrated with a strain of Bacillus sphaericus [139, 210-214]. Scheme (27) in Sec. III, F, 1 has shown that the reaction yields the corresponding aniline, carbon dioxide, and N, O-dimethylhydroxylamine. The identity of the latter metabolite suggested that the amide bond was the site of enzymic cleavage. Cell-free extracts derived from the bacterium did not demonstrate activity toward linuron unless the organism was precultured in the presence of the herbicide. This would indicate that linuron induced formation of the hydrolytic enzyme in Bacillus sphaericus. Examination of a partially purified cell-free preparation showed that different acylanilides were degraded at rates at least 10 times higher than those determined for methoxy-substituted phenylureas. Since the highest specific activity was observed with aminoacylanilides, the enzyme was classified as an acylamidase. Apparently the activity increased with increasing hydrophilic properties of the substrates. This phenomenon was suggested as the reason for the inability of the enzyme to hydrolyze 1, 1-dimethylureas. It would be interesting to examine induction of hydrolytic activity on a larger scale. Such a mechanism might be responsible for the acknowledged more rapid degradation of methylmethoxyphenylureas when compared to dimethyl derivatives.

In contrast to the limited information on direct cleavage of the ureido grouping, the process of stepwise N-demethylation has not been demonstrated for a number of phenylureas in soil, plant, and animal systems and therefore represents the major initial pathway in urea herbicide degradation.

The N-demethylation of substituted ureas resembles the N-dealkylation as observed for alkylated amines [185, 244], amides [245, 246], and carbamates [247, 248]. This common oxidative, enzymatic reaction has been most extensively investigated with mammalian liver microsomal systems and in general requires molecular oxygen and reduced pyridine nucleotides [249, 250].

Apparently the velocity of the reaction depends on the type of nonalkyl nitrogen substituents. Thus amides are less rapidly dealkylated than ionizable amines [246]. Unfortunately, no systematic kinetic studies on urea-type compounds as compared to amines, amides, and carbamates are available at this time. Hodgson and Casida [247] examined the dealkylation of a large number of carbamates in vitro with an enzyme system derived from rat liver microsomes and included monuron and diuron in their ancillary group of noncarbamates. These authors observed that the ureas were less readily dealkylated than the carbamates. They concluded that the ureas, which are quite insoluble in water, might not be easily available to the enzymes. On the other hand, high lipid solubility has been demonstrated to be a significant factor in permitting the dealkylation of alkylated amines [251]. Apparently additional properties or structural features of urea compounds, which are as yet undetermined, might be more important than their water solubility in controlling their rates of dealkylation by mammalian liver microsomal systems.

A plant microsomal enzyme that is able to successively N-demethylate 1,1-dialkylated phenylureas was isolated from etiolated cotton seedling hypocotyls by Frear et al. [170, 171]. Active enzyme preparations were also derived from leaf tissues of several plant species. By using ^{14}C-methyl-labeled substrate, it was demonstrated that the enzyme formed 1 mole of formaldehyde for each mole of substrate [Scheme (31)]:

$$(31)$$

The mixed function oxidase type of enzyme required molecular oxygen and either NADPH or NADH as cofactors. It appeared to be quite specific for the 1,1-dialkylphenylureas since linuron and a number of additional compounds which have been reported to be N-dealkylated in vivo by plants (phenylacetamides, methylcarbamates, triazines) did not function as sub-strates for the cotton microsomal system [252]. Inhibition of the enzyme by several specific inhibitors (carbon monoxide, ionic detergents, sulfhydryl reagents, chelating agents) provided indirect evidence for a microsomal electron transport system in plants similar to that reported for animals. However, several differences between cotton N-demethylase and the corre-sponding mammalian systems were also noted. These included greater instability and decreased sensitivity to such specific inhibitors or syner-gists as SKF 525-A and Sesamex.

During the isolation and the characterization of the demethylated reaction products of the cotton microsomal N-demethylase system, an unstable me-tabolite was detected by thin-layer chromatography and autoradiography. The formation and the identity of this metabolite were examined in a series of experiments in which 1-methyl-3-(p-chlorophenyl)urea was used as a substrate [253]. Acid hydrolysis of the unstable product yielded 4-chloro-phenylurea and formaldehyde. This suggested the structure to be 1-hydroxy-methyl-3-(p-chlorophenyl)urea (14):

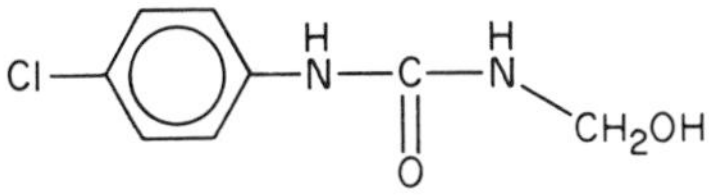

(14)

Since stabilization of this labile intermediate by known derivatization reac-tions failed, its identity had to be confirmed by other means. More definite proof for its structure was obtained by demonstrating that both the corre-sponding β-D-glucoside conjugate [structures (12) and (13) in Sec. III,F,2) and the metabolite itself were converted to 1-methoxymethyl-3-(p-chloro-phenyl)urea (15) when they were stored in anhydrous methanol. The identity

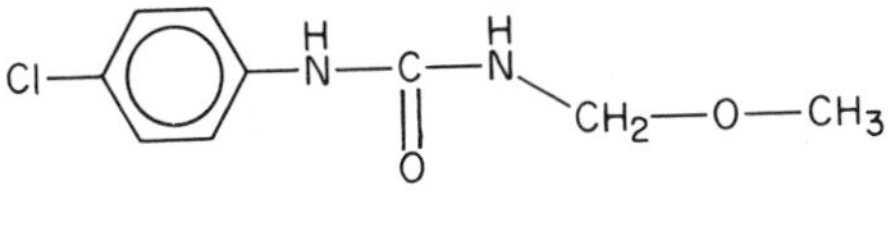

(15)

of the methoxymethyl derivative was established by infrared and mass-spectral analyses.

At present there is only indirect evidence for the formation of a hydroxy-methyl metabolite in the initial degradation of dialkylated phenylureas [253]. However, from the occurrence of the corresponding glucoside conjugate (see Sec. III,F,2) it may be deduced that both N-demethylation reactions proceed by intermediate formation of the hydroxymethyl derivatives. The mechanism of N-dealkylation of phenylureas, including the trapping of intermediates by glucosyl- or glucuronyl-transferase systems, thus resembles the pathways established for phenylacetamides, phosphoramides, and methylcarbamates [245-248, 254, 255].

Additional information on the nature of the active site and the reaction mechanisms of the cotton microsomal N-demethylase system was obtained by inhibition and stable isotope experiments [252, 256]. The description of these specialized enzyme reaction studies is beyond the scope of this review on transformation pathways.

The second major pathway postulated for phenylureas is ring hydroxylation. Since the normally preferred para position is either substituted or sterically hindered, as for example in fluometuron, ring hydroxylation seems to occur mainly in the ortho positions. Like the process of N-dealkylation, ring hydroxylation of foreign compounds has been most extensively investigated with mammalian liver microsomal systems [257]. The mechanism again requires oxygen and reduced pyridine nucleotides. However, the applicability of this aromatic oxidation reaction to phenylureas remains to be established.

H. Conclusions

When compared with the data available a few years ago [258, 259] the present review demonstrates that considerable progress has been made in elucidating the pathways of transformation of urea herbicides in biological environments. Table 3 summarizes the information available today and evaluates the significance of the various pathways with regard to the three main biological substrates. A number of reactions (N-dealkylation, N-dealkoxylation, ring hydroxylation, carbohydrate conjugation) have been established with enough certainty to define in qualitative and quantitative terms their role in urea herbicide transformations. However, the table demonstrates that several gaps remain to be closed before a complete description of the metabolic pathways of urea herbicides can be given. Thus, the extent of aniline formation and subsequent transformations in soil substrates has not yet been clearly defined, although several types of experiments have demonstrated that aniline-containing metabolites do disappear from soil. Additional data are needed to describe the polypeptide and protein complexes that are apparently formed in plants and the different water-soluble conjugates that are eliminated by mammals. Metabolism experiments should be extended to phenyl derivatives with N-alkyl chains other than methyl and methoxy and also to ureas substituted with saturated and heterocyclic ring systems.

In spite of the detailed data now available on the N-dealkylation mechanism of phenylureas in plant microsomal systems, further efforts are

TABLE 3

Present Status of Biochemical Pathways and
Enzymatic Mechanisms Involved in the Transformation of
Phenylurea Herbicides by Plants, Mammals, and Soil (Microbial) Systems

Pathway or Mechanism	Status and significance
N-Dealkylation	Verified as major initial pathway in plants, mammals, and soil microorganisms
Hydroxyalkyl intermediates	Demonstrated with cotton microsomes and by plant studies in vivo; assumed for mammals
Conjugation of intermediates with carbohydrates	Demonstrated by plant experiments in vivo and in vitro; assumed in mammals
N-Dealkoxylation	Demonstrated in plants, mammals, and fungi
Hydroxylamide intermediates	Indicated by preliminary experiments
Ring hydroxylation	Demonstrated in rats (major pathway for alkylalkoxy compounds); indicated in plants by preliminary characterization
Conjugation of hydroxy metabolites	Demonstrated in rats by enzyme hydrolysis
Oxidation of ring substituents	Demonstrated for 4-methyl in rats and plants; minor pathway for $3-CF_3$ in soil
Hydrolysis (cleavage of ureido grouping to aniline)	Demonstrated as minor pathway in plants and mammals; apparent varying rates in soil; direct cleavage of alkylalkoxy compounds demonstrated with one bacterial strain
Aniline Transformations Oxidation to NO_2	Indicated by plant studies but not confirmed
Acetylation	Postulated for soil microorganisms
Formylation	Assumed from studies with free anilines
Azobenzene formation	Demonstrated as absent in soils and plants
Ring hydroxylation	Assumed
Sorption/conjugation	Assumed from soil studies with free anilines
Complexing with peptide-protein structures	Postulated from plant studies and from model experiments with serum albumin

required to characterize the different metabolic reactions at the enzyme and molecular levels. More extensive comparative studies would help to establish the relationship between chemical structure, enzymatic activity, and the ease of degradation. The advantages of in vitro enzyme systems for such a comparative approach have been described by Frear et al. [252]. On the practical side, this type of work will help to develop efficient herbicides which create increasingly fewer contamination problems for the total environment.

IV. MODE OF ACTION

A. Biochemical Mechanisms

1. The Inhibition of Photosynthesis

Soon after the discovery of the herbicidal properties of substituted phenylureas their potent effect on the photosynthetic mechanism was recognized. Thus Cooke [260] and Wessels and van der Veen [261] reported that these compounds inhibited the Hill reaction, i.e., the evolution of oxygen in the presence of living chloroplasts and a suitable hydrogen acceptor; see Fig. 7 for a scheme of the photosynthetic mechanism.

The photosynthetic process is based on a light-induced electron transfer from water to carbon dioxide, which results in the formation of oxygen and a carbohydrate. In the light-dependent part of the photosynthetic mechanism two coupled photoreactions act in series. They are driven by two different pigment systems (see Fig. 7, PS I and PS II) capable of absorbing light energy.

The transfer of the absorbed light energy from PS II to the photochemical reaction center P_x results in the excitation of the latter. This provides electrons for the reduction of compound Q. At the same time another compound (Y) is oxidized, which is the substance that is supposed to be responsible for the oxidation of water and thus for the final photosynthetic oxygen production. This light-induced transfer of electrons and the evolution of oxygen in the presence of an added hydrogen acceptor is known as the Hill reaction.

Photosystems I and II are interconnected by an electron-transfer complex which includes several cytochromes (cyt.), plastoquinone (PQ), and plastocyanine (PC). The electron transfer occurring from the quencher Q along this cytochrome chain is associated with a phosphorylation reaction, i.e., the formation of ATP. At the end of the chain the redox potential has reached a level that is no longer sufficient for the reduction of carbon dioxide. The further electron transfer is then mediated by the excitation of the photochemical reaction center P 700 by uptake of light energy from PS I. The redox potential of compound Z is now sufficient for ferredoxin (Fd) and NADP

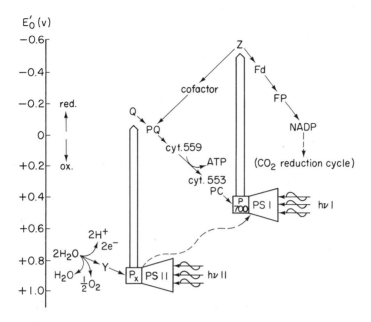

FIG. 7. Diagram of the electron-transfer chain in photosynthesis. (Reprinted with the permission of van Rensen [270].)

reduction, the latter being catalyzed by a flavoprotein enzyme (FP). The resulting NADPH is finally utilized for carbon dioxide reduction (dark reac reaction).

Another electron-transfer complex with an unknown "cofactor" exists between Z and the cytochrome chain. ATP formation due to this type of electron transfer is known as cyclic photophosphorylation. It requires the presence of a suitable catalyst, such as phenazine methosulfate (PMS). Pseudocyclic photophosphorylation occurs when reduced ferredoxin is oxi-dized by oxygen. This results in a lack of oxygen production and NADP reduction as observed in cyclic photophosphorylation. Photosystem II, how-ever, participates in pseudocyclic photophosphorylation, since electrons are transported from water by the electron carriers and ferredoxin to oxygen.

Monuron and diuron were found to inhibit the light-dependent part of the photosynthetic process much more drastically than phenylurethane, which was used as a standard [261]. These studies indicated the possible mech-anism of action of herbicidal substituted ureas, and at the same time, initiated further basic research in photosynthesis whereby the well-known compounds CMU (monuron) and DCMU (diuron) served as important model inhibitors. Many fundamental aspects of the photosynthetic process were subsequently investigated, but it would be far beyond the scope of this chapter to present a detailed description of the scientific findings available today. Therefore, the reader is referred to several excellent review articles published by other workers [262-271].

Some early experimental data supported the assumption that herbicidal inhibitors of the Hill reaction kill plants by depressing the photosynthetic formation of carbohydrates. Thus, after the addition of glucose to the culture medium, the concentration of phenylureas required to inhibit growth of Chlorella vulgaris increased [272]. In addition, sucrose supplied to barley plants through cut leaf tips kept plants alive and growing in the presence of lethal concentrations of fenuron, monuron, diuron, neburon, and 1-methyl-3-(3,4-dichlorophenyl)urea [273]. The starvation hypothesis, however, was soon contradicted by data which demonstrated that the toxicity of monuron to Chlorella and Euglena fed with glucose or succinate, respectively [274, 275], did not result solely from depletion of nutrient supply. It was suggested that the buildup of a phytotoxic compound which originated from blocking the oxygen-liberating pathway was involved [274]. This hypothesis was also supported by Ashton [276] who described a relationship between light intensity and toxicity symptoms of monuron in bean plants. The author assumed that phytotoxicity was not caused by monuron itself but rather by a secondary substance formed by some mechanism involving the interaction of the herbicide and light.

A correlation between the inhibitory effect of various phenylureas in algal cultures and the degree of their photochemical reactivity with flavin mononucleotide (FMN) led to the conclusion that inhibition of photosynthesis by ureas was due to an interaction with FMN or a flavoprotein [277]. At the same time this interaction was found to inactivate phenylureas since FMN sprayed on the leaves of bean plants prevented the herbicidal action of monuron when applied to the roots. However, no indications for the postulated flavin-urea complex were obtained by Homann and Gaffron [278] in their studies on the effect of monuron on various photooxidation reactions. Urea herbicides were also demonstrated to stimulate fluorescence in Chlorella pyrenoidosa [279]. The authors concluded that the observed correlation between the inhibition of oxygen evolution and the increase in fluorescence suggested a close relationship between energy requirements for oxygen evolution (oxidation) and fluorescent energy.

The discovery that urea herbicides inhibited the Hill reaction was soon followed by studies undertaken to locate the inhibition site. Bishop [280] observed that diuron did not inhibit carbon dioxide fixation in Scenedesmus cultures which were adapted to photoreduction with hydrogen instead of water. In addition Rhodospirillum rubrum, a photosynthetic bacterium that lacks the oxygen-liberating pathway associated with photosystem II (see Fig. 7), was relatively insensitive to monuron [275]. Diuron prevented photosystem II from reducing cytochrome in <u>Porphyridium cruentum</u> [281] , and produced a very high level of fluorescence in <u>Ankistrodesmus braunii</u> under aerobic conditions but not in hydrogen-adapted algae [282] . The same herbicide inhibited endogenous oxygen consumption, which depends on photoreaction II, in chloroplasts of normal pea plants. No such inhibition was observed in a mutant strain with a defective electron transport chain which made the plants unable to assimilate carbon dioxide and liberate oxygen [283] . Other authors [284, 285] investigated the effects of ureas on adenosine triphosphate (ATP) formation and found that monuron and diuron inhibited pseudocyclic photophosphorylation (Fig. 7), which is catalyzed by FMN to a much greater extent than cyclic photophosphorylation (Fig. 7) which is catalyzed by phenazine methosulfate (PMS). In addition, inhibition of NADP photoreduction by phenylureas was overcome by an artificial electron donor system such as ascorbate-DCPIP (2, 6-dichlorophenol-indophenol) [286] . The oxidation of ascorbate and the reduction of nicotinamide adenine dinucleotide phosphate (NADP) is presumed to involve only light reaction I and circumvents the requirement for hydroxyl ions as electron donors. All the findings just given favor the hypothesis that the site of photosynthesis inhibition is located close to photosystem II. However, at high concentrations ureas can exhibit an inhibitory effect on cyclic photophosphorylation as well, which is only dependent on photosystem I [287-288] .

Most workers today accept the view that ureas interfere with the reducing side rather than with the oxidizing side of photosystem II. From several sources of evidence van Rensen [270] recently concluded that diuron forms a complex with the oxidized form of an unknown component located in the electron-transfer pathway close to photosystem II, and this component also takes part in cyclic electron transport. The advantage of this hypothesis is that it explains the inhibition of noncyclic electron transport as well as that of cyclic electron transport which occurs with higher concentrations of phenylureas. Since photosynthesis inhibition by ureas is assumed to be due to hydrogen bond formation of the NH group and/or the carbonyl oxygen with the active centers in the chloroplasts [285] , van Rensen proposed the formation of such bonds between diuron and the oxidized state of plastoquinone which is thought to be a possible candidate for the still unknown substance mentioned above. Structure (16) illustrates that the proposed complex formation is only possible when plastoquinone is in its oxidized state. A mechanism for phytotoxicity caused by diuron was recently proposed by Stanger and Appleby [289] . They found that DCPIP protected functional spinach chloroplasts from diuron-induced toxicity even in the presence of methylamine hydrochloride, a known inhibitor of photophosphorylation. ATP deficiency was therefore not considered to be the primary cause of diuron toxicity. The authors observed, however, that in diuron-poisoned plants carotenoid pigments began to degrade before chlorophyll. This finding was interpreted as follows: As a result of the interruption of the electron flow diuron induces phytotoxicity by increasing the concentration of oxidized

(16)

chlorophyll and by inhibiting NADPH formation. Under normal photosynthetic conditions the latter compound is necessary for the reduction of
carotenoids which in turn inactivate the lethal chlorophyll-oxygen complex.

2.   Structure-Activity Correlations

The molar concentration of phenylureas required for 50% inhibition of
photosynthetic activity (I_{50} values) of isolated chloroplasts and algae varies
widely. This is demonstrated by Table 4 which summarizes the values
obtained by different investigators. It must be kept in mind, however, that
all data reported refer to the herbicide concentration in the reaction mixture
and not to the amount of inhibitor inside the chloroplast. This quantity is
actually much smaller, as estimated for monuron and diuron: one inhibitor-
sensitive site for approximately 2500 chlorophyll molecules [290].

Early attempts to correlate chemical structure with biological activity
had already indicated the need for a free and sterically unhindered amido
hydrogen, which is believed to be responsible for significant inhibitory action.
Halogen substituents in the ortho position of 1, 1-dimethylphenylureas reduce
inhibition and biological effects, since they prevent the amido hydrogen from
interacting with adjacent molecules. The formation of hydrogen bonds with
the active center(s) of the chloroplast could also occur through the carbonyl
oxygen, since replacement of this oxygen with sulfur caused a considerable
loss of inhibition potency [285]. The observation that diuron can be removed
from Scenedesmus cells by washing favors the assumption that the mentioned
bonds are rather weak [270].

The values presented in Table 4 indicate that the in vitro activity of
substituted dimethylphenylureas can be arranged according to the following

Inhibitory Activity of Various Substituted Phenylurea Derivatives as Determined with Isolated Chloroplasts of Several Plant Species or Chlorella[a]

| Common name | Substituent[b] | | | References | | | | | |
	R$_1$	R$_2$	R$_3$	261	279[c]	285	291	292	293[c]
Fenuron	Me	Me	—	4.4		5.2	4.6		5.1
	Me	Me	2-Cl			3.3			
	Me	Me	3-Cl	5.7		6.3			6.1
Monuron	Me	Me	4-Cl	5.4	6.2	6.3	5.4		6.1
Monolinuron	MeO	Me	4-Cl						5.7
Metobromuron	MeO	Me	4-Br						6.0
	Me	H	3,4-di-Cl			7.0			
	Et	H	3,4-di-Cl			6.2			
	Pr	H	3,4-di-Cl			5.8			
Diuron	Me	Me	3,4-di-Cl	6.7	6.8	7.5	6.5	5.8	7.0
	Et	Et	3,4-di-Cl			6.8			
	Pr	Pr	3,4-di-Cl			4.7			
Neburon	Me	Bu	3,4-di-Cl					6.9	
Linuron	MeO	Me	e,4-di-Cl					6.7	6.4
Chlorbromuron	MeO	Me	3-Cl,4-Br						7.0
	Me	Me	3,5-di-Cl			6.0			
Chlortoluron	Me	Me	3-Cl,4-Me						6.3
	Me	Me	4-Me	4.5					
	Me	Me	4-MeO	4.3					
Fluometuron	Me	Me	3-CF$_3$	5.2					5.8
	Me	Me	4-CF$_3$	5.4					
Chloroxuron	Me	Me	4-(4'-Cl-phenoxy)					6.8	7.0

[a]The results are expressed as pI$_{50}$ values (negative logarithms of I$_{50}$ values).

[b]Me, methyl; MeO, methoxy; H, hydrogen; Et, ethyl; Pr, n-propyl; Bu, n-butyl.

[c]Experiments with Chlorella.

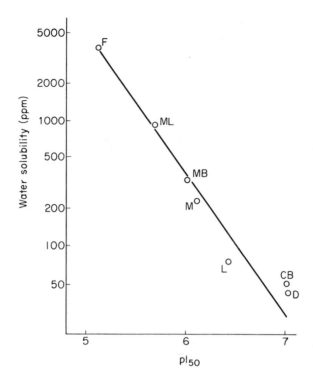

FIG. 8. Correlation between water solubility and pI_{50} values of fenuron (F), monolinuron (ML), metobromuron (MB), monuron (M), linuron (L), chlorbromuron (CB), and diuron (D). The test organism was Chlorella pyrenoidosa [293].

sequence: 3,4- > 3,5- > 3- and 4- > unsubstituted 2-. The majority of the commercial phenylureas are substituted in the 3,4, 3, or 4 position, with most substituents being halogen atoms. The inhibitory activity is reduced if the halogen substituents are replaced by trifluoromethyl, alkyl, or nitro groups. The finding of Good [285] that 3-(3,4-dichlorophenyl)-1-methyl-urea is a strong inhibitor of the Hill reaction demonstrated that the N-demethylated metabolites (see Sec. III, B) are likely to contribute to the phytotoxic action of urea herbicides.

The lipophilic nature of the ring substituent(s) is of primary importance for inhibition potency of phenylureas. This is the major conclusion to which Hansch and Deutsch [294, 295] arrived when studying the effect of various ring substituents in several groups of Hill-reaction inhibitors. The authors evaluated the pI_{50} values of Wessels and van der Veen [261] by multiple regression analysis, introducing substituent parameters such as partition coefficients, electron distribution, and steric factors. They concluded that phenyl substituents enhanced inhibition by affecting the lipophilic properties of the chemical and postulated that the inhibitors acted in a very lipophilic compartment or on a lipophilic enzyme. Since the lipophilic nature of a compound is normally reflected in its water solubility (see Table 1) an attempt was undertaken to relate this parameter with pI_{50} values obtained from experiments with Chlorella [293]. Figure 8 demonstrates that a plot of water solubility (log scale) versus pI_{50} values of several halogenated ureas resulted in satisfactory correlation. The number of compounds evaluated in this respect, however, is still too limited to suggest a general interdependence.

The lipophilic nature of the 1,1-dialkyl side chain contributes much less to the inhibitory effect of ureas. From the data of Good [285], who included diuron and the corresponding diethyl and dipropyl analogs in his experiments, it appeared that an increase in length of the alkyl groups resulted in a decrease of inhibitory activity (Table 4). Apparently it is not the overall lipophilic character of the herbicide that is important for inhibition, but only the hydrophobic properties of the ring and its substituents [294].

B.   Selectivity

A number of review articles on herbicide selectivity [172, 296-299] demonstrate that this feature is determined by several parameters whose contribution is quite variable and often based on complex interactions. The major biological and environmental factors involved are adsorption and leaching properties of the compounds in soil, uptake, translocation, and metabolism in plants, and behavior at the site of action. Knowledge of these factors can certainly help to explain selective activity, but its application to the design of specific weedkillers has been limited. The search for and the development of a new selective herbicide still depend on empirical screening, although some progress has been made in predicting selectivity of substituted phenylureas. This became possible by studying the relationship between chemical structure and selective activity as described in Sec. I,A,1 for such herbicide pairs as diuron-neburon diuron-monuron, and diuron-metoxuron.

This section is confined to two mechanisms that have been postulated to be involved in urea herbicide selectivity: differential uptake and/or translocation, and differential metabolism. Selectivity caused by special formulation and application techniques is not considered [296]. The complex problem of selectivity due to the placement of substituted ureas in soil has not been investigated sufficiently to justify its inclusion in this chapter. In addition, it is not regarded as a phenomenon specific for urea herbicides.

One of the early approaches toward an understanding of selectivity of herbicides that inhibit photosynthesis was that of van Oorschot who followed the exchange of carbon dioxide and the rate of transpiration of whole plants in the presence of various herbicides. In a first set of experiments he demonstrated that different ureas exhibited different effects in a single plant species. When diuron and fluometuron were added to a nutrient solution in a concentration of 2×10^{-5} M, they severely inhibited carbon dioxide uptake in carrot plants, whereas an equimolar concentration of linuron produced only a very slight effect [300]. In a second series of experiments he investigated the physiological response of two different plant species toward a single herbicide, i.e., monuron. This compound caused almost complete inhibition of carbon dioxide absorption in both corn and plantain (Plantago lanceolata). Removal of the herbicide from the nutrient solution resulted in fast recovery of carbon dioxide uptake in plantain but not in corn [301]. These results demonstrated some physiological inactivation of the herbicides in the tolerant plant species. However, no indication was obtained at that time as to the nature of such inactivation. More recent studies by Swanson and Swanson [169] with leaf discs of several plant species confirmed these findings by demonstrating that the rate of metabolic breakdown of monuron in plantain (Plantago major) was very high when compared to that in corn.

The physiological and biochemical mechanisms which have been postulated to be involved in urea herbicide selectivity are summarized in Table 5. It appears that differential metabolism, often coupled with differential uptake and translocation, represents an important factor in determining selectivity. Differences in uptake and/or translocation alone seem to be exceptional. In addition, such experiments require careful interpretation, since differential absorption in two plant species can be the reverse in nutrient solution when compared with soil. This has been demonstrated for chloroxuron with Galinsoga parviflora and Polygonum convolvulus [153]. The selective behavior of the two weed species under field condition could only be explained by using the data obtained from the normal soil environment in connection with subsequent metabolic studies [124].

The problem of differential uptake and/or translocation becomes even more complex when pre- and postemergence applications are compared. Hogue [302] applied six urea herbicides (monuron, monolinuron, metobromuron, diuron, linuron, chlorbromuron), both pre- and postemergent, to tomato (Lycopersicum esculentum) and coriander plants (Coriandrum sativum). He found that both types of treatments resulted in equal sensitivity in tomatoes. Similar results were also obtained for coriander with the monohalogenated ureas; however, the dihalogenated derivatives were much more toxic after preemergence application than after postemergence application. From these results Hogue concluded that root uptake and/or translocation which dominate under preemergence conditions were limited with diuron, linuron, and chlorbromuron.

TABLE 5

Physiological and Biochemical Mechanisms Involved in Selectivity of Urea Herbicides

| Herbicide | Plant species compared | | Selectivity[a] | Reference |
	Susceptible	Tolerant		
Chlorbromuron	Cucumber	Corn	Metabolism	195
	Amaranthus retr.	Parsnip	Uptake/translocation	304
Chloroxuron	Galinsoga parviflora	Polygonum convolvulus	Metabolism	124
	Morningglory	Soybean	Uptake/translocation	305
Diuron	Soybean, oats, corn	Cotton	Metabolism	161
	Soybean, corn	Cotton, plantain	Uptake/translocation, metabolism	169
	Hawaiian sugarcane varieties		Uptake/translocation, metabolism	306
Fluometron	Cucumber	Cotton	Metabolism	160
	Setaria, Amaranthus	Cotton	Metabolism, phytotoxicity of metabolites	193, 200
Linuron	Amaranthus retr.	Parsnip	Uptake/translocation	304
	Tomato	Parsnip	Uptake/translocation, metabolism	164
	Ambrosia artemisiifol.	Carrot	Uptake/translocation, metabolism, phytotoxicity of metabolites	196
Metobromuron	Sinapis arvensis	Galium aparine	Metabolism	159
	Sinapis arv., Veronica pers.	Galium aparine	No differential mechanism detected	158
Monuron	Soybean, oats, corn	Cotton	Metabolism	161
	Corn, soybean	Cotton	Metabolism	169

[a]Based mainly on differential.

A rather unique selectivity mechanism was recently proposed by Strang and Rogers [163]. They found that root-applied [^{14}C]diuron accumulated in the lysigenous glands of cotton plants and assumed that this phenomenon could contribute to the resistance of cotton to substituted phenylureas by preventing them from reaching the chloroplasts. However, the same mechanism failed to explain the intermediate tolerance of cotton to atrazine since no difference in phytotoxicity was observed when glanded and nonglanded isogenic lines of a special cotton variety were compared [303].

Table 5 presents sufficient evidence that differential metabolism is one of the major factors responsible for the selectivity of phenylurea herbicides. In particular, cotton exhibited a remarkable ability to degrade monuron, diuron, and fluometuron [160, 161, 169, 193, 200] in contrast to certain monocotyledonous plant species, such as corn, oats, and Setaria. Differences in metabolism, however, were always based on different rates of degradation and not on differences in biochemical transformation pathways.

The presence of phytotoxic N-monomethyl intermediates in plants can also contribute to selectivity as demonstrated by Rubin and Eshel [200] and Kuratle et al [196]. The monomethyl derivative of fluometuron was more phytotoxic to Amaranthus retroflexus than to cotton. In addition, both the N-methyl and N-methoxy derivatives of linuron were more herbicidal to common ragweed (Ambrosia artemisiifolia) than to carrot plants. The question of whether or not differential phytotoxicity is based on differential metabolism of the herbicidal intermediates remained unanswered.

Another mechanism suggested as contributing to the selectivity of substituted phenylurea herbicides is binding or conjugation [194, 195] of the active parent compounds and/or their partially detoxified metabolites. According to Frear et al. [299] this process may result in immobilization of the toxicant which prevents it from reaching the primary site of action in the chloroplast.

The scientific data are insufficient at the present time to explain completely the selectivity of certain ureas. The search for a new selective herbicide is still based on biological testing. Any success in this direction is more the result of trial and error than that of objective synthesis of molecules based on acquired biological knowledge. The findings reviewed above, however, have contributed to a partial understanding of the phenomenon of selectivity. Future studies on the relation between chemical structure of herbicides and their selectivity should not be confined to the terminal biological effects. These studies should also take into account the relation between chemical structure of herbicides and their physical and biochemical characteristics. The latter factors have a significant effect on the behavior of herbicides in soil and plants and therefore contribute to the final expression of herbicide selectivity.

The literature review for this chapter had been concluded in September 1972. This addendum covers the period from October 1972 to March 1974. During this time no publications came to the attention of the authors which would make significant new contributions to Secs. I, II, and IV, or which would alter the views expressed therein. Recent reference numbers have therefore been inserted into appropriate parts of the text. On the other hand, a number of papers appeared that provide supporting evidence or some new leads for Sec. III. This additional information is reviewed below under the same subheadings as used in Sec. III.

A. N-Dealkylation and N-Dealkoxylation

The occurrence of the mechanism of N-dealkylation has been confirmed for several 1,1-dimethyl- and 1-methyl-1-methoxy-subsituted ureas in a number of additional biological materials including various crop and weed species [310-314], soil samples [308, 328], and microbial cultures [315-317]. The fungus Rhizopus japonicus has been demonstrated to remove the 1-(1-methyl-2-propinyl) moiety from buturon to yield 3-(p-chlorophenyl)-1-methylurea according to scheme (32) [316]. The identity of the latter metabolite is said to have been verified by nuclear magnetic resonance and mass spectrometry.

$$ (32) $$

Buturon

N-Demethylation has now been shown to also occur with the heterocyclic urea derivative methabenzthiazuron [3-(2-benzothiazolyl)-1,3-dimethylurea] in wheat and other plant species [326, 327]. The corresponding N-hydroxymethyl intermediate which occurred as a prominent metabolite in the chloroform-soluble fraction derived from young wheat seedlings was identified by comparison with a reference standard and by ir, nmr, and mass spectroanalysis. This metabolite seems to be much more stable than the N-hydroxymethyl intermediates of phenylureas.

Apparently no advances have been made in describing the mechanism of N-demethoxylation of 1-methyl-1-methoxy-substituted ureas. However, it seems that in soil environments the significance of this pathway may be limited. Ross and Tweedy [317] followed the fate of metobromuron and chlorbromuron in microbial suspensions derived from two different soils. Under the short-term conditions applied (three weeks) formation of the N-demethoxylated metabolites and/or the unsubstituted phenylureas was either not detectable or restricted to small quantities. From ancillary observations, the authors concluded that direct hydrolysis to the corresponding anilines was functioning as an alternative pathway in these experiments.

B. Ring Hydroxylation

The tentative evidence provided by Lee and Fang [198] that monuron is ring-hydroxylated in the ortho (2) position by corn and bean tissues (see page 247) has meanwhile been confirmed by mass-spectrometric analysis [318] . The following two of the three postulated aglycones were identified after β-glucosidase hydrolysis: 1, 1-dimethyl-3-(2-hydroxy-4-chlorophenyl)-urea and 1-methyl-3-(2-hydroxy-4-chlorophenyl)urea. Although the mass spectra do provide proof for the presence of an aromatic hydroxyl group, its exact positioning would have to be verified by nuclear magnetic resonance analysis. In a recent report by Frear and Swanson [319] the same type of arylhydroxylation was shown to occur with the monuron metabolite 4-chloro-phenylurea in cotton. In these experiments the aglycone from β-glucosidase or hesperidinase hydrolysis was identified as 2-hydroxy-4-chlorphenylurea.

The reports mentioned above definitely demonstrate that ring hydroxylation of phenylureas does occur in plants. From the present data it would seem that the hydroxyl group may be introduced at any stage of the N-dealkylation pathway. However, this assumption as well as the relative importance of the two processes in different plant species remains to be verified. Some preliminary evidence was recently obtained with fluometuron showing that aromatic hydroxylation in the 6 position also occurs in soil [320] .

C.   Oxidation of Ring Substituents

In a survey type of degradation study, Kaufman and Blake [321] incubated a variety of carbanilate, acylanilide, and phenylurea herbicides with soil-enrichment solutions and pure cultures of active microbial isolates. In these experiments it was shown that in the presence of the fungus Fusarium oxysporum, halogen ions were released into the culture medium from the following phenylureas (listed in order of decreasing release): DMU [3-(3,4-dichlorophenyl)-1-methylurea] , chloroxuron, neburon, metobromuron, chlorbromuron. No halide ion release was observed with diuron. Although no dehalogenated metabolites were searched for in this survey, the data provide preliminary evidence that oxidative removal of halogen ring substituents may potentially occur for some substituted ureas in certain soil substrates.

D.   Aniline Formation and Further Pathways

Some new fragments have been added to present knowledge on the formation of anilines from substituted ureas in soils and their subsequent transformations. In the above-mentioned survey by Kaufman and Blake [321] it was shown that a number of fungal and bacterial isolates which were able to liberate anilines from carbanilates and acylanilides, did not hydrolyze fenuron and diuron. However, in a more extended experiment with Fusarium oxysporum, small amounts of anilines were formed from chlorbromuron, metobromuron, and noruron, but not from fenuron, diuron, fluometuron, neburon, and chloroxuron. These data give additional support to the assumption that direct microbial hydrolysis may be more prevalent with 1-methoxy-1-methylureas than with 1,1-dialkyl derivatives.

Direct hydrolysis of monolinuron to form N, O-dimethylhydroxylamine was reported to occur in soil [315]. However, in these experiments, which were carried out with the O-methyl-^{14}C-labeled compound, no more than 8% of the original radioactivity was recovered as the hydroxylamine. The same pathway was observed in sterile soil samples, indicating that the reaction was chemical rather than microbial.

As regards the further transformation of anilines, yet another mechanism has been added to the growing list of potential degradation or removal processes. Ross and Tweedy [322] observed that anilines were conjugated with malonic acid by mixed populations of soil microbes. The significance of this pathway, and the others listed in the original article, for the transformation of substituted ureas remains to be determined.

E. Conjugation and Complex Formation

The ring-hydroxylated metabolites of monuron recovered from different plant tissues (see above) were found to be present as O-(β-D-)glycosides [318, 319]. A glycoside of the N-hydroxymethyl intermediate of monolinuron has been described to be formed by the green alga Chlorella [315, 329]. However, it is difficult to conceive that such a conjugate would be an intermediate in N-demethylation of monolinuron, as postulated by the author. The N-hydroxymethyl metabolite of methabenzthiazuron was also observed to undergo conjugation to form an O-(ß-D-)glycoside [326, 327].

F. Miscellaneous Reactions

The N-acylated phenylurea fenobenzuron (see Table 1) was shown to be converted to diuron [Eq. (33)] by the fungus Fusarium solani and by an unidentified bacterium [323]. This conversion was described as being involved in the prolongation and/or retardation of the herbicidal effect of fenobenzuron.

$$\text{(33)}$$

Fenobenzuron   Diuron

Extensive metabolism data have recently been published for the herbicide methazole [2-(3,4-dichlorophenyl)-4-methyl-1,2,4-oxadiazolidine-3,5-
dione], which is not a substituted urea as defined for the present chapter,
but which is considered to be a heterocyclic derivative (see Sec. I). In cotton
methazole was found to be transformed first to DMU [3-(3,4-dichlorophenyl)-
1-methylurea] and then to 1-(3,4-dichlorophenyl)urea [324]. After initial
hydrolysis of the exadiazolidine moiety, the herbicide was thus observed to
follow the usual N-demethylation pathway of phenylureas. In these experiments, significant quantities of more polar transformation products remained
unidentified. Further studies by Dorough et al. [325] with cotton and bean
tissues, demonstrated that upon acid treatment, extractable polar compounds
and nonextractable residues released the monomethyl and unsubstituted
phenyl derivates. In addition, evidence was presented to show that 1-hydroxy-
methyl-3-(3,4-dichlorophenyl)urea was both an intermediate in N-demethylation and a hydrolysis product of polar metabolite(s). The composite metabolic pathway of methazole was therefore proposed to follow scheme (34):

(34)

REFERENCES

1. H. E. Thompson, C. P. Swanson, and A. G. Norman, Bot. Gaz., 107, 476 (1946).

2. H. C. Bucha and C. W. Todd, Science, 114, 453 (1951).

3. C. W. Todd (to E. I. du Pont de Nemours & Co.), U. S. Pat. 2,655,444-447 (1953).

4. H. E. Cupery, N. E. Searle, and C. W. Todd (to E. I. du Pont de Nemours & Co.), Ger. Pat. 935,165 (1949).

5. Herbicide Handbook of the Weed Society of America, 2nd ed., 1970, p. 163.

6. A. L. Abel, Chem. Ind. (London), 1957, 1106.

7. A. Fischer, Meded. Landbouwhogesch. Gent, 29, 719 (1964).

8. R. Wegler, Chemie der Pflanzenschutz- und Schädlingsbekämpfungs- mittel, Vol. 2, Springer-Verlag, New York, 1970, p. 238.

9. O. Scherer and G. Hörlein (to Farbwerke Hoechst), Fr. Pat. 1,531,212 (1966).

10. O. Scherer, G. Hörlein, and H. Schönowsky (to Farbwerke Hoechst), Fr. Pat. 1,520,220 (1966).

11. H. Martin, D. Dürr, P. S. Janiak, G. Pissiotas, O. Rohr, and W. Töpfl (to Ciba Ltd.), Swiss Pat. 491,600 (1968).

12. O. Scherer, G. Hörlein, and K. Härtel, Angew. Chem., 75, 852 (1963).

13. M. Schuler (to Sandoz Ltd.), Swiss Pat. 466,943 (1965).

14. A. Fischer, G. Scheuerer, O. Schlichting, and H. Stummeyer (to BASF), Ger. Pat. 1,027,930 (1956).

15. W. R. Diveley and M. M. Pombo (to Hercules Powder Co.), U.S. Pat. 3,150,179 (1962).

16. N. E. Searle (to E. I. du Pont de Nemours & Co.), U. S. Pat. 2,756,135 (1955).

17. H. Hack, W. Schäfer, R. Wegler, and L. Eue (to Bayer), Brit. Pat. 1,004,469 (1963).

18. H. Hack, L. Eue, and W. Schäfer (to Bayer), Brit. Pat. 1,085,430 (1965).

19. W. Siefken, Ann. Chem., 562, 99 (1949).

20. Houben-Weyl, Methoden der Organischen Chemie, 4th ed., Vol. VIII, Thieme, Stuttgart, 1952, pp. 149-195.

21. D. Dürr (to Ciba Ltd.), Swiss Pat. 490,796 (1967).

22. A. Rohm (to I. G. Farbenindustrie AG), Fr. Pat. 872,920 and add. 51,638 (1938/1939).

23. American Cyanamid Co., New Product Bulletin, Coll. Vol. I, 1949, p. 108.

24. E. E. Gilbert and J. Rumanowski (to Allied Chemical Corp.), Fr. Pat. 1,320,068 (1961).

25. O. Scherer and G. Hörlein (to Farbwerke Hoechst), Ger. Pat. 1,140,925 (1961).

26. T. P. Johnston, G. S. McCaleb, and J. A. Montgomery, J. Med. Chem., 6, 669 (1963).

27. A. Koshiro, Chem. Pharm. Bull., 7, 725 (1959).

28. E. Kühle, Angew. Chem., 74, 861 (1962).

29. H. W. Grimmel, A. Günther, and I. F. Morgan, J. Am. Chem. Soc., 68, 539 (1946).

30. H. Lautenschlager, Ph.D. Thesis, Institute of Technology, Munich, 1953.

31. O. Sedlmaier, Ph.D. Thesis, Institute of Technology, Munich, 1958.

32. A. Michaelis, Ann., 326, 169 (1903).

33. R. A. Franz, J. Org. Chem., 27, 4341 (1962).

34. R. A. Franz, M. D. Barnes, and F. Appelgath (to Monsanto), U.S. Pat. 2,857,430 (1956).

35. E. S. Wallis and F. S. Lane, in Organic Reactions, 7th ed., Vol. 3, Wiley, New York, 1959, p. 267.

36. P. A. S. Smith, in Organic Reactions, 7th ed., Vol. 3, Wiley, New York, 1959, p. 337.

37. E. W. Brunnett and C. W. McCarthy, J. Heterocycl. Chem., 5, 417 (1968).

38. E. Kühle, L. Eue, and R. Wegler (to Bayer), U.S. Pat. 3,280,190 (1959).

39. E. E. Gilbert, J. A. Otto, and S. A. Pellerano (to Allied Chemical Corp.), Ger. Pat. 951,181 (1954).

40. W. K. Lowen, W. E. Bleidner, J. J. Kirkland, and H. L. Pease, in
    Analytical Methods for Pesticides, Plant Growth Regulators and Food
    Additives, Vol. IV (G. Zweig, ed.), Academic Press, New York,
    1959, p. 157.

41. R. L. Dalton and H. L. Pease, J. Assoc. Off. Agr. Chem., 45, 377
    (1962).

42. H. Martin, H. Aebi, and L. Ebner (to Ciba Ltd.), Swiss Pat. 398,543
    (1961).

43. D. Dürr and L. Ebner (to Ciba Ltd.), Swiss Pat. 478,523 (1966).

44. I. Baunok and H. Geissbühler, Bull. Environ. Contam. Toxicol., 3, 7
    (1968).

45. T. Fujita, H. Tsuli, H. Deura, and M. Nakajiama, Agr. Biol. Chem.
    (Tokyo), 33, 785 (1969).

46. O. Scherer and G. Hörlein (to Farbwerke Hoechst), Fr. Pat. 1,531,212
    (1966).

47. K. Goliasch, R. Heiss, L. Eue, and H. Hack (to Bayer), Fr. Pat.
    1,516,527 (1966).

48. P. Poignant (to Péchiney-Progil), Fr. Pat. 1,250,422 (1958).

49. A. Didier and J. Lehureau (to Péchiney-Progil), Fr. Pat. 1,544,745
    (1967).

50. P. Poignant, D. Pillon, and R. Gaffiero (to Péchiney-Progil), Fr. Pat.
    1,535,465 (1967).

51. K. R. Wilson and K. L. Hill (to FMC Corp.), U.S. Pat. 3,434,822
    (1965).

52. S. T. Young (to FMC Corp.), U.S. Pat. 3,383,194 (1964).

53. P. E. Drummond, K. L. Hill, and K. R. Wilson (to FMC Corp.),
    U.S. Pat. 3,383,195 (1964).

54. R. J. Gobeil and R. W. Luckenbaugh (to E. I. du Pont de Nemours & Co.),
    U.S. Pat. 3,020,144 (1965).

55. H. Martin, P. S. Janiak, and G. Pissiotas (to Ciba Ltd.), Ger. Offen.
    2,044,735 (1969).

56. R. W. Luckenbaugh (to E. I. du Pont de Nemours & Co.), U.S. Pat.
    2,902,356 (1956).

57. M. Carmack and R. W. Luckenbaugh (to E. I. du Pont de Nemours &
    Co.), U.S. Pat. 2,985,663 (1958).

58. D. Mayer, H. Hack, and L. Eue (to Bayer), Ger. Offen. 1,817,119 (1968).

59. W. Siefken, O. Bayer, and L. Eue (to Bayer), Ger. Pat. 1,039,302 (1957).

60. P. J. Stoffel (to Monsanto), U.S. Pat. 3,418,334 (1963).

61. R. W. Luckenbaugh (to E. I. du Pont de Nemours & Co.), U.S. Pat. 3,895,817 (1956).

62. D. Rücker, C. Metzger, and L. Eue (to Bayer), Ger. Offen. 1,910,895 (1969).

63. W. Schäfer, R. Wegler, L. Eue, and H. Hack (to Bayer), Ger. Pat. 1,217,693 (1964).

64. J. A. Baskakow, M. J. Faddeewa, L. A. Bakumenko, W. G. Kaskakowa, L. S. Astaf'cwa, and L. W. Winogradowa (to Plant Protection Research Institute, Moscow), Ger. Offen. 1,817,119 (1968).

65. G. Zimmer, Arch. Pharm., 294, 765 (1961).

66. A. Zschocke and A. Fischer (to BASF), Fr. Pat. 1,560,971 (1967).

67. J. Krenzer (to Velsicol), U.S. Pat. 3,437,644 (1966).

68. D. L. Smathers (to E. I. du Pont de Nemours & Co.), Brit. Pat. 1,208,112 (1966).

69. G. Zumach, L. Eue, W. Weiss, E. Kühle, and H. Hack (to Bayer), Fr. Pat. 1,523,876 (1966).

70. W. Schäfer, R. Wegler, and L. Eue (to Bayer), Ger. Pat. 1,249,003 (1962).

71. G. Steinbrunn, A. Fischer, and Z. Zschocke (to BASF), Fr. Pat. 1,530,030 (1966).

72. G. Scheuerer, A. Zeidler, and A. Fischer (to BASF), Fr. Pat. 1,498,024 (1965).

73. T. J. Sheets, J. Agr. Food Chem., 12, 30 (1964).

74. T. J. Sheets and D. D. Kaufman, Tech. Papers FAO Int. Conf. Weed Control, Weed Sci. Soc. Am., 1971, pp. 513-538.

75. T. J. Sheets, Proc. Brit. Weed Control Conf., 8, 842 (1966).

76. G. S. Hartley, in The Physiology and Biochemistry of Herbicides (L. J. Audus, ed.), Academic Press, New York, 1964, pp. 111-116.

77. S. S. Sharp, M. C. Swingle, G. L. McCall, M. B. Weed, and L. E. Cowart, Agr. Chem., 8 (9), 56 (1953).

78. A. J. Lostalot, T. J. Muzik, and H. J. Cruzado, Agr. Chem., 8, (11), 52 (1953).

79. S. Dallyn, Proc. Northeast. Weed Control Conf., 8, 13 (1954).

80. H. R. Sherburne and V. H. Freed, J. Agr. Food Chem., 2, 937 (1954).

81. H. R. Sherburne, V. H. Freed, and S. C. Fang, Weeds, 4, 50 (1956).

82. R. P. Upchurch, Weeds, 6, 161 (1958).

83. T. J. Sheets, Weeds, 6, 413 (1958).

84. T. J. Sheets and A. S. Crafts, Weeds, 5, 53 (1957).

85. Q. H. Yuen and H. W. Hilton, J. Agr. Food Chem., 10, 386 (1962).

86. H. W. Hilton and Q. H. Yuen, J. Agr. Food Chem., 11, 230 (1963).

87. H. Geissbühler, C. Haselbach, and H. Aebi, Weed Res., 3, 140 (1963).

88. C. I. Harris and G. F. Warren, Weeds, 12, 120 (1964).

89. C. I. Harris and T. J. Sheets, Weeds, 13, 215 (1965).

90. C. I. Harris, Weeds, 14, 6 (1966).

91. R. J. Hance, Weed Res., 5, 98 (1965).

92. R. J. Hance, Weed Res., 5, 108 (1965).

93. R. J. Hance, Weed Res., 7, 29 (1967).

94. R. J. Hance, Can. J. Soil Sci., 49, 357 (1969).

95. R. J. Hance, Weed Res., 9, 108 (1969).

96. R. J. Hance, Weed Res., 11, 106 (1971).

97. F. Arndt, Z. Pflanzenkrankh., Pflanzenschutz, 67, 25 (1960).

98. N. Taubel, Z. Pflanzenkrankh., Pflanzenschutz, Sonderheft IV, 1968, 187.

99. G. MacNamara and S. J. Toth, Soil Sci., 109, 234 (1970).

100. R. Grover and R. J. Hance, Can. J. Plant Sci., 49, 378 (1969).

101. L. C. Liu, H. Cibes-Viadé, and F. K. S. Koo, Weed Sci., 18, 470 (1970).

102. P. D. Jordan and L. W. Smith, Weed Sci., 19, 541 (1971).

103. J. B. Weber, Adv. Am. Chem. Soc., 111, 55 (1972).

104. C. W. Coggins and A. S. Crafts, Weeds, 7, 349 (1969).

105. J. M. Davidson and P. W. Santelmann, Weed Sci., 16, 544 (1968).

106. Ciba-Geigy Ltd., unpublished data, 1972.

107. A. S. Crafts and H. R. Drever, Weeds, 8, 12 (1960).

108. C. A. Shadbolt and F. L. Whiting, Calif. Agr., 15 (11), 10 (1961).

109. P. A. Frank, Weeds, 14, 219 (1966).

110. H. Börner, Z. Pflanzenkrankh., Pflanzenschutz, 72, 516 (1965).

111. J. C. Majumdar and W. Koch, Z. Pflanzenkrankh., Pflanzenschutz, Sonderheft IV, 1968, 201.

112. J. C. Majumdar, Z. Pflanzenkrankh., Pflanzenschutz, 76, 95 (1968).

113. M. Horowitz, Weed Res., 9, 314 (1969).

114. E. E. Schweizer and J. T. Holstun, Weeds, 14, 22 (1966).

115. G. D. Hill, J. W. McGahen, H. M. Baker, D. W. Finnerty, and C. W. Bingeman, Agron. J., 47, 93 (1955).

116. R. J. Hance, J. Sci. Food Agr., 18, 544 (1967).

117. L. W. Weldon and F. L. Timmons, Weeds, 9, 111 (1961).

118. R. D. Comes and F. L. Timmons, Weeds, 13, 81 (1965).

119. L. S. Jordan, C. W. Coggins, B. E. Day, and W. A. Clerx, Weeds, 12, 1 (1964).

120. L. S. Jordan, J. D. Mann, and B. E. Day, Weeds, 13, 43 (1965).

121. J. D. Rosen, R. F. Strusz, and C. C. Still, J. Agr. Food Chem., 17, 206 (1969).

122. J. D. Rosen and R. F. Strusz, J. Agr. Food Chem., 16, 568 (1968).

123. D. G. Crosby and C. S. Tang, J. Agr. Food Chem., 17, 1041 (1969).

124. H. Geissbühler, C. Haselbach, H. Aebi, and L. Ebner, Weed Res., 3, 277 (1963).

125. T. J. Muzik, H. J. Cruzado, and A. J. Loustalot, Bot. Gaz., 116, 65 (1954).

126. R. E. Ogle and G. F. Warren, Weeds, 3, 257 (1954).

127. E. M. Rahn and R. E. Baynard, Weeds, 6, 432 (1958).

128. H. D. Dubey and J·. F. Freeman, Soil Sci., 5, 334 (1964).

129. D. D. Kaufman, in Pesticides in the Soil, Ecology, Degradation and Movement, Int. Symp. Michigan State University, East Lansing, February 1970, pp. 73-86.

130. G. D. Hill and J. W. McGahen, Proc. South. Weed Control Conf., 8, 284 (1955).

131. J. E. Bowen, Weed Sci., 15, 317 (1967).

132. D. S. Murray, W. L. Rieck, and J. Q. Lind, Weed Sci., 17, 52 (1969).

133. H. Börner, H. Burgemeister, and M. Schroeder, Z. Pflanzenkrankh., Pflanzenschutz, 76, 385 (1969).

134. P. Wallnöfer, Mitt. Biol. Bundesanst. Land- und Forstw., 132, 69 (1969).

135. B. G. Tweedy, C. Loeppky, and J. A. Ross, J. Agr. Food Chem., 18, 851 (1970).

136. M. Weinberger, M.Sc. Thesis, Pennsylvania State University, University Park, March 1972.

137. L. L. McCormick and A. E. Hiltbolt, Weeds, 14, 77 (1966).

138. G. A. Bozarth and H. H. Funderburk, Weed Sci., 19, 691 (1971).

139. P. Wallnöfer, Weed Res., 9, 333 (1969).

140. L. J. Audus, in The Physiology and Biochemistry of Herbicides (L. J. Audus, ed.), Academic Press, New York, 1964, pp. 163-206.

141. R. J. Hance and C. E. McKone, Pestic. Sci., 2, 31 (1972).

142. A. Süss, Bayer. Landw. Jahrb., 47, 425 (1970).

143. J. M. Bollag, CRC Critical Reviews in Microbiology, 2 (1), 35 (1972).

144. W. H. Minshall, Can. J. Bot., 32, 785 (1954).

145. T. J. Muzik, H. J. Cruzado, and A. J. Loustalot, Bot. Gaz., 116, 65 (1954).

146. T. J. Muzik, H. J. Cruzado, and M. P. Morris, Weeds, 5, 133 (1957).

147. J. R. Haun and J. H. Peterson, Weeds, 3, 177 (1954).

148. A. S. Crafts and S. Yamaguchi, Hilgardia, 27, 421 (1958).

149. A. S. Crafts and S. Yamaguchi, Am. J. Bot., 47, 248 (1960).

150. A. S. Crafts, Plant Physiol., 34, 613 (1959).

151. A. S. Crafts, Int. J. Appl. Radiat. Isot., 13, 407 (1962).

152. S. C. Fang, V. H. Freed, R. H. Johnson, and D. R. Coffee, J. Agr. Food Chem., 3, 400 (1955).

153. H. Geissbühler, C. Haselbach, H. Aebi, and L. Ebner, Weed Res., 3, 181 (1963).

154. H. Börner, Z. Pflanzenkrankh., Pflanzenschutz, 72, 449 (1965).

155. D. Penner, Weed Sci., 19, 571 (1971).

156. D. E. Bayer and S. Yamaguchi, Weeds, 13, 232 (1965).

157. G. Voss and H. Geissbühler, Proc. 8th Brit. Weed Control Conf., 1, 266 (1966).

158. J. C. Majumdar and F. Müller, Weed Res., 9, 322 (1969).

159. H. Fykse, Acta Agr. Scand., 19, 97 (1969).

160. R. L. Rogers and H. H. Funderburk, J. Agr. Food Chem., 15, 577 (1967).

161. J. W. Smith and T. J. Sheets, J. Agr. Food Chem., 15, 577 (1967).

162. R. Goren, Weed Res., 9, 121 (1969).

163. R. H. Strang and R. L. Rogers, Weed Sci., 19, 355 (1971).

164. E. J. Hogue and G. F. Warren, Weed Sci., 16, 51 (1968).

165. G. D. Hill, I. J. Belasco, and H. L. Ploeg, Weeds, 13, 103 (1965).

166. D. E. Bayer and H. R. Drever, Weeds, 13, 222 (1965).

167. C. G. McWhorter, Weeds, 11, 265 (1963).

168. H. Chandler, Proc. Calif. Weed Conf., 15, 99 (1963).

169. C. R. Swanson and H. R. Swanson, Weed Sci., 16, 137 (1968).

170. D. S. Frear, Science, 162, 674 (1968).

171. D. S. Frear, H. R. Swanson, and F. S. Tanaka, Phytochemistry, 8, 2157 (1969).

172. H. Geissbühler, Meded. Landbouwhogesch. Gent, 29, 704 (1964).

173. H. R. Schütte, H. H. Kühnau, and G. Siegel, Tagungsbericht Nr. 109, Dtsch. Akad. Landw. Wissenschaften, Berlin (DDR), 1970, pp. 31-39.

174. Ciba-Florida, Research Report CF-2564, 1968.

175. Ciba-Florida, Research Report CF-5487, 1969.

176. Ciba-Florida, Research Report CF-3101, 1968.

177. W. Mucke, R. E. Menzer, K. O. Alt, H. O. Esser, and W. Richter, in preparation.

178. H. C. Hodge, W. L. Downs, B. S. Panner, D. W. Smith, and E. A. Maynard, Food Cosmet. Toxicol., 5, 513 (1967).

179. H. C. Hodge, W. L. Downs, D. W. Smith, and E. A. Maynard, Food Cosmet. Toxicol., 6, 171 (1968).

180. H. Geissbühler and G. Voss, in Pesticide Terminal Residues (A. S. Tahori, ed.), Suppl. Pure and Applied Chemistry, Butterworths, London, 1971, pp. 305-322.

181. D. V. Parke, The Biochemistry of Foreign Compounds, Int. Series of Monographs in Pure and Applied Biol., Division Biochem., Vol. 5, Pergamon Press, Oxford, 1968.

182. D. V. Parke and R. T. Williams, Brit. Med. Bull., 25, 256 (1969).

183. R. T. Williams, in Handbook of Experimental Pharmacology, Vol. 18 (B. B. Brodie and J. R. Gillette, eds.), Springer-Verlag, New York, 1971, Pt. 2.

184. R. T. Williams, Detoxification Mechanisms, 2nd ed., Chapman & Hall, London, 1959.

185. B. B. Brodie, J. R. Gilette, and B. N. La Du, Ann. Rev. Biochem., 27, 427 (1958).

186. R. L. Dalton, A. W. Evans, and R. C. Rhodes, Weeds, 14, 31 (1966).

187. W. Ernst and C. Böhme, Food Cosmet. Toxicol., 3, 789 (1965).

188. C. Böhme and W. Ernst, Food Cosmet. Toxicol., 3, 797 (1965).

189. Ciba-Florida, Research Report CF-2614, 1968.

190. Sandoz Ltd., Basle, Dosanex, general data, 1971.

191. V. F. Boyd and R. W. Fogleman, 153rd American Chemical Society Meeting, Miami, April 1967.

192. M. Geldmacher-v. Mallinckrodt and F. Schüssler, Arch. Toxikol., 27, 187 (1971).

193. Y. Eshel and B. Rubin, Proc. 2nd Int. IUPAC Congr. Pestic. Chem., Tel Aviv, 5, 113-123 (1971).

194. R. B. Nashed and R. D. Ilnicki, Weed Sci., 18, 25 (1970).

195. R. B. Nashed, S. E. Katz, and R. D. Ilnicki, Weed Sci., 18, 122 (1970).

196. H. Kuratle, E. M. Rahn, and C. W. Woodmansee, Weed Sci., 17, 216 (1969).

197. J. H. Onley, G. Yip, and M. H. Aldridge, J. Agr. Food Chem., 16, 426 (1968).

198. S. S. Lee and S. C. Fang, Weed Res., 13, 59 (1973).

199. G. Voss and H. Geissbühler, 2nd Int. IUPAC Congr. Pestic. Chem.,
     4, 525-536 (1971).

200. B. Rubin and Y. Eshel, Weed Sci., 19, 592 (1971).

201. C. Klotzsche, Proc. 7th Int. Congress Plant Protection, Paris, 1970,
     pp. 764-765.

202. H. G. Bray, H. J. Lake, and W. V. Thorpe, Biochem. J., 44, 136
     (1949).

203. W. Ernst, J. S. Afr. Chem. Inst., 22, 879 (1969).

204. I. J. Belasco and R. W. Reiser, J. Agr. Food Chem., 17, 1000 (1969).

205. I. J. Belasco and W. P. Langsdorf, J. Agr. Food Chem., 17, 1004
     (1969).

206. W. E. Splittstoesser and H. J. Hopen, Weed Sci., 16, 305 (1968).

207. D. T. Gibson, Science, 161, 1093 (1968).

208. P. H. Mazzocchi and M. P. Rao, J. Agr. Food Chem., 20, 957 (1972).

209. R. E. Hoagland and G. Graf, Weed Sci., 20, 303 (1972).

210. P. R. Wallnöfer and J. Bader, Appl. Microbiol., 19, 714 (1970).

211. P. Wallnöfer and G. Engelhardt, Arch. Mikrobiol., 80, 315 (1971).

212. G. Engelhardt, Ph.D. Thesis, Technische Universität, Munich, 1972.

213. G. Engelhardt, P. R. Wallnöfer, and R. Plapp, Appl. Microbiol., 22,
     284 (1971).

214. G. Engelhardt, P. R. Wallnöfer, and R. Plapp, Appl. Microbiol., 23,
     664 (1972).

215. C. Böhme and W. Grunow, Food Cosmet. Toxicol., 7, 125 (1969).

216. W. Grunow, C. Böhme, and B. Budczies, Food Cosmet. Toxicol., 8,
     277 (1970).

217. A. Bobik, G. M. Holder, and A. J. Ryan, Food Cosmet. Toxicol., 10,
     163 (1972).

218. B. G. Tweedy, C. Loeppky, and J. A. Ross, Science, 168, 482 (1970).

219. P. C. Kearney and J. R. Plimmer, J. Agr. Food Chem., 20, 584
     (1972).

220. R. Bartha, H. A. B. Linke, and D. Pramer, Science, 161, 582 (1968).

221. J. D. Rosen, M. Siewiersky, and G. Winnett, J. Agr. Food Chem.,
     18, 494 (1970).

222. R. Bartha and D. Pramer, Science, 156, 1617 (1967).

223. R. Bartha, J. Agr. Food Chem., 16, 602 (1968).

224. R. Bartha and D. Pramer, Adv. Appl. Microbiol., 13, 317 (1970).

225. R. Bartha, Science, 166, 1299 (1969).

226. R. Bartha and D. Pramer, Bull. Environ. Contam. Toxicol., 4, 240 (1969).

227. H. A. B. Linke, Naturwiss., 57, 307 (1970).

228. P. C. Kearney, J. R. Plimmer, and F. G. Guardia, J. Agr. Food Chem., 17, 1418 (1969).

229. J. R. Plimmer, P. C. Kearney, H. Chisaka, J. B. Yount, and U. I. Klingebiel, J. Agr. Food Chem., 16, 602 (1968).

230. D. D. Kaufman, J. R. Plimmer, J. Iwan, and U. I. Klingebiel, J. Agr. Food Chem., 20, 916 (1972).

231. L. M. Bordeleau, J. D. Rosen, and R. Bartha, J. Agr. Food Chem., 20, 2573 (1972).

232. I. J. Belasco and H. L. Pease, J. Agr. Food Chem., 17, 1414 (1969).

233. H. Maier-Bode, personal communication, 1972.

234. H. Maier-Bode, Herbizide und ihre Rückstände, Verlag Eugen Ulmer, Stuttgart, 1971, pp. 175-241.

235. J. A. Guth, H. Geissbühler, and L. Ebner, Meded. Rijksfaculteit Landbouw. Gent, 34, 1027 (1969).

236. J. A. Guth, G. Voss, and L. Ebner, Z. Pflanzenkrankh., Pflanzen-schutz, Sonderheft V, 1970, 51.

237. H. Chisaka and P. C. Kearney, J. Agr. Food Chem., 18, 854 (1970).

238. R. Bartha, J. Agr. Food Chem., 19, 385 (1971).

239. V. H. Freed, M. Montgomery, and M. Kief, Proc. Northeast. Weed Control Conf., 15, 6 (1961).

240. N. D. Camper and D. E. Moreland, Weed Sci., 19, 269 (1971).

241. D. S. Frear and H. R. Swanson, Abstr. 162nd Meeting Am. Chem. Soc., Washington, D.C., 1971.

242. D. S. Frear and H. R. Swanson, Phytochemistry, 11, 1919 (1972).

243. J. B. Sumner and G. F. Somers, Chemistry and Methods of Enzymes, 3rd ed., Academic Press, New York, 1953.

244. J. Axelrod, J. Pharmacol. Exp. Ther., 117, 322 (1956).

245. R. E. McMahon, Biochem. Pharmacol., 12, 1225 (1963).

246. R. E. McMahon and H. R. Sullivan, Biochem. Pharmacol., 14, 1085 (1965).

247. E. Hodgson and J. E. Casida, Biochem. Pharmacol., 8, 179 (1961).

248. H. W. Dorough and J. E. Casida, J. Agr. Food Chem., 12, 294 (1964).

249. L. Shuster, Ann. Rev. Biochem., 33, 571 (1964).

250. T. E. Gram, in Handbook of Experimental Pharmacology, Vol. 18 (B. B. Brodie and J. R. Gillette, eds.), Springer-Verlag, New York, Pt. 2, pp, 334-348, 1971.

251. L. E. Gaudette and B. B. Brodie, Biochem. Pharmacol., 2, 89 (1959).

252. D. S. Frear, H. R. Swanson, and F. S. Tanaka, in Structural and Functional Aspects of Phytochemistry, Academic Press, New York, 1972, pp. 225-246.

253. F. S. Tanaka, H. R. Swanson, and D. S. Frear, Phytochem., 11: 2709 (1972).

254. R. E. Menzer and J. E. Casida, J. Agr. Food Chem., 13, 102 (1965).

255. G. W. Lucier and R. E. Menzer, J. Agr. Food Chem., 18, 698 (1970).

256. F. S. Tanaka, H. R. Swanson, and D. S. Frear, Phytochem., 11: 2709 (1972).

257. J. Daly, in Handbook of Experimental Pharmacology, Vol. 18 (B. B. Brodie and J. R. Gillette, eds.), Springer-Verlag, New York, 1971, Pt. 2, pp. 285-311.

258. J. E. Casida and L. Lykken, Ann. Rev. Plant Physiol., 20, 618 (1969).

259. C. M. Menzie, Metabolism of Pesticides, U.S. Fish Wildlife Serv. Spec. Sci. Rep. Wildlife, 1969, p. 127.

260. A. R. Cooke, Weeds, 4, 397 (1956).

261. J. S. C. Wessels and R. van der Veen, Biochim. Biophys. Acta, 19, 548 (1956).

262. H. Gaffron, in Plant Physiology, Photosynthesis and Chemosynthesis, Vol. IB, Academic Press, New York, 1960.

263. J. L. Hilton, L. L. Jansen, and H. M. Hull, Ann. Rev. Plant Physiol., 14, 353 (1963).

264. J. van Overbeek, in Physiology and Biochemistry of Herbicides (L. J. Audus, ed.), Academic Press, New York, 1964, pp. 387-400.

265. D. E. Moreland, Ann. Rev. Plant Physiol., 18, 365 (1967).

266. A. Barth and H.-J. Michel, Pharmazie, 24, 11 (1969).

267. D. C. Fork and J. Amesz, Ann. Rev. Plant Physiol., 20, 305 (1969).

268. G. M. Cheniae, Ann. Rev. Plant Physiol., 21, 467 (1970).

269. A. Trebst, Ber. Dtsch. Bot. Ges., 83, 373 (1970).

270. J. J. S. van Rensen, Meded. Landbouwhogesch. Wageningen (Netherlands), 1971, 71-79.

271. K. H. Büchel, Pestic. Sci., 3, 89 (1972).

272. M. J. Geoghegan, New Phytol., 56, 71 (1957).

273. W. A. Gentner and J. L. Hilton, Weeds, 8, 413 (1960).

274. P. B. Sweetser and C. W. Todd, Biochim. Biophys. Acta, 51, 504 (1961).

275. C. E. Hoffman, R. T. Hersh, P. B. Sweetser, and C. W. Todd, Proc. 14th Northeast. Weed Control Conf., New York, 1960.

276. F. M. Ashton, Weeds, 13, 164 (1965).

277. P. B. Sweetser, Biochim. Biophys. Acta, 66, 78 (1963).

278. P. Homann and H. Gaffron, Science, 141, 905 (1963).

279. G. Zweig, I. Tamas, and E. Greenburg, Biochim. Biophys. Acta, 66, 196 (1963).

280. N. I. Bishop, Biochim. Biophys. Acta, 27, 205 (1958).

281. L. N. M. Duysens, J. Amesz, and B. M. Kamp, Nature, 190, 510 (1961).

282. E. Kessler, in Progress in Photosynthesis Research, Vol. 2 (H. Metzner, ed.), Tübingen, Germany, 1969, pp. 938-942.

283. G. S. Grishna and N. P. Voskresenskaya, in Progress in Photosynthesis Research, Vol. 3 (H. Metzner, ed.), Tübingen, Germany, 1969, pp. 1262-1267.

284. A. T. Jagendorf and M. Margulies, Arch. Biochem. Biophys., 90, 184 (1960).

285. N. E. Good, Plant Physiol., 36, 788 (1961).

286. L. P. Vernon and W. P. Zaugg, J. Biol. Chem., 253, 2728 (1960).

287. T. Asahi and A. T. Jagendorf, Arch. Biochem. Biophys., 100, 531 (1963).

288. W. Tanner, L. Dächsel, and O. Kandler, Plant Physiol., 40, 1151 (1965).

289. C. E. Stanger and A. P. Appleby, Weed Sci., 20, 357 (1972).

290. S. Izawa and N. E. Good, Biochem. Biophys. Acta, 102, 20 (1965).

291. D. E. Moreland and K. L. Hill, Weeds, 10, 229 (1962).

292. D. E. Moreland, in Progress in Photosynthesis Research, Vol. 3 (H. Metzner, ed.), Tübingen, Germany, 1969, pp. 1693-1711.

293. Ciba-Geigy, unpublished data, 1967.

294. C. Hansch and C. W. Deutsch, Biochim. Biophys. Acta, 112, 381 (1966).

295. C. Hansch, in Progress in Photosynthesis Research, Vol. 3 (H. Metzner, ed.), Tübingen, Germany, 1969, pp. 1685-1692.

296. K. Holly, in The Physiology and Biochemistry of Herbicides (L. J. Audus, ed.), Academic Press, New York, 1964, pp. 423-464.

297. R. L. Wain, in The Physiology and Biochemistry of Herbicides (L. J. Audus, ed.), Academic Press, New York, 1964, pp. 465-482.

298. J. L. P. van Oorschot, Pestic. Sci., 1, 33 (1970).

299. D. S. Frear, R. H. Hodgson, R. H. Shimabukuro, and G. G. Still, Adv. Agron., 24, 327 (1972).

300. J. L. P. van Oorschot, Meded. Landbouwhogesch. Gent, 29, 683 (1964).

301. J. L. P. van Oorschot, Weed Res., 5, 84 (1965).

302. E. J. Hogue, Weed Sci., 18, 580 (1970).

303. R. H. Shimabukuro and H. R. Swanson, Weed Sci., 18, 231 (1970).

304. H. L. Palm, Diss. Abstr. Int., B32, No. 6, 3113 (1971).

305. R. W. Feeny, Diss. Abstr., 29, No. 7, 2261-B (1969).

306. R. V. Osgood, Diss. Abstr. Int., 30, No. 12, 5330-B (1970).

307. R. J. Hance, Soil Biol. Biochem., 6, 39 (1974).

308. K. A. Savage, Weed Sci., 21, 416 (1973).

309. M. Weinberger and J. M. Bollag, Appl. Microbiol., 24, 750 (1972).

310. L. S. Jordan and A. J. Dawson, Res. Prog. Rep. West. Soc. Weed Sci., Meeting 166 (1973).

311. J. Menashe and R. Goren, Weed Res., 13, 158 (1973).

312. R. W. Feeny, J. V. Parochetti, and S. R. Colby, Weed Sci., 22, 143 (1974).

313. R. V. Osgood, R. R. Romanowsky, and H. W. Hilton, Weed Sci., 20, 537 (1972).

314. A. Walker and R. M. Featherstone, J. Exp. Bot., 24, 450 (1973).

315. I. Schuphan, Mitt. Biol. Bundesanstalt Land- und Forstwirtsch.,
     151, 193 (1973).

316. P. R. Wallnoefer, S. Safe, and O. Hutzinger, Pestic. Biochem.
     Physiol., 3, 253 (1973).

317. J. A. Ross and B. G. Tweedy, Soil Biol. Biochem., 5, 739 (1973).

318. S. S. Lee, D. A. Griffin, and S. C. Fang, Weed Res., 13, 234 (1973).

319. D. S. Frear and H. R. Swanson, Phytochemistry, 13, 357 (1974).

320. Ciba-Geigy Ltd., unpublished data, 1974.

321. D. D. Kaufman and J. Blake, Soil Biol. Biochem., 5, 297 (1973).

322. J. A. Ross and B. G. Tweedy, Bull. Environ. Contam. Toxicol., 10,
     234 (1973).

323. J. C. Fournier, G. Soulas, and G. Catroux, C. R. Acad. Sci. Paris,
     Ser. D, 276, 2993 (1973).

324. D. W. Jones and G. L. Foy, Pestic. Biochem. Physiol., 2, 8 (1972).

325. H. W. Dorough, D. M. Whitacre, and R. A. Cardona, J. Agr. Food
     Chem., 2, 797 (1973).

326. V. Pont, H. J. Jarczyk, G. F. Collet, and R. Thomas, Phytochem-
     istry, 13, 785 (1974).

327. G. F. Collet and V. Pomp, Weed Res., 14, 151 (1974).

328. I. Schuphan, Chemosphere, 3, 127 (1974).

329. I. Schuphan, Chemosphere, 3, 131 (1974).

Chapter 4

SUBSTITUTED URACIL HERBICIDES

JOHN A. GARDINER

Biochemicals Department, Research Division
E. I. du Pont de Nemours & Co.
Wilmington, Delaware

I. INTRODUCTION: HISTORY AND DEVELOPMENT

The discovery of the substituted uracils, a family of highly effective herbicides, was first announced by Du Pont in December 1961 [1]. Bucha et al. [2] reported in the following year that the combination of the high phytotoxicity and the low mammalian toxicity of these compounds held promise for their future development and use. The first material introduced on a commercial scale was isocil, which has since been superseded by other members of the class. In 1972, two substituted uracil herbicides were available commercially in the United States. The structural formulas and names of these products are shown in Table 1. Isocil is included in the table for reference.

Bromacil is used on railroad rights-of-way and other industrial, noncropland areas at rates up to 24 lb active ingredient/acre for general nonselective weed control, particularly of perennial grasses. It is also used in citrus, pineapple, and sisal for selective control of weeds at rates of about 2-6 lb active ingredient/acre. Terbacil is useful against many annual and some perennial weeds in a number of crops such as sugarcane, apples, peaches, citrus, and mint. These compounds are also useful in combination with other herbicides, for example, bromacil with diuron for selective control of weeds in citrus and for general weed control in noncropland areas.

A. Synthesis and Selected Properties

Methods for preparing the substituted uracil herbicides are given in the U.S. patents listed in Table 1. The syntheses of 2-^{14}C-labeled bromacil and terbacil have been described by Gardiner et al. [3]. In the latter procedures, which are suitable for small-scale laboratory preparations, the appropriate alkylureas were reacted under acidic conditions with methyl or ethyl acetoacetate to form methyl or ethyl 3-(3-alkylureido)crotonates. These materials were then cyclized to the uracils with sodium methoxide, followed by the appropriate ring-halogenation steps. The ^{14}C label in both cases was introduced by way of the urea carbonyl carbon.

Merkle et al. [4] reported that isocil exists almost entirely in the keto form based on nuclear magnetic resonance spectra. The structures of the substituted uracil herbicides in Table 1 are normally written this way.

All of the uracil herbicides in pure form are white, crystalline solids and are temperature stable up to their melting points. Table 2 gives additional selected information. For comparison with the LD_{50} values in the table, that of aspirin is 1750 mg/kg. Sherman [5, 6] has presented other portions of the toxicity information on bromacil and terbacil. The K values in Table 2 are indicative of soil adsorption characteristics and are discussed in Sec. II, B.

TABLE 1

Substituted Uracil Herbicides

Structural formula	Chemical name; empirical formula	Common name	Introductory code name; year of commercial introduction	Trade name of formulated product[a]	Patent coverage
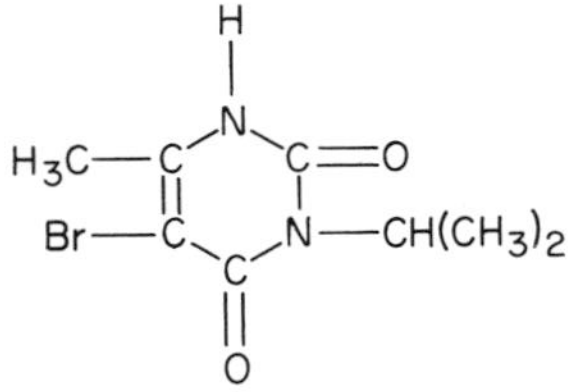 (uracil ring structure with substituents: H_3C–, Br–, –CH(CH$_3$)$_2$)	5-Bromo-3-isopropyl-6-methyluracil $C_8H_{11}BrN_2O_2$	Isocil	Du Pont Herbicide 82 1962	—[b]	U.S. 3,235,357 (use) U.S. 3,352,862 (cpd)
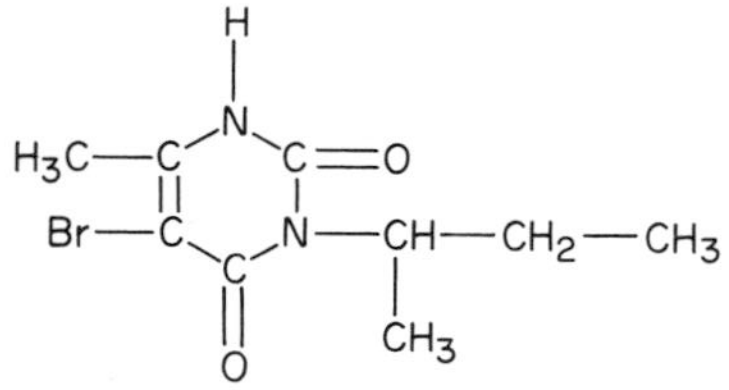 (uracil ring structure with substituents: H_3C–, Br–, –CH–CH$_2$–CH$_3$ with CH$_3$)	5-Bromo-3-sec-butyl-6-methyluracil $C_9H_{13}BrN_2O_2$	Bromacil	Du Pont Herbicide 976 1963	Hyvar X	U.S. 3,235,357 (use) U.S. 3,352,862 (cpd)
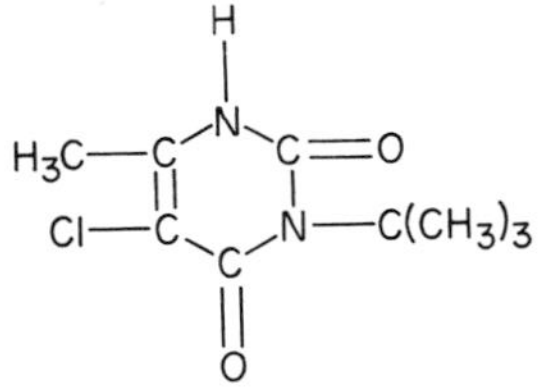 (uracil ring structure with substituents: H_3C–, Cl–, –C(CH$_3$)$_3$)	3-tert-Butyl-5-chloro-6-methyluracil $C_9H_{13}ClN_2O_2$	Terbacil	Du Pont Herbicide 732 1966	Sinbar	U.S. 3,235,357 (use) U.S. 3,352,862 (cpd)

[a]Registered trademarks of E. I. du Pont de Nemours & Co., Wilmington, Delaware.
[b]Isocil is no longer commercially available.

295

TABLE 2

Properties of the Substituted Uracil Herbicides

Common name	Mol wt	Melting point (°C)	Solubility in H_2O at 25°C (ppm)	K value (Keyport silt loam)	Acute oral LD_{50}, rats (mg/kg)
Isocil	247.1	158–159	2150	1.0	3250
Bromacil	261.1	158–159	815	1.5	5200
Terbacil	216.7	175–177	710	1.7	>5000

B.   Chemical Reactions and Analytical Methods

Bromacil and terbacil are stable in water, aqueous bases (in which they form salts), and common organic solvents. They decompose slowly in strong acids. Terbacil is more susceptible to decomposition by acids because the tert-butyl group on the 3 position of the uracil ring is relatively more labile than the sec-butyl group of bromacil. In hot caustic, both materials decompose.

Jordan et al. [7] have studied the effects of ultraviolet (uv) light on bromacil and isocil. The greatest changes and losses occurred under far ultraviolet (240-260 nm) and the least under near ultraviolet irradiation (320-450 nm). About 70% of the bromacil was recovered after 500 hr of exposure to near ultraviolet, which is the closest to natural sunlight.

Residue methods for the intact parent materials have been possible and desirable for the substituted uracil herbicides. Table 3 summarizes several published procedures that have been developed. Certain thin-layer chromatography (tlc) systems for separating isocil, bromacil, and terbacil are also available [19]. A method for analysis of bromacil formulations has also been described [20].

C.   Formulations

Bromacil is formulated by Du Pont for specific uses as follows:

1.   Hyvar X bromacil weedkiller, a wettable powder containing 80% bromacil
2.   Hyvar X-L bromacil weedkiller, a water-soluble liquid containing 2 lb bromacil/gal (present as lithium salt)
3.   Hyvar X-P bromacil brushkiller, pellets containing 10% bromacil

Other formulations of bromacil, such as granules, solutions, etc., alone and in combination with other herbicides, are available from additional suppliers for specialized uses.

Terbacil is formulated by Du Pont as Sinbar terbacil weedkiller, a wettable powder containing 80% terbacil.

The formulations of the substituted uracil herbicides, except for Hyvar X-L, are compatible with most crop protection chemicals. They may be combined with many other herbicides to control a broader spectrum of weeds.

TABLE 3

Published Residue Methods for the Substituted Uracil Herbicides

Compound	Tissues	Initial extraction solvent	Method of measurement	Claimed sensitivity or detectability (ppm)	Reference
Bromacil	Soil, plants, animals	1% NaOH	Microcoulometric gas chromatography	0.04	8[a]
	Soil, plants	1.5% NaOH (soil) Chloroform (plants)	Electron-capture gas chromatography	0.01 (soil) 0.1 (plants)	9
	Soil	1.5% NaOH	Thin-layer chromatography with silver nitrate spray	0.03	10
	Soil	Ethyl acetate	Electron-capture gas chromatography	0.02	11
	Soil	1.5% NaOH	Thin-layer chromatography with brilliant green/ bromine detection	0.1	12
	Soil, water	1% NaOH	Electron-capture gas chromatography	0.005	13

	Cow milk, urine, and feces	Acetone/chloroform (milk and urine) Ethyl acetate (feces)	Electron-capture gas chromatography	0.005 (milk) 0.04 (urine) 0.01 (feces)	14
	Soil	—	Bioassay	—	15
Terbacil	Soil, plants, animals	1% NaOH	Microcoulometric gas chromatography	0.04	16[a]
	Soil, alfalfa	Ethyl acetate	Electron-capture gas chromatography	0.5	11
	Soil	1.5% NaOH	Thin-layer chromatography with brilliant green/ bromine detection	0.1	12
	Cow milk, urine, and feces	Acetone/chloroform (milk and urine) Ethyl acetate (feces)	Electron-capture gas chromatography	0.01 (milk) 0.03 (urine) 0.02 (feces)	17
	Nutsedge	1% NaOH	Electron-capture gas chromatography	0.1	18

[a]Manufacturer's recommended method.

II. DEGRADATION PATHWAYS

A. Degradation in Animals

The most definitive study on the fate of the substituted uracil herbicides in animals was reported by Rhodes et al. [21] using ^{14}C techniques. These workers conditioned a male beagle dog over a three-week period to a diet containing 2500 ppm terbacil at the time of testing. The dog was then given a single dose of 2-^{14}C-labeled terbacil by gelatin capsule. The animal was maintained on the diet of 2500 ppm unlabeled terbacil while daily urine and feces samples were collected and pooled for 3 days after the ^{14}C treatment. At the end of 3 days, the dog was sacrificed for tissue analyses. No attempt was made to monitor for ^{14}C-containing offgases in this work and the authors reported that a portion of the first day's urine sample was lost by splashing. These two factors may account for the less than quantitative overall recovery (77. 3%) of the original ^{14}C dose.

Data from the study just mentioned are shown in Table 4. The fact that essentially all of the detected ^{14}C was eliminated from the system within 72 hr, predominantly in the urine, is of particular significance to the fate of terbacil in the dog. Less than 0. 1% of the total was retained in the tissues at sacrifice.

In additional work reported in the same paper [21], urinary metabolites of unlabeled terbacil from female beagle dogs were isolated by combinations of various chromatographic techniques and identified by mass, infrared (ir), and nuclear magnetic resonance (nmr) spectroscopy and elemental analysis. Structures of the pertinent materials are shown in Fig. 1.

In the ^{14}C portion of the work described above, the urinary metabolites were characterized as being the same materials shown in Fig. 1, based on tlc R_f values. Relative concentrations of the various metabolites, after enzyme hydrolysis to liberate possible glucuronide and/or sulfonate conjugates, are given in Table 5. The major urinary excretion product of terbacil was metabolite (1), where the 6-methyl group has been oxidized to a 6-hydroxymethyl substituent. Since the appearance of the original paper [21], the structures of terbacil metabolites (1), (2), and (3) have been confirmed by synthesis and direct comparison of chromatographic behavior and mass spectra [22].

The metabolism of bromacil in rats has been studied by Gardiner et al. [23]. These workers examined the urine from male rats maintained on a diet containing 1250 ppm bromacil for one month. Structural assignments for the detected metabolites shown in Fig. 2 were based on isolations by tlc techniques followed by characterization by ir, mass, and nmr spectroscopy. The two principal urinary excretion products were metabolite (6) (major), 5-bromo-3-sec-butyl-6-hydroxymethyluracil, and metabolite (8) (minor),

TABLE 4

[2-^{14}C]Terbacil Dog Study[a]

Tissue	Percent of original dose found	Percent of total activity detected
Urine (0–24 hr)	50.14	64.8
Urine (24–48 hr)	11.04	14.3
Urine (48–72 hr)	1.16	1.5
Feces (0–24 hr)	10.57	13.7
Feces (24–48 hr)	3.92	5.1
Feces (48–72 hr)	0.41	0.5
Liver	0.06	0.1
Kidney	<0.005	—
Blood	0.001	—
Stomach	0.01	—
GI tract	0.02	—
Brain	<0.005	—
Lung	<0.005	—
Heart	<0.005	—
Bladder and prostate	<0.005	—
Testes	<0.005	—
Thyroid	<0.005	—
Adrenals	<0.005	—
Pancreas	<0.005	—
Spleen	<0.005	—
Bone marrow	<0.005	—
Fat	<0.001	—
Muscle	<0.001	—
Total	77.34	100.0

[a]Reprinted from Rhodes et al. [21, p. 977], by courtesy of the Journal of Agricultural and Food Chemistry.

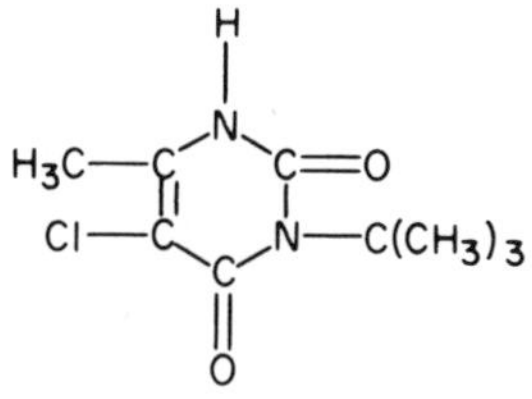

Terbacil

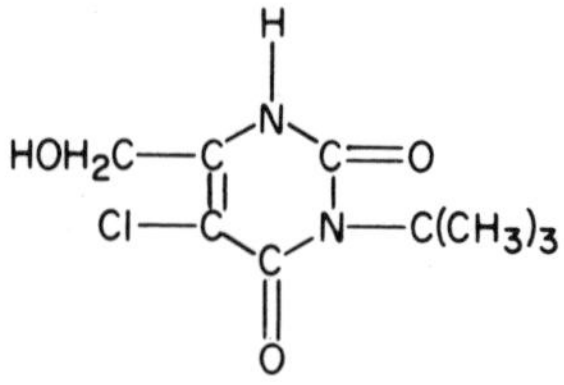

(1)

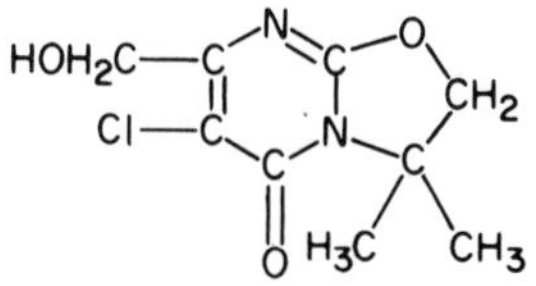

(2)

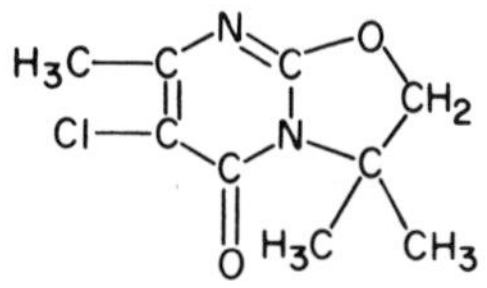

(3)

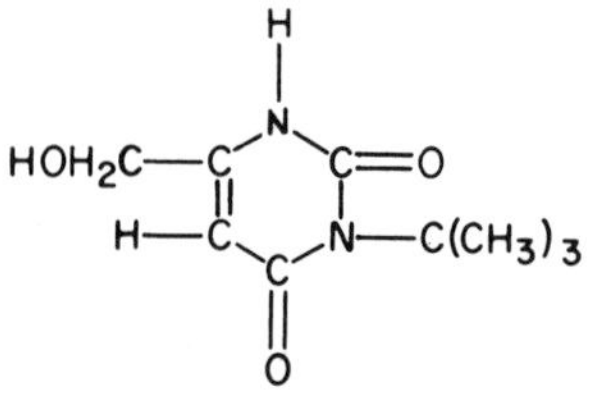

(4)

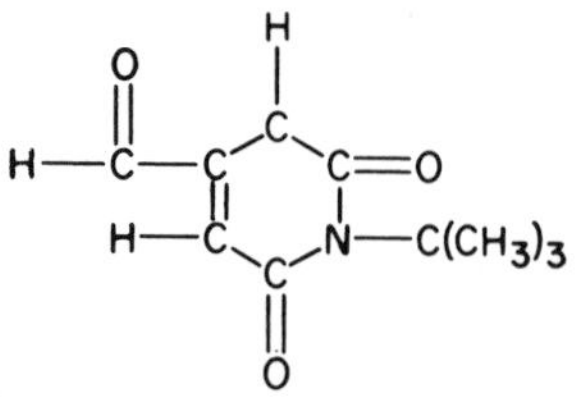

(5)

FIG. 1. Terbacil metabolites detected in dog urine. (Reprinted from Rhodes et al. [21, p. 975], by courtesy of the Journal of Agricultural and Food Chemistry.)

TABLE 5

Relative Concentrations of
[2-^{14}C]Terbacil and Metabolites in Dog Urine Extract[a]

Compound	Relative concentration
Terbacil	1
Metabolite	
(1)	100
(2)	10
(3)	1
(4)	5
(5)	2
(6)	>1
Unknown[b]	7

[a]Reprinted from Rhodes et al. [21, p. 978], by courtesy of the Journal of Agricultural and Food Chemistry.

[b]Activity remained at origin of tlc plate.

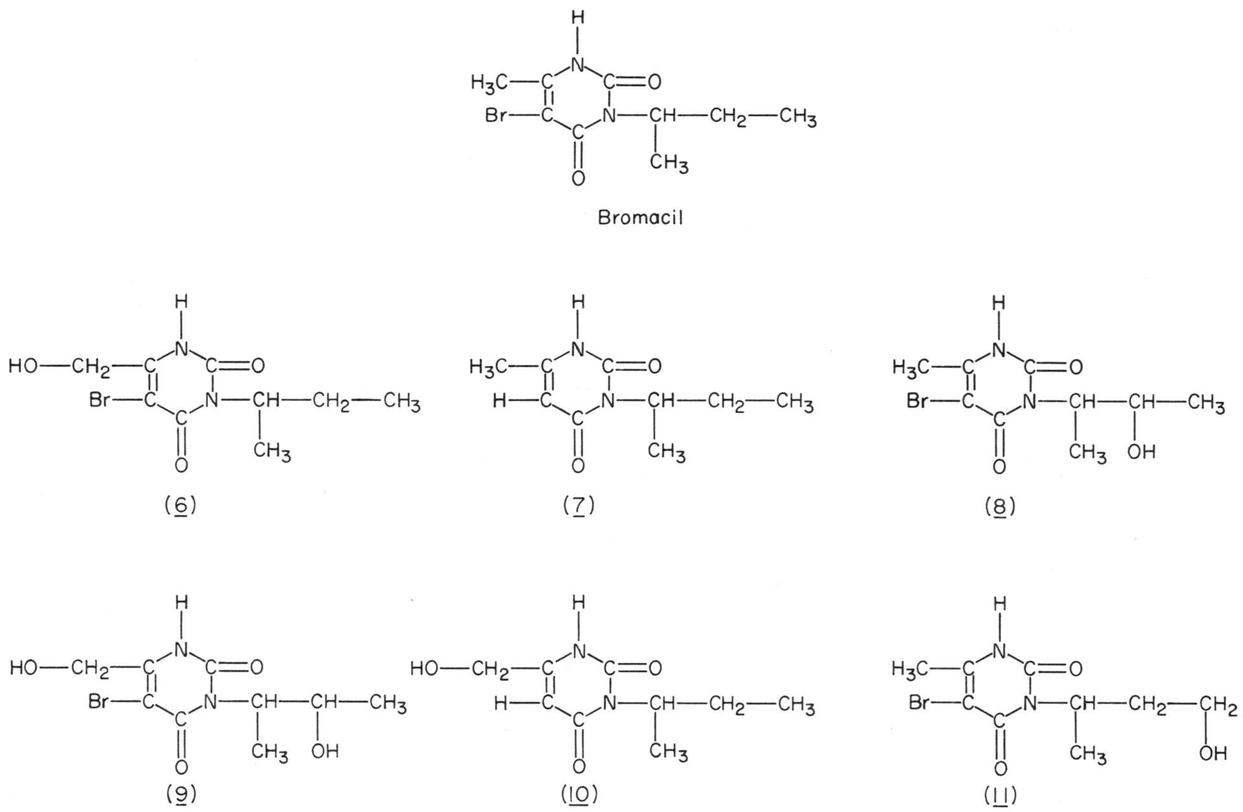

FIG. 2. Bromacil metabolites detected in rat urine. (Reprinted from Gardiner et al. [23, p. 968], by courtesy of the Journal of Agricultural and Food Chemistry.)

5-bromo-3-(2-hydroxy-1-methylpropyl)-6-methyluracil. These are produced
by oxidation of the 6-methyl group and the 3-sec-butyl group of bromacil,
respectively. Structure identifications were confirmed by synthesis of reference compounds and direct comparisons of mass, nmr, and ir spectra. A
residue method for metabolite (6) was developed, based on microcoulometric
gas chromatography, and was used to show that this metabolite is excreted
from the rat primarily in conjugated form. Fresh rat urine contained 21 ppm
metabolite, whereas after enzyme hydrolysis with β-glucuronidase/aryl
sulfatase, 146 ppm was detected, an increase of about sevenfold.

Much of the identification work on isolated metabolites of the substituted
uracil herbicides has involved mass spectroscopic techniques. Reiser [24]
has published a separate paper on some of the background information useful
in the interpretation of mass spectra from substituted uracils.

The presence of only minute amounts of bromacil (0.019 and 0.13 ppm)
and terbacil (0.03 and 0.08 ppm) in milk from dairy cows maintained for
4 days on diets containing 5 and 30 ppm for each compound, respectively,
has been reported by Gutenmann and Lisk [14, 17]. Kutches et al. [25]
found that bromacil and several nonrelated pesticides had negligible effects
on rumen digestibility and other associated rumen functions when ingested
in amounts equivalent to no more than 250 ppm in feedstuffs. Use patterns
and residue tolerances for the substituted uracil herbicides are such that
no more than 0.1 ppm residue is likely to be in livestock forage. Thus,
these herbicides should not be of concern to the world's milk and meat
supply.

In summary, animals ingesting residues of the substituted uracil herbicides can be expected to excrete the major portion of these materials by the
urine and feces within a relatively short time. Most of the material is
excreted in the form of conjugated urinary metabolites where oxidative
attack (hydroxylation) has occurred on one or more alkyl side groups. This
conversion appears to be the major pathway of degradation. With both
bromacil and terbacil, dehalogenated minor metabolites have also been
detected (Figs. 1 and 2), indicating a secondary route of degradation. The
major urinary metabolite of bromacil in rat urine was the 6-hydroxymethyl
metabolite (6) and the next most prevalent species was metabolite (8) with
an -OH group in the sec-butyl side chain. Terbacil was principally converted
to a 6-hydroxymethyl metabolite also (1), and other metabolites indicated
hydroxylation in the tert-butyl side chain, followed by elimination of water
and ring closure.

Similar hydroxylation of alkyl-substituted pyrimidine derivatives is
reported in the literature, i.e., thymine is converted in part to 5-hydroxy-
methyluracil, and 5-n-butyl-5-ethylbarbituric acid is converted to 5-ethyl-
5-(3'-hydroxybutyl)barbituric acid [26]. Other studies [27, 28] on the in vitro
metabolism of uracil and thymine by rat liver slices or extracts indicate that

these uracils can undergo ring opening and subsequent degradation down to β-alanine and β-aminoisobutyric acid. Loss of $^{14}CO_2$ from both [2-^{14}C]uracil and [2-^{14}C]thymine was noted. β-Aminoisobutyric acid has also been detected in rat urine from animals ingesting thymine [29]. However, presumably because of the rapid conversion to hydroxylated and easily excretable metabolites, the substituted uracil herbicides do not undergo extensive ring degradation in living animals.

B.   Degradation in Soils

Steady loss of substituted uracil herbicide residues has been observed in most soils, based both on practical field use and on direct chemical analysis. Work with specific microorganisms has shown that this class of herbicides is biodegradable, and this mode of dissipation appears to play a prime role as discussed later. However, in general, several factors may contribute to loss of soil residues. These are discussed in the following order: volatilization, leaching and mobility, photodecomposition, and biological and/or chemical degradation.

Volatilization is one mechanism by which herbicide residues might be removed from treated soils. This route for the substituted uracil herbicides is very minor. The vapor pressure of bromacil at 100°C is reported as 0.8×10^{-3} mm Hg [20]. Hill [30] has described studies where the rate of volatilization of Hyvar bromacil was measured for two weeks in an air circulation oven. Losses were less than 0.1% per week at a temperature (49°C) selected to approximate high soil temperatures. In laboratory studies with 2-^{14}C-labeled terbacil and bromacil applied to soil in enclosed systems, Gardiner et al. [3] found no evidence for volatile ^{14}C-containing materials other than $^{14}CO_2$. These studies lasted for 6.5 (terbacil) and 9 weeks (bromacil) and when concluded, the gas trapping system contained 32 and 25%, respectively, of the applied ^{14}C as $^{14}CO_2$. In spite of this extensive breakdown and evolution of ^{14}C, no intact [2-^{14}C]bromacil or [2-^{14}C]terbacil was found in the trapping systems. Thus, loss of the substituted uracil herbicides by volatilization is regarded as a minor factor.

The effects of soil adsorption as they relate to leaching and mobility of the uracils have also been examined. The K values in Table 2 are indicative of soil adsorption characteristics. A K value is the ratio at 25°C of parts per million of herbicide adsorbed to soil (25 g) at the point where 1.0 ppm is in an equilibrated aqueous phase (30 ml). The higher the K value, the stronger is the adsorption to soil. For the uracils, these values range from 1.0 for isocil to 1.7 for terbacil on Keyport silt loam. For comparison, K values of diuron, linuron, and atrazine on the same soil type are 5.0, 3.2, and 2.8, respectively, when determined under similar conditions [31]. Thus, this type of soil-water-herbicide equilibrium measurement indicates that the uracil herbicides are less tightly adsorbed to this particular soil than other familiar herbicides.

Several researchers have noted that detectable amounts of bromacil and terbacil can leach in soil if applied at high enough rates [32-37] . The adsorption of isocil and bromacil from aqueous solution onto certain mineral surfaces has been examined by Haque and Coshow [38] . A humic acid surface adsorbed considerably more chemical than illite, montmorillonite, silica gel, and kaolinite surfaces.

Rhodes et al. [39] studied the mobility and adsorption of bromacil and terbacil by the soil-tlc technique developed by Helling and Turner [40] . Based on the classification system suggested by the latter workers, bromacil is between class 3 and 4, whereas terbacil is in class 3 of a scale ranging from class 1, essentially immobile materials, to class 5, the very mobile [39, 41, 42] . Thus, the substituted uracil herbicides should be considered neither excessively mobile nor immobile. From an efficacy viewpoint, a degree of downward movement into germination zones is desirable.

Photodecomposition is yet another route by which residues of herbicides can be dissipated in the increments near the soil surface. The work of Jordan et al. [7] has been described in Sec. I, B. This work plus several practical field tests indicate that bromacil, and probably the other members of the class, are not affected adversely by sunlight. However, certain substituted uracils (not bromacil, terbacil, or isocil) are known to undergo photochemical reactions [43-46] . In particular, thymine and uracil formed dimers on filter paper and quartz fiber sheets when irradiated at 254 nm [43] . It is therefore possible that sunlight might play a small role in degradation but probably not a major role.

The last two general routes of degradation in soil are chemical decomposition and biological degradation. None of the foregoing processes, i. e., volatility, leaching, and photodecomposition, appears to account for observed losses of substituted uracil herbicide residues from soils. There is positive evidence that these compounds can be broken down by microorganisms, yet it is possible that chemical decomposition also plays an important role.

Gardiner et al. [3] have published the results of a field study where $2-{}^{14}C$-labeled bromacil was applied at 4 lb active ingredient/acre to a Butlertown silt loam in an agricultural area in Delaware. An identical study was performed using $2-{}^{14}C$-labeled terbacil. In order to isolate undisturbed field soil and yet provide for quantitative recovery of the ${}^{14}C$, these workers simply drove the required number of 12-in. sections of 4-in. diameter stainless steel tubing into the ground. Radiolabeled herbicides were applied to the soil surfaces inside the rims of the tubing. Just enough of the tubing (0.5-in.) was left protruding from the ground surface to protect against loss by runoff. The tubing walls underground prevented any sideways diffusion

TABLE 6

Percentage of Original ^{14}C Activity in Various Soil Fractions[a]

| Soil depth[b] (in.) | Exposure to | | | | | |
| | Bromacil | | | Terbacil | | |
	5 weeks	14 weeks	1 year	5 weeks	14 weeks	1 year
0-1	34.2	25.2	4.3	17.4	46.2	4.5
1-3	24.0	17.7	7.1	24.8	17.5	3.2
3-5	9.6	12.5	5.9	11.8	5.6	2.7
5-8	0.7	5.8	4.6	12.1	1.3	3.3
8-12	0.3	1.8	1.6	6.3	0.2	0.8
Total	68.8	63.0	23.5	72.4	70.8	14.5

[a]Reprinted from Gardiner et al. [3, p. 983], by courtesy of the Journal of Agricultural and Food Chemistry.

[b]Soil type: Butlertown silt loam (Delaware, United States); application rate: 4 lb active ingredient/acre, surface applied.

and loss of the ^{14}C chemicals. The open ends of the tubing permitted free passage of water, etc. At various sampling intervals, a selected tube was removed from the field in toto and brought back to the laboratory where the treated soil was pushed out in increments and analyzed for ^{14}C content.

Results are shown in Table 6. Total ^{14}C residues from both compounds decreased by 50% after five to six months under the conditions studied. The appreciable loss of ^{14}C activity from the samples was presumably due to evolution of $^{14}CO_2$. This assumption is based on laboratory studies with both compounds on a similar soil type in enclosed systems equipped with gas traps [3]. These tests were discussed earlier in connection with volatility. No evidence for volatile ^{14}C fragments other than $^{14}CO_2$ was found.

Various soil increments from the field samples were extracted with 80% ethanol and analyzed for residues of the intact herbicides. About 90% of the total activity in the extracts was bromacil and terbacil, even after field exposures of up to one year. The soil extract from the 0- to 1-in. soil increment of the five-week bromacil sample was cochromatographed by tlc with metabolites isolated from rat urine. Although the bulk of the ^{14}C activity coincided with bromacil (89.5%), positive evidence for hydroxylated bromacil metabolites as minor residues in soil was obtained. Bromacil metabolites tentatively identified in soil by the tlc procedures were metabolites (6), (8), and (11) in Fig. 2.

In separate experiments at the same test area with unlabeled bromacil and terbacil, larger scale field plots were treated at the rate of 4 lb/acre. The soil (0- to 4-in. depth) from these plots was analyzed for bromacil and terbacil residues after one year of exposure by the gas-chromatographic methods of Pease [8, 16] . These analyses showed that 0.44 ppm bromacil and 0.21 ppm terbacil remained in the soil after one year. These results compared favorably with values calculated from the radiochemical data in Table 6. Calculated ^{14}C residues of 0.49 and 0.30 ppm, respectively, remained in the 0- to 5-in. depth. These latter values were calculated from the specific activity of the 2-^{14}C-labeled compounds by assuming that the entire ^{14}C residue was either bromacil or terbacil.

The studies just mentioned are the only published experiments conducted under field conditions using radiolabeled substituted uracil herbicides. They show moderately rapid breakdown of bromacil and terbacil and demonstrate that the major residues left after extended exposures are those of the parent herbicides. The presence of hydroxylated metabolites in the bromacil work is evidence for degradation. However, whether the materials are first hydroxylated, followed by ring opening to liberate $^{14}CO_2$, or whether there is an alternate chemical route to open the ring and evolve $^{14}CO_2$ has not been resolved.

The rate of disappearance of the substituted uracil herbicides in soil has also been studied by other workers (bromacil [9, 32, 33, 47, 48] , terbacil [32, 34, 36, 48-50]). Postulated routes of degradation include attack on the halogen-substituted position [48] and microbial degradation [32, 51] . The reported rate of disappearance of these herbicides varies considerably with geographical location, as might be expected. Tucker and Phillips [32] found that in citrus orchards in Florida (Leon Immokalee fine sand) where a total of 100 lb of either bromacil or terbacil had been applied separately at rates of 20 lb/acre annually over a period of five years, approximately 1 lb was still present in the soil (0- to 18-in. depth) at 13 months after the last application. Analyses were performed by microcoulometric gas chromatography. This represents over a 95% loss of residue in the Florida soil over the five-year period. In contrast, Migchelbrink [49] has reported that in Oregon soils, an interim period of two years may be necessary for crop rotation purposes where sensitive species such as beans are planted after mint.

Microbiological degradation appears to play a very important role in the disappearance of the substituted uracils from soils. Reid [52] studied isocil and reported that soil diphtheroids and Pseudomonas spp. attacked this molecular structure. These species are present in a wide variety of agricultural soils. Bryant [53] found that isocil could serve as an energy source for bacteria of both the Pseudomonas and Arthrobacter spp. Torgeson and Mee [54] reported that a soil isolate of Penicillium paraherquei Abe. was particularly active in the degradation of bromacil. For example, no herbicidal effects on buckwheat could be detected 28 days after treatment with 12.5 lb/acre (about 12 ppm) of bromacil in sterile soil inoculated with P. paraherquei. Buckwheat was sensitive to bromacil at 1 ppm in the standard bioassay. However, Zimdahl et al. [48] did not observe detectable lag periods in the rates of degradation of bromacil and terbacil applied to Chehalis silt loam, and thus suspected nonenzymatic degradation. With regard to this latter work, it seems possible that there may be chemical routes of degradation in soil that could obscure the biological pathway in certain situations. From an environmental standpoint, however, the cited work with microorganisms provides direct and positive evidence that the substituted uracil herbicides are biodegradable.

C. Degradation in Plants

The fate of the substituted uracil herbicides in plants had received less attention as of early 1972, possibly because the class in general does not display a great deal of selectivity which might generate interest in plant degradation studies. The most active member of this class of herbicides is bromacil, which is used mainly as a nonselective weedkiller for nonagricultural areas. However, there is some evidence available that this compound can undergo oxidation (hydroxylation) in plants.

Gardiner et al. [3] grew young orange seedlings for four weeks in sand watered with nutrient solution containing 10 ppm [2-^{14}C]bromacil. Less than 5% of the total ^{14}C was taken up by the plants. Approximately 83% (8.5 ppm) of the material absorbed remained in the roots and about 17% moved up into the stem and leaves (1.1 ppm). Although the air chambers around the aerial portions of the plants were monitored for ^{14}C-containing gases, no ^{14}C evolution was found. Extracts of the plant roots, stem, and leaves contained three ^{14}C-labeled materials in a relative ratio of 10:5:1. The first two components, respectively, were tentatively identified by tlc R_f values as bromacil and the 6-hydroxymethyl metabolite (6), the same major metabolite found in rat urine. The minor metabolite was unidentified.

Barrentine and Warren [55] applied 2-^{14}C-labeled terbacil in an isoparaffinic oil to the leaves of susceptible ivyleaf morningglory and to peppermint, a tolerant species. After 9 days, the leaves of both species were extracted with 80% ethanol which was evaporated. The residue was then partitioned between water and chloroform. Based on tlc studies of these fractions, terbacil was metabolized in both species but at a higher rate in peppermint. In morningglory, about 90% of the ^{14}C activity was terbacil in both chloroform and water fractions. One or two metabolites were seen. In peppermint, only about 50% of the ^{14}C was terbacil in the chloroform fraction with about 40% contained in a single metabolite spot that had the same R_f value as a metabolite in morningglory. When [2-^{14}C]terbacil was fed to the roots of both species for 4 days by nutrient solution techniques, the herbicide was taken up and translocated to the foliage. Again, it was metabolized in both plants, but at a higher rate in peppermint. About 50% of the ^{14}C in the peppermint chloroform extract was terbacil and 50% was an unidentified metabolite, based on tlc procedures. In morningglory, about 70% of the translocated material was terbacil.

Herholdt [56] fed [2-^{14}C]terbacil to bean (susceptible) and citrus (tolerant) seedlings by nutrient solution techniques. Extensive degradation of the compound in both species was reported based on tlc analyses, but metabolites were not characterized. More residual terbacil was found in bean than in citrus.

In future studies on the fate of substituted uracil herbicides in plants, the following work should be considered as background information. The reductive pathway of pyrimidine degradation appears widespread in nature [57]. Uracil is degraded via dihydrouracil and β-ureidopropionic acid to β-alanine, carbon dioxide, and ammonia. The reductive pathway has been described from bacteria, animals, and tissues of higher plants [58-63]. For example, Tsai and Axelrod [64] found that externally added [2-^{14}C]-thymine and [2-^{14}C]uracil were catabolized by germinating rape seedlings through the 5,6-dihydropyrimidines and β-ureidoamino acids to give β-aminoisobutyric acid and β-alanine, respectively. β-[2-^{14}C]Alanine gave rise to labeled carboxylic acids, including malic, pyruvic, and citric acids. Carbon-14 was also incorporated into lipids and sterols. In contrast, [^{14}C]-urea was formed from both [2-^{14}C]uracil and [2-^{14}C]orotic acid in fruit bodies of Agaricus bisporus and Lycoperdon pyriforme [65]. This body of information suggests possible pathways of ultimate degradation of the substituted uracil herbicides in plants. It also suggests that metabolite identification must be done cautiously in view of the demonstrated reincorporation of ring ^{14}C into natural products.

D. Enzymatic Studies

There was essentially no work reported by 1972 on the role of specific plant enzyme systems in the degradation of the substituted uracil herbicides. Remaining portions of this section are devoted to describing certain in vitro and in vivo studies which showed that the substituted uracil herbicides do not substitute for thymine in DNA synthesis.

The work of Freese [66, 67] demonstrated that 5-bromouracil could substitute for thymine in a thymine-requiring mutant of Escherichia coli and that 5-bromouracil markedly increased the back-mutation rate of several γII mutants of coliphage T4. Because of their structural relationship to 5-bromouracil, the possibility that the substituted uracil herbicides could also show this effect has been carefully examined.

5—Bromouracil   Thymine

In 1963, McGahen and Hoffmann [68] reported results of tests in juvenile mice which showed that bromacil is not recognizable as a thymine analog by the mouse. In these studies, weanling white mice were injected by oral intubation with either [2-^{14}C]bromacil or 5-[2-^{14}C]bromouracil. Dose rates were 100 mg/kg twice daily for 2 days and 50 mg/kg for an additional 8 days. Sixteen hours after the final injection, the mice were sacrificed and DNA isolated from the livers and spleens. Labeled bromacil was not detectable in the DNA from either tissue. 5-[2-^{14}C]Bromouracil was incorporated into DNA from both tissues as suspected. Also in 1963, McGahen and Hoffmann [69] reported studies on the action of bromacil on a thymine-requiring strain (15T-) of Escherichia coli. Positive evidence for the incorporation of 5-[2-^{14}C]bromouracil into the DNA was found. Under the conditions used, this compound substituted for 21% of the replaceable thymine when thymine was absent from the growth medium, and 7.5% when thymine was present. [2-^{14}C]Bromacil, on the other hand, did not substitute for thymine in DNA either in the presence or absence of thymine in the growth medium.

The studies just mentioned demonstrated that bromacil does not incorporate into DNA under conditions in which 5-bromouracil can be incorporated. This work, however, did not rule out the possibility of indirect mutagenic activity, which was the subject of a later paper by McGahen and Hoffmann [70]. These workers found no evidence of mutagenic effects for the substituted uracil herbicides with E. coli K12 and mutant AP72, an aminopurine-induced γII mutant of coliphage T4 used earlier by Freese [67]. Included in the study for comparison were 5-bromouracil, bromacil, isocil, 5-bromo-6-methyluracil, and 5-bromo-3-isopropyluracil. The results showed that alkyl substitution of 5-bromouracil at either the 3 or the 6 position, or both, blocks the strong mutagenic action which was observed as expected in the test with 5-bromouracil. Presumably, substitution of 5-bromouracil prevents incorporation into DNA or its precursors according to these workers.

In the degradation pathways discussed earlier for the substituted uracil herbicides, specific attempts were made to determine if 5-bromouracil is a degradation product of bromacil and to detect 5-chlorouracil as a possible residue from terbacil. No evidence was found that these materials are metabolites of bromacil or terbacil in soil [3], plants [3], or animals [21,

23]. Rather, in the metabolites that had been firmly identified as of 1972, there were positive indications that the halogen substituent in the 5 position leaves the molecule before side groups at the 3 and 6 positions are eliminated. All identified metabolites in Figs. 1 and 2 retain substituents in the 3 and 6 positions, whereas several degradation products contain no halogen component.

Somewhat related work done in Czechoslovakia [71] has shown that 5-bromo[2-14C]uracil is rapidly degraded by cucumber seedlings when fed by nutrient solution techniques; $^{14}CO_2$ was detected as an end product. Also, 5-bromouracil was not incorporated into nucleic acids in the plants and did not form the corresponding nucleosides and nucleotides.

III. MODE OF ACTION

It is generally accepted that the substituted uracil herbicides are potent direct inhibitors of photosynthesis acting at the chloroplast level. They are taken up by the roots and move through plants in the transpiration stream. Limited amounts may also enter through the stem or leaves when applications are made directly to the plant. Both positional and physiological factors may come into play when this class of herbicides is used for selective weed control.

A. Uptake and Translocation

Direct visual evidence of the uptake and subsequent behavior of 2-^{14}C-labeled terbacil has been provided by Barrentine and Warren [55]. This paper contains several excellent radioautograms obtained with peppermint (tolerant) and ivyleaf morningglory (susceptible). As discussed in Sec. II,C, when [2-^{14}C]terbacil was fed to the roots of both species for 4 days by nutrient solution techniques, ^{14}C was readily taken up and translocated to the foliage. The ^{14}C concentrated in the vascular tissue of the peppermint leaves, but in the susceptible morningglory species, there was essentially no restriction of ^{14}C movement to all parts of the plant. Terbacil was absorbed by both types of leaves when small amounts of [2-^{14}C]terbacil was applied in an isoparaffinic oil. However, there was little or no movement out of the treated tolerant peppermint leaves in 9 days, whereas in the susceptible morningglory, ^{14}C activity translocated to all plant parts above the treated leaves.

Herholdt [56] has also studied the uptake and translocation of [2-^{14}C]-terbacil. This worker used bean (susceptible) and citrus (resistant) seedlings with a variety of experimental techniques. The labeled compound was fed to the seedlings by nutrient solution. Terbacil was found to be rapidly absorbed by the roots of both plant species. In bean, ^{14}C was rapidly translocated from the roots to the leaves, where it was uniformly distributed in

the mesophyll. In contrast, the ^{14}C was retained to a much greater extent
in the citrus roots and the rate of accumulation in the leaves was slower.
In the citrus leaves, ^{14}C tended to concentrate mainly in and around the
veins. Translocation of ^{14}C occurred mainly acropetally in the xylem, but
some evidence for movement in the apoplast, both laterally and basipetally,
was also obtained.

In other work related to uptake, translocation, and/or penetration of
the substituted uracil herbicides, Ray et al. [72] found that intact terbacil
could translocate in purple nutsedge. Analyses of treated and untreated
pairs of plants, connected by rhizomes, showed that after 9 days the
untreated plants contained about 10% as much terbacil as the plants treated
directly.

Gardiner et al. [3] reported that young orange seedlings, grown for
four weeks in sand watered with nutrient solution containing 10 ppm [2-^{14}C]-
bromacil, took up less than 5% of the total ^{14}C available. Of the amount
in the plants, 83% remained in the roots with only 17% going into the foliar
parts. This distribution of [2-^{14}C]bromacil in citrus appears to be in agreement with the results of Herholdt [56] , who found that ^{14}C from [2-^{14}C]-
terbacil was retained in citrus roots. Also in the former study, [2-^{14}C]-
terbacil was injected into the fifth node of sugarcane plants. Three weeks
later, 90% of the ^{14}C activity was found in the leaves but only 3 and 7% in
the juice and pulp of the plant, respectively.

Barrentine and Warren [73] demonstrated that terbacil can penetrate
into plants more rapidly when applied in an isoparaffinic oil rather than in
water. The ability of various surfactants to enhance the activity of bromacil
as a foliar spray on weeds has been reported by Hill et al. [74] .

B.   Biochemical Processes Affected

The knowledge of the mode of action of bromacil and related uracils
was reviewed and expanded by Hoffmann in 1971 [75] . In essence, the substituted uracil herbicides appear to act in much the same manner as the
substituted phenylureas to inhibit photosynthesis.

In 1964, Hilton et al. [76] and Hoffmann et al. [77] reported that bromo-
cil and isocil were strong inhibitors of photosynthesis in a variety of laboratory systems. The former workers concluded that the degree of inhibition
of the Hill reaction by bromacil and isocil was comparable with that produced
by known photosynthetic inhibitors. The respective molar concentrations
required for 50% inhibition of the Hill reaction manifested by isolated turnip
green chloroplasts were 1.4×10^{-6} and 2.7×10^{-6} . They also demonstrated
that nonphotosynthetic growth systems were only about 0.001 times as sensitive to isocil as the Hill reaction, suggesting that this material does not

interfere appreciably with pyrimidine metabolism. In addition, exogenous carbohydrates could keep plants alive and growing in the presence of growth-inhibitory concentrations of isocil. The low activity against nonphotosynthesizing plant tissue was confirmed by Jordan et al. [78], who found that bromacil had no effect on the dark growth of tobacco callus tissue supplied with an organic energy source, except at much greater concentrations than required for inhibition of photosynthesis.

The early results obtained with the substituted uracil herbicides in laboratory test systems were similar to those obtained with the substituted phenylurea herbicides and implied the same type of inhibition, possibly even at the same site. In 1971, Hoffmann [75] showed that the effects of combinations of monuron and bromacil on Chlorella cells growing in the light in an inorganic medium were additive rather than synergistic or nonadditive. This finding provided further evidence that the compounds acted at the same site and not sequentially. The most convincing evidence, according to this worker, that the urea and uracil herbicides have a similar activity was shown by tests with a monuron-resistant Euglena strain. By alternate passage of the wild strain of Euglena in 5 and 2 ppm of monuron, a strain resistant to 80 ppm monuron was isolated. Resistance was measured by photosynthesis and growth. The isolated strain was not affected in the light by 40 ppm bromacil and was only partially inhibited by 100 ppm bromacil, whereas the wild strain was markedly inhibited by 2 ppm and completely inhibited by 10 ppm. Neither strain was affected in the dark. These results provide strong evidence that the basic inhibitory activity of the substituted uracil herbicides is due to a block along the oxygen-liberating site of photosynthesis in the same manner as the urea herbicides.

In other closely related work, Hogue [79] found that bromacil was a strong Hill-reaction inhibitor of isolated tomato chloroplasts. Swann and Buchholtz [80] reported carbohydrate depletion in the rhizomes of quackgrass by the action of isocil and bromacil. Bromacil reduced $^{14}CO_2$ fixation in the light but had no effect on fixation in the dark, according to Couch and Davis [81]. Barrentine and Warren [55] showed that foliar-applied terbacil reduced $^{14}CO_2$ fixation in both peppermint and ivyleaf morningglory, but that peppermint could inactivate the herbicide as indicated by a recovery in the rate of CO_2 uptake. Van Oorschot [82] has also studied effects on CO_2 fixation by various substituted uracils.

Experiments on the wavelength dependency of the inhibition of the Hill reaction have been reported by Nishimura et al. [83]. These studies included isocil, whose inhibition was not found to be wavelength dependent. In later studies, Nishimura and Takamiya [84] said that both isocil and monuron were strongly inhibitory to the reduction of cytochrome-553 by photochemical system II.

In some additional, nonrelated studies dealing with effects on plants, Hewitt and Notton [85] found that bromacil and isocil could inhibit the induction of nitrate reductase in plants. It was not clear if the effect was direct or indirect. Oat seeds contained a higher protein content as a result of terbacil applications and yielded more grain in the subsequent growing season, according to Schweizer and Ries [86]. Several compounds, including isocil, inhibited transpiration when sprayed on bean plants at concentrations ranging from 1.0 to 1.7×10^{-3} M [87]. Finally, total elongation of oat and radish roots is reported to be inhibited by bromacil [88]. The effects noted appeared to be associated with loss of membrane integrity.

C.   Selectivity

The substituted uracils constitute a highly active, broad-spectrum class of herbicides whose primary mode of action depends on the inhibition of photosynthesis. Laboratory studies with in vitro plant fractions have indicated that all these materials will inhibit photosynthesis if they can reach the active sites in high enough concentration. Selectivity therefore appears to arise from both positional and physiological factors to prevent transport of the intact materials to these sites.

Both terbacil and bromacil are used for selective weed control in citrus. These uses appear to depend mainly on positional factors. Jordan et al. [51] reported that citrus seedlings germinating in the presence of bromacil or terbacil were killed, whereas deeper-lying roots of trees received subtoxic dosages. They concluded that the positional differences, coupled with some degree of physiological tolerance, form the basis of selectivity in citrus.

The degree of physiological tolerance in citrus apparently is due to a combination of absorption, translocation, and degradation effects. As discussed in Sec. II,C, Gardiner et al. [3] found that of the small amount of [2-^{14}C]bromacil taken up from nutrient solution by young citrus seedlings, 83% remained in the roots. Evidence for degradation of the compound in the plant was also reported.

Herholdt [56] found that [2-^{14}C]terbacil is retained to a much greater extent in citrus roots than in bean roots, a susceptible species. While both plants appeared capable of metabolizing terbacil, more residual terbacil was found in beans. Of the ^{14}C translocated into foliar portions of the plants, there was fairly uniform distribution into the mesophyll in bean leaves but a noticeable accumulation around the veins of the leaves in citrus. All of these factors tend to limit the amount of intact herbicide that can reach the photosynthesis sites. Herholdt confirmed that selectivity resides in the combination of these factors by demonstrating that isolated bean and citrus chloroplasts were equally sensitive to terbacil, thus indicating that the mechanism of selectivity did not reside in the oxygen evolution of photosystem II.

The work of Barrentine and Warren [55] has also been discussed in detail in Secs. II,C and III,A, but their conclusions bear repeating here. These workers felt that greater translocation plus slower rates of degradation accounted for the much higher susceptibility of ivyleaf morningglory versus peppermint to foliarly applied terbacil. In root applications to the same species, the selectivity appeared to arise from binding in the vascular system of peppermint plus a slightly higher rate of metabolism.

REFERENCES

1. R. W. Varner, Abstr. Weed Soc. Am., 1961, 19.

2. H. C. Bucha, W. E. Cupery, J. E. Harrod, H. M. Loux, and L. M. Ellis, Science, 137, 537 (1962).

3. J. A. Gardiner, R. C. Rhodes, J. B. Adams, Jr., and E. J. Soboczenski, J. Agr. Food Chem., 17, 980 (1969).

4. M. Merkle, A. Danti, and R. Hall, Weed Res., 5, 27 (1965).

5. H. Sherman, Toxicol. Appl. Pharmacol., 12, 313 (1968).

6. H. Sherman, Toxicol. Appl. Pharmacol., 14, 657 (1969).

7. L. S. Jordan, J. D. Mann, and B. E. Day, Weeds, 13, 43 (1965).

8. H. L. Pease, J. Agr. Food Chem., 14, 94 (1966).

9. V. A. Jolliffe, B. E. Day, L. S. Jordan, and J. D. Mann, J. Agr. Food Chem., 15, 174 (1967).

10. D. J. Hamilton, J. Agr. Food Chem., 16, 142 (1968).

11. W. H. Gutenmann and D. J. Lisk, J.A.O.A.C., 51, 688 (1968).

12. F. G. Von Stryk and G. F. Zajacz, J. Chromatogr., 41, 125 (1969).

13. A. Bevenue and J. N. Ogata, J. Chromatogr., 46, 110 (1970).

14. W. H. Gutenmann and D. J. Lisk, J. Agr. Food Chem., 18, 128 (1970).

15. C. Parker, Weed Res., 5, 181 (1965).

16. H. L. Pease, J. Agr. Food Chem., 16, 54 (1968).

17. W. H. Gutenmann and D. J. Lisk, J. Agr. Food Chem., 17, 1011 (1969).

18. W. B. Wheeler, N. P. Thompson, B. R. Ray, and M. Wilcox, Weed Sci., 19, 307 (1971).

19. W. Ebing, J. Chromatogr., 65, 533 (1972).

20. H. L. Pease and J. F. Deye, in Analytical Methods for Pesticides, Plant Growth Regulators, and Food Additives, Vol. V (G. Zweig, ed.), Academic Press, New York, 1967, Chap. 18.

21. R. C. Rhodes, R. W. Reiser, J. A. Gardiner, and H. Sherman, J. Agr. Food Chem., 17, 974 (1969).

22. R. C. Rhodes, K. Lin, J. B. Adams, and R. W. Reiser, unpublished work, 1968-1970.

23. J. A. Gardiner, R. W. Reiser, and H. Sherman, J. Agr. Food Chem., 17, 967 (1969).

24. R. W. Reiser, Org. Mass Spectrom., 2, 467 (1969).

25. A. J. Kutches, D. C. Church, and F. L. Duryee, J. Agr. Food Chem., 18, 430 (1970).

26. R. T. Williams, Detoxication Mechanisms, 2nd ed., Chapman & Hall, London, 1959, pp. 590, 599.

27. K. Fink, R. E. Cline, R. B. Henderson, and R. M. Fink, J. Biol. Chem., 221, 425 (1956).

28. E. S. Canellakis, J. Biol. Chem., 221, 315 (1956).

29. K. Fink, R. B. Henderson, and R. M. Fink, J. Biol. Chem., 197, 441 (1952).

30. G. D. Hill, Proc. Calif. Weed Conf., 1971, 158.

31. H. L. Pease and T. F. Blanchfield, unpublished work, 1965, 1969.

32. D. P. Tucker and R. L. Phillips, Citrus Ind., 51, 11 (1970).

33. W. G. McCully, C. W. Robinson, and J. G. Darroch, U.S. Clearinghouse Fed. Sci. Tech. Inform. No. 723402, 1971; C.A., 75, 97463n (1971).

34. L. C. Liu, H. R. Cibes-Viadé, and J. González-Ibáñez, J. Agr. Univ. Puerto Rico, LV, 147 (1971).

35. L. C. Liu, H. Cibes-Viadé, and F. K. S. Koo, J. Agr. Univ. Puerto Rico, LV, 451 (1971).

36. W. A. Skroch, T. J. Sheets, and J. W. Smith, Weed Sci., 19, 257 (1971).

37. G. Liefstingh and G. Blink, Landbouwvoorlichting, 26, 433 (1969); C.A., 73, 86815m (1970).

38. R. Haque and W. R. Coshow, Environ. Sci. Technol., 5, 139 (1971).

39. R. C. Rhodes, I. J. Belasco, and H. L. Pease, J. Agr. Food Chem., 18, 3 (1970).

40. C. S. Helling and B. C. Turner, Science, 162, 562 (1968).

41. C. S. Helling, P. C. Kearney, and M. Alexander, in Advances in Agron-
    omy, Academic Press, New York, 1971, Vol. 23, p. 163.

42. C. S. Helling, Soil Sci. Soc. Am. Proc., 35, 737 (1971).

43. H. Ishihara, Photochem. Photobiol., 2, 455 (1963).

44. H. Ishihara and S. Y. Wang, Biochemistry, 5, 2302 (1965).

45. H. Ishihara and S. Y. Wang, Biochemistry, 5, 2307 (1965).

46. H. Ishihara and S. Y. Wang, Nature (London), 210, 1222 (1966).

47. M. Horowitz, Weed Res., 9, 314 (1969).

48. R. L. Zimdahl, V. H. Freed, M. L. Montgomery, and W. R. Furtick,
    Weed Res., 10, 18 (1970).

49. K. Migchelbrink, Bett. Fruit Bett. Veg., 63, 33 (1969); Weed Abstr.,
    20, 170 (1971).

50. D. J. Allott, Rep. Hort. Centre, Loughgall, N. Ire., 26-9 (1968);
    Weed Abstr., 19, 2608 (1970).

51. L. S. Jordan, B. E. Day, and R. C. Russell, Proc. Int. Citrus Symp.,
    1, 463 (1968).

52. J. J. Reid, Science for the Farmer (Penn. State Univ. Agr. Exp. Sta.),
    Vol. X, No. 4, 1963.

53. J. B. Bryant, Diss. Abstr., 24, 4915 (1964).

54. D. C. Torgeson and H. Mee, Proc. Northeast. Weed Control Conf.,
    1967, 584.

55. J. L. Barrentine and G. F. Warren, Weed Sci., 18, 373 (1970).

56. J. A. Herholdt, Ph.D. Thesis, University of California, Riverside,
    1968; Diss. Abstr., 30, 1978-B (1969).

57. M. P. Schulman, in Metabolic Pathways (D. M. Greenberg, ed.), Aca-
    demic Press, New York, 1961, Vol. 2, Chap. 18.

58. W. R. Evans, C. S. Tsai, and B. Axelrod, Nature (London), 190, 809
    (1961).

59. W. R. Evans and B. Axelrod, Plant Physiol., 36, 9 (1961).

60. R. L. Barnes and A. W. Naylor, Plant Physiol. Suppl., 36, XVIII
    (1961).

61. R. L. Barnes and A. W. Naylor, Plant Physiol., 37, 171 (1962).

62. S. T. Takats and R. M. S. Smellie, J. Cell. Biol., 17, 59 (1963).

63. J. Bauerova, K. Sebesta, and Z. Sormova, Coll. Czech. Chem. Commun., 29, 807 (1964).

64. C. S. Tsai and B. Axelrod, Plant Physiol., 40, 39 (1965).

65. H. Reinbothe, Tetrahedron Lett., 37, 2651 (1964).

66. E. Freese, Molecular Genetics, Academic Press, New York, 1963, Pt. I, Chap. 5.

67. E. Freese, Proc. Natl. Acad. Sci. (U.S.), 45, 622 (1959).

68. J. W. McGahen and C. E. Hoffmann, Nature (London), 199, 810 (1963).

69. J. W. McGahen and C. E. Hoffmann, Nature (London), 200, 571 (1963).

70. J. W. McGahen and C. E. Hoffmann, Nature (London), 209, 1241 (1966).

71. K. Sebesta, J. Bauerova, F. Sorm, and Z. Sormova, Coll. Czech. Chem. Commun., 25, 2899 (1960).

72. B. R. Ray, M. Wilcox, W. B. Wheeler, and N. P. Thompson, Weed Sci., 19, 306 (1971).

73. J. L. Barrentine and G. F. Warren, Weed Sci., 18, 365 (1970).

74. G. D. Hill, Jr., I. J. Belasco, and H. L. Ploeg, Weeds, 13, 103 (1965).

75. C. E. Hoffmann, Proc. 2nd Int. IUPAC Congr. Pestic. Chem., 5, 65-85 (1971).

76. J. L. Hilton, T. J. Monaco, D. E. Moreland, and W. A. Gentner, Weeds, 12, 129 (1964).

77. C. E. Hoffmann, J. W. McGahen, and P. B. Sweetser, Nature (London), 202, 577 (1964).

78. L. S. Jordan, T. Murashige, J. D. Mann, and B. E. Day, Weeds, 14, 134 (1966).

79. E. J. Hogue, Thesis, Purdue University, 1967; Diss. Abstr., 28, 414-B (1967).

80. C. W. Swann and K. P. Buchholtz, Weeds, 14, 103 (1966).

81. R. W. Couch and D. E. Davis, Weeds, 14, 251 (1966).

82. J. L. P. Van Oorschot, Weed Res., 5, 84 (1965).

83. M. Nishimura, H. Sakurai, and A. Takamiya, Biochim. Biophys. Acta, 79, 241 (1964).

84. M. Nishimura and A. Takamiya, Biochim. Biophys. Acta, 120, 45 (1966).

85. E. J. Hewitt and B. A. Notton, Biochem. J., 101, 39C (1966).

86. C. J. Schweizer and S. K. Ries, Science, 165, 73 (1969).

87. P. E. Waggoner and I. Zelitch, Science, 150, 1413 (1965).

88. F. M. Ashton, E. G. Cutter, and D. Huffstutter, Weed Res., 9, 198 (1969).

Chapter 5

THIOCARBAMATES

S. C. FANG

Department of Agricultural Chemistry
Oregon State University
Corvallis, Oregon

I. INTRODUCTION

The first thiocarbamate, S-ethyl dipropylthiocarbamate (EPTC), was introduced by Stauffer Chemical Company in 1956 for use as an experimental herbicide for the control of annual grasses and many broadleaved weeds. Low rates of EPTC were found to have marked inhibitory effects on nutgrass with no injury to bean plants or potatoes. Since then, several thiocarbamate herbicides have been introduced.

A. Chemical and Physical Properties

All compounds in this group are clear liquids with an aromatic odor. They are miscible with most organic solvents including benzene, toluene, xylene, acetone, and alcohols. Their solubilities in water are generally low. Their structural formulas and physical properties are listed in Table 1.

B. Synthesis and Analytical Methods

Two methods, with good yield, have been used by Tilles [1] for the synthesis of numerous substituted thiocarbamates. They are outlined in the following subsections.

1. Sodium Dispersion Method

In this procedure, an anhydrous alkoxide-free sodium alkylmercaptide is refluxed with the appropriate dialkylcarbamoyl chloride in xylene [Eq. (1)]. After removal of most of the solvent, the remaining products are then subjected to fractional distillation. The yields of thiocarbamate prepared by this method ranged from 30 to 90%. Compounds prepared by this procedure include EPTC, vernolate, and butylate.

$$R_1SNa + ClC{\overset{\displaystyle O}{\overset{\displaystyle \|}{-}}}N{\overset{R_2}{\underset{R_2}{\big<}}} \xrightarrow[\text{xylene}]{} R_1{-}S{-}\overset{\displaystyle O}{\overset{\displaystyle \|}{C}}{-}N{\overset{R_2}{\underset{R_2}{\big<}}} + NaCl \qquad (1)$$

2. Chlorothiolformate Method

In this method, an amine is reacted with an alkyl chlorothiolformate in ether to form a thiocarbamate according to the following reaction:

$$\underset{\text{ether}}{R_1SC\text{-}Cl} + 2HN\overset{R_2}{\underset{R_3}{\diagdown}} \xrightarrow{\text{ether}} R_1SC\text{-}N\overset{R_2}{\underset{R_3}{\diagdown}} + \overset{R_2}{\underset{R_3}{\diagdown}}NH\text{-}HCl \qquad (2)$$

The yields obtained by this method, which includes the preparation of pebulate and cycloate, were in the range of 53-84%.

2-Chloroallyl diethyldithiocarbamate (CDEC) may be synthesized in the laboratory by reacting 2,3-dichloropropene with sodium diethyldithiocarbamate.

3. Analytical Methods

A variety of techniques have been utilized for the determination of thiocarbamate herbicides. Hughes and Freed [2] used gas-liquid chromatography for the measurement of microamounts of EPTC in crops. At present, this method has been employed for the analysis of other thiocarbamates. A colorimetric procedure based on determination of the amine after hydrolysis of the thiocarbamate with concentrated sulfuric acid is also used [3].

C. Formulation, Chemical Reactions,
and Toxicological Properties

Most thiocarbamate herbicides are available either as emulsifiable concentrates, which are diluted in water and applied in a volume of 20-100 gal/acre, or as 10-20% granules. For optimum weed control under normal soil and climatic conditions, thiocarbamate herbicides, such as EPTC, butylate, pebulate, and vernolate, must be mechanically incorporated into the soil to a depth of 2-3 in. immediately after application.

EPTC and other thiocarbamates are stable and apparently noncorrosive. However, some may be hydrolyzed in aqueous solution. The half-life of CDEC at pH 5 is 47 days and at pH 8 is 30 days [4].

The acute toxicity of this group of compounds is relatively low. The LD_{50} values for experimental animals range from 395 mg/kg for diallate to 3998 mg/kg for butylate. No-effect levels were established as 125 mg/day for diallate at the higher range and 8 mg/day for molinate at the lower range.

TABLE 1

Physical and Chemical Properties of the Thiocarbamate Herbicides

Chemical name	Common name	Structural formula (molecular weight)	Specific gravity or density (°C)
S-Ethyl dipropylthio-carbamate	EPTC	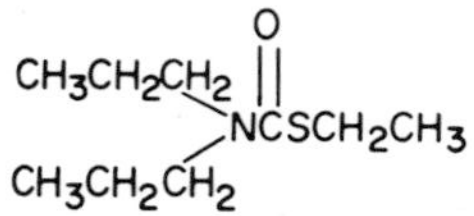 $CH_3CH_2CH_2$ $CH_3CH_2CH_2$$NCSCH_2CH_3$ (O) (189.3)	0.955 (30)
Sodium methyldithio-carbamate	Metham	$CH_3NHCSSNa \cdot 2H_2O$ (165.2)	
S-Propyl dipropylthio-carbamate	Vernolate (R-1607)	$CH_3CH_2CH_2CH_2SCN$$CH_2CH_2CH_3$ $CH_2CH_2CH_3$ (O) (203.4)	0.954 (20)
S-Propyl butylethylthio-carbamate	Pebulate (PEBC) (Tillam)	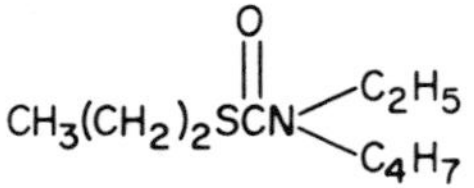 $CH_3(CH_2)_2SCN$$C_2H_5$ C_4H_7 (O) (203.3)	0.945 (30)
2-Chloroallyl diethyldithio-carbamate	CDEC	CH_3CH_2 $CH_3CH_2$$NCSCH_2CClCH_2$ (S) (223.8)	

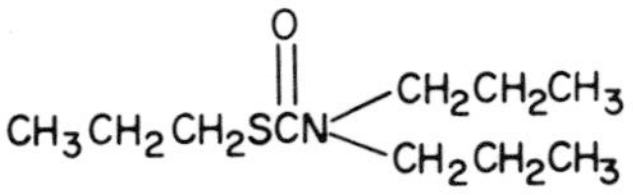

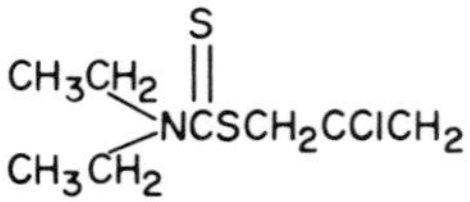

326

Diallate

S-(2,3-Dichloroallyl)
diisopropylthiocarbamate

(270.2)

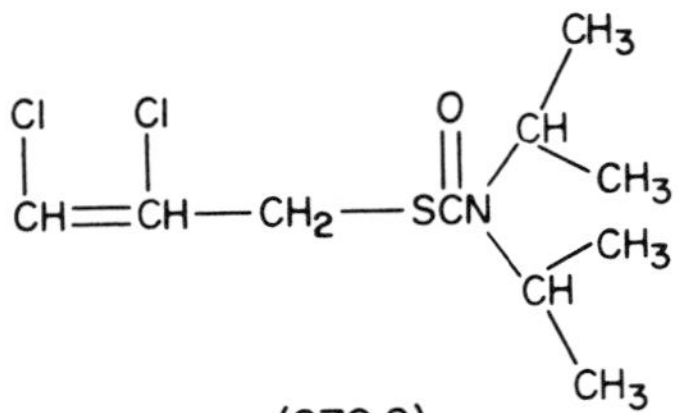

Molinate
(Ordram)

S-Ethyl hexahydro-1H-azepine-
1-carbothioate

(187.3)

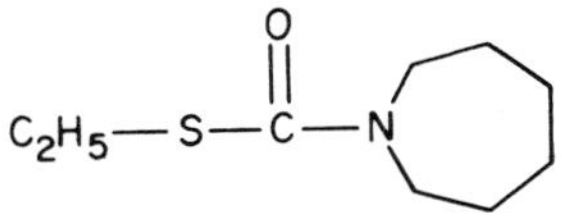

1.06
(20)

Butylate
(Sutan)

S-Ethyl diisobutylthio-
carbamate

(217.4)

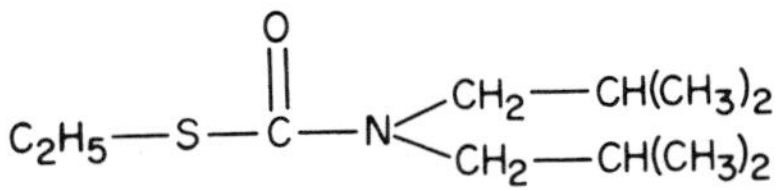

0.930
(30)

Cycloate
(RO-Neet)

S-Ethyl N-ethylcyclohexyl-
thiocarbamate

(215.4)

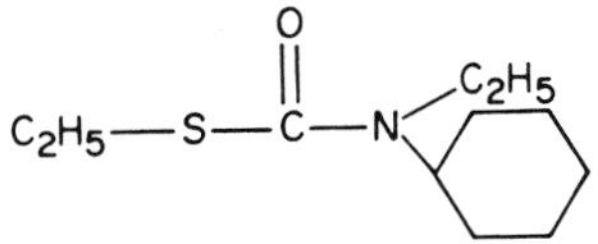

1.016
(30)

Triallate

S-(2,3,3-Trichloroallyl)
diisopropylthiocarbamate

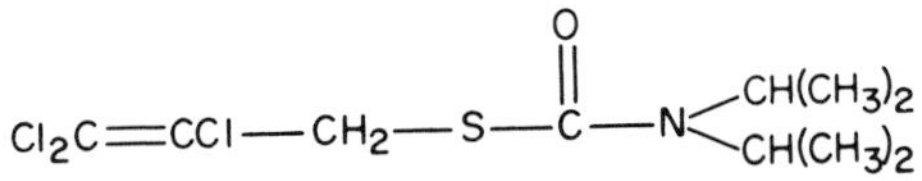

(304.7)

(continued)

327

TABLE 1 (continued)

Chemical name	Melting or boiling point (°C)	Vapor pressure, mm Hg (°C)	Solubility, g/100 ml (°C) Water	Organic	N_D^{30}	H Solution (kcal/mole)	Acute oral LD_{50} (mg/kg)
S–Ethyl dipropylthio-carbamate	127 (20 mm)	1.55×10^{-1} (25)	375 (20)	Soluble acetone, benzene, xylene, toluene, methanol isopropanol	1.4755	-3.93	1630 (rat) 3160 (mice)
Sodium methyldithio-carbamate			72.2 (20)				820 (rat) 50 (mice)
S–Propyl dipropylthio-carbamate	140 (20 mm)	10.4×10^{-3} (25)	107 ppm (21)	Soluble xylene, kerosene	1.4736		1780 (rat)
S–Propyl butylethylthio-carbamate	142 (20 mm)	4.8×10^{-3} (25)	92 ppm (21)	Soluble acetone, benzene, toluene, xylene, methanol, isopropanol	1.4750	-2.74	1120 (rat) 1652 (mice)

Compound						
2-Chloroallyl diethyldithiocarbamate	128–130 (1 mm)	2.2×10^{-3} (200)	100 ppm (25)			850 (rat)
S-(2,3-Dichloroallyl) diisopropylthiocarbamate	150 (9 mm)		14 ppm	Soluble acetone, alcohols, benzene, xylene		395 (rat) 510 (dog)
S-Ethyl hexahydro-1H-azepine-1-carbothioate	202 (10 mm)	5.6×10^{-3} (25)	800 ppm (20)	Soluble	1.516	720 (rat) 795 (mice)
S-Ethyl diisobutylthiocarbamate	138 (21.5 mm)	1.3×10^{-3} (25)	45 ppm (22)	Soluble	1.470	3998 (rat) 1659 (guinea pig)
S-Ethyl N-ethylcyclohexylthiocarbamate	146 (10 mm)	6.2×10^{-3} (25)	85 ppm (22)	Soluble		3190 (rat)
S-(2,3,3-Trichloroallyl) diisopropylthiocarbamate	148 (9 mm)		4 ppm (25)	Soluble		1570 (rat)

II. DEGRADATION PATHWAYS

A. Degradation in Plants

Fang and Yu [5] reported on the rate of [^{35}S]EPTC degradation in germinating seeds. Data in Table 2 show a greater degradation of EPTC in seeds of those species which are resistant to this herbicide than in susceptible species. The radiosulfur from ^{35}S-labeled EPTC was incorporated into cysteic acid, cystine, methionine, methionine sulfone, and two unidentified compounds. The ratio of ^{35}S labeling in these compounds varied with plant species. In all cases only a small amount of radioactivity was found in inorganic sulfate.

Using [ethyl-1-^{14}C]EPTC, Nalewaja et al. [6] reported the degradation of EPTC to CO_2 by alfalfa plants (Medicago sativa). The ^{14}C was incorporated into fructose, glucose, and several amino acids (aspartic acid, asparagine, glutamic acid, glutamine, serine, threonine, and alanine). A lack of radioactivity in cystine, cysteine, and methionine (sulfur-containing amino acids) from [ethyl-1-^{14}C]EPTC [6] and an incorporation of sulfur-35 from [^{35}S]EPTC in these compounds [5] suggests a cleavage between the sulfur atom and the ethyl group.

Fang and George [7] determined the rate of degradation of [propyl-1-^{14}C]pebulate in germinating mung bean (Phaseolus aureus) and wheat (Triticum aestivum) seeds. In mung bean seeds, which are quite resistant to pebulate, degradation was higher and increased with temperature (Fig. 1),

TABLE 2

Absorption of [^{35}S]EPTC and Percent Incorporation of Radiosulfur in Different Fractions of Plant Seedlings after Exposure[a] to [^{35}S]EPTC

Plant species	EPTC absorption (μg/seed)	Percent recovery in various fractions			
		EPTC residue	Inorganic sulfate	Hot water extract	H_2O-insoluble residue
Kidney bean	0.34	10.7	0.4	80.1	8.8
Mung bean	0.21	4.3	0.2	89.7	5.9
Pea	0.16	2.6	1.2	87.4	8.7
Corn	0.30	8.7	0.4	70.7	20.1
Wheat	0.11	8.0	0.6	74.2	17.2
Oat	0.25	23.0	0.3	62.8	13.9

[a]For 48 hr.

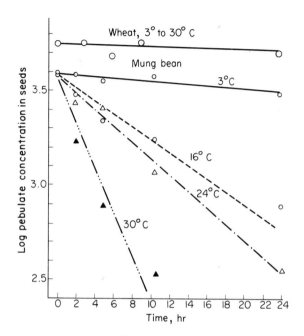

FIG. 1. Kinetic plot of [^{14}C]pebulate degradation by mung bean and wheat seedlings at different temperatures. Initial concentration w = 2.44 μg/g; MB = 1.70 μg/g.

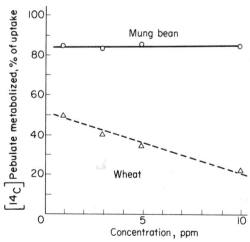

FIG. 2. Effect of pebulate concentration on its metabolism in mung bean and wheat seedlings.

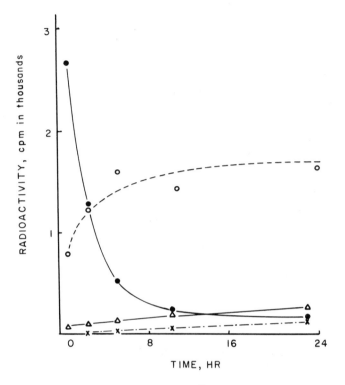

FIG. 3.   Time course conversion of [^{14}C]pebulate in mung bean seedlings   (×, CO_2; •, pebulate; Δ, EtOH insoluble; o, EtOH soluble).

whereas less pebulate degradation was evident in the susceptible wheat seeds.
Furthermore, the rate of pebulate degradation by mung beans remained quite
constant and was not affected by concentrations below 10 ppm. In contrast,
the rate of degradation of pebulate in wheat decreased as its concentration
increased (Fig. 2). This observation suggests the absence of an active pebulate detoxication system in wheat seedlings.

The time course conversion of [^{14}C]pebulate (Fig. 3) illustrates a
gradual disappearance of pebulate and an increase in other fractions, particularly the alcohol-soluble compounds. The incorporation of radioactivity
into the alcohol-insoluble fraction and CO_2 was small, suggesting that the
young seedlings did not have sufficient oxidative enzymes. However, the
catabolic oxidation of pebulate to CO_2 is more rapid in both root and shoot
tissues (Fig. 4). From 18.3 to 39.6% of the radioactivity was recovered as
CO_2 from root and shoot tissues (Table 3).

Bourke and Fang [8] examined the influence of age on the metabolism of
S-[propyl-1-^{14}C]vernolate in soybean (Glycine max L. Merr.). Figure 5

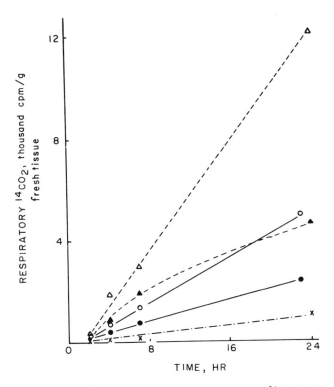

FIG. 4. In vitro $^{14}CO_2$ production from absorbed [^{14}C]pebulate by excised plan tissues and seedlings (o, pea shoot; ●, pea root; Δ, corn shoot; ▲, corn root; ×, mung bean and wheat seeds).

TABLE 3

In Vitro Metabolism of [^{14}C]Pebulate in Plant Tissues

| Plant tissue | Fresh wt (mg) | Total uptake (cpm) | Percent recovery of ^{14}C in various fractions | | | |
			Pebulate residue	CO_2	EtOH soluble	EtOH insoluble
Pea shoot	1,425	17,840	14.6	39.6	43.8	2.1
Pea root	648	4,270	25.1	37.8	26.2	10.8
Corn shoot	743	23,744	34.3	24.9	39.7	1.1
Corn root	530	5,793	13.1	36.9	43.0	6.9
Mung bean shoot	460	10,667	21.9	18.3	50.1	9.7
Mung bean root	480	2,612	30.9	22.2	39.0	8.0
Mung bean seed	723	30,773	26.0	2.7	64.2	7.1
Wheat seed	528	21,768	63.5	2.1	30.0	4.4

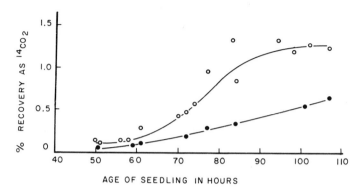

FIG. 5.   Influence of age and pretreatment with vernolate on the rate
of vernolate degradation by soybean seedlings (o—o, germinated in water;
●—●, germinated in 5 ppm vernolate).

shows $^{14}CO_2$ production as a function of age of seedlings germinated either
in water or in a 5-ppm solution of carrier vernolate. As can be seen from
Fig. 5, $^{14}CO_2$ production is low during the early stages of germination but
increases gradually with age. Preexposure to vernolate resulted in reduced
$^{14}CO_2$ production. This observation suggests that inhibition of enzyme synthesis may be the primary site of action of thiocarbamate herbicides in germinating seeds. The overall absorptions are similar: 20.98 µg of vernolate for
water-germinated and 19.47 µg for carrier-germinated seedlings. Seedlings
germinated in vernolate produced 3.25 µg equiv. of $^{14}CO_2$, whereas water-
germinated seedlings produced a total of 7.40 µg equiv. The difference in
$^{14}CO_2$ production is reflected in the amount of radioactivity recovered in the
ethanol-soluble fraction, of which 9.90 and 14.74 µg were found in the water-
and vernolate-germinated seedlings, respectively. Separation of the ethanol-
soluble fraction by paper chromatography revealed the presence of four
radioactive spots, two of which contained most of the radioactivity. One of
the highly labeled spots has been tentatively identified as citric acid by silica
column chromatography.

Using [ethyl-1-^{14}C]molinate or [azepine-2-^{14}C]molinate, Gray [9] reported
degradation of the ethyl and azepine moieties to CO_2 by rice (Oryza sativa)
plants following root uptake. Productions of $^{14}CO_2$ were 4.0% in 6 days from
chain-labeled molinate and 11.4% in 13 days from ring-labeled molinate.
Paper chromatography of the cell sap from the rice shoots harvested 3 days
after treatment with chain-labeled molinate showed the presence of 16 radioactive metabolites. Seven of these radioactive spots coincided with the amino
acids asparagine, glycine, threonine, alanine, tryptophan, phenylalanine,
and isoleucine. Five metabolites were organic acids, two of which were identified as lactic and glycolic acids. The insoluble cellulose residue and certain

protein fractions were also labeled with carbon-14. Several metabolites from rice shoots treated with [azepine-2-^{14}C]molinate were also identified as organic acids and amino acids.

Antognini et al. [10] studied the fate of ring-labeled cycloate in sugarbeets and reported that the ^{14}C was incorporated into CO_2 and a number of natural plant constituents including glycine, asparagine, proline, alanine, valine, leucine, phenylalanine, sucrose, glucose, and cellulose. No unchanged cycloate could be found in the sugarbeets 3 days after applicatication.

Jaworski [11] reported that the 2-chloroallyl moiety of 2-chloroallyl [2-^{14}C]diethyldithiocarbamate (CDEC) was extensively degraded in cabbage. Radioactivity associated with the major component corresponded to lactic acid by cochromatography.

B. Degradation in Soils

Several factors are known to determine the persistence of herbicides in soil. These include uptake and degradation by soil microorganisms, loss through physical processes (volatilization, leaching), and chemical changes (photodecomposition, chemical reaction). Volatilization is an important mechanism for the loss of thiocarbamate herbicides from soil. The loss of EPTC is greater in moist soils than in dry soils [12-14]. Significant correlations exist between the loss of EPTC by evaporation of soil moisture and the amount of organic matter, clay content, or both [12]. During the first 15 min after spraying on the soil surface, 20% of the applied EPTC disappeared from dry soil, 27% from moist soil, and 44% from wet soil [15]. The loss was 23, 49, and 69% after 1 day and 44, 68, and 90% after 6 days on dry, moist, and wet soils, respectively. Incorporation to a depth of 2-3 in. prevented large losses of EPTC from soils.

Danielson et al. [16] demonstrated with ryegrass bioassays that equivalent rates of EPTC persisted longer in soils of relatively high organic matter content and under subirrigation conditions than in soils of relatively low organic matter content and under conditions of top irrigation. In comparing the loss by vaporization of several thiocarbamate herbicides, Gray [13] reported that vernolate and pebulate were much less volatile than EPTC, and that molinate was the least volatile. Gray and Weierich [14] reported that volatilization from dry soil was much less and was not affected significantly by temperature in further studies on the effect of temperature and soil moisture on the loss of five thiocarbamates after surface application. Of the five herbicides tested, cycloate was the least volatile, followed by molinate, pebulate, vernolate, and EPTC in order of increasing volatility. Increasing the temperature from 1.7° to 37.7° caused an increase in loss of vernolate from moist and wet soils. The effect was more pronounced as the soil moisture content increased. Horowitz [17] also indicated that the loss

TABLE 4

Effect of Soil Characteristics on the Vaporization Loss of Pebulate from Soils and on the Uptake of Pebulate by Mung Bean Seedlings

Soil[a]	Organic matter (%)	Clay (%, < 20μ)	Loss of pebulate from soil after evaporation of soil water (%)			Pebulate retention in dry soil, room temperature (%)		Pebulate uptake[b] by mung bean seedlings (cpm/g fresh tissue)
			33%	50%	60%	1.5 months	4 months	
Peat	83.3	—	0.0	4.7	8.5	51.9	55.5	245
Knappa	18.3	25.1	8.2	12.8	14.7	69.5	28.0	1330
Quillayutte	17.6	36.3	5.9	9.9	14.3	91.1	22.5	1120
Astoria	12.8	34.6	5.8	8.3	13.4	53.9	42.3	1140
Wingville	7.5	25.1	0.0	3.5	5.5	88.3	43.9	310
Aiken	6.8	57.3	18.6	18.8	26.6	62.8	63.4	2625
Barron	5.4	21.3	14.8	21.1	32.7	92.9	65.1	2780
Athena	5.0	29.0	12.1	14.3	23.6	57.8	70.6	3180
Williamette	4.0	29.1	15.3	26.8	28.5	80.8	38.1	3900
Medford	3.3	10.2	21.7	31.2	41.7	84.3	52.5	3720
Sams	2.9	30.6	19.7	28.3	35.8	89.7	67.3	3525
Baker	2.8	16.3	26.5	29.5	36.3	67.1	54.4	4290
Chehalis	2.2	18.6	13.9	12.8	19.5	100.0	56.7	1780
Walla Walla	2.2	18.5	17.5	31.2	30.8	87.1	51.9	4610
Powder	2.1	12.8	24.9	35.4	40.1	86.5	56.0	4740
Deschutte	1.6	8.1	27.3	37.5	48.8	84.8	37.0	3940

[a] General characteristics and some chemical and mineralogical properties of the soils were published in Soil Sci. Soc. Am. Proc., 26, 27 (1962).

[b] Pebulate concentration in soil was 3 ppm. The uptake was carried out at room temperature for 25 hr and calculated as counts per minute per gram of fresh tissue.

of pebulate by volatilization from dry soil did not seem to constitute an impor-
than mechanism of dissipation. However, the organic matter and clay content
of soil seem to affect the herbicidal activity of pebulate. Under conditions
where pebulate was applied to dry soil without incorporation, its activity persisted for one or two months. Fang [18] (Table 4) noted that the amount of
pebulate loss from 15 Oregon soils is linearly related to the time of storage,
but the rate of loss from an individual soil is not correlated with the organic
matter content in that soil. The loss of pebulate from wet soil is related to
the amount of moisture loss and is correlated to the amount of soil organic
matter. The adsorption process probably plays an important role in preventing the loss of thiocarbamates from soils and also their availability for
plant uptake (Table 4). Although there are no available data to prove directly
that volatilization of pebulate and vernolate is reduced by soil incorporation,
an increase of herbicidal activity by this practice suggests a reduction in
vaporization and, therefore, more persistence. The addition of nonvolatile
solvents or solid materials reduces the loss of EPTC [13, 19]. In contrast,
Danielson et al. [16] reported a reduction of EPTC persistence when it was
dissolved in kerosene and an increase of persistence by the addition of specific surfactants. At present, whether this additive will also modify the
availability of herbicide to plants is not yet known and the advantage of this
practice is difficult to evaluate. Other factors, such as leaching, photodecomposition, and chemical reaction, will undoubtedly have some effects
on the persistence of thiocarbamates in soils. However, with the exception
of leaching [20] little research has been done to evaluate separately their
influence on the persistence of these herbicides.

Soil microorganisms contribute significantly to the disappearance of
thiocarbamate herbicides when incorporated in soil [21-23]. Sheets [23]
reported that EPTC was inactivated about one-third as rapidly in autoclaved
as in nonautoclaved soil, suggesting that microbial breakdown is a major
pathway of EPTC loss from soils. Therefore, EPTC persistence may be
greatly prolonged in the field when conditions for microbial activity are not
favorable. MacRae and Alexander [22] measured the microbial degradation
of EPTC by determining the release of $^{14}CO_2$ from labeled EPTC applied to
soil and by plant bioassay. They reported that EPTC was metabolized micro-
biologically, but the rate of $^{14}CO_2$ release from the ethyl moiety of the molecule was slow in comparison to the rate of inactivation as determined by
plant bioassay. A similar observation was reported by Kaufman [21] in a
Hagerstown silty clay loam. In this experiment, approximately 25% [^{14}C]-
EPTC applied to soil was recovered as $^{14}CO_2$ in 35 days, whereas a complete
inactivation of EPTC was revealed by bioassay. Since vaporization of EPTC
takes place either in dry or in wet soil, the difference observed between the
bioassay and the measurement of $^{14}CO_2$ could be due partially to vaporization
loss of [^{14}C]EPTC during this period. Also, it is conceivable that the ^{14}C
from the ethyl moiety of EPTC could be utilized anabolically by soil microorganisms as it is in plants; thereby the label would not be immediately

released as $^{14}CO_2$. Gray and Weierich [14] reported that the disappearance
of butylate, cycloate, molinate, and vernolate from nonautoclaved soil was
much faster than from autoclaved soil. To ascertain whether the disappearance of thiocarbamate from soil is due solely to the microbial breakdown,
and not from volatilization, Antognini et al. [10] stored the soil samples in
sealed containers and analyzed for cycloate at monthly intervals. The rate
of decomposition was about 20% in autoclaved soil and 60% in nonautoclaved
soil after three months of storage.

Isolation or identification of soil microorganisms capable of degrading
thiocarbamate and dithiocarbamate herbicides has not been reported, nor
has the actual mechanism of microbial degradation been determined. Several
sites of attack are possible, e.g., the alkyl groups, the amide linkage, or
the ester linkage. Based on available data obtained from studies in plants
and in animals, the thiocarbamate molecule is probably hydrolyzed at the
ester linkage with formation of mercaptan, CO_2, and amine [reaction (3)]:

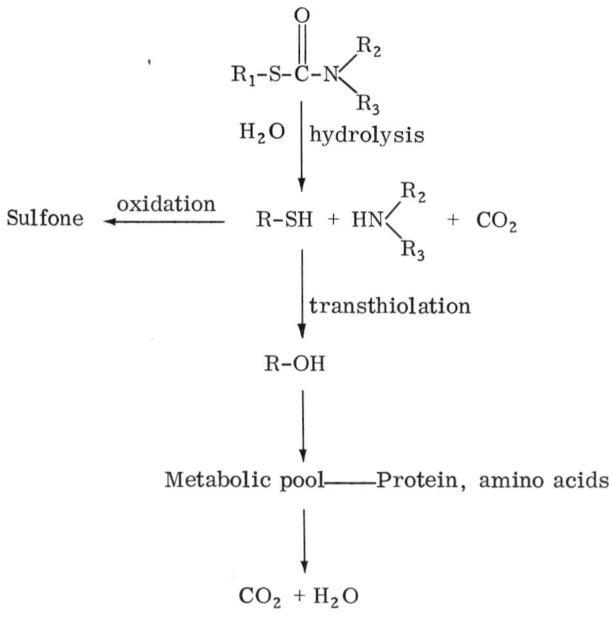

$$
\begin{array}{c}
\overset{\displaystyle O}{\underset{\displaystyle \|}{}} \\
R_1\text{-S-C-N} \diagdown \begin{array}{l} R_2 \\ R_3 \end{array} \\
H_2O \; | \; \text{hydrolysis} \\
\downarrow
\end{array}
$$

Sulfone $\xleftarrow{\text{oxidation}}$ R-SH + HN$\diagdown \begin{array}{l} R_2 \\ R_3 \end{array}$ + CO_2

$$\downarrow \text{transthiolation} \qquad\qquad (3)$$

R-OH

$$\downarrow$$

Metabolic pool——Protein, amino acids

$$\downarrow$$

$CO_2 + H_2O$

The mercaptan could then be converted to an alcohol and further oxidized.
The fate of the aliphatic amines formed in the degradation of thiocarbamates
in soil is also unknown. Early workers reported that certain bacteria utilize
amines as a source of nitrogen, based on their ability to support growth [24].
Later, Gale [25] studied the oxidation of several amines by washed cells of
Pseudomonas aeruginosa and reported that the amines were completely

oxidized to ammonia, CO_2, and water. Williams [26] indicated that short-chain aliphatic amines were mainly degraded to the corresponding carboxylic acid and urea, whereas the intermediate amines are degraded to the corresponding aldehyde and ammonia.

The degradation of dithiocarbamates would yield carbon disulfide (CS_2). This compound is highly toxic and volatile; therefore, utilization of CS_2 by soil microorganisms appears unlikely.

C. Degradation in Animals

Fang et al. [27] examined the distribution, elimination, and metabolism of S-[propyl-1-^{14}C]pebulate in adult rats and reported that approximately 55% of radioactivity was found in the expired air as $^{14}CO_2$, 23% was excreted in the urine, and less than 5% in the feces. Tissue accumulation studies showed that pebulate was absorbed and distributed throughout the whole body with the highest concentrations found in the liver and blood. The accumulation of radioactivity in most internal organs increased with the increasing dosages. A higher dosage resulted in a greater accumulation in the tissues. This relationship is approximately linear between the dosages used. The biological half-life of the radioactivity in these organs was between 2 and 3.6 days. Analysis of urine samples revealed an extensive labeling of urinary constituents, such as urea, and many amino acids. Using steam distillation, fractional distillation, and isotope-dilution techniques for the separation and identification of labeled volatile compounds in the urine revealed the presence of 0.3-2.8% as unchanged pebulate, and also a small amount of [^{14}C]propyl-mercaptan and [^{14}C]propanol. These results indicated a hydrolytic cleavage of the thiocarbamate molecule to propylmercaptan which was further converted to propanol by transthiolation. The propanol was then oxidized to a C-3 acid and entered a metabolic pool. Hydrolysis of the urine sample with strong acid converted some nonvolatile labeled metabolites to steam-volatile, isooctane-extractable metabolites, indicating that the pebulate molecule had undergone conjugation. Complete oxidation of the ethyl moiety of the EPTC molecule to CO_2 by rats was reported by Ong and Fang [28]. Increasing the dose of EPTC from 0.63 to 103 mg per rat led to a decrease in $^{14}CO_2$ output with a corresponding increase in urinary excretion of radioactivity. Generally, $^{14}CO_2$ elimination was complete within 15 hr for lower doses, but extended to approximately 35 hr for higher doses. The time for complete urinary elimination of radioactivity was likewise extended. Six major labeled metabolites were found in the urine, and the relative amounts of three metabolites changed with the dose. One of the major metabolites was identified as urea, and three others, which accounted for 50-70% of the total radioactivity in the urine, were conjugates of EPTC or EPTC-like compounds. Due to the labile nature of these conjugates, attempts at isolation by ion-exchange column chromatography and by solvent extraction have met with failure. Using ^{14}C-ring-labeled cycloate, Antognini et al. [10] reported that about

82% of the radioactivity was excreted in the rat urine and the remainder in the feces. No significant amounts of radioactivity were found in the respiratory CO_2, indicating that the cyclohexyl ring was not broken. Ethylcyclohexylamine was found as a metabolite in rat urine, providing direct evidence for cleavage at the carbonyl carbon and nitrogen bond in the thiocarbamate molecule.

III.   MODE OF ACTION

A.   Uptake and Translocation

Most thiocarbamate herbicides, when used for preemergence control, are applied as sprays to the soil surface and then mechanically incorporated into the soil immediately after application. It is possible that the chemical may be absorbed by the seed during the early stages of germination. In early experiments, Fang and Yu [5] covered the seeds of kidney bean, pea, corn, and oat with dry soils containing various levels of [35S]EPTC, ranging from 10 to 100 ppm, to determine whether or not EPTC was absorbed under these conditions. No significant amount of radioactivity was found in the seeds after 6 days, which indicates no absorption of EPTC by seeds in dry soil. However, a significant amount of radioactivity was taken up when the germinating seeds were covered with wet soil containing [35S]EPTC. EPTC was also taken up by seeds that were germinated between two sheets of filter paper moistened with an EPTC solution. In a study on the absorption from soil of a closely related herbicide, S-propyl butylethylthiocarbamate (pebulate), Fang and George [7] reported that the amount of uptake and the rate of breakdown varied with seed type and herbicide concentration. The uptake of pebulate from soil was increased with time of germination and proportionally more with increasing concentration (Fig. 6). Bourke [29] studied the absorption of S-propyl dipropylthiocarbamate (vernolate) by soybean seedlings and reported that the absorption of [14C]vernolate reached its maximum in 48 hr. During this stage, no absorbed herbicide was degraded until the seedlings were 6 or 7 days old.

Root absorption of EPTC was reported by Appleby et al. [30] in oat (Avena sativa) and by Nalewaja et al. [6] in alfalfa. The absorbed EPTC was readily moved upward to the foliage. This chemical is also absorbed by the coleoptiles and can be translocated downward to the roots.

Studies of the translocation of thiocarbamate herbicides have been conducted by Fang and co-workers [5, 8, 31-34], Yamaguchi [35], Nalewaja [36], and Chen et al. [37] using radioautographic and direct counting techniques. [35S]EPTC was absorbed from a preemergence soil application by a variety of horticultural crops and distributed throughout the entire plants [31, 32, 34]. The radiosulfur from labeled EPTC accumulates in growing stem and

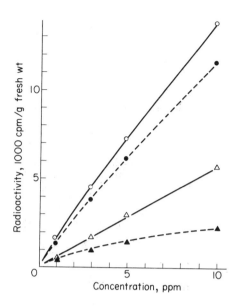

FIG. 6.   Uptake and metabolism of [^{14}C]pebulate by mung bean and wheat
seedlings. (Mung bean: o, uptake; •, metabolized. Wheat: Δ, uptake;
▲, metabolized.)

root tips after application to the leaves. When the application was made to
the roots, the distribution was more uniform [35]. Nalewaja et al. [6]
reported that [^{14}C]EPTC was readily taken up from nutrient solution by the
roots of growing alfalfa plants. The amount of radioactivity accumulated in
plants during the 5-day uptake period was approximately twice the amount
accumulated in 2 days. The ^{14}C was distributed throughout the entire plant,
with greater accumulation in the youngest tissue. The uptake of [^{14}C]pebulate
by tomato plants (Lycopersicon esculentum) from soil receiving either pre-
or postemergence treatment has been reported [33]. When pebulate was
applied to soil, it was quickly absorbed by the root system and the radioactivity translocated upward to the foliage and fruits at the bloom stage. At
1-lb/acre treatment, the ^{14}C in the foliage reached a maximum concentration
in 7 days and remained at a fairly constant level during the next 70 days. At
4 and 8 lb/acre, the maxima were not reached until the sixth or eighth week
after treatment. The concentration of pebulate residue and total ^{14}C in the
foliage correlated with the rate of treatment. Bourke and Fang [34] reported
the root uptake of [^{14}C]vernolate by soybean and peanut plants (Arachis
hypogaea L.) after a preemergence treatment and upward translocation of
^{14}C throughout the entire plant. Higher concentrations of [^{14}C]vernolate were
found in the root and stem than in the foliage. Nalewaja [36] compared the

uptake and translocation of ^{14}C from labeled diallate by roots and coleoptiles of wild oat (Avena fatua L.), wheat (Triticum aestivum L. var. Selkirk), barley (Hordeum vulgare L. var. Trail), and flax (Linum usitatissimum L. var. Bolley), and concluded that the pattern of ^{14}C uptake and movement was similar in these species.

Selectivity of diallate for wild oat is not due to the uptake and translocation of this herbicide. Chen et al. [37] reported that leaf-applied [^{14}C]molinate moved almost exclusively in an acropetal direction in both barnyardgrass (Echinochloa crusgalli L. Beauv.) and rice, whereas the root-applied molinate moved rapidly throughout the plants of both species. However, excised organs of barnyardgrass absorbed more molinate than did corresponding organs of rice. Oliver et al. [38] and Prendeville et al. [39] compared the site of EPTC uptake in relation to tolerance of several plant species, and concluded that differences in tolerance can be associated, in part, with differential sites of uptakes.

B.   Biochemical Processes

Research on the mechanism of action of thiocarbamate herbicides lagged behind technical development in their application. Studies in this area have been very limited. The precise actions of these compounds at the cellular level are not certain. Ashton et al. [40] reported the effect of several herbicides on the growth and proteolytic activity of 3-day-old squash (Cucurbita maxima Duchesne) seedlings. CDEC at 1×10^{-4} M showed a complete inhibition of root growth and approximately 75% reduction in proteolytic activity. Moreland et al. [41] showed that both EPTC at 6×10^{-4} M and CDEC at 2×10^{-4} M inhibited the development of α-amylase activity of germinating barley seeds, with CDEC being the stronger inhibitor. Inhibition of enzymic activity during the germination of seedlings is generally linked to inhibition of enzyme synthesis. Using ^{14}C-labeled leucine, ATP, or orotic acid, Moreland et al. reported a moderate inhibition of ATP incorporation into RNA by EPTC and of leucine incorporation into protein by CDEC in corn (Zea mays L.) mesocotyl and soybean hypocotyl sections, which suggests differences in their sites of action. Chen et al. [37] reported the inhibition of growth of barnyardgrass by molinate. This inhibitive effect can be overcome by addition of an equimolar concentration of gibberellic acid. The authors also found that molinate treatment reduced the level of soluble RNA in barnyardgrass. They concluded that treatment with molinate resulted in a limited amount of gibberellic acid in the plant which, in turn, caused a reduced level of transfer RNA. Antagonistic interaction between 2,4-D ([2,4-dichlorophenoxy]acetic acid) and EPTC or other thiocarbamates has been reported in several plant species [42-44]. Generally, the inhibition of growth is related to the concentrations of thiocarbamates, and this inhibition can be reversed by 2,4-D. Beste and Schreiber [44] investigated the effect of the interaction of EPTC and 2,4-D, either used single or in combination, on the synthesis of ribosomal RNA (r-RNA), DNA-like RNA (D-RNA), and tenaciously bound RNA (TB-RNA) in soybean hypocotyl sections. They reported that EPTC alone inhibited the synthesis of these RNAs, whereas 2,4-D alone stimulated synthesis. When 2,4-D was combined with EPTC, an enhancement of D-RNA and TB-RNA was observed. The inhibition of r-RNA synthesis by EPTC was not reversed by 2,4-D. Therefore, the authors concluded that the 2,4-D-enhanced synthesis of D-RNA and TB-RNA in the presence of EPTC appears to be the basis for antagonism between these two herbicides.

The formation of foliar wax in cabbage (Brassica oleracea L. var. capitata) was inhibited by EPTC, and this inhibition was correlated directly with the rate of application [45]. The reduction of foliar wax resulted in a higher rate of transpiration and altered the susceptibility of treated plants to subsequently applied herbicide. Inhibition of cuticle deposition by EPTC was also observed in sicklepot (Cassia obtysifolia L.) [46, 47], which was attributed to the inhibition of long-chained hydrocarbons. Neither the fatty acid content nor its composition was significantly affected. In an incorporation study utilizing [^{14}C]malonic acid, Mann and Pu [48] reported that CDEC at 20 mg/liter gave a 60% inhibition of ^{14}C incorporation into lipid, whereas EPTC at the same concentration had no effect.

The effect of EPTC on oxidative phosphorylation in cucumber mitochondria was first reported by Ashton [49]. Both phosphorus esterification and oxygen consumption were inhibited at 1×10^{-2} M. Since the concentration of EPTC required to produce these effects was quite high, the physiological significance of these results is questionable. Lotlikar et al. [50] reported that EPTC at 1×10^{-3} M severely inhibited oxidative phosphorylation of cabbage mitochondria. This would indicate that cabbage mitochondria are more susceptible to the action of EPTC. Winely and San Clemente [51] reported that EPTC exerted an uncoupling effect on oxidative phosphorylation linked to nitrite oxidation in cell-free extracts of Nitrobactor agilis. EPTC at 1×10^{-3} M depressed oxygen uptake by 41% but reduced phosphate esterification by 94%. EPTC apparently does not affect electron transport since NADH$_2$ oxidase activity was not affected by the addition of 1.57×10^{-3} M of EPTC.

Effects of EPTC on other physiological and biochemical processes, such as photosynthesis [49], respiration [49], acetate metabolism [52], and glucose metabolism [53], have been reported. The effect on photosynthesis in kidney bean plants was moderate and was inhibited only after long exposures to high concentrations of EPTC. The relative distribution of ^{14}C in the various labeled compounds was not altered. The respiration of excised embryos of corn and mung bean was stimulated after 24 or 48 hr of exposure in 10 or 100 ppm EPTC, and the percent stimulation was greater in 100 ppm solution and at longer exposure. Pretreatment of pea roots with 1×10^{-4} M EPTC for 2.5 hr resulted in no change of acetate uptake, but an approximately 10% increase in $^{14}CO_2$ production from either the C-1 or C-2 carbon of acetate [52]. Pretreatment of pea roots with 1×10^{-4} M EPTC resulted in a slight inhibition of glucose oxidation [53]. During the first hour, a 20 and 34% inhibition of C-1 and C-6 carbon utilization were observed, respectively. However, this inhibition diminished after the third hour. The quick recovery of the plant following EPTC treatment was probably due to rapid metabolism of the herbicide.

C. Selectivity

For herbicides to be effective in killing plants, the chemicals must achieve the following steps: (a) they must enter the plant either through the foliage or through the root system; (b) they must be translocated through the plant and reach the site(s) of action; and (c) they must effect the toxic action. The processes of absorption, translocation, and degradation, as well as the ultimate or primary action at the cellular level, may all be expected to play a role in determining effectiveness and selectivity. At present, basic knowledge on the selectivity of thiocarbamate herbicides is very limited. It has been shown by several workers [37-39] that greater absorption of thio-carbamates in barnyardgrass and sorghum (susceptible species) may be the reason for the lack of tolerance. On the other hand, Nalewaja [36] did not consider that the selectivity of diallate for wild oat was due to absorption or translocation. Degradation of the absorbed herbicide in the plant is another factor contributing to the selectivity, thereby reducing the concentration in the cell. Higher rates of pesticide degradation are effective in preventing the chemical from reaching a toxic level at the target sites. Fang and George [7] have shown a difference in the rate of pebulate degradation in mung bean and wheat, and suggested that the lack of a rapid degrading system for pebu-late in wheat may be responsible for the selectivity of this chemical. In the study of vernolate metabolism in soybean, Bourke and Fang [8] reported that no degradation of vernolate was observed in germinating soybean seedlings during the first 5 days. This observation seems to contradict the proposed concept on selective action. Although many physiological and biochemical processes have been linked to the action of thiocarbamates, it is uncertain that these processes contribute to the selective action.

IV. SUMMARY AND CONCLUSIONS

Current knowledge on the metabolism of thiocarbamate herbicides and their mode of action is rather limited. The data collected thus far may be summarized as follows:

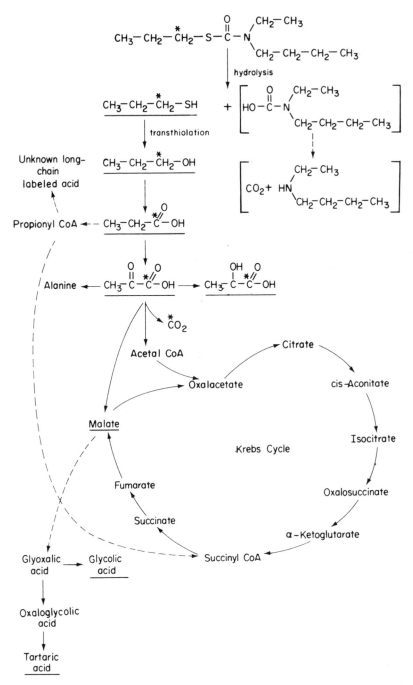

FIG. 7. Proposed pathways for pebulate degradation in plants. (The bracket encloses hypothetical intermediate. The underlined compounds have been identified in tomatoes treated with $[^{14}C]$pebulate.)

Thiocarbamates are readily absorbed by plants, but do not remain as a residue for a very long time. Relatively little work has been done to investigate their possible metabolic products. It is generally believed that upon hydrolysis, thiocarbamates yield a molecule of mercaptan, carbox dioxide, and dialkylamine [7]. Evidence suggests a further cleavage of the sulfur atom from the mercaptan. Labeling experiments with [^{35}S]EPTC revealed that the sulfur-35 is incorporated into many sulfur-containing amino acids in plants. Use of ^{14}C-labeled thiocarbamates (ethyl-1-^{14}C; n-propyl-1-^{14}C; 2,3-dichloroallyl-2-^{14}C; [ring-^{14}C]cyclohexylethylamine; and [1-^{14}C]diisobutylamine) has revealed that all label carbons gave rise to ^{14}CO$_2$ or are incorporated into many natural plant constituents such as sugars, amino acids, and plant acids. Cleavage may take place between the sulfur and carbon bond to yield an alcohol, carbonyl sulfide, and dialkylamine. However, there is no direct evidence to support this scheme.

Recent reports describing degradation of phenylurea herbicides in soil show that dealkylation precedes hydrolysis [54, 55]. Demethylation of a number of closely related N, N-dimethylcarbamates has also been reported by an enzyme system from rat liver microsomes [55]. However, Bourke [29] was not able to detect volatile labeled metabolites by steam distillation and gas-liquid chromatography after treating soybean seedlings with [^{14}C] - vernolate. If dealkylation of vernolate takes place before hydrolysis in soybean seedlings, volatile metabolic products should result. The scheme given in Fig. 7 shows the proposed pathways for thiocarbamate degradation. It is based on the unpublished work with tomatoes [56]. The underlined compounds have been identified in tomatoes treated with [^{14}C]pebulate.

It is evident that thiocarbamates are subjected to hydrolytic degradation and a wide variety of biotransformations in animals and soils. The hydrolytic pathway has received the major attention. Several conjugates of thiocarbamates have been demonstrated in rat urine, which indicates that hydroxylations of aliphatic side chains or N-alkyl groups must take place. At present, the precise nature of the conjugates is not known; pathways of degradation of these compounds in biological systems require further intensive study.

REFERENCES

1. H. Tilles, J. Am. Chem. Soc., 81, 714 (1959).

2. R. E. Hughes, Jr., and V. H. Freed, J. Agr. Food Chem., 9, 381 (1961).

3. G. H. Batcheder and G. G. Patchett, J. Agr. Food Chem., 8, 214 (1960).

4. Herbicide Handbook of the Weed Science Society of America, 2nd ed., W. F. Humphrey Press, Geneva, New York, 1970.

5. S. C. Fang and T. C. Yu, West. Weed Control Conf. Res. Prog. Rep., 1959, 91.

6. J. D. Nalewaja, R. Behrens, and A. R. Schmid, Weeds, 12, 269 (1964).

7. S. C. Fang and M. George, Plant Physiol. Suppl., 37, XXVI (1962).

8. J. B. Bourke and S. C. Fang, Weed Sci., 16, 290 (1968).

9. R. A. Gray, Weed Sci. Soc. Am. Abstr., 1969, 174.

10. J. Antognini, R. A. Gray, and J. J. Menn, Proc. Selective Weed Control in Beetcrops, 2nd Int., 1, 293 (1970).

11. E. G. Jaworski, J. Agr. Food Chem., 12, 33 (1964).

12. S. C. Fang, P. Theisen, and V. H. Freed, Weeds, 9, 569 (1961).

13. R. A. Gray, Weeds, 13, 138 (1965).

14. R. A. Gray and A. J. Weierich, Proc. 9th Brit. Weed Control Conf., 1968, 94.

15. R. A. Gray and A. J. Weierich, Weeds, 13, 141 (1965).

16. L. L. Danielson, W. A. Gentner, and L. L. Jansen, Weeds, 9, 463 (1961).

17. M. Horowitz, Weed Res., 6, 1 (1966).

18. S. C. Fang, unpublished data, 1967.

19. V. H. Freed, J. Vernetti, and M. Montgomery, Proc. West. Weed Control Conf., 16, 21 (1962).

20. R. A. Gray and A. J. Weierich, Weed Sci., 16, 77 (1968).

21. D. D. Kaufman, J. Agr. Food Chem., 15, 582 (1967).

22. I. J. MacRae and M. Alexander, J. Agr. Food Chem., 13, 72 (1966).

23. T. J. Sheets, Weeds, 7, 442 (1959).

24. J. R. Porter, Bacterial Chemistry and Physiology, 2nd ed., Wiley, New York, 1947, p. 1073.

25. E. F. Gale, Biochem. J., 36, 64 (1962).

26. R. T. Williams, Detoxication Mechanisms, 2nd ed., Wiley, New York, 1959, p. 796.

27. S. C. Fang, M. George, and V. H. Freed, J. Agr. Food Chem., 12, 37 (1964).

28. V. Y. Ong and S. C. Fang, Toxicol. Appl. Pharmacol., 17, 418 (1970).

29. J. B. Bourke, Ph.D. Thesis, Oregon State University, Corvallis, 1964.

30. A. P. Appleby, W. R. Furtick, and S. C. Fang, Weed Res., 5, 115 (1965).

31. S. C Fang and P. Theisen, J. Agr. Food Chem., 7, 770 (1959).

32. S. C. Fang and P. Theisen, J. Agr. Food Chem., 8, 295 (1960)

33. S. C. Fang and E. Fallin, Weeds, 13, 152 (1965).

34. J. B. Bourke and S. C. Fang, J. Agr. Food Chem., 13, 340 (1965).

35. S. Yamaguchi, Weeds, 9, 374 (1961).

36. J. D. Nalewaja, Weed Sci., 16, 309 (1968).

37. T. M. Chen, D. E. Seaman, and F. M. Ashton, Weed Sci., 16, 28 (1968).

38. L. R. Oliver, G. N. Prendeville, and M. M. Schreiber, Weed Sci., 16, 534 (1968).

39. G. N. Prendeville, L. R. Oliver, and M. M. Schreiber, Weed Sci., 16, 538 (1968).

40. F. M. Ashton, D. Penner, and S. Hoffman, Weed Sci., 16, 169 (1968).

41. D. E. Moreland, S. S. Malhotra, R. D. Gruenhagen, and E. H. Shokrah, Weed Sci., 17, 556 (1969).

42. C. E. Beste and M. M. Schreiber, Weed Sci., 18, 484 (1970).

43. C. E. Beste and M. M. Schreiber, Weed Sci., 20, 4 (1972).

44. C. E. Beste and M. M. Schreiber, Weed Sci., 20, 8 (1972).

45. W. A. Gentner, Weeds, 14, 27 (1966).

46. R. E. Wilkinson and W. S. Hardcastle, Weed Sci., 17, 335 (1969).

47. R. E. Wilkinson and W. S. Hardcastle, Weed Sci., 18, 125 (1970).

48. J. D. Mann and M. Pu, Weed Sci., 16, 197 (1968).

49. F. M. Ashton, Weeds, 11, 295 (1963).

50. P. D. Lotlikar, L. F. Remmert, and V. H. Freed, Weed Sci., 16, 161 (1968).

51. C. L. Winely and C. L. San Clemente, Can. J. Microbiol., 17, 47 (1971).

52. V. L. Stevens, J. S. Butts, and S. C. Fang, Plant Physiol., 37, 215 (1962).

53. J. B. Bourke, J. S. Butts, and S. C. Fang, Weeds, 12, 272 (1964).

54. R. L. Dalton, A. W. Evans, and R. C. Rhodes, Proc. South. Weed Conf., 18, 72 (1965).

55. H. Geissbühler, C. Haselbach, H. Aebi, and L. Ebner, Weed Res., 3, 277 (1963).

56. S. C. Fang and E. Fallin, unpublished data, 1967.

Chapter 6

CHLOROACETAMIDES

ERNEST G. JAWORSKI

Monsanto Agricultural Products Co.
Monsanto Company
St. Louis, Missouri

I.   INTRODUCTION: CHEMICAL AND PHYSICAL PROPERTIES

The metabolism of pesticides has received particularly wide and intensive study during the last 18 years. Although these investigations have been

349

initiated in many instances because of a desire and need to know the persistence, metabolism, and fate of subsequent degradation products of a given
pesticide in crop plants, in many cases such studies have resulted in collateral findings regarding detoxication mechanisms, possible modes of action,
basis of selectivity, basis of resistance, and other facts of physiological and
biochemical interest. Some of these collateral findings are mentioned here
in the subsequent discussions of preemergence chloroacetamide herbicides
possessing unique qualities of specificity.

The first material is N,N-diallyl-2-chloroacetamide (CDAA) (1). This
grass-specific, preemergent herbicide was commercially introduced in 1956
for use in corn and soybeans to control foxtail, bromegrass, cheatgrass,
crabgrass, and certain broadleaf weeds [1].

$$\underset{\text{ClCH}_2\text{C}}{\overset{\overset{\displaystyle O}{\|}}{}}\text{—N}\overset{\diagup\text{CH}_2\text{CH}=\text{CH}_2}{\diagdown\text{CH}_2\text{CH}=\text{CH}_2}$$

(1)

Propachlor, 2-chloro-N-isopropylacetanilide (2), was introduced in
1965 and was designed to perform under a broad set of climatic and geo-
graphic conditions. It has been particularly effective in light, sandy soils
where the water solubility of CDAA precluded complete weed control due to

$$\underset{\underset{\underset{\text{CH}_3 \quad \text{CH}_3}{\diagup \quad \diagdown}}{\text{CH}}}{\underset{|}{\overset{\overset{\displaystyle O}{\|}}{\text{ClCH}_2\text{C}}}\text{—N}}\!\!-\!\!\bigcirc$$

(2)

rapid leaching. It was also designed as a preemergent herbicide for use in
corn and soybeans to control a broader spectrum of broadleaf weeds as well
as to control barnyardgrass, foxtail, bromegrass, cheatgrass, and crabgrass [2, 3]. As pointed out by Selleck et al. [3], this herbicide is effective
against most annual grass weeds, Chenopodiaceae, and most species of the
Compositae.

CDAA (molecular weight 173.6) is an amber liquid having a specific
gravity of 0.990 at 25°C. Its vapor pressure is 9.4×10^{-3} mm Hg at 20°C.
The water solubility is 1.97% at 22°C, and it has a high degree of solubility
in ethanol, acetone, xylene, and hexane.

Propachlor (molecular weight 211.7) is a white, crystalline solid having
a melting point of 78°-79°C. It is very soluble in ether (18%), acetone (31%),

benzene (50%), ethanol (29%), and chloroform (38%). Its water solubility, however, is only 0.07%.

Both CDAA and propachlor are prepared by the reaction of chloroacetyl chloride with the appropriate secondary amine, diallylamine, and N-isopropylaniline, respectively. Each compound has an active halogen that can undergo nucleophilic displacement reactions and hydrolysis.

Since the publication of the first edition, two additional chloroacetamide herbicides have been developed. 2-Chloro-2',6'-diethyl-N-(methoxymethyl) acetanilide (alachlor) (3) was introduced in 1969 as a promising new herbicide for corn, soybeans, peanuts, and cotton. It also possesses selectivity

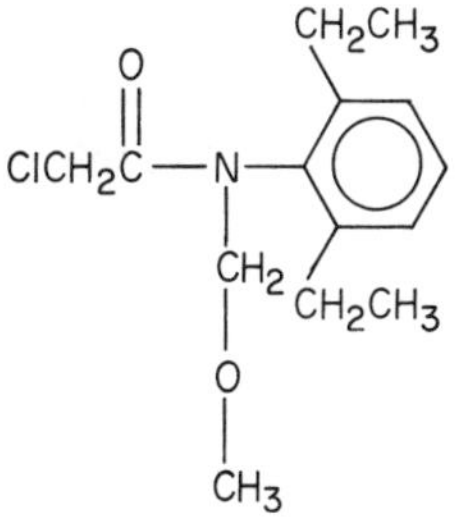

$$ClCH_2\overset{\overset{\displaystyle O}{\|}}{C}-N\!\!\left\langle \begin{array}{c} CH_2CH_3 \\ \\ \underset{CH_2CH_3}{} \end{array} \right.$$

CH₂
|
O
|
CH₃

(3)

for cabbage, dry beans, radish, rapeseed, tree fruits, and nuts while controlling the annual grasses and many broadleaf species [4, 5].

Alachlor (molecular weight 269.5) is a crystalline solid having a melting point of 40°-41°C. It has a specific gravity of 1.133 (25/15.6°C), and is soluble in ether, acetone, benzene, ethanol, and ethyl acetate. Its solubility in water is 240 ppm at 24°C.

In 1970 Baird and Upchurch [6] reported on the characteristics of N-(butoxymethyl)-2-chloro-2',6'-diethylacetanilide (butachlor) (4) as a herbicide for rice. This compound exhibited good performance in selectively

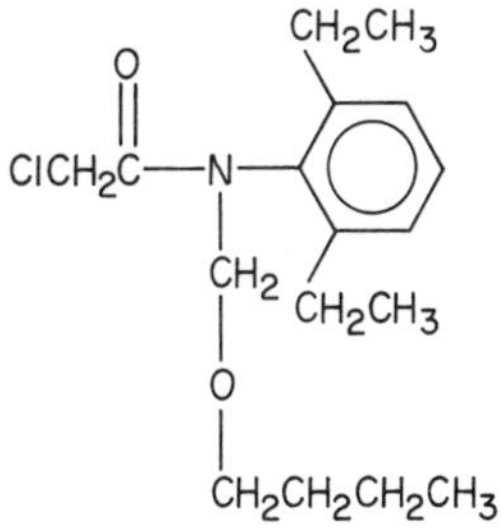

$$ClCH_2\overset{\overset{\displaystyle O}{\|}}{C}-N\!\!\left\langle \begin{array}{c} CH_2CH_3 \\ \\ \underset{CH_2CH_3}{} \end{array} \right.$$

CH₂
|
O
|
CH₂CH₂CH₂CH₃

(4)

controlling grass and specific broadleaf weeds in transplanted rice grown under flooded conditions and in seeded rice which was drilled directly into soil.

Butachlor (molecular weight 311.9) is an amber oil. It has a specific gravity of 1.65 at 25°C. It is soluble in ether, acetone, benzene, ethanol, and ethyl acetate but is soluble in water only to the extent of 23 ppm at 24°C.

II.   DEGRADATION OF CDAA

In 1954, when the first CDAA degradation studies were undertaken, no real precedent had been established for a study of metabolic degradation of pesticides, although a considerable literature existed with regard to the general reaction of haloacetamides such as iodoacetamide [7]. Two types of studies were therefore undertaken as described by Jaworski [8], one in which the carbonyl carbon was labeled with ^{14}C and the other in which the allyl moiety was labeled with ^{14}C in the 2-carbon.

A.   Chloroacetyl Moiety

Corn and soybeans were capable of rapid uptake of radioactivity from soil treated with ^{14}C-carbonyl-labeled CDAA. Uptake reached a maximum for both crops at 4-5 days following emergence, with a maximum specific activity of 85,000 cpm/g dry weight for soybeans at 5 days and 30,000 cpm/g dry weight for corn at 4 days. In subsequent harvests the specific radioactivity of both crops decreased at a rapid rate.

After the CDAA was shown to be readily taken up by plants, experiments were conducted to determine the nature of the radioactivity in the plants. Prior to doing this, the recovery procedure was established, and the results indicated that recoveries were greater than 90% at a 0.1 ppm concentration of CDAA. A bioassay technique was developed to analyze the plant extracts for CDAA-like activity. Table 1 shows the standard results for CDAA responses, as well as those from corn plant extracts. This bioassay was described by Jaworski [8] and involved using germinating ryegrass seed. The data indicate that essentially complete metabolism of the molecule took place prior to the fourth day following emergence of the corn seedlings. Similar results were also obtained for soybean seedlings.

Chromatographic analyses were made of these extracts to determine the number and types of radioactive components present. Standard R_f values are shown in Table 2 for CDAA, as well as possible expected metabolites and degradation products. The analyses of corn and soybean extracts are shown in Tables 3 and 4, respectively. Four days following emergence of corn seedlings, the primary radioactive material was at R_f 0.00 in solvent A and

TABLE 1

Ryegrass Bioassay of Corn Extracts[a]
([Carbonyl-^{14}C]CDAA Treatment)

Concentration of CDAA (ppm)	Concentration of CDAA based on radioactivity (ppm)	Ryegrass growth (mm)	Concentration of CDAA based on bioassay (ppm)
Control
—	—	21.3	—
CDAA
0.1	—	3.0	0.1
0.05	—	4.8	0.05
0.01	—	8.6	0.01
Plant extract (4 days after emergence)
—	0.12	16.6	0.01
—	0.06	20.8	0
Plant extract (7 days after emergence)
—	0.06	21.5	0
—	0.03	19.9	0

[a]The standards contained plant extracts from untreated plants equivalent to those of treated plant extracts.

TABLE 2

R_f Value of CDAA and Possible Degradation Products

Compound	Solvent system[a]		
	A	B	C
CDAA	0.89	0.91	0.93
α-Hydroxy analog of CDAA	0.89	0.76	0.89
Glycolic acid	0.00	0.61	0.59
Glyoxylic acid	0.00	0.47	0.15
Lactic acid	0.02	0.67	0.71
Glyceric acid	0.00	0.75	0.80
Oxalic acid	0.00	0.00	0.49
Chloroacetic acid	0.48	0.83	0.86

[a]Solvent system: A, benzene-methanol-H_2O (10:5:5); B, 80% aqueous phenol; C, butanol-acetic acid-H_2O (4:1:5).

TABLE 3

Chromatographic Analysis of Corn Extracts

Days after emergence	Solvent A[a]		Solvent B[a]		Solvent C[a]	
	R_f	Radioactivity (%)	R_f	Radioactivity (%)	R_f	Radioactivity (%)
4	0.00	92	0.64	98	0.63	97
	0.64	5	0.89	2	0.94	3
7	0.00	88	0.64	68	0.55	14
	0.21	4	0.88	20	0.69	68
	0.65	4			0.86	15
	0.90	5				

[a]For solvents, see Table 2.

TABLE 4

Chromatographic Analysis of Soybean Extracts

Days after emergence	Solvent A[a]		Solvent B[a]	
	R_f	Radioactivity (%)	R_f	Radioactivity (%)
2	0.00	70	0.42	64
	0.40	17	0.54	18
	0.77	13	0.66	5
5	0.00	93	0.44	1
	0.71	6	0.66	98
10	0.00	87	0.10	7
	0.44	7	0.25	7
	0.68	3	0.42	13
	0.92	3	0.59	73

[a]For solvents, see Table 2.

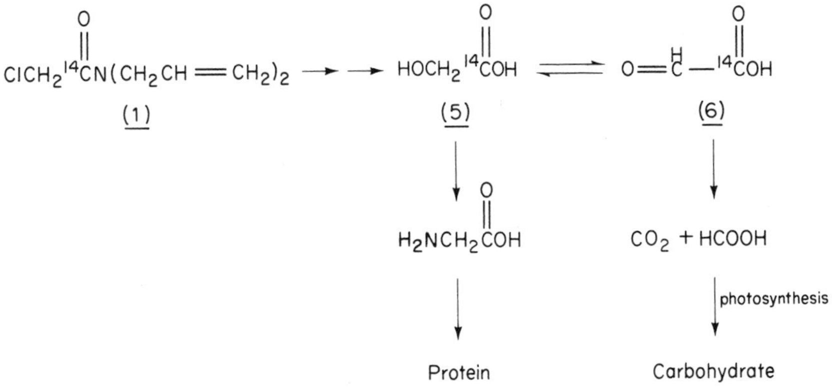

$$\text{ClCH}_2{}^{14}\text{CN(CH}_2\text{CH}=\text{CH}_2)_2 \longrightarrow \longrightarrow \text{HOCH}_2{}^{14}\text{COH} \rightleftharpoons \text{O}=\overset{\text{H}}{\underset{}{\text{C}}}-{}^{14}\text{COH}$$

(with carbonyl O on the amide of (1), and C=O groups)

(1)   (5)   (6)

(5) → H$_2$NCH$_2$COH

(6) → CO$_2$ + HCOOH

H$_2$NCH$_2$COH → Protein

CO$_2$ + HCOOH → (photosynthesis) → Carbohydrate

FIG. 1. Metabolism of CDAA.

R_f 0.64 in solvent B. These R_f values coincided with those for glycolic acid (see Fig. 1). A similar pattern was noted 7 days following emergence of the corn plants. The results further indicated the absence of unmetabolized CDAA, its 2-hydroxy analog, and 2-chloroacetic acid. Glycolic acid and possibly lactic or glyceric acids were therefore the most likely metabolite candidates. A number of other solvent systems were investigated and coupled with cochromatographic analyses to verify the presence of one of these components. Confirmation of glycolic acid as the major radioactive product formed was obtained by cochromatography of the labeled products with unlabeled glycolic acid in seven different solvent systems.

Examination of the results in Table 4 of soybean extracts 2 days after emergence indicated the presence of a compound behaving like glyoxylic (6) (see Fig. 1) rather than glycolic acid. In 5- and 10-day extracts this component was modified to one that behaved cochromatographically as glycolic acid. Zelitch and Ochoa [9] have shown that glyoxylic acid is a natural constituent of plants and is in equilibrium with glycolic acid. Zelitch has also pointed out [10] that glycolic and glyoxylic acids may function in tandem in plant respiration. On the basis of this information, it was not considered unusual to find the labeled glyoxylic acid. Furthermore, the relative proportions of these acids could vary widely, depending on such factors as light, temperature, and time of harvest.

Additional information was obtained by growing [carbonyl-^{14}C]CDAA-treated plants in a closed system, in which CO$_2$-free air was introduced into the system and CO$_2$ evolution from the plant was assayed for radioactivity. Table 5 shows that more than 50% of the radioactivity absorbed through the root system of corn seedlings was evolved as ^{14}CO$_2$, thus

TABLE 5

CO$_2$ Train Results with [Carbonyl-^{14}C]CDAA-Treated Corn
(24 days)

Fraction	Total activity (cpm × 10^{-7})	Radioactivity (%) based on:	
		Total	Plant
CO$_2$ collected	0.395	16.5[a]	54.4[a]
Plant	0.331	13.8	45.6
Soil	1.675	69.7	

[a] Data not corrected for ^{14}CO$_2$ liberated by soil microflora.

indicating that the plant readily degraded the CDAA molecule. This was also consistent with the finding of glycolic acid in the system.

On the basis of these studies, the generalized metabolic route for a portion of the CDAA molecule may be written as shown in Fig. 1. A mechanism involving the hydrolysis of both the α-chloro grouping and the amide linkage is postulated to account for the generation of glycolic acid. Glycolic acid may be in equilibrium with glyoxylic acid and the concentration of each will depend on the circumstances of plant respiration and photosynthesis at the time of harvest. Once glycolic and glyoxylic acid are formed, ^{14}CO$_2$ would be generated and photosynthetic fixation of the CO$_2$ as well as incorporation of labeled glycolic acid into a variety of biosynthetic sequences would lead to the generation of numerous labeled products in the plant. Chromatographic analyses of plant extracts from harvests made more than 20 days postemergence always contained numerous radioactive components.

B. Allylic Moieties

The allylic radicals of CDAA were tagged in the C-2 position with ^{14}C and plants were treated in a manner similar to that described for the carbonyl studies. The enclosed CO$_2$-free system was utilized to measure ^{14}CO$_2$ production and Table 6 summarizes the results obtained for a variety of crops including soybeans and corn. All plants studied were capable of metabolizing the allylic moiety to the extent where the C-2 atom was liberated as CO$_2$. In such a system, there is undoubtedly some photosynthetic reincorporation of respiratory CO$_2$ into the plants, because of the removal of CO$_2$ from air entering the system. The amount of ^{14}CO$_2$ determined therefore could be expected to represent a minimal picture of the CO$_2$ evolved.

TABLE 6

CO$_2$ Results with [Allyl-^{14}C]CDAA-Treated Crops

Crop, treatment level (lb/acre)	Duration of study (days)	Radioactivity liberated[a] as ^{14}CO$_2$ from plants (%)
Corn (6)	13	14
Soybeans (6)	13	28
Peas (10)	12	23
Onions (6)	12	19
Sugarbeets (6)	16	60
Cabbage sets (6)	10	46
Potato sets (6)	12	76
Tomato sets (6)	10	19
Strawberry sets (6)	15	24
Flax (10)	13	25
Barley (10)	14	60
Sorghum (6)	9	35

[a]Data are corrected for microbial breakdown of CDAA.

Chromatographic analyses of extracts of these plants did not elucidate the route of breakdown, possibly because of the lability and volatility of some of the intermediate products. None of the expected breakdown products arising from the cleavage of the amide linkage could be found in extracts of the plants. The type of degradation products sought included diallylamine, monoallylamine, allyl alcohol, acrylic acid, and acrolein. From the data in Table 6, however, it is apparent that both dicotyledonous and monocotyledonous plants can readily metabolize the allylic moiety of CDAA.

To establish that the breakdown of the CDAA molecule was complete and that accumulation of some specific product derived from the allylic moiety was not taking place, plants were treated with [allyl-^{14}C]CDAA and grown to maturity. Following maturation, the crops were harvested and fractionated as shown in Table 7. The table illustrates the various fractions isolated and shows their corresponding specific radioactivities. A low level of radioactivity was distributed among all fractions isolated. Since each fraction is morphologically, as well as in many instances chemically, different from

TABLE 7

Radioactivity of Corn Seed Fractions

Fraction	Specific radioactivity (cpm/g)	Concentration of radioactivity (ppm CDAA)
Hulls		
Whole hulls	33.0	3.7
Petroleum ether extract	8.3	0.9
Aqueous extract	16.1	1.8
Residue	30.5	3.4
Germs		
Whole germs	26.9	3.0
Petroleum ether extract	6.4	0.7
Santomerse soluble	4.6	0.5
Protein	10.1	1.1
Residue	4.7	0.5
Endosperm		
Whole endosperm	22.6	2.5
Petroleum ether extract	5.0	0.6
Aqueous extract	5.7	0.8
Protein	4.4	0.5
Starch	16.9	1.9

the other, the random distribution of radioactivity throughout the fractions suggests that no unusual metabolite is formed as a result of the degradation of the allylic moiety of CDAA. Rather, the carbon atoms are presumed to be randomized and incorporated through normal metabolic pathways into many natural products. These results would be anticipated in the light of the CO_2 train studies with ^{14}C-allyl-labeled CDAA. Similar fractionation studies were conducted with all the other crops shown in Table 6 and the results indicated the same type of general nonspecific distribution of radioactivity in all fractions.

Unfortunately, no mechanistic scheme can be proposed for the metabolism of the diallylamine portion of CDAA. It would not be unreasonable, however, to speculate that an amine oxidase could oxidize the amine to two acrolein molecules. These would be highly unstable and could be oxidized very rapidly.

C. Degradation in Soil

The degradation of CDAA in the soil has not been studied with labeled compounds, with the exception that studies using the CO_2 train showed $^{14}CO_2$ liberation from soil treated with either ^{14}C-carbonyl- or ^{14}C-allyl-labeled CDAA. The amounts generated represented approximately 20% of the total $^{14}CO_2$ trapped. Thus some decomposition of CDAA occurs in the soil.

Dissipation studies in soil using chemical analyses for the determination of CDAA indicated a rapid loss of the parent compound. The half-life was estimated to be approximately 16-18 days and was fairly representative over a broad range of soil types and climatic conditions. Some of this loss would be attributable to volatility, as shown by Deming [11].

III. DEGRADATION OF PROPACHLOR

The uptake and metabolism of uniformly ring-labeled (^3H) propachlor in corn and soybeans have been studied by Jaworski and Porter [12]. Both plants were capable of rapid uptake of radioactivity from soil treated with the acetanilide. As shown in Table 8, plant growth was sufficiently rapid to cause a continued dilution of the amount of radioactivity in the plants on a per gram fresh weight basis. Residues of radioactivity in corn decreased from 46 ppm in plants harvested 5 days after planting to 0.79 ppm in plants harvested 74 days after planting. In soybeans a similar dilution of radioactivity was noted. Since the extraction of radioactivity from treated plants with 80% aqueous acetone was essentially quantitative and the insoluble fraction from plant tissues was extremely low in radioactivity, virtually all the radioactivity in the plants remained in a soluble form and was not fixed into polymeric, insoluble products.

Paper and thin-layer chromatographic analyses of the 80% aqueous acetone extracts of corn plants harvested at various intervals of time illustrated that the metabolism of the chloroacetanilide must be extremely rapid, since no tritiated propachlor was detected even at the earliest harvest. The chromatographic data suggested that the product or products formed were highly polar materials. A harvest made 18 days after planting indicated that the conversion of the chloroacetanilide to the hydrophilic material was virtually complete in 18 days. This was borne out by the appearance of a single radioactive peak upon cochromatography of extracts from 18-day plants with extracts from plants harvested 34 days after planting.

The partitioning characteristics of the corn metabolite indicated that more than 90% of the metabolite(s) was water soluble and would not partition into ether.

TABLE 8

Specific Radioactivity of Corn and Soybean Plants
Treated with Propachlor

Days after planting	80% Acetone-soluble fraction	Insoluble fraction	Fresh wt/ plant, g	ppm expressed as the herbicide based on total radioactivity in plants
		Corn		
5	5.4×10^5	0.02×10^5	0.4	46
18	0.9×10^5	0.01×10^5	3.7	7.8
34	0.7×10^5	0.01×10^5	8.5	5.8
74	0.9×10^4		29	0.79
		Soybean		
5	1.5×10^5	0.06×10^5	0.8	12.8
18	0.6×10^5	0.01×10^5	2.3	5.5
34	0.2×10^5	0.01×10^5	4.8	1.4

TABLE 9

Fractionation of Corn Metabolite(s) by
Ion-Exchange Procedures

Fraction	Total radioactivity (%)
80% Acetone extract	100
Cation (Dowex 50, H^+)	10
Anion (Dowex 1, $HCOO^-$)	87
Neutral	0

TABLE 10

Vapor-Phase Chromatography (VPC) of Ether Extract from
Base Hydrolysis of Corn Metabolite with
N-Isopropylaniline and Aniline

VPC fraction no.	DPM in fraction
1	20
2 (aniline)	199
3	5
4 (N-isopropylaniline)	5377
5	118

To study the nature of the metabolites, ion-exchange fractionation was performed on the crude 80% aqueous acetone extract; these results are shown in Table 9. Eighty-seven percent of the total radioactivity was found to be anionic in nature, suggesting the presence of an acidic functional group.

Since this water-soluble anionic radioactive metabolite could contain either N-isopropylaniline or aniline, it was isolated and subjected to rigorous base hydrolysis in a sealed tube. If an amide linkage existed, it would be cleaved by this procedure to liberate a moiety such as aniline or N-isopropylaniline. Vapor-phase chromatography of an ether extract from the base hydrolysis was conducted by adding cold carrier aniline and N-isopropylaniline to the extract. As shown in a vapor-phase chromatographic trapping experiment (Table 10), most of the radioactivity cochromatographed with carrier N-isopropylaniline. The recovery of radioactive N-isopropylaniline represented 84% of the total radioactivity injected into the column. It is therefore reasonable to assume that the corn metabolite contains N-isopropylaniline as an integral moiety. Further corroboration of the presence of N-isopropylaniline was obtained by thin-film chromatography of the ether extract from base hydrolysis.

Similar hydrolysis and vapor-phase chromatography trapping experiments were conducted using acidic rather than basic hydrolysis conditions. The conditions of the experiments were such that if the chloroacetanilide had been conjugated through its active chlorine to some natural product to form a glycosidic linkage, acidic hydrolysis would liberate the 2-hydroxy analog of 2-chloro-N-isopropylacetanilide. Vapor-phase chromatographic analysis of an ether extract from acid hydrolysis of the corn metabolite fortified with cold 2-hydroxy-N-isopropylacetanilide as a carrier indicated that the major radioactive peak coincided with the hydroxy analog (Table 11). These results again were verified by thin-film chromatography.

TABLE 11

VPC of Ether Extract from Acid Hydrolysis of Corn Metabolite
with 2-Hydroxy-N-isopropylacetanilide

VPC fraction no.	DPM in fraction
1	30
2	164
3 (2-hydroxy-N-isopropylacetanilide)	2327
4	620
5	227

Similar studies with soybeans reported by Porter and Jaworski [13] were in complete agreement with the results found with corn. Thus these two highly resistant plant species metabolize propachlor in identical fashions.

While it is known that propachlor is rapidly metabolized by corn, soybeans, and a variety of other resistant crop plants to a water-soluble acidic metabolite, the absolute structure of this metabolite has not been defined. It is known that the metabolite contains essentially the entire structure of the original herbicide, with the exception that the chloro group appears to have been displaced, probably by some nucleophilic endogenous substrate in the plant.

Frear and Swanson [14] reported that propachlor reacts nonenzymatically in vitro with glutathione. Lamoureux et al. [15] investigated the possible in vivo conjugation of propachlor with glutathione. When ^{14}C-ring-labeled propachlor was used, the metabolism of propachlor in corn seedlings and in excised leaves of corn, sorghum, sugarcane, and barley was found to be similar during the first 6-24 hr following treatment. A minimum of three water-soluble metabolites was produced in each species. Two of these were isolated and characterized. One of the metabolites was identified as the glutathione conjugate of propachlor (7) and the other as the γ-glutamylcysteine conjugate or propachlor (8). Compounds (7) and (8) were found to be transitory metabolites in corn seedlings and were not detected in significant concentrations 72 hr after treatment. Thus, the primary mode of metabolism appeared to involve a nucleophilic displacement of the α-chloro group of propachlor by the sulfhydryl group of a peptide. It was not established whether the reaction was enzymatic. The transitory nature of these metabolites was consistent with those reported earlier by Jaworski and Porter [12].

Kaufman et al. [16] found that [carbonyl-^{14}C]propachlor-treated Hagerstown silty clay loam liberated little $^{14}CO_2$ (8.5%). A pure culture of Fusarium

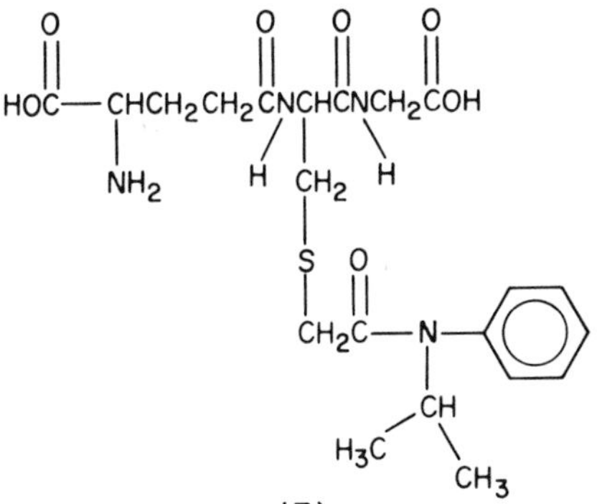

(7)

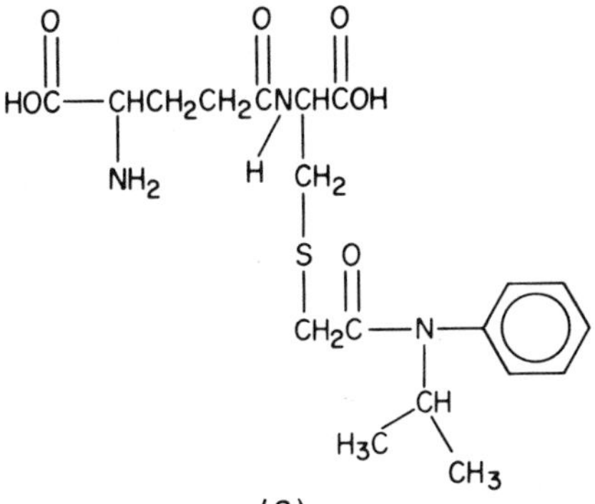

(8)

oxysporum Schlecht liberated 5.7% $^{14}CO_2$ from carbonyl-^{14}C propachlor. Dehalogenation appeared to be the major degradation mechanism. Nearly quantitative release of Cl^- occurred when pure cultures of F. oxysporum were exposed to propachlor. 2-Hydroxy-N-isopropylacetanilide and bis(N-isopropylanilino-N-carboxymethylene)oxide were identified as metabolites. A third unidentified metabolite more polar than propachlor was also isolated and comprised nearly 60% of the ^{14}C label. Only $^{14}CO_2$ and 2-hydroxy-N-isopropylacetanilide were observed as products of degradation in propachlor-treated soil.

Soil degradation experiments were also conducted using Nixon sandy loam (pH 5.5) treated with propachlor at a level of 500 ppm [17]. Bartha [17] indicated that this concentration was almost 100 times the usual field application level. The soil was moistened to 60% of holding capacity and incubated at 27°C in beakers covered with thin polyethylene film. On the basis of respirometric data, it appears that of the several herbicides studied, propachlor was the least subject to oxidation. Gas-chromatographic analyses showed a 10% decrease in the propachlor content of treated soil after 21 days. No evidence for the production of aniline or azobenzene was obtained.

IV. UPTAKE AND METABOLISM OF OTHER 2-CHLOROACETAMIDES

Studies by Smith et al. [18] defined the possible relationship between the degree of susceptibility of various plant species to selected 2-chloro-acetamides and differential uptake or metabolism of the compounds by the plants. The seeds used in these studies were corn (Zea mays L. CV US 13), oats (Avena sativa L.), soybean (Glycine max L. Merr CV Clark), and cucumber (Cucumis sativus L. CV Straight B). Oat and cucumber seeds were representative of susceptible monocotyledonous and dicotyledonous plant species, respectively, and corn and soybeans were representative of

TABLE 12

Growth Inhibition by 2-Chloroacetamides[a,b]

$$CICH_2\overset{\overset{\displaystyle O}{\|}}{C}N\underset{R_2}{\overset{R_1}{<}}$$

Derivative					
R_1	R_2	Corn	Oats	Soybean	Cucumber
$-CH_2CH=CH_2$	$-CH_2CH=CH_2$	-	++	-	+
H	$-CH_2CH=CH_2$	-	++	-	+
$-CH_2CH_2CH_3$	$-CH_2CH_2CH_3$	-	+	-	+
H	$-CH_2CH_2CH_3$	-	++	-	+
H	$-CH{<}^{CH_3}_{CH_3}$	-	+	-	+
H	$-(CH_2)_5CH_3$	-	+	-	+
H	(S-cyclohexyl ring)	-	+	-	+
H	(cyclohexenyl ring)	-	+	-	+

[a] No inhibition, -; slight inhibition, +; marked inhibition, ++.

[b] Inhibitor concentrations 10 ppm, except for diallyl derivative, where concentration was 100 ppm.

resistant plant species. The compounds studied are shown in Table 12, along with the relative sensitivities of the plant species to the chloroacetamides. The seeds used in these studies were surface sterilized with aqueous sodium hypochlorite, rinsed, dried, and placed in Scientific Product's seedpacks. The outer plastic pouch contained the chloroacetamide solutions and the inner paper wick held the seeds. All chloroacetamides were labeled at the carbonyl carbon with ^{14}C and the concentration of the compounds was 10 ppm, except for the diallyl derivative which was used at 100 ppm because of its low specific activity. At these concentrations, all four types of seeds germinated well. Following various times of germination, the seeds were removed from the pouches, rinsed with distilled water and diethyl ether, air-dried

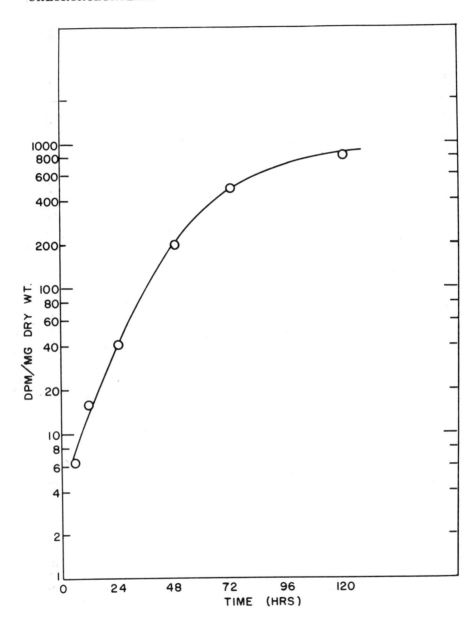

FIG. 2.   Uptake of ^{14}C-labeled N,N-diallyl-2-chloroacetamide by germinating oat seeds.

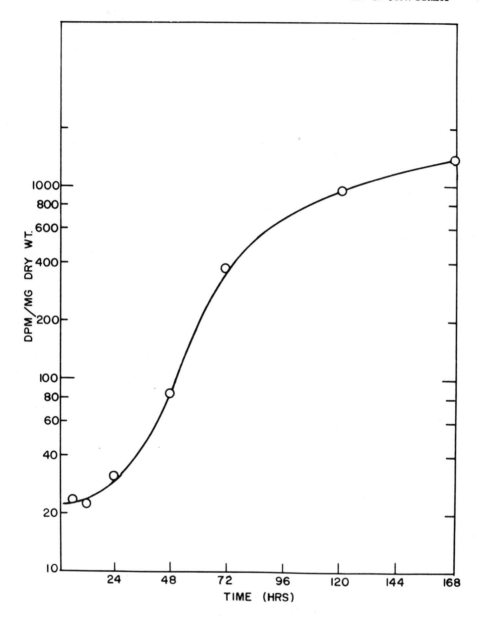

FIG. 3.   Uptake of ^{14}C-labeled N,N-diallyl-2-chloroacetamide by germinating cucumber seeds.

for 20 min, and weighed. The tissue was homogenized in 80% acetone, fil-
tered, and reextracted with more acetone. Extracts were then assayed for
radioactivity by liquid scintillation counting.

Two types of uptake curves were observed (as shown in Figs. 2 and 3).
Corn, soybeans, and oats yielded a parabolic uptake curve, as shown in
Fig. 2. While the uptake of only the diallyl derivative by oats is shown, the
curve was typical for all derivatives. The uptake curve for cucumbers
(Fig. 3) was sigmoidal in nature and was the same for all derivatives. A
statistical analysis of the data from six sampling times indicated that all
chloroacetamide derivatives were taken up to the same extent by a given
seed. Corn, one of the more resistant species, invariably took up less
chemical than the other seeds, but soybeans, the resistant species, took up
more than any other seed or about three times that of corn. Since oats and
cucumbers are susceptible and corn and soybeans very resistant, it appears
that the susceptibility is not determined by the amount of chemical absorbed.

The extent of metabolism of the chloroacetamides was studied by determining the amount taken up in 6 and 48 hr. The radioactivity of the plants
was extracted and partitioned between chloroform and water since the chloroacetamides were found to have very high chloroform-water partition coefficients, ranging from 5 to 25. Suspected breakdown products, such as
chloroacetic acid and glycolic acid, had chloroform-water partition coefficients of less than 0.01. Thus, it was assumed that all radioactivity found
in the chloroform fraction was due to nonmetabolized chemicals, and the
balance of the radioactivity remaining in the aqueous phase was attributed
to metabolites. Both corn and oats metabolized a large fraction of the chemicals absorbed in 48 hr. Thus, both the resistant and susceptible species
clearly have the ability to metabolize 2-chloroacetamides. At 6 hr there was
a definite difference between the amount of metabolism by corn and that by
oats. Corn metabolized significant amounts within a short time, but oats,
the more susceptible species, metabolized essentially none.

The same relationship was observed between soybean and cucumbers,
as shown in Fig. 4. Again, both metabolized large fractions in 48 hr, but
only the resistant soybeans metabolized a significant amount in 6 hr.

The degree of susceptibility of various weeds to chloroacetamides could
be directly related to the length of time required to initiate metabolism of
these chemicals. Those species that are able to metabolize the chemical as
soon as it enters or within a short time thereafter have only small amounts
of the chemical present internally at any time. On the other hand, those
species with delayed or slow metabolic capabilities accumulate relatively
higher and therefore lethal concentrations. The basis for selectivity of the
chloroacetamide herbicides could therefore be attributed to the ability of
resistant plants to metabolize them at a rate sufficient to keep cellular
levels below that required for growth inhibition.

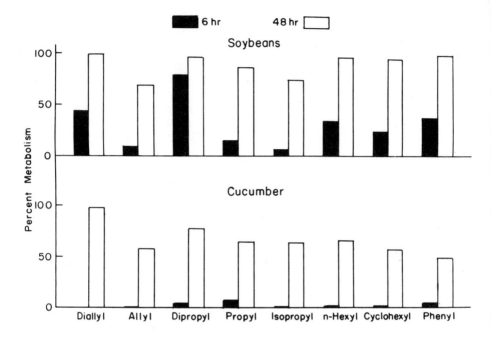

FIG. 4.   Metabolism of 2-chloroacetamides.

Some selectivity based on the kinetics of absorption could also be expected and might explain intermediate ranges of toxicity, as shown between cucumbers and oats. The latter are somewhat more susceptible to CDAA than the former. In this case the metabolism rates appear to be comparable in both species, but due to the parabolic uptake kinetics for oats, they would be expected to absorb greater amounts of the herbicides per unit time (compare Figs. 2 and 3) and build up higher intracellular concentrations.

Baird [19] studied the influence of soil moisture, pH, temperature, and organic matter on the detoxication rate of butachlor in soil. Emphasis was placed on the role of microorganisms. The recovery of radioactivity from [^{14}C]butachlor-treated soils (24 ppm) as $^{14}CO_2$ was 13% at the end of 11 weeks at 32°C from carbonyl-labeled material and 4.5% from ring-labeled material under nonflooded conditions. Less of both labeled materials was recovered as $^{14}CO_2$ from flooded, autoclaved, and KN_3-treated (400 ppm) soil and at temperatures of 16° and 24°C. Nevertheless, plant bioassays indicated that the detoxication of butachlor was favored by flooding and high temperatures. It was suggested that the initial detoxified product of butachlor is not necessarily a decarbonylated product but may be an N-dealkylated or an α-hydroxylated product. Thus the detoxication of butachlor can proceed and occur under conditions not especially suitable for total degradation to CO_2.

Electrophoretic and thin-layer chromatograms of soil produced aqueous and organic soluble metabolites which emphasized the ability of the environment to degrade butachlor. Three organic soluble and five aqueous soluble metabolites of butachlor were separated. Two of the organic soluble metabolites were tentatively identified as 2-hydroxy-2',6'-diethyl-N-(butoxymethyl)acetanilide and 2-chloro-2',6'-diethylacetanilide. High temperatures, ample moisture, and long-term exposure enhanced the relative production and concentration of these degradation products.

While no attempt was made to identify the aqueous soluble metabolites, relative electrophoretic mobilities at pH 5.4 and 2.2 suggested they were acidic in nature. The evidence in hand did not support the possible formation of glutathione conjugates as reported by Lamoureux et al. [15] for propachlor. However, conjugations with various soil amino acids and/or carbohydrates could not be excluded.

The degradation of alachlor in soil and by isolated soil microorganisms has been examined. Information from soil incubation studies suggests that alachlor is biodegraded rapidly in soils [20-22] but that very little ring-labeled [^{14}C]alachlor is converted to $^{14}CO_2$ [22]. The majority of the radioactivity could be recovered from soil only after alkaline hydrolysis, suggesting that herbicide metabolites were bound to soil organic matter. Only one alachlor degradation product has been identified from soil. 2-Chloro-2',6'-diethylacetamide [3] was formed in alachlor-treated air-dried soils incubated at 46°C [23]. This product was believed to result from acid-catalyzed hydrolysis of alachlor on mineral surfaces. Some loss of propachlor, alachlor, and butachlor by volatilization from moist soils was also reported [20, 23].

Soil fungi degrade alachlor and release chloride ion [22, 24]. Four additional organic metabolites were also identified in cultures of Chaetomium globosum: 2-chloro-2',6'-diethylacetanilide (9), 2,6-diethyl-N-(methoxymethyl)aniline (10), 2,6-diethylaniline (11), and 1-chloroacetyl-2,3-dihydro-7-ethylindole (12). Incubation of C. globosum with 2-chloro-2',6'-diethylacetanilide, 2,6-diethylaniline, and monochloroacetate demonstrated further degradation of these products. Six other soil fungi studied were unable to effectively degrade alachlor. On the basis of the products isolated the degradation scheme shown in Fig. 5 was proposed.

Although no studies have been reported on the degradation of alachlor in plants, the effective sites of uptake were examined by Groenwold [25] in two monocots (susceptible oats and resistant corn) and in two dicots (susceptible cucumbers and resistant soybean). In oats, uptake by the shoot region was almost twice that by the root system. Root uptake was more important than shoot uptake for soybean and cucumber.

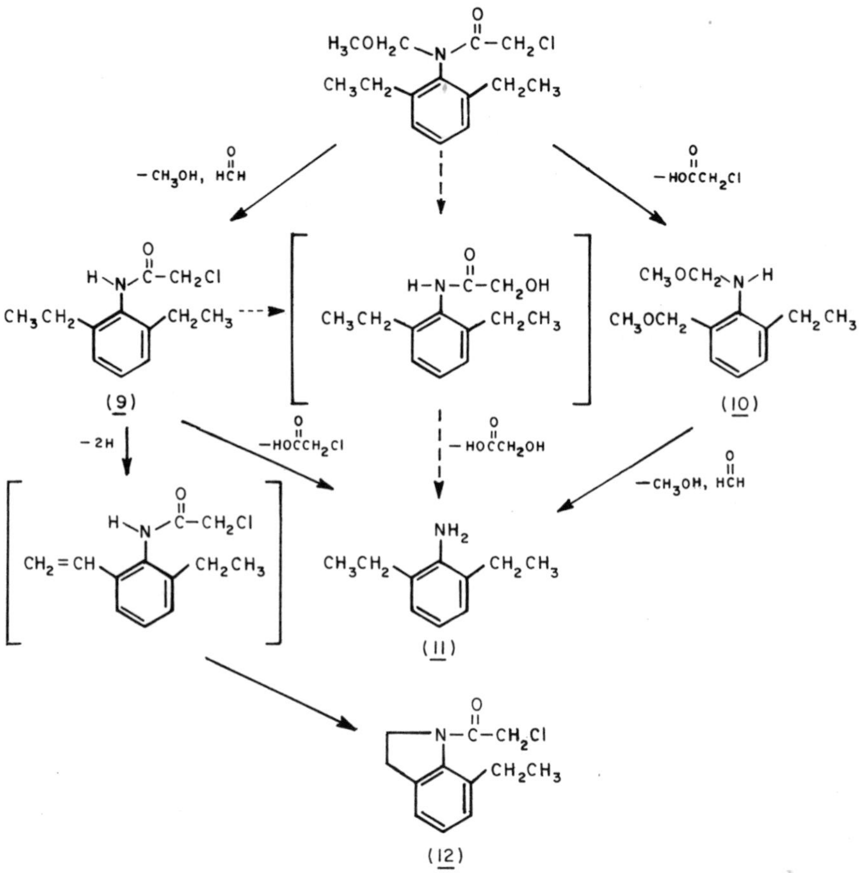

FIG. 5.   Alachlor metabolism by C. globosum [21].

Studies on the effects of light intensity on the toxicity of alachlor to cucumber, corn, and oats revealed that herbicidal activity in corn and cucumber increased with an increase in light intensity from 60 to 1600 ft-c. Light intensity had little effect on the herbicidal activity of alachlor for oats.

The absolute amounts of alachlor taken up by the four plant species studied appeared related to their relative tolerance to the herbicide. Uptake increased from corn to soybeans, to cucumber, to oats. This selectivity may be due to different rates of uptake even though the uptake mechanism appeared to be a primarily passive process.

V. MODE OF ACTION

Mechanistic studies on the mode of action of CDAA have been very limited despite the fact that CDAA is a most interesting herbicide from the aspect of its high degree of selectivity and high unit activity. Studies by Jaworski [26] were conducted to evaluate the effects of CDAA upon respiration of germinating wheat and ryegrass seeds. Ryegrass respiration was strongly inhibited by CDAA, whereas it was only moderately affected in germinating wheat, a much less susceptible plant species. This was also demonstrated by a marked decrease in respiratory quotient (RQ) in ryegrass seed. The RQ of wheat was only slightly affected. In both instances strong growth inhibition of the coleoptiles was achieved. Reversal studies demonstrated that glutathione, calcium pantothenate, and α-lipoic acid could reverse the respiratory inhibition of ryegrass by CDAA, but growth inhibition persisted. Although these data suggested that CDAA could inhibit certain sulfhydryl-containing enzymes involved in respiration, the basic lethal effect must involve some mechanism more intimately connected with growth.

The inhibition of seed germination and subsequent development of red pine seedlings by CDAA were studied by Sasaki and Kozlowski [27]. In a subsequent paper [28], they evaluated the effects of a commercial formulation of CDAA as well as pure CDAA and the inert ingredients in the formulated product for effects on the respiration of young pine seedlings. Warburg respirometer measurements indicated that the inert ingredients in the CDAA formulation may cause inhibition of respiration if young pine seedlings are exposed to them for a long time (150 min). However, it was evident that CDAA alone (4000 ppm) was much more inhibitory to respiration and that this inhibition was enhanced by the inert ingredients.

CDAA has also been shown to inhibit the oxidative phosphorylation of isolated cabbage mitochondria [29]. The P:O ratio was decreased by 46% at a CDAA concentration of 9×10^{-3} M. The inhibition of phosphorylation was greater than the inhibition of oxygen uptake.

In a more recent study by Mann et al. [30], a survey of a variety of herbicides and their effects upon protein synthesis was described. In these studies barley coleoptiles or Sesbania hypocotyls were preincubated with CDAA for 1 hr. L-[1-^{14}C]Leucine was then added and incubation was continued for an additional 2 hr. The tissues were extracted with hot ethanol, and residual ^{14}C was assayed by liquid scintillation counting. The results of these studies with barley indicated a 51 and 70% inhibition of [^{14}C]leucine incorporation in the protein at 2 and 5 ppm of CDAA, respectively. Results for Sesbania were approximately the same. Mann et al. [30] concluded that the inhibition of protein synthesis by CDAA is probably due to a more fundamental action, since even the uptake of amino acids (α-amino-n-butyric acid) was somewhat inhibited by CDAA. This latter effect could be attributed, however, to the inhibition of respiration and subsequent effects on active uptake of amino acids by the plant tissue.

Further studies by Mann and Pu [31] involved the examination of 30 herbicides for possible effects upon the incorporation of radioactivity from [2-^{14}C]malonic acid into lipid by excised hypocotyls of hemp Sesbania. CDAA inhibited incorporation of malonic acid into lipid by 65 and 85% at 1 and 20 ppm, respectively. Auxins (2,4-D and 2,4,5-T) stimulated lipid synthesis.

The mechanisms that may be applicable in accounting for the detoxication of the chloroacetamides may also be involved in the mode of action of these compounds. That is, the haloacetamides can act as alkylating reagents, particularly with sulfhydryl groups which are highly effective nucleophiles. The studies of Lindley [32] on the reaction of thiol compounds with chloro-acetamides are particularly illustrative of the potential of these compounds for interacting with sulfhydryl groups and hence various enzyme systems.

Very little work has been conducted on the investigation of the mode of action of 2-chloro-N-isopropylacetanilide. Duke et al. [33] reported that the inhibition of growth of cucumber roots by propachlor was closely corre-lated with the inhibition of protein synthesis in root tips and suggested that this was probably the primary site of action of the herbicide. He further suggested that the primary site of action was at the level of nascent protein formation and was probably due to the prevention of the transfer of aminoacyl-tRNA to the growing polypeptide chain. These results are consistent with studies conducted by Smith and Jaworski [34] where the inhibition of gibber-ellic acid-induced (GA$_3$) amylase production in barley endosperm was demon-strated. Varner [35] demonstrated that GA$_3$ stimulated the de novo synthesis of α-amylase in the aleurone layers of barley. A 67% inhibition of amylase production resulted from the treatment of barley half-seeds with 3×10^{-4} M CDAA or propachlor (70 versus 205 units of α-amylase per 10 half-seeds). This inhibitory response was concluded to be an effect on protein synthesis or some molecular level below that of protein synthesis, since the effect of CDAA was not reversed at higher levels of GA$_3$. Similar experiments using intact barley seeds indicated that 3×10^{-4} M CDAA resulted in a 27% inhibi-tion of amylase production, and 10^{-2} M CDAA completely inhibited the for-mation of amylase in intact barley seeds. These levels of CDAA inhibited growth of barley by 78% when shoot lengths were measured over a 73-hr growing period. The growth of barley seed shoots was stimulated approxi-mately 13% by GA$_3$ at 10^{-5} M; however, CDAA caused approximately the same degree of inhibition of growth in this instance as with non-GA$_3$-treated barley seeds.

This work and that reported by Duke et al. [33] suggest a role for CDAA and propachlor at the level of protein or nucleic acid synthesis. Interference at the site of action of plant growth hormones, such as GA$_3$, could also be involved, but a direct interference with GA_3 activity cannot be involved, since the inhibition was irreversible even at very high levels of GA_3. It has also been demonstrated that CDAA and propachlor do not inhibit α-amylase activity per se even at 10^{-3} M.

Studies by Devlin and Cunningham [36] have confirmed the inhibition of GA_3-induced α-amylase production in barley half-seeds. They found 82% inhibition at 4.7×10^{-4} M propachlor. No inhibition of the amylase-starch reaction was found. Alachlor was also found to inhibit GA-induced amylase production, although to a lesser degree. Alachlor at the maximum concentration tested (3.7×10^{-4} M) inhibited enzyme production by 37%.

Propachlor was tested by Duke [37] for its capacity to alter the rate of growth of the elongating zone of excised cucumber hypocotyl tissue. Significant inhibition (23%) of cell elongation was noted at levels of 1 ppm propachlor. 2,4-D in similar experiments was found to stimulate elongation at concentrations of 0.1-5 ppm. Auxin (2,4-D)-induced elongation was more susceptible to propachlor inhibition than control growth, particularly at low auxin levels. The inhibition of auxin-induced hypocotyl growth by actinomycin D, cycloheximide, and propachlor was also compared. The effects of propachlor and chloramphenicol were greatest at low auxin levels. The expansion of cells elongating at a constant maximum rate in response to auxin was blocked by propachlor or actinomycin D. Pretreatment of tissue with propachlor or actinomycin D caused a loss in sensitivity to auxin. [^{14}C]Leucine incorporation into cucumber hypocotyl protein was inhibited by propachlor to approximately the same extent as growth (23 versus 30% at 5 ppm). Cycloheximide was more effective as a protein synthesis and growth inhibitor (45 and 38%, respectively, at 0.1 ppm). Duke concluded that propachlor acted somewhat like actinomycin D and cycloheximide.

In contrast to Duke's [33] studies with propachlor, Edmonson [38] showed that alachlor acted as an auxin herbicide stimulating growth of excised cucumber hypocotyl sections. However, alachlor did inhibit induced growth of these sections. Alachlor also stimulated [^{14}C]leucine incorporation by cucumber seedlings even up to the point where root growth ceased. Treated tissues were found to contain more RNA, DNA, intact polyribosomes, and functional enzymes. Thus, none of Edmonson's data pointed to a basis for the herbicidal effects of alachlor.

The effects of herbicides and inorganic phosphate on phytase activity in seedlings were reported by Penner [39] and included data on CDAA and propachlor. CDAA had no effect upon barley phytase per se. Phytase activity increased two- and threefold in barley, sixfold in corn, and 12-fold in squash seedlings 2 days after the initiation of germination in the dark. Most of the phytase was found in the embryo. The presence of propachlor (10^{-4} M) in the culture solutions during germination partially inhibited the development of phytase activity in the system studied. This could result from a general suppression of protein synthesis as noted earlier [33, 37].

Studies by Lanzilotta and Pramer [40] on herbicidal transformation included data on the effects of propachlor upon an acylamidase from <u>Fusarium solani</u>. The fungal acylamidase (E.C. 3.5.1., an aryl acylamine aminohydrolase) was detected using acetanilide as the substrate (K_m = 0.195 mM). Propachlor was found to be a competitive inhibitor of acetanilide hydrolysis (K_i = 0.167 mM). This fungal enzyme was highly specific for N-acetylarylamine and differed from acylamidases isolated from rice, rat liver, chick kidney, and <u>Neurospora</u>. Thus, it is difficult to relate these findings to the possible mode of herbicidal activity of propachlor.

VI. CONCLUSIONS

Based on the present knowledge of the metabolism of chloroacetamide and chloroacetanilide herbicides, it is apparent that a generalization regarding their metabolic fates cannot be made. Whether the substitution of a phenyl group on the nitrogen results in a completely different detoxication mechanism than when the nitrogens are substituted with aliphatic moieties, such as in CDAA, must await further research. However, it would appear that crop species resistant to these herbicides are capable of rapid detoxication by one of two major mechanisms. One mechanism involves a reaction of the α-halogen with exogenous substrates leading to the formation of water-soluble acidic metabolites, and the other involves the cleavage of the amide linkage and hydrolysis of the α-halogen, as is the case with CDAA.

The published works to date suggest that the chloroacetamides are subject to extensive degradation in plants and soil and represent an attractive class of herbicidal chemistry when considerations are given to minimizing the environmental impact of agricultural products.

While the mode of action of chloroacetamides remains to be resolved, the elegant studies of Duke [33, 37, 41] seem to have at least localized the areas for future consideration. Many of the reported biochemical and physiological effects of the chloroacetamides can be interpreted on the basis of protein synthesis inhibition. The exact molecular level of effect requires further investigation. Duke's work suggests an effect similar to that of cycloheximide involving the inhibition of the transfer of aminoacyl-tRNA during polypeptide formation. Jaworski [42] proposed a model involving a more specific interaction of the α-chloroacetamides with the amino group of methionyl-tRNA involved in protein synthesis initiation. Such a nucleophilic displacement between the herbicide and the amino group of methionyl-tRNA would be consistent with Duke's theory which clearly indicates an inhibition of nascent protein synthesis.

It is always tempting to have a single explanation for the mode of action of a given chemical. However, it is possible that the α-chloroacetamides may be acting by a variety of mechanisms whose sum total results in growth inhibition and death of susceptible plant species. For example, at least 30 enzymes have been inhibited by iodoacetamide [7]. Reversal studies on the inhibition of respiration [26] suggest that the α-haloacetamides can generally react with sulfhydryl reagents both in vitro and in vivo. Since numerous enzyme systems contain sulfhydryl groups vital to enzyme activity, it is conceivable that a large variety of enzymes might be inhibited. The levels of α-haloacetamides involved in such reactions are comparable to those involved in the inhibition of protein synthesis, and thus a multivalent rather than specific inhibition or mode of action may be involved.

REFERENCES

1. P. C. Hamm and A. J. Speziale, J. Agr. Food Chem., 4, 518 (1965).

2. D. D. Baird, R. F. Husted, and C. L. Wilson, Proc. South. Weed Conf., 18, 653 (1965).

3. G. W. Selleck, P. L. Berthet, D. M. Evans, and P. M. Vincent, Symp. New Herb., 2nd, Paris, pp. 277-285, 287-296 (1965).

4. G. William Selleck, Proc. Ann. Calif. Weed Conf., 21, 157 (1969).

5. D. M. Evans, Chem. Ind. (London), 19, 615 (1969).

6. D. D. Baird and R. P. Upchurch, Proc. South. Weed Sci. Soc., 23, 101 (1970).

7. J. Leyden Webb, Enzyme and Metabolic Inhibitors, Academic Press, New York, 1966, pp. 1-270.

8. E. Jaworski, J. Agr. Food Chem., 12, 33 (1964).

9. I. Zelitch and S. Ochoa, J. Biol. Chem., 201, 707 (1953).

10. I. Zelitch, J. Biol. Chem., 201, 719 (1953).

11. J. M. Deming, Weeds, 11, 91 (1963).

12. E. G. Jaworski and C. A. Porter, American Chemical Society, 148th Meeting, Detroit, 1965.

13. C. A. Porter and E. G. Jaworski, American Society of Plant Physiologists, Urbana, Illinois, 1965.

14. D. S. Frear and H. R. Swanson, Phytochemistry, 9, 2123 (1970).

15. G. L. Lamoureux, L. E. Stafford, and F. S. Tanaka, J. Agr. Food Chem., 19, 346 (1971).

16. D. D. Kaufman, J. R. Plimmer, and J. Iwan, American Chemical Society, 162nd Meeting, Washington, D.C., 1971.

17. R. Bartha, J. Agr. Food Chem., 16, 602 (1968).

18. G. R. Smith, C. A. Porter, and E. G. Jaworski, American Chemical Society, 152nd Meeting, New York, 1966.

19. D. Baird, Ph.D. Dissertation, The University of Tennessee, Knoxville, 1971.

20. G. B. Beestman and J. M. Deming, Agron. J., 66, 308 (1974).

21. J. M. Tiedje and M. L. Hagedorn, J. Agr. Food Chem., 23, 77 (1975).

22. S. F. Chou and J. M. Tiedje, unpublished data on alachlor degradation in soils, Michigan State University, East Lansing, 1973; cited by J. M. Tiedje and M. L. Hagedorn, J. Agr. Food Chem., 23, 77 (1975).

23. R. S. Hargrove and M. G. Merkle, Weed Sci., 19, 652 (1971).

24. D. D. Kaufman and J. Blake, Soil Biol. Biochem., 5, 297 (1973).

25. B. E. Groenwold, Ph.D. Dissertation, Oregon State University, Corvallis, 1971.

26. E. G. Jaworski, Science, 123, 847 (1956).

27. S. Sasaki and T. T. Kozlowski, Bot. Gaz., 129, 238 (1968).

28. S. Sasaki and T. T. Kozlowski, Bot. Gaz., 129, 286 (1968).

29. P. D. Lotlikar, L. S. Jordan, and B. E. Day, Plant Physiol., 40, 840 (1965).

30. J. D. Mann, L. S. Jordan, and B. E. Day, Plant Physiol., 40, 840 (1965).

31. J. D. Mann and M. Pu, Weed Sci., 16, 197 (1968).

32. H. Lindley, Biochem. J., 82, 418 (1962).

33. W. B. Duke, F. W. Slife, and J. B. Hanson, Weed Soc. Am. Abstr., 1967, 50.

34. G. R. Smith and E. G. Jaworski, unpublished work, 1966.

35. J. E. Varner, Plant Physiol., 37, 413 (1964).

36. R. M. Devlin and R. P. Cunningham, Proc. Northeast. Weed Control Conf., 24, 149 (1970).

37. W. B. Duke, Proc. Northeast. Weed Control Conf., 22, 504 (1968).

38. J. B. Edmonson, Ph.D. Dissertation, University of Illinois, Urbana, 1966.

39. D. Penner, Weed Sci., 18, 360 (1970).

40. R. P. Lanzilotta and D. Pramer, Appl. Microbiol., 19, 307 (1970).

41. W. Duke, Ph.D. Dissertation, University of Illinois, Urbana, 1967.

42. E. Jaworski, J. Agr. Food Chem., 17, 165 (1969).

Chapter 7

AMITROLE

MASON C. CARTER

Department of Forestry
Auburn University
Auburn, Alabama*

---

*Present address: Department of Forestry and Natural Resources,
Purdue University, West Lafayette, Indiana.

I. INTRODUCTION

A. History

Amitrole (3-amino-s-triazole) was patented as a herbicide and plant growth regulator in 1954 [1], although research reports concerning the chemical had appeared the preceding year [2-4]. Since its introduction, amitrole has been widely used for a variety of agricultural and industrial applications [5]. Residues of this pesticide led to withdrawal of certain lots of cranberries from the market in 1959 when it was proposed that the pesticide could induce thyroid tumors in rats.* All registered uses of amitrole on food crops were canceled in 1971 [6].

B. Physical and Chemical Properties

Amitrole is a white, crystalline solid with a molecular weight of 84 and a melting point of 150°-153°C [17]. It is soluble in water, methanol, ethanol, and chloroform; sparingly soluble in ethyl acetate; and insoluble in ether and acetone [7]. Amitrole forms neutral aqueous solutions but acts as a weak base with a K_b of 10^{-10} [8].

Although amitrole may be thought to exist in two tautomeric forms (Fig. 1), it is probably more accurate to consider the imino hydrogen to exist as a charged atom that is bound to the triazole nucleus and stabilized by resonance [9]. Amitrole behaves chemically as a typical aromatic amine [9]; hence, it will diazotize and couple with several dyes. Methods of detecting amitrole based on azo dye formation have been reported by several workers [10-12]. Sund [13] reported an assay method utilizing a nitroprusside-ferrocyanide reagent.

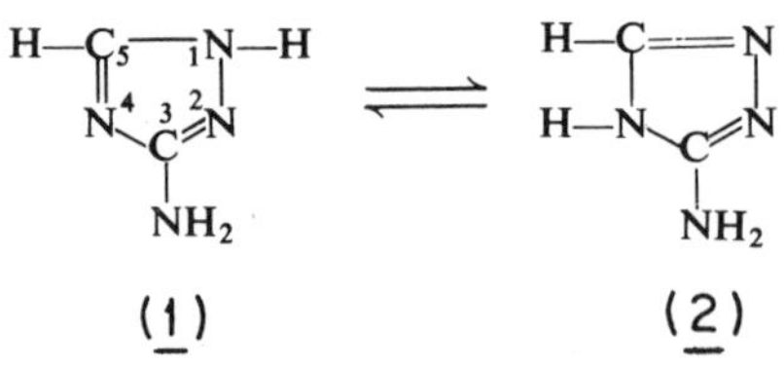

FIG. 1.   Two possible tautomeric forms of 3-amino-1,2,4-triazole: (1) 1H; (2) 4H.

---

*Report in the New York Times, International Edition, Weekly Review, November 15, 1959.

C. Synthesis

Amitrole may be prepared in nearly quantitative yield by heating an aminoguanidine salt with formic acid in an inert solvent at 100°-120°C [16, 17]. It may also be prepared from formylguanine nitrate and sodium carbonate but with a much lower yield [9]. Heating s-triazine to 120°C in the presence of aminoguanidine also yields amitrole [9].

D. Chemical Reactions

As mentioned earlier, amitrole behaves as a typical aromatic amine as well as an s-triazole. Potts [9] has reviewed the chemistry of the s-triazoles in general, while Cooper [18] and Hutchinson [19] have presented detailed discussion of the reactions of amitrole in particular.

E. Formulations

Amitrole has been marketed alone and in combination with other herbicides. Trade names Amizon, Woodazol, and Amino Triazole Weed Killer are used for the 50% powder. Amitrol-T formulation is a liquid containing equal amounts of amitrole and ammonium thiocyanate. Amizine is the trade name for a mixture of amitrole and simazine [2-chloro-4,6-bis(ethylamino)-s-triazine], and Fenamine refers to a combination of amitrole, fenac [2,3,6-trichlorophenyl)acetic acid], and atrazine [2-chloro-4-(ethylamino)-6-(isopropylamino)-s-triazine] [20].

II. DEGRADATION PATHWAYS

A. Degradation in Plants

The s-triazole nucleus is very stable [9]; hence, it is not surprising that few workers have reported evidence of ring cleavage under physiological conditions. Yost and Williams [21] reported the disappearance of [5-^{14}C]-amitrole from corn plants in approximately six weeks with a half-life of about 8 days. Disappearance was also observed from soybean but at a much slower rate [21]. The possibility that amitrole was lost from the roots, similar to the situation with certain other herbicides [22, 23], was not investigated. Reports of disappearance from treated plants do not necessarily indicate ring cleavage.

Miller and Hall [24] could not detect amitrole in cotton 4 days after treatment. However, large quantities of metabolic products were present. These derivatives were probably conjugates.

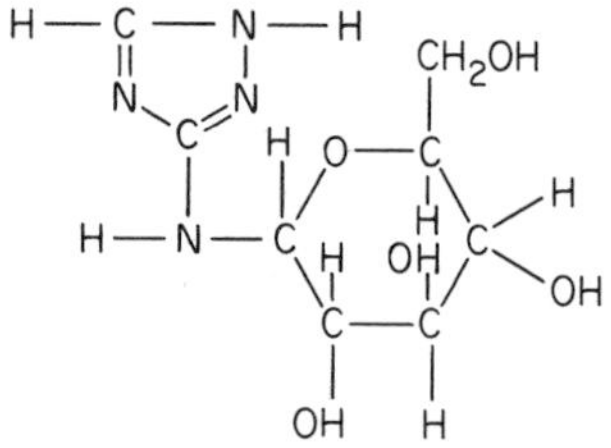

FIG. 2.   Glucose adduct of amitrole [39].

Freed et al. [25] reported evolution of $^{14}CO_2$ from [5-^{14}C]amitrole-treated oat and barley plants, indicating ring cleavage. Resistant oats released $^{14}CO_2$ more readily than sensitive barley. Lund-Hoie [26] also reported a very low rate of $^{14}CO_2$ release from oats. Massini [27] found no loss of $^{14}CO_2$ from beans or tomatoes. Muzik [28] observed chlorosis in scions grafted on tomato plants 103 days after treatment of the stock with amitrole, indicating long persistence of the toxic moiety.

Hilton [29] and Castelfranco et al. [30] reported photodecomposition of amitrole in the presence of riboflavin. Ring cleavage occurred with the loss of ^{14}C as CO_2 or formic acid [30]. This mechanism is discussed in Sec. II,B.

Studies of amitrole degradation in plants have been complicated by the fact that the radioactive material is available commercially only with the 5-carbon labeled. Apparently, the 5-carbon is quickly lost (as CO_2 or formate) when ring rupture occurs and the remaining fragment(s) is thus unlabeled. However, if significant amounts of $^{14}CO_2$ or [^{14}C]formate were produced by higher plants from [^{14}C]amitrole, one would expect to find some incorporation of ^{14}C into normal metabolites. Such is not the case. The vast majority of the literature indicates that the extractable ^{14}C from plants treated with [5-^{14}C]amitrole remains in the intact ring as free amitrole or conjugates. Considerable amounts of amitrole are attached to protein [31, 32] or somehow bound in an insoluble form [11]. Conclusive evidence for rapid and extensive ring cleavage by higher plants has not been reported.

Rogers [33, 34] reported a derivative of amitrole from several plants which was "chromatographically identical" with an aminoglucoside derivative that forms quite readily in vitro [35]. Its occurrence in plant extracts is probably an artifact, since numerous attempts by other workers to detect the compound in plant extracts have failed [11, 24, 27, 36-38]. However, Fredricks and Gentile [35, 39-41] have published a series of studies on the properties and metabolism of the glucose derivative (Fig. 2). In one paper [41] these authors suggest that the glucose adduct gives rise to a triose derivative through the action of aldolase; this would require isomerization of the glucose to fructose. Furthermore, they suggest [41] that the triose

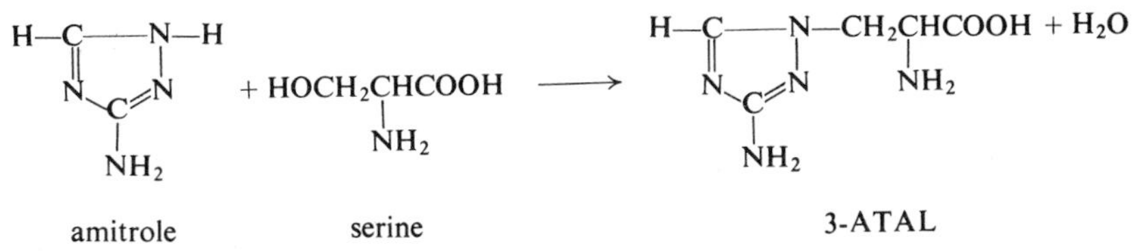

FIG. 3.   Formation of 3-ATAL from amitrole and serine [45] .

derivative represents the true structure of the amitrole derivatives reported
by other workers [11, 24, 27, 37, 38, 42] when, according to their own work
[35, 41] , the glucose adduct will not diazotize and does not give a positive
reaction with ninhydrin, two characteristics of the amitrole derivative
reported by other workers [11, 27, 42] . Fredricks and Gentile [41] suggest
that the ninhydrin reaction found by others may have resulted from a con-
taminant, an unlikely possibility. The fact that the amitrole derivatives
reported by most workers [11, 24, 27, 37, 38, 42] are readily diazotizable
and stable in strong acid makes it highly unlikely that the 3-amino group is
substituted or that a glucose or a triose is part of the molecule. The existing
evidence indicates that the glucose adduct of amitrole is simply an interesting
artifact.

Racusen [11] reported the first comprehensive studies of amitrole
metabolism. Two major metabolites were isolated from plants exposed to
[5-^{14}C]amitrole. The more abundant metabolite, termed X by Racusen,
exhibited positive reactions with azo dyes and ninhydrin reagents and
behaved as a zwitterion during electrophoresis at different pH levels. The
second metabolite, called Y, exhibited only acidic properties above pH 4.5,
gave no ninhydrin reaction, but did form azo dyes. Neither X nor Y appeared
as phytotoxic as amitrole. Both compounds were stable in 6 N HCl for 5 hr
at 100°C, indicating that neither compound is a simple amide or an aminoglucoside.

Shortly after Racusen's work, Massini [8, 27] and Carter and Naylor
[42, 43] reported studies of amitrole metabolism. The principal metabolites
reported by these workers exhibited azo dye reactions, ninhydrin sensitivity,
and zwitterion behavior. In all probability X [11] , ATX [8] , 1 [42] , and
similar compounds reported by other workers [24, 37, 38] are the same
compound which Massini [27] identified as 3-(3-amino-s-triazole-1-yl)-2-
aminopropionic acid (3-ATAL) (Fig. 3). The suggestion by Carter and Naylor
[42] that the alanine chain is attached to the triazole through the 3-amino
group is not consistent with the azo-dye-forming properties of the compound.
The existence of the C-N bond between the 3-carbon of alanine and the 1-
nitrogen in the triazole ring has been confirmed by synthesis [19] .

Castelfranco and Brown [32] suggested that amitrole, in the presence of a free-radical generating system, accepts electrons and becomes a free radical, which is capable of alkylating proteins and possibly amino acids, indiscriminately. However, Carter and Naylor [44] and Carter [45] have shown that ^{14}C from serine or glycine readily enters 3-ATAL, but the incorporation of radiocarbon from glucose, succinate, alanine, glyoxylate, and formate is much less rapid. Most likely 3-ATAL is formed by the condensation of amitrole and serine [27, 45] in a manner analogous to the formation of β-pyrazol-1-yl-α-alanine (BPA) from pyrazole and serine [46] (see Fig. 3).

The formation of 3-ATAL apparently represents detoxication, since the derivative does not appear to be as toxic as amitrole [11, 27, 42] nor as mobile [27]. Furthermore, ammonium thiocyanate, which synergizes the action of amitrole, inhibits the formation of 3-ATAL [38, 45]. Donnalley [47] suggested that the synergism resulted from increased transport of amitrole, but increased mobility is probably a secondary result of reduced formation of the less mobile 3-ATAL [45]. Other divalent sulfur compounds reduce 3-ATAL formation but none of those tested is as effective as thiocyanate (Table 1).

In addition to 3-ATAL, another compound has been reported under various designations: Y [11, 24], unknown I [37], and compound II [36]. None of these workers reported sufficient chemical and physical properties to conclude that the compounds were identical. Chromatographic properties were similar and all compounds gave a positive azo dye reaction but a negative ninhydrin reaction. Miller and Hall [24] reported that Y was the most abundant metabolite of amitrole in cotton leaves, whereas X (i.e., 3-ATAL) predominated in the seed. In beans, Canada thistle, bindweed, silver maple, honeysuckle, and alfalfa, 3-ATAL was equal to or greater than all other products [11, 27, 36-38].

Racusen [11] reported that Y was stable in 6 N HCl for 5 hr at 100°C. Herrett and Linck [37] reported the disappearance of their unknown I following similar treatment. 3-ATAL (X [11], unknown II [37]) was stable. Thus, it would appear that compound Y of Racusen [11] and unknown I of Herrett and Linck [37] were not the same. Verification of the structure of products other than 3-ATAL must await further studies. Carter and Naylor [36] observed more than a dozen radioactive compounds derived from [5-^{14}C]amitrole, none of which appeared to be normal metabolites arising from $^{14}CO_2$ fixation. Hence, most were probably conjugates.

Herrett and Bagley [48] reported a metabolic product of amitrole which they designated unknown III. Their material was found to be five to eight times as effective as amitrole in suppressing the growth of tomato and lettuce roots [48]. Unknown III did not react with ninhydrin and would not form azo dyes, suggesting substitution of the 3-amino group, but it was not the aminoglucoside.

TABLE 1

Inhibition of 3-ATAL Formation by Organic Sulfur Compounds[a]

[14]C compound in extract	Sulfur compound				
	None	Thiocyanate	Thiourea	Thioacetamide	Cysteine
Amitrole	17.4	93.2	38.3	57.7	46.9
3-ATAL	63.1	6.7	57.2	36.8	45.8
ATY[b]	19.3	< 1	4.4	5.5	7.4

[a]Bean trifoliates given 5.0 μCi of [5-^{14}C]amitrole in equimolar solution of sulfur compounds for 24 hr. Data expressed as percent total radioactivity in 80% ethanol extract [J. C. Brown and M. C. Carter, unpublished data].
[b]Compound possessing characteristic of Y [11, 24], unknown I [37], and compound 2 [42].

Lund-Hoie [26] reported a phytotoxic metabolite from oats which was chromatographically similar to 3-ATAL.

It is difficult to assess the significance of these phytotoxic metabolites. Both Herrett and Bagley [48] and Lund-Hoie [26] maintained their extracts at room temperature or above for several hours during extraction and concentration. Both report their toxic metabolites were sensitive to hydrolysis. Attempts to detect these compounds in our laboratory using bean plants have been unsuccessful, and Smith et al. [38] were unable to detect such compounds in three ecotypes of Canada thistle. The lack of confirmation casts serious doubts about the existence of toxic amitrole metabolites in vivo. They may be formed during extraction and, like the aminoglucoside, they may be artifacts. However, artifacts that are several times as toxic as amitrole are well worth further study.

B. Degradation in Soils

Amitrole disappears rapidly from soils [49-52]. Disappearance has been attributed to adsorption [13, 532] and microbial degradation, although attempts to isolate organisms capable of degrading amitrole have not been successful [49, 52, 54].

Kaufman and co-workers [54, 55] proposed that most of the amitrole degradation occurring in soils proceeds by nonbiological reactions. Approximately 69% of the 5-carbon from amitrole was released as $^{14}CO_2$ in 20 days by nonsterilized soil [54]. Autoclaved soil released only 2% in a comparable period, whereas soil treated with potassium azide or ethylene oxide released

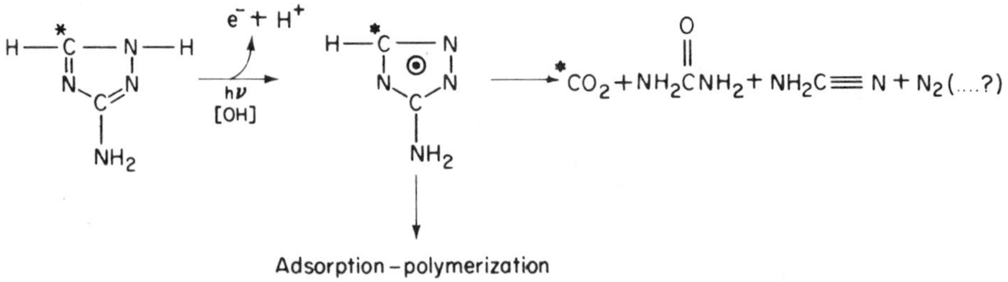

Adsorption – polymerization

FIG. 4.   Activation and degradation of amitrole. (After Plimmer et al. [57] and Castelfranco and Brown [32].)

46 and 35%, respectively [54]. Reinoculation of autoclaved soil did not restore the capacity to metabolize amitrole [54]. These authors propose that amitrole is degraded in soil by an oxidative mechanism involving an attack on the triazole nucleus by (OH·) or other free radicals.

Plimmer et al. [56, 57] have studied the degradation of amitrole by free-radical generating systems. They further investigated the riboflavin-mediated photodecomposition of amitrole reported earlier [29, 30]. Carbon dioxide arising from the 5-carbon and cyanamid and urea arising from the 3-carbon and associated nitrogens were liberated from amitrole in the presence of riboflavin [57]. Fenton's reagent produced the same products. However, the reagent of Castelfranco et al. [30], consisting of ascorbic acid, cupric sulfate, and molecular oxygen, liberated no CO_2, although amitrole was degraded [57]. Plimmer et al. [57] concluded that riboflavin (and light) or Fenton's reagent promotes oxidation of amitrole, resulting in ring cleavage, the loss of CO_2, and the production of urea, cyanamid, and possibly molecular nitrogen (Fig. 4). The ascorbate-copper reagent evidently reduces amitrole to a free radical which then polymerizes.

This work provides strong evidence for nonbiological destruction of amitrole in soil; however, microbial attack cannot be totally discounted. Treatment with potassium azide failed to reduce amitrole degradation to the level of autoclaved soil but it produced a 40% or greater reduction of amitrole degradation compared to nonsterilized soil [51, 54]. Soil moisture, temperature, and pH markedly affected amitrole degradation [51, 53], indicating possible microbial involvement. In Riempa's studies [51], amitrole degradation in soil exhibited a lag phase typical of microbial degradation, but the studies by Ercegovich and Frear [53] and Kaufman et al. [54] showed that amitrole degradation obeys first-order kinetics, suggesting a chemical reaction. No one has reported an investigation of the possible involvement of extracellular enzymes.

Whatever the mechanism whereby the triazole ring is opened, there appears to be little doubt that ring opening does occur rapidly in soils and the resulting products (urea, cyanamid, and nitrogen) should be readily metabolized by soil microorganisms.

C. Degradation in Animals

Fang et al. [58] administered [^{14}C]amitrole orally to rats and determined the fate of the radioactivity. Insignificant amounts of radioactivity were lost as respiratory CO_2. After 3 days, from 80 to 104% of the administered ^{14}C was recovered in urine and feces, the majority, by far, being present in the urine. The liver was the only internal organ containing significant amounts of ^{14}C after 3 days. Amitrole and one unidentified metabolite were recovered from the liver. Geldmacher and Mallinckrodt [59] also observed rapid excretory loss of amitrole from an adult, female human subject.

III. MODE OF ACTION

A. Uptake

Hall et al. [60] were the first to report that amitrole is readily absorbed by both roots and leaves of higher plants and rapidly translocated. More detailed studies by Crafts and Yamaguchi [61, 62] demonstrated that amitrole was freely transported in both xylem and phloem. They concluded that amitrole possessed the right combination of chemical and physical properties for extensive translocation and that the effectiveness of amitrole as a herbicide was largely dependent on its movement and accumulation in the growing regions of plants [62]. Subsequent work by other workers confirmed the high mobility of amitrole and the importance of mobility to herbicide effectiveness [8, 37, 38, 47].

B. Biochemical Processes Affected

1. Amino Acid and Protein Metabolism

Amitrole blocks histidine biosynthesis in bacteria [63], yeast [64, 65], algae [66], and higher plants [67] by inhibiting the enzyme imidazoleglycerol phosphate dehydrase [63, 68] and imidazoleglycerol (IG) accumulates in the microbial growth medium. Inhibition is competitive with substrate [69] and may be potentiated by PO_4 [67].

Inhibition of histidine synthesis appears to be the major growth-controlling action in microorganisms but it does not appear to be the primary site of action of amitrole in higher plants [70, 71]. McWhorter and Hilton [72] observed no reduction in histidine content of amitrole-treated corn, nor were they able to demonstrate IG accumulation.

Schroeder [73] found that amitrole reduced amino acid levels in the algae Poteriochromones stipitata, but in light-grown higher plants, amitrole increased free amino acids and decreased proteins [72, 74, 75].

Bartels and Wolf [74] concluded that the effect of amitrole on protein synthesis is indirect, arising from interferences with purine metabolism. Brown and Carter [31] also found no evidence for a direct effect of amitrole on protein synthesis. McWhorter and Hilton [72] found that while total free amino acids increased in amitrole-treated corn, free glycine and serine declined to very low levels. These authors suggested that formation of 3-ATAL may have depleted glycine-serine pools as suggested earlier by Carter and Naylor [44]. However, McWhorter and Hilton [72] could not detect 3-ATAL in treated corn tissue. Castelfranco and Brown [32] discounted amino acid conjugation and removal as a toxic mechanism on the basis of the large quantities of amitrole that would be required. Possibly amitrole interferes with C_1 metabolism and disrupts glycine-serine interconversion [72]. Such a mechanism might also disrupt purine metabolism.

Williams et al. [76] suggested that amitrole was converted to 3-ATAL by E. coli and that 3-ATAL, acting as a histidine analog, was incorporated into protein. Production of "lethal" protein could be responsible for part of amitrole's toxicity. Brown and Carter [31] were unable to demonstrate the incorporation of 3-ATAL into bean plant protein. Confirmation of the theory of Williams et al. [76] has not appeared.

Hilton [70] emphasized the fact that young seedlings, which are very sensitive to amitrole, have adequate supplies of amino acids from storage tissues in the seed. Hence, growth inhibition and reduced protein content in treated plants must result from some action other than a direct effect upon the availability of amino acids for protein synthesis.

2. Nucleotide and Nucleic Acid Metabolism

While histidine biosynthesis appears to be the most sensitive site of amitrole action in yeast and bacteria, it is not the only site. At high amitrole concentrations, addition of histidine does not totally nullify amitrole inhibition of growth [63, 77, 78]. A combination of histidine plus adenine does totally nullify growth inhibition [63, 77, 78]. Klopotowski and Bagdasarian [78] found that histidine plus cytosine was nearly as effective as histidine plus adenine in nullifying amitrole inhibition of Saccharomyces cerevisiae but only histidine plus adenine was effective with Salmonella typhimirium. Wolf [79] reported that adenine, alone, reversed amitrole inhibition of Chlorella.

Thus, the hypothesis arose that amitrole interferes with some phase of purine metabolism. Hulanicka et al. [80] concluded that amitrole inhibits the step in purine synthesis in which formylglycineamidine ribotide is cyclized to form 4-aminoimidazole ribotide (4-AI). Earlier, Sund et al. [81] reported that 4-AI partially reversed amitrole inhibition of the growth of tomato plants, and Rabinowitz and Pricer [82] found that amitrole inhibited enzymic degradation of 4-AI.

Hilton [70] attempted unsuccessfully to demonstrate accumulation of purine precursors by feeding [^{14}C]glycine to Salmonella. He attributed his failure to the possibility that amitrole inhibits the formation of one-carbon units in Salmonella [83], and this inhibition prevented the buildup of purine precursors.

In algae grown autotrophically, adenine will reverse amitrole's inhibition [79, 84], but in mixotrophically or heterotrophically grown algae, histidine is more effective [66, 85, 86]. Thus, amitrole appears to have a different action on light- and dark-grown organisms. This conclusion is substantiated by work with higher plants.

Bartels and Wolf [74] grew wheat plants with and without amitrole in both light and darkness. After 7 days, amitrole had no significant effect on nucleic acid or soluble nucleotides in plants grown in the dark. But amitrole significantly reduced the RNA content of light-grown plants. Most of the reduction occurred in the microsomal RNA, but a general reduction of all RNA fractions was observed. Base ratios in RNA were not affected by amitrole, but the incorporation of [^{14}C]glycine and [^{14}C]formate into ribonucleotides and RNA was reduced.

DNA levels were not reduced by amitrole in either light- or dark-grown plants [74], but incorporation of [^{14}C]glycine into DNA was reduced by amitrole [74], indicating possible interference with cell division as suggested by other workers [84, 87].

Bartels and Wolf [74] concluded that amitrole interferes with some aspect of purine metabolism, thus leading to a shortage of nucleotides for both DNA and RNA synthesis. In later work, Bartels et al. [88] found that amitrole treatment of light-grown plants resulted in a complete absence of 70 S ribosomes but did not affect 80 S ribosomes. Neither 70 S nor 80 S ribosomes were reduced in dark-grown plants. However, amitrole-treated dark-grown plants that contained normal complements of 70 S and 80 S ribosomes did not develop normally when illuminated [89, 90]. This raises the question of whether amitrole in the light inhibits formation or promotes destruction of 70 S ribosomes. These possibilities are explored further in the ensuing discussion on pigment synthesis and plastid development. The different effects of amitrole in light and darkness indicate that the action of the herbicide on higher plants differs appreciably from its action on microorganisms.

3.   Riboflavin Metabolism

Nullification of amitrole toxicity by several flavins and the reduced riboflavin concentration in treated plants led Sund et al. [81, 91] to suggest that amitrole inhibited riboflavin synthesis. Castelfranco [30] discounted Sund's suggestion, since riboflavin mediates photodecomposition of amitrole (see Sec. II, B). Therefore, Castelfranco [30] concluded that the reversal of amitrole toxicity was due to nonbiological photodestruction occurring in vitro. Hilton [29] found that while both riboflavin and isoriboflavin nullified amitrole's action in the light, only riboflavin was capable of nullification in the dark. Hilton [70] suggested that the action of riboflavin on amitrole differed from the photooxidation mechanism. Evidence for this belief is derived from the fact that the riboflavin nullification of [5-^{14}C]amitrole in the light results in the liberation of $^{14}CO_2$, but nullification of [5-^{14}C]amitrole in the dark does not [70]. Nullification in darkness requires the presence of a living organism [70] and when riboflavin reacts with amitrole in the presence of protein, amitrole is activated and attacks the protein without loss of the 5-carbon as CO_2 [31, 32]. Thus, nullification of amitrole toxicity by riboflavin in darkness could still involve oxidation of the triazole, but the reaction would be coupled to some redox reaction in the cell rather than the photoactivation of riboflavin.

4.   Inhibition of Catalase

Inhibition of catalase and several other metalloprotein enzymes was one of the first biochemical actions of amitrole reported [92-94]. Catalase and fatty acid peroxidase are irreversibly inhibited [32, 93]. Inhibition of catalase requires that the enzyme, substrate (H_2O_2), and amitrole be incubated together [93]. Apparently, amitrole is oxidized and then reacts with the enzyme. Agrawal and co-workers [95, 96] reported that amitrole reacts with catalase in a ratio of one amitrole per hematin. The C-5 of amitrole becomes attached to an imidazole nitrogen in histidyl residue 74 of catalase [96].

Castelfranco [32] suggested that amitrole undergoes a one-electron oxidation and attacks protein indiscriminately. At high concentrations, when amitrole causes rapid desiccation and necrosis [97], such widespread nonspecific enzyme inactivation may occur. At lower concentrations, however, growth may not be noticeably affected while other processes are strongly inhibited [97, 98], suggesting a specific site of action.

Inhibition of catalase and other peroxidases may play a role in the light-induced changes in amitrole-treated plants discussed below.

5.   Pigment Synthesis and Plastid Development

Perhaps the most obvious effect of amitrole upon higher plants is the pronounced absence of chlorophyll in new growth arising after amitrole application [2, 34, 97]. Chlorophyll content declines with increasing amitrole in a linear relationship [99].

TABLE 2

Effect of Amitrole on the
Incorporation of [2, 3-^{14}C]Succinate into Heme[a]

Amitrole (μM)	Reaction ingredients[b]			Specific activity of heme (cpm/mM)
	[2, 3-^{14}C]Succinate (μCi)	Glycine (μM)	Glucose (mg)	
0	2	400	20	97,300
400	2	400	20	124,750

[a]M. C. Carter and B. Jacobson, unpublished data.
[b]Washed red cells taken from 20 ml of duck blood suspended in 10 ml of isotonic buffer. (Procedures taken from Shemin and Kumin [101] .)

Since amitrole does not promote chlorophyll degradation in extracts or in plants treated and placed in darkness [97, 98] , a disruption of chlorophyll synthesis might be suspected. Naylor [100] reported that amitrole did not inhibit the formation of protochlorophyll in the dark nor the conversion of protochlorophyll to chlorophyll in the light. Amitrole does not inhibit the formation of heme in avian erythrocytes (Table 2) nor the incorporation of δ-aminolevulinic acid and porphobilinogen into porphyrins in plant systems [102, 103] . Doerfling et al. [104] reported the accumulation of coproporphyrin in Porteriochromones stipitata cultures grown in the presence of amitrole, but the existing evidence indicates that an inhibition of porphyrin synthesis does not occur in higher plants [100, 102, 103, 105] . Most investigators concluded that amitrole disrupts plastid development in some way which indirectly reduces pigment synthesis [36, 102, 106, 107] .

Bartels and his co-workers [88, 90, 108] found that proplastids from dark-grown, amitrole-treated wheat seedlings have a normal ultrastructure and a normal complement of fraction I protein, 70 S ribosomes, and plastid DNA. However, plastids from amitrole-treated plants grown in the light are quite abnormal in appearance, contain few 70 S ribosomes, and have reduced amounts of fraction I protein and plastid DNA [88, 90, 108] . Moreover, when dark-grown treated plants are exposed to light all plastid structure seems to disintegrate and these changes appear to be directly related to the intensity and duration of light exposure [90] .

Recently, Burns et al. [105] suggested that amitrole induced plastid degeneration and chlorosis by preventing normal carotenoid synthesis. Other workers also have reported a reduction in carotenoids in amitrole-treated plants [98, 109] but Burns et al. [105] emphasized the fact that amitrole

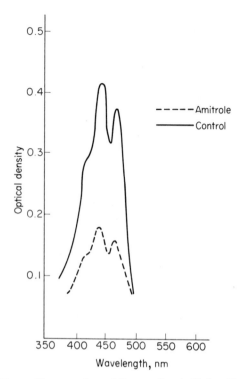

FIG. 5.   Absorption spectra of the xanthophyll fraction of 6-day-old
etiolated wheat seedlings; 10 seedlings per 6.5 ml of diethyl ether. (Redrawn
from Burns et al. [105].)

altered carotenoid synthesis in the dark at an amitrole concentration which,
according to Bartels [88, 90, 108], does not affect proplastid structure or
fraction I protein, 70 S ribosome, and DNA contents.

Burns [105, 110] grew wheat plants in darkness in the presence of
10^{-4} M amitrole. After 6 days, the fresh weight and protochlorophyll content
were not affected by amitrole, but xanthophyll levels were sharply reduced
(Fig. 5) and qualitative and quantitative changes were noted in the carotenes
(Fig. 6). β-Carotene was the principal carotene in untreated plants but
amitrole-treated plants contained mainly the precursors: ζ-carotene, phyto-
ene, and phytofluene [109, 110]. Illumination at low light intensities caused
some increase in carotene (Fig. 7), but high light intensities resulted in
almost complete loss of carotenes from amitrole treated seedlings (Fig. 8).
Similarly, chlorophyll accumulated in amitrole-treated plants exposed to
low light intensities but declined sharply at higher intensities.

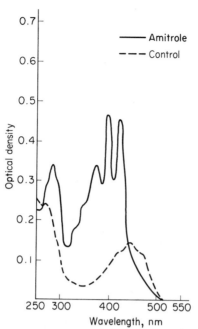

FIG. 6. Absorption spectra of the carotene fraction of 6-day-old etiolated wheat seedlings; 10 seedlings per 6.5 ml hexane. (Redrawn from Burns
et al. [105].)

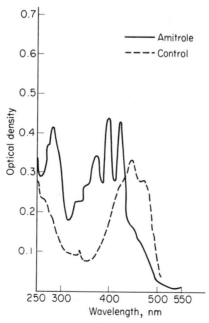

FIG. 7. Absorption spectra of the carotene fraction from wheat seedlings grown 6 days in darkness followed by 36 hr at 60 ft-c illumination;
10 seedlings in 10 ml hexane. (Redrawn from Burns et al. [105].)

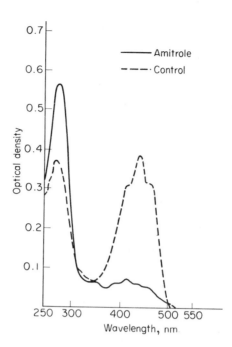

FIG. 8. Absorption spectra of the carotene fraction from wheat seedlings grown 6 days in darkness, 36 hr at 60 ft-c, and 8 hr at 3000 ft-c; 10 seedlings in 10 ml hexane. (Redrawn from Burns et al. [105].)

Chlorosis in a variety of mutant organisms is believed to result from the inability of the organisms to synthesize carotenoids [111-114]. In the absence of carotenoids, chlorophyll is photooxidized and destroyed at high light intensities [111-115]. In a similar manner, amitrole prevents carotenoid accumulation; consequently, photooxidation of chlorophyll and probably other plastid components occurs upon illumination. Apparently, several other herbicides act in a similar way [105, 116, 117].

The mechanism whereby amitrole prevents the accumulation of β-carotene and xanthophylls remains to be elucidated. However, it is not a light-requiring reaction and it must take place very early in plastid development. Both Burns [110] and Bartels [90] emphasized the fact that wheat seed must be presoaked in amitrole solution and continuously exposed to the herbicide during germination to obtain full suppression of carotenoids and plastid development.

Interference with carotenoid accumulation provides the best explanation of amitrole action in higher plants but microorganisms do not appear to be similarly affected. Habercom and Carter [118] found that while amitrole inhibited growth and carotenoid synthesis in the fungus <u>Phycomyces blakeslee-</u><u>anus</u>, both inhibitions were reversed by the addition of histidine. At present, no close relationship is known to exist between histidine biosynthesis and carotenoid biosynthesis. Thus, one must conclude that the principal site of action of amitrole in higher plants differs from that in microorganisms.

C. Selectivity

Amitrole effectiveness is largely dependent on foliar uptake and trans-location [5, 8, 37, 38, 47, 60-62]; therefore, selectivity could result from the many variations in leaf structure which affect wetting and penetration [119]. Differences in amitrole activity, which have been attributed to differences in translocation [120], are difficult to separate from differences in metabolism, since metabolic alteration leads to products less mobile than amitrole [8, 27]. Herrett and Linck [37] found that differences in the rate of metabolism of amitrole could account for the differences in sensitivity between Canada thistle and field bindweed. Smith et al. [120] observed different rates of amitrole metabolism between susceptible and resistant ecotypes of Canada thistle. Freed et al. [25] and Lund-Hoie [26] found that amitrole-resistant plans released $^{14}CO_2$ from [^{14}C]amitrole at a faster rate than sensitive plants. Thus, it would appear that the rate.of detoxication is probably largely responsible for the herbicide's selectivity in some plants.

REFERENCES

1. W. W. Allen, U.S. Pat. 2,670,282 (1954).

2. W. C. Hall, S. P. Johnson, and C. L. Leinweber, <u>Texas Agr. Exp.</u><u>Sta. Bull.</u>, <u>1953</u>, 759.

3. R. Behrens, <u>Proc. North Central Weed Conf.</u>, <u>10</u>, 61 (1953).

4. W. C. Shaw and C. R. Swanson, <u>Weeds</u>, <u>2</u>, 43 (1953).

5. R. A. Alverson, <u>Annotated Bibliography of Amino Triazole</u>, American Cyanamid Co., 1957.

6. W. A. Meissner, <u>Report of the Amitrole Advisory Committee</u>, EPA, 21 pp., 1971.

7. American Cyanamid Co., <u>Tech. Data Sheet: Amino Triazole</u>, 1956.

8. P. Massini, <u>Acta Bot. Neerl.</u>, <u>7</u>, 524 (1958).

9. K. T. Potts, <u>Chem. Rev.</u>, <u>61</u>, 87 (1961).

10. F. D. Aldrich and S. R. McLane, <u>Plant Physiol.</u>, <u>32</u>, 153 (1957).

11. D. Racusen, <u>Arch. Biochem. Biophys.</u>, <u>74</u>, 106 (1958).

12. R. W. Storherr and J. Burke, J.A.O.A.C., 44, 196 (1961).

13. K. A. Sund, J. Agr. Food Chem., 4, 57 (1956).

14. H. W. Hilton and G. K. Uyehara, J. Agr. Food Chem., 14, 90 (1966).

15. J. R. Bishop, J.A.O.A.C., 50, 568 (1967).

16. J. Thiele and W. Manchot, Ann., 303, 33 (1898).

17. C. F. H. Allen and A. Bell, Org. Synth., 26, 11 (1946).

18. E. A. Cooper, Ph.D. Dissertation, Michigan State University, East Lansing, 1959.

19. J. H. Hutchinson, Ph.D. Dissertation, Auburn University, Auburn, Alabama, 1968.

20. Anonymous, Herbicide Handbook, WSSA, 1970, p. 12.

21. J. F. Yost and E. F. Williams, Proc. Northeast. Weed Conf., 1958, 9-15.

22. J. B. Hanson and F. W. Slife, Ill. Res., 1961, 3-4.

23. C. L. Foy and W. Hurtt, Weed Soc. Am. Abstr., 1967, 40.

24. C. S. Miller and W. C. Hall, J. Agr. Food Chem., 9, 210 (1961).

25. V. H. Freed, M. Montgomery, and M. Kief, Proc. Northeast. Weed Conf., 1961, 6-16.

26. K. Lund-Hoie, Weed Res., 10, 367 (1970).

27. P. Massini, Acta Bot. Neerl., 12, 64 (1963).

28. T. J. Muzik, Weed Res., 5, 207 (1965).

29. J. L. Hilton, Plant Physiol., 37, 238 (1962).

30. P. Castelfranco, A. Oppenheim, and S. Yamaguchi, Weeds, 11, 111 (1963).

31. J. C. Brown and M. C. Carter, Weed Sci., 16, 222 (1968).

32. P. Castelfranco and M. S. Brown, Weeds, 11, 116 (1963).

33. B. J. Rogers, Weeds, 5, 5 (1957).

34. B. J. Rogers, Hormolog., 1, 10 (1957).

35. J. F. Fredricks and A. C. Gentile, Physiol. Plant., 13, 761 (1960).

36. M. C. Carter and A. W. Naylor, Bot. Gaz., 122, 138 (1960).

37. R. A. Herrett and A. J. Linck, Physiol. Plant., 14, 767 (1961).

38. L. W. Smith, D. E. Bayer, and C. L. Foy, Weed Sci., 16, 523 (1969).

39. J. F. Fredricks and A. C. Gentile, Biochim. Biophys. Acta, 92, 356 (1961).

40. J. F. Fredricks and A. C. Gentile, Physiol. Plant., 15, 186 (1962).

41. J. F. Fredricks and A. C. Gentile, Phytochemistry, 4, 851 (1966).

42. M. C. Carter and A. W. Naylor, Physiol. Plant., 14, 20 (1961).

43. M. C. Carter and A. W. Naylor, Plant Physiol. Suppl., 34, 6 (1959).

44. M. C. Carter and A. W. Naylor, Physiol. Plant., 14, 62 (1961).

45. M. C. Carter, Physiol. Plant., 18, 1054 (1965).

46. P. M. Dunnill and L. Fowden, J. Exp. Bot., 14, 237 (1963).

47. W. F. Donnalley, Ph.D. Thesis, Michigan State University, East Lansing, 1964.

48. R. A. Herrett and W. P. Bagley, J. Agr. Food Chem., 12, 17 (1964).

49. C. D. Ercegovich, Ph.D. Thesis, Pennsylvania State University, University Park, 1958.

50. D. D. Bondarenko, Proc. North Central Weed Conf., 1958.

51. P. Riempa, Weed Res., 2, 41 (1962).

52. F. M. Ashton, Weeds, 11, 167 (1963).

53. C. D. Ercegovich and D. E. H. Frear, J. Agr. Food Chem., 12, 26 (1964).

54. D. D. Kaufman, J. R. Plimmer, P. C. Kearney, J. Blake, and F. S. Guardia, Weed Sci., 16, 226 (1968).

55. D. D. Kaufman, Weed Sci. Soc. Am., 1967, 78-79.

56. P. C. Kearney and J. R. Plimmer, Weed Sci. Soc. Am., 1967, 76-77.

57. J. R. Plimmer, P. C. Kearney, D. D. Kaufman, and F. S. Guardia, J. Agr. Food Chem., 15, 996 (1967).

58. S. C. Fang, M. George, and Te Chang Yu, J. Agr. Food Chem., 12, 219 (1964).

59. M. Geldmacher and V. Mallinckrodt, Arch. Toxikol., 27, 13 (1970).

60. W. C. Hall, S. P. Johnson, and C. L. Leinweber, Texas Agr. Exp. Sta. Bull., 1954, 789.

61. A. S. Crafts and S. Yamaguchi, Hilgardia, 27, 421 (1958).

62. S. Yamaguchi and A. S. Crafts, Hilgardia, 29, 171 (1959).

63. J. L. Hilton, P. C. Kearney, and B. N. Ames, Arch. Biochem. Biophys., 112, 544 (1965).

64. T. Klopotowski and D. Hulanicka, Acta Biochim. Pol., 10, 209 (1963).

65. J. L. Hilton, Weeds, 8, 392 (1960).

66. J. N. Siegel and A. C. Gentile, Plant Physiol., 41, 670 (1966).

67. A. Wiater, K. Krajewska-Grynkiewicz, and T. Klopotowski, Acta Biochim. Pol., 18, 299 (1971).

68. T. Klopotowski and A. Wiater, Arch. Biochem. Biophys., 112, 562 (1965).

69. A. Wiater, D. Hulanicka, and T. Klopotowski, Acta Biochim. Pol., 18, 289 (1971).

70. J. L. Hilton, J. Agr. Food Chem., 17, 192 (1969).

71. A. Wiater, T. Klopotowski, and G. Bagdasarian, Acta Biochim. Pol., 18, 309 (1971).

72. C. G. McWhorter and J. L. Hilton, Physiol. Plant., 20, 30 (1967).

73. I. Schroeder, Z. Allg. Mikrobiol., 10, 295 (1970).

74. P. G. Bartels and F. T. Wolf, Physiol. Plant., 18, 805 (1965).

75. C. G. McWhorter, Physiol. Plant., 16, 31 (1963).

76. A. K. Williams, S. T. Cos, and R. G. Eagon, Biochem. Biophys. Res. Commun., 18, 250 (1965).

77. F. W. Weyter and H. P. Broquist, Biochim. Biophys. Acta, 40, 567 (1960).

78. T. Klopotowski and G. Bagdasarian, Acta Biochim. Pol., 13, 153 (1966).

79. F. T. Wolf, Plant Physiol. Suppl., 36, XXXIX (1961).

80. D. Hulanicka, T. Klopotowski, and G. Bagdasarian, Acta Biochim. Pol., 16, 127 (1969).

81. K. A. Sund, E. C. Putala, and H. N. Little, J. Agr. Food Chem., 8, 210 (1960).

82. J. C. Rabinowitz and W. E. Pricer, J. Biol. Chem., 222, 537 (1956).

83. J. Boguslawski, W. Walczak, and T. Klopotowski, Acta Biochim. Pol., 14, 133 (1967).

84. P. Castelfranco and T. Bisalputra, Am. J. Bot., 52, 222 (1965).

85. P. J. Casselton, Nature, 204, 93 (1964).

86. P. J. Casselton, Physiol. Plant., 19, 411 (1966).

87. M. G. Srivastava, Trans Bose Res. Inst. Calcutta, 21, 119 (1956-1959).

88. P. G. Bartels, K. Matsuda, A. Siegel, and T. E. Weier, Plant Physiol., 42, 736 (1967).

89. P. G. Bartels, Plant Cell Physiol., 6, 361 (1965).

90. P. G. Bartels and T. E. Weier, Am. J. Bot., 56, 1 (1969).

91. K. A. Sund and H. N. Little, Science, 132, 622 (1960).

92. W. G. Heim, D. Appleman, and H. T. Pyfrom, Am. J. Physiol., 186, 19 (1956).

93. E. Margoliash and A. Novogrodsky, Biochem. J., 68, 468 (1958).

94. P. Castelfranco, Biochim. Biophys. Acta, 41, 485 (1960).

95. B. B. L. Agrawal and E. Margoliash, Fed. Proc., 28, 405 (1969).

96. B. B. L. Agrawal, E. Margoliash, M. I. Levenberg, R. S. Egan, and M. H. Studier, Fed. Proc., 29, 732 (1970).

97. W. C. Hall, S. P. Johnson, and C. L. Leinweber, Texas Agr. Exp. Sta. Bull., 1954, 789.

98. F. T. Wolf, Nature, 188, 164 (1960).

99. E. B. Minton, W. H. Preston, Jr., and W. H. Orgell, Plant Physiol. Suppl., 33, XLVIII (1958).

100. A. W. Naylor, J. Agr. Food Chem., 12, 21 (1964).

101. D. Shemin and S. Kumin, Fed. Proc., 11, 285 (1952).

102. P. G. Bartels, Proc. Assoc. South. Agr. Workers, 60, 305 (1963).

103. L. Bogorad, J. Biol. Chem., 233, 501 (1958).

104. P. Doerfling, W. Dummler, and D. Mucke, Experientia, 26, 728 (1970).

105. E. R. Burns, G. A. Buchanan, and M. C. Carter, Plant Physiol., 47, 144 (1971).

106. H. Linser and O. Kiermayer, Planta, 49, 498 (1957).

107. P. G. Bartels, Plant Cell Physiol., 6, 361 (1965).

108. P. G. Bartels and A. Hyde, Plant Physiol., 46, 825 (1970).

109. T. Guillot-Saloman, R. Douce, and M. Signol, Bull. Soc. Fr. Physiol.-Veget., 13, 63 (1967).

110. E. R. Burns, Ph.D. Dissertation, Auburn University, Auburn, Alabama, 1970.

111. M. Griffiths and R. Y. Stanier, J. Gen. Microbiol., 14, 698 (1956).

112. H. M. Haberman, Physiol. Plant., 13, 718 (1960).

113. B. Walles, Hereditas, 53, 247 (1965).

114. A. Faludi-Daniel, F. Lang, and L. D. Fradkin, Symposium on Bio-
      chemistry of Chloroplasts (T. W. Goodwin, ed.), Academic Press,
      New York, 1965.

115. J. M. Anderson and N. K. Boardman, Aust. J. Biol. Sci., 17, 93
      (1964).

116. J. L. Hilton, A. L. Scharen, J. B. St. John, D. E. Moreland, and
      K. H. Morris, Weed Sci., 17, 541 (1969).

117. P. G. Bartels and A. Hyde, Plant Physiol., 45, 807 (1970).

118. M. Habercom and M. C. Carter, Proc. SWSS, 25, 419 (1972).

119. K. Holly, Physiology and Biochemistry of Herbicides (L. J. Audus,
      ed.), Academic Press, New York, 1964.

120. L. W. Smith, D. E. Bayer, and C. L. Foy, Weed Sci., 16, 523
      (1968).

Chapter 8

THE CHLORINATED ALIPHATIC ACIDS

CHESTER L. FOY

Department of Plant Pathology and Physiology
Virginia Polytechnic Institute and State University
Blacksburg, Virginia

I. INTRODUCTION

A. History

The halogenated aliphatic acids have been known for many years, but their use as herbicides is fairly recent. The patent history of trichloroacetic acid (TCA) and dalapon has been reviewed and summarized by Mukula and Ruuttenen [1]. In 1944, Bousquet [2], assignor to E. I. du Pont de Nemours & Co., applied for a patent for the ammonium salt of TCA. About nine months later, Virtanen [3] presented a patent application in Finland "for the use of those organic compounds in which the carbon atom adjacent to CO group was completely chlorinated, such as TCA and chloral, against perennial weeds, especially couchgrass (quackgrass)"; the next year he presented a similar application in Sweden. The American patent was weak, because it covered only the ammonium salt of this compound. On the other hand, the Finnish and Swedish patents were not sufficiently comprehensive to include dalapon, which was patented a few years later in the United States by Barrons [4]. Patent citations were also made for 2,2,3-trichloropropionic acid in 1957 by Barrons [5] and for 2,2-dichlorobutyric acid in 1959 by Toornman [6].

In 1950, in a comprehensive review on herbicides by Norman et al. [7], only four references were quoted in relation to halogenated aliphatic acids. All referred to TCA, and the earliest was in 1948. However, other chlorinated aliphatic acids have become well known as herbicides or growth regulators within the past 25 years. One literature survey [8] revealed over 700 papers on dalapon alone between 1958 and 1962. Also, mono- and trichloro-substituted acetates, propionates, butyrates, and other analogs have been investigated. Numerous additional references on the chlorinated aliphatic acids have appeared since 1962.

Most citations are concerned with relatively few of these derivatives. The major ones, in order of their appearance, are TCA, dalapon, 2,2,3-trichloropropionic acid, 2,2-dichlorobutyric acid, and 2,3-dichloroisobutyric acid. Alpha chlorination appears to be a major requirement for activity, with TCA, dalapon, and 2,2-dichlorobutyric acid being herbicidally most active. The unsubstituted aliphatic acids, including formic, acetic, acrylic, butyric, and oleic (among others), are essentially inactive as herbicides.

The herbicidal action of a 1% chloroacetic acid was described in 1951 by Zimmerman and Hitchcock [9]. They claimed that several weed species were killed in 1-24 hr. A selective action was also found since no damage was sustained by corn plants, roses, and carnations. Subsequently TCA and dalapon found a place in agricultural practices. The now well-known grass-killing properties of TCA were observed early by Virtanen (cf. Mukula and Ruuttenen [1]) and by McCall and Zahnley [10]. The announcement of dalapon in 1953 as a systemic grass-selective herbicide [11] represented a promising new approach to chemical week control. Sodium 2,2-dichloropropionate (dalapon, sodium salt) was first made available by the Dow Chemical Company for industrial weed control; it later showed considerable promise for certain selective agricultural uses [11] . Both sodium TCA and dalapon are somewhat selectively toxic to annual and perennial grasses in much the same way that 2,4-D is to broadleaved species. Several such uses for each compound, in various food or feed crops, are listed in the current EPA Summary of Registered Agricultural Pesticide Chemical Uses [12] .

Sodium 2,3-dichloroisobutyrate has received considerable research attention since the late 1950s as a male gametocide in cotton, tomatoes, and several other crops [13, 14] .

B. Physical and Chemical Properties

Replacement of hydrogen with halogen in the aliphatic acids yields derivatives which ionize to a greater extent if unsubstituted. The halogens, being strongly electronegative, tend to attract electrons. Thus, the inductive effect of the chlorines makes the chlorinated aliphatic compounds stronger acids than the corresponding acetic or propionic acids. The effect of chloro substitution on the pK_a values of several acids is shown in Table 1.

The strongly electronegative chlorine atoms in 2,2-dichloropropionate greatly influence other properties of the molecule such as the rotation of the methyl group around the α-carbon.

Chloroformic acid, chloroacetic acid, and dichloroacetic acid are all inactive as selective grass-control herbicides. Within the propionic acid

TABLE 1

Effect of Chloro Substitution on Acidity
of Acetic and Propionic Acids at 25°C [15, 16]

Acetic acid	pK_a	Propionic acid	pK_a
Acetic	4.76	Propionic	4.88
Monochloroacetic	2.81	2-Chloropropionic	2.80
Dichloroacetic	1.29	3-Chloropropionic	4.10
Trichloroacetic[a]	0.08	2,3-Dichloropropionic[b]	1.71

[a]Refers to a 0.03 M solution that is 89.5% ionized [15].
[b]pK_a value reprinted by permission of Kearney et al. [16]; all others reprinted by permission of Reinhold Book Corporation from Fieser and Fieser [15].

and the butyric acid series, α chlorination results in herbicidal activity, with the 2-chloro compounds being highly active. Chlorination in other positions alone does not result in activity and in combination with α chlorination may weaken the effects of α chlorination [8].

Increasing chain length reduces activity, even with α chlorination. The 2,2-dichlorovaleric acid is only weakly active, and the 2,2-dichlorohexanoic acid is inactive. The substitution of other halogens for chlorine generally reduces activity throughout the series [8].

Freed [17] conducted certain physical and chemical examinations on pure 2,2-dichloropropionic acid and 2,2,3-trichloropropionic acid, as shown in Table 2.

Table 3 presents another interesting property of chlorinated aliphatic acids investigated by Freed and Montgomery [18], the relative molar volume (V_m).

The relationship of structure, including three-dimensional configuration, to biological activity is very complex. Since dalapon is probably in an aqueous phase in most biological systems, the authors suggest that the partial molar volume (or the volume contribution of the constituent at infinite dilution) might be pertinent to activity. They regarded it significant that dalapon, a systemic herbicide, is the only acid showing no appreciable volume change in solution. Actually, the position of dalapon relative to molar volumes is not inconsistent with known physical chemical principles. Rather, this

TABLE 2

Selected Physical Properties of
Dalapon and 2,2,3-Trichloropropionic Acid [17]

Physical property	2,2-Dichloropropionic acid	2,2,3-Trichloropropionic acid
Melting point (°C)	20	65-66
Density	1.399	1.491
pH (aqueous)	1.32 (0.099 N, 23°C)	1.21 (0.10 N, 24°C)
Ka	2.94×10^{-2}	9.94×10^{-2}
pK_a	1.53	1.00
Reference index	1.453	1.485
Molar volume (ml)	102.20	119.00
Molecular weight	142.98	177.43

TABLE 3

Molar Volumes and Partial Molar
Volumes of Several Chlorinated Aliphatic Acids [18]

Acid	Molar volume (ml)	Partial molar volume (ml)
Monochloroacetic	59.8	65.0
Trichloroacetic	101.1	97.5
2,2-Dichloropropionic	102.2	102.2
2,2,3-Trichloropropionic	118.9	131.0

property may be considered a logical consequence of its other properties based on strength of acids and hydrogen bonding among different species in solution.

The chlorinated aliphatic acids are water-soluble, anionic compounds that do not possess functional groups generally associated with hydrogen bonding. These facts suggest little or no adsorption of these compounds on soil colloids [16]. Likewise, one should expect little adsorption or binding in plant tissues, except through metabolic accumulation and/or mechanical trapping. Recently, however, Kemp et al. [196] studied the hydrogen bonding of dalapon to N-methylacetamide as a model protein using infrared spectral analysis. Changes in the spectra indicated that the dalapon carboxyl-OH hydrogen bonds significantly to the model amide carboxyl. The biological significance of these findings has not yet been determined.

Some of the most important physical properties of dalapon and TCA are compared in Table 4.

Since dichloropropionic acid or trichloroacetic acid can exist in the acid form only at very low pH values, Kearney et al. [16] have suggested that the ionic species encountered under most biological conditions will be the anions dichloropropionate or trichloroacetate. They concluded that the appearance of the acid or undissociated form would be extremely rare under most physiological conditions.

Investigation of several derivatives of dalapon has failed to reveal any with greater biological activity than the sodium salt [11, 20]. However, activity has been considerably enhanced by changing the pH or by adding any of several suitable surfactants. The latter alter the physiologically important ionic and polar properties of dalapon solutions [20-28] (see Sec. III).

The foregoing conclusions were substantiated further by two recent studies [197, 198], as follows:

TABLE 4

Physical Properties of Dalapon and TCA[a]

Property	Dalapon	TCA
Empirical formula	$C_3 H_3 Cl_2 NaO_2$ $(C_2 H_4 Cl_2 O_2)$	$C_2 Cl_3 NaO_2$ $(C_2 HCl_3 O_2)$
Molecular weight	165.0 (143.0)	185.4 (163.4)
Physical state and color	White powder (colorless liquid)	White granules or pellets (white deliquescent solid)
Density	1.0049^b at 25°C 1.2978^c at 25°C (1.389 at 22.8°C)	— (1.6013 at 60°C) (1.62 at 25°C)
Melting point	Decomposes before melting at 193–197°C	Decomposes before melting
Decomposition temperature (°C)	166.5	165–200
Boiling point (°C)	(185–190)	

Solubility at 25°C:

Solvent	g/100 g solvent	g/100 g solvent
Acetone	0.014	0.76
Benzene	0.002	0.007
Carbon tetrachloride	Nil	0.004
Ether	0.016	0.02
n–Heptane	0.002	0.002
Methanol	17.9	23.2
Water	50.2	83.3

Vapor pressure	—	Temp (°C)	mm Hg
		(76.99)	(5)
		(88.67)	(10)
		(101.52)	20
		(109.65)	30
		(115.72)	40
		(124.75)	60
		(136.98)	100
		(155.30)	200
		(175.97)	400
		(197.97)	760

[a]Values listed are for the sodium salts; values in parentheses refer to the acids. (Adapted by permission of Kearney et al. [16] and the Weed Science Society of America [19].)

[b]One percent solution.

[c]Fifty percent solution.

McWhorter [197] found that the addition of many alkali metal salts to dalapon sprays increased johnsongrass control 20-40%. Maximum effectiveness was obtained when surfactant was included. Salts of sodium, potassium, and lithium were nearly equal in effectiveness, but the anion was also important in affecting herbicide activity. Sulfates, phosphates, and nitrates were most effective; carbonates and chlorides intermediate; while nitriles, sulfites, and hydroxides were ineffective. Ammonium aluminum sulfate was effective in reducing the activity of dalapon and MSMA. Also, Wills [198] reported that the addition of the monovalent ammonium and potassium ions to dalapon increased nutsedge control [193]. Phosphate appeared to increase the toxicity of dalapon.

Sodium 2,2-dichloropropionate is a white to tan-white, free-flowing powder prepared by neutralizing 2,2-dichloropropionic acid with sodium hydroxide. An anionic wetting agent (a polyglycol of Dow manufacture) is also formulated in the commercial product.

The chemistry of such compounds as TCA and dalapon is deceptively simple. The compounds are easily prepared and the products undergo standard reactions [8]. In the laboratory, α-chlorinated acids can undergo dehydrochlorination, yielding (according to reaction conditions) hydroxy acids, amino acids, or cyano acids. However, these reactions apparently do not take place in plant tissue. Freed et al. [29] indicated that dalapon undergoes the typical reactions of salt formation, esterification, and acyl chloride formation.

In aqueous solutions these acids decompose at room temperature [8]. TCA breaks down in solution to form chloroform and carbon dioxide. Dry dalapon sodium salt is stable, but aqueous solutions are subject to decomposition, undergoing the probable reaction shown in Eq. (1):

$$
\begin{array}{ccc}
\text{Cl} & & \text{O} \\
| & & \| \\
\text{H}_3\text{C-C-COONa} + \text{H}_2\text{O} & \rightarrow & \text{H}_3\text{C-C-COOH} + \text{NaCl} + \text{HCl} \\
| & & \\
\text{Cl} & &
\end{array}
\qquad (1)
$$

Dalapon   Pyruvic acid

The reaction does not take place readily under acid conditions and proceeds very slowly at temperatures below 20°C. The sodium salt of dalapon is almost as hygroscopic as TCA. The two chlorine atoms on the second carbon are of considerable interest. As shown, these can be removed by alkaline hydrolysis to yield pyruvic acid and inorganic chloride.

Another reaction involving the chlorine in dalapon has been demonstrated in vitro [29]. This reaction, with sulfhydryl groups of organic compounds, presumably forming thioether linkages, may prove to be of biological importance. Preliminary experiments showed this reaction to proceed at a physiological pH. Reaction of halogen-substituted alkyl acids with sulfhydryl-containing metabolites is not new. For example, monochloroacetic acid and TCA react with glutathione. By comparison, dalapon reacted much more slowly [30]. It has been suggested, without further explanation, that this difference in reaction rate with the SH groups may, in part, account for differences in their effects on plants. This particular reaction with the ubiquitous, physiologically active compound glutathione may or may not be important. However, the fact that the SH group is an active functional group in certain enzymes seems relevant. For example, it is by such an alkylation reaction that iodoacetic acid exerts its toxic action [31].

2,2,3-Trichloropropionate is stable in water solutions and also in weakly acidic or weakly basic solutions [32]. In very strongly acidic solutions, propionic acid is formed. Loss of chlorine also occurs by prolonged exposure to strong alkaline media (pH 9 or higher).

C.   Analytical Methods

A gas-chromatographic method has been used successfully for the detection of dalapon residues as low as 0.1 ppm in cranberries and bananas [19]. Kratky and Warren [33] have developed three bioassays: respiratory activity of Chlorella, root growth of sorghum and cucumber, and shoot growth of sorghum and oats at different dalapon concentrations. TCA or sodium TCA can be determined in flax seed and other crops by a colorimetric method, sensitivity 0.1-2.5 ppm [19]. Hardcastle and Wilkinson [34] used a bioassay to evaluate the activity of dalapon and TCA in combinations with other herbicides or each other on rice root growth.

D.   Toxicological Properties of Dalapon and TCA

1.   Dalapon

Goldfish mortality at the end of 24 hr exposure to dalapon was 0% at 100 ppm and 100% at 500 ppm and above [19]. Bluegill, green sunfish, lake chubsucker, and smallmouth bass fry survived a concentration of 500 ppm of dalapon for 8 days [35].

The 48-hr LC_{50} for waterfleas (Daphnia magna) exposed to dalapon was 6000 ppb [36]. Stonefly nymphs (Pteronarcys californica) exposed to dalapon for 96 hr at 100 ppm were not affected [37].

The LD_{50} for the rat was 7570-9330 mg/kg; for the female mouse, >4600 mg/kg; for the female rabbit, 3860 mg/kg for dalapon administered orally. When calves were fed 1000 mg/kg/day for 10 days, females exhibited nonspecific symptoms during the last days of the experiment but completely recovered upon cessation of dosing. No adverse effects on the male calves were observed during dosing, but autopsy showed a possible slight kidney involvement. Dogs fed from 50 to 1000 mg/kg/day, 5 days a week, for 80 days showed no adverse effects other than vomiting. A slight kidney weight increase occurred when dogs were fed 100 mg/kg/day of dalapon, but no adverse effects were observed with 50 mg/kg/day after one year. After two years, rats showed a slight kidney weight increase when fed 50 mg/kg/day but no adverse effects were observed with 15 mg/kg/day [19].

The LD_{50} for chicks was 5660 mg/kg [19]. The LC_{50} for mallards was >5000 ppm; for pheasants, >5000 ppm; and for coturnix, >5000 ppm of dalapon in diets of two-week-old birds when fed treated feed for 5 days followed by untreated feed for 3 days [38].

Dalapon is moderately irritating to skin and eyes and there is no antidote except to flush with plenty of water. Dusts may be irritating to the upper respiratory tract, but not likely to cause systemic injury [19].

2. TCA

TCA and its sodium salt have a low order of toxicity to wildlife and fish. The oral LD_{50} for the rat was 5000 mg/kg; for the mouse, 3640 mg/kg; and for the rabbit, 4000 mg/kg. The LD_{50} for chicks was 4280 mg/kg. Rats fed TCA at 0.1% in diets for 126 days showed no adverse effects, and only growth depression due to reduced food intake was observed when fed 0.3% TCA in diets [19].

TCA is capable of causing burns on skin upon prolonged contact of 1 hr or more and is very painful in eyes and may cause injury. Dust is very irritating to nose and throat, but is not likely to cause systemic injury [19].

II. DEGRADATION

A. Degradation in Plants

Higher plants do not readily attack or metabolize dalapon [39-43] or TCA [44, 45]. Barrons and Hummer [46] and Tibbetts and Holm [47] demonstrated the presence of TCA in plants grown in TCA-treated soils. The colorimetric method used by Tibbetts and Holm [47], however, did not distinguish between TCA and its possible degradation products. Blanchard [44] treated pea and corn plants with ^{14}C-labeled TCA, extracted the sap. and found only a single radioactive spot which cochromatographed with labeled TCA. No physiological variables were introduced. The presence of trichloromethyl compounds, in addition to TCA, in treated tomato and tobacco plants has been reported [45]. Only intact TCA could be detected in flax, black radish, maize, barley, and dandelion plants from soil applications of the herbicide. Thus, the virtual lack of metabolic alteration of TCA in plants is based on just three reports. Conclusive as these reports may appear, it would seem that more complete study is needed on the metabolic fate of TCA in plants.

Dalapon is absorbed, translocated, and accumulated in plants largely as the original chemical [21, 40-43, 48-51]. The most intensive series of studies on dalapon metabolism in plants employed autoradiography, extraction and fractionation, counting, and paper-chromatographic techniques to analyze for the herbicide and its possible metabolic products.

Absorption of dalapon by roots or leaves of cotton resulted in extensive distribution throughout the plant [41, 42, 49], and accumulation was greatest in regions of high metabolic activity [41-43, 50]. Similar accumulations have been noted for sugarbeets [39, 48], barley [49], corn [21], sorghum, and wheat [42, 50].

Foy [41, 42, 50] in studies on cotton (a tolerant species), sorghum (a susceptible species), and wheat found that dalapon was absorbed, translocated, redistributed, and accumulated in higher plants, principally as the intact molecule or its dissociable salt. It remained essentially nonmetabolized for long periods, especially in dormant or quiescent tissues. The metabolic stability and persistence of dalapon is emphasized by the fact that it accumulated in the seeds of cotton and wheat and was transmitted from one generation to the next [42, 50, 52]. Indeed, dalapon stimulus was traced to the third generation in wheat; it was carried over in the seeds after exposure of first-generation seedlings to preplant applications of dalapon at the rate of 4.48 kg/ha in the field (Fig. 1).

In an intensive study of dalapon metabolism in cotton, Foy [42] incubated tissue homogenates with [2-^{14}C]dalapon and [^{36}Cl]dalapon for short periods; no metabolic changes were detected by extraction, fractionation, and chromatography. Moreover, 7 days after foliar application to intact cotton and sorghum, dalapon, and no other prominent radioactive species, was recoverable with water or ethanol from all plant parts (Fig. 2).

Loss of radioactivity has been found to occur from the roots of cotton [41, 43] and sorghum [41]. However, the excreted radioactive material was chromatographically indistinguishable from authentic dalapon [41]. Exudation of dalapon from the roots was increased by creating an unfavorable ionic balance in the nutrient medium.

In long-term studies, however, a small amount of nonextractable radioactivity appeared in both cotton and sorghum. In one phase of the cotton study [42], for example, dried ground fruits of various ages were extracted and fractionated according to the scheme shown in Fig. 3. Nine to 10 weeks after treatment with [2-^{14}C]dalapon (through severed petioles), approximately 85-90% of the radioactivity was recoverable as dalapon. The remainder was associated with the ether-soluble portion, the neutral and cationic fractions

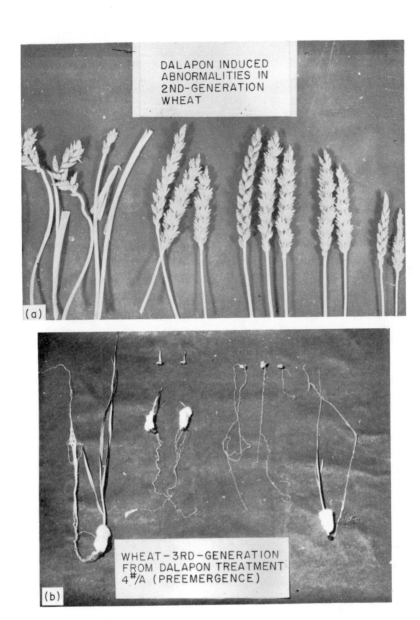

FIG. 1.   Carryover effect of characteristic dalapon symptoms in
second-generation (a) and third-generation (b) wheat following inhibition
of first-generation plants by preplanting application of dalapon at 4.48
kg/ha. Groups of heads in (a) are arranged left to right in order of occurrence on the same plant. In (b) the seedling at left is normal and others show
inhibition or other anomalies. (Reprinted from Foy [42], with permission.)

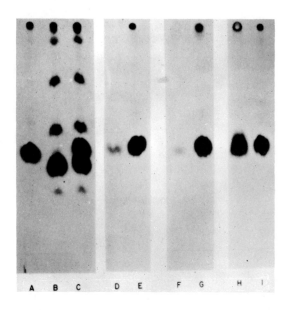

FIG. 2.   Radioautograms of representative cochromatograms. (A) Five microliters of 96% [2-^{14}C]dalapon stock solution; (B) 5 μl of impure 2,2,3-trichloro[2-^{14}C]propionate (TCP); (C) dalapon plus TCP, 5 μl of each; (D) aqueous extract from roots of dalapon-treated cotton plants; (E) extract in D plus dalapon; (F) aqueous drop (from ether extract) of nutrient solution in which dalapon-treated cotton was grown; (G) extract in F plus dalapon; (H) aqueous extract from dalapon-treated leaves of sorghum; (I) extract in H plus dalapon. (Reprinted from Foy [42] , with permission.)

of the ethanol extract, and the insoluble plant residue. Some dalapon may be slowly degraded and the ^{14}C incorporated metabolically into the plant constituents. The presence of nonextractable radioactive residues in cotton has been interpreted differently by Smith and Dyer [43] . They suggested that this was occluded or trapped, but chemically unaltered, dalapon. Quantitatively it accounted for only a very small percentage of the applied chemical.

Very small counts of radioactivity (possible ^{36}Cl) were detected in the water of guttation from hydathodes of sorghum 6-8 days after treatment [42, 50] . None was detected in the case of plants treated comparably with [2-^{14}C]-dalapon. As in the preceding situation, however, the low levels of radioactivity precluded further characterization of the chemical substance(s). The degradation of dalapon in higher plants is very slight. Eventual breakdown, if indeed it does occur, may possibly involve an initial dehalogenation followed by normal or modified propionate oxidation. Giovanelli and Stumpf [53] stated

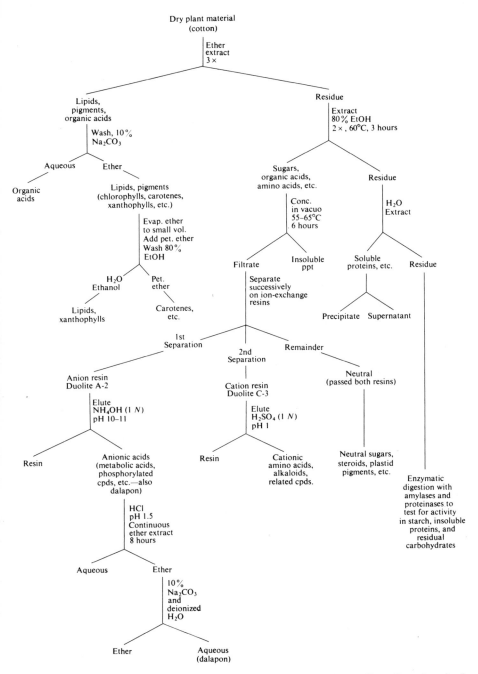

FIG. 3.   Analytical scheme followed in categorizing radioactive chemical constituents from fruits and leaves of cotton 10 weeks after treatment with 96% sodium 2,2-dichloro[2-^{14}C]propionate.   (Reprinted from Foy [42] by permission of the American Society of Plant Physiologists.)

that the oxidation of propionate in animal tissues occurs by a carboxylation pathway through methyl malonate to succinate. The same reaction might perhaps be acceptable for dalapon and 2,2,3-trichloropropionate in plant tissue after dehalogenation.

Most studies with dalapon have been of relatively short duration. Blanchard et al. [40] agreed that dalapon resisted degradation based on their studies with soybean and corn in which no metabolic products were found 4 days after herbicide application to roots or foliage. Care must be exercised in long-term studies, however, to interpret correctly the decline of herbicide content. As Schreiber [54] indicated, what seems to be loss of a chemical may actually be dilution due to growth. Also, in long-term studies, microbial contamination of injured plant tissue and subsequent degradation by microorganisms may sometimes become a factor. McIntyre [55] has suggested that differences in the rate of translocation of dalapon in the light and in the dark could be attributed to the formation of a chemical or physical combination of the herbicide with a photosynthate. However, no direct experimental evidence exists to support this hypothesis.

No detailed degradation studies of other chlorinated aliphatic acids in higher plants were found in the literature.

B. Degradation in Animals

Leasure [8], in a literature review, found certain similarities in the behavior of dalapon in plant, soil, and animal systems. Dalapon fed to animals was quickly excreted (nonmetabolized) in herbicidal concentrations in the urine. According to Leasure, limited evidence indicated that the relatively small amount of dalapon remaining in the animal system was decomposed along the same general lines as in soil systems.

Dalapon residues reported in milk were much less than 1% of the amount ingested in the feed in one study [56]. In another report, cited by Leasure [8], Redemann and Hamaker found only two labeled compounds in milk from a cow whose feed contained ^{36}Cl-labeled dalapon. These were dalapon and chloride ion, and the latter was present in much larger quantities than was the dalapon. It was postulated that dalapon hydrolyzes to pyruvate, which then breaks down to acetate and carbon dioxide.

C. Degradation in Soils

Some portions of practically all herbicides used, whether soil- or foliar-applied compounds, eventually contact the soil. Herbicides reaching the soil, accidentally or by design, normally become dissipated or removed with time, in one or more of several ways, as follows: (a) volatilization; (b) photodecomposition; (c) soil absorption-inactivation; (d) leaching; (e) chemical breakdown; (f) microbiological degradation; (g) plant uptake, followed by metabolic degradation and/or physical removal at harvest. The ways in which herbicides may encounter the soil and the possible ensuing courses of events are depicted schematically in Fig. 4 [57] .

In most critical studies conducted thus far on herbicide disappearance from within soils, degradation by soil microorganisms has proved to be very significant, perhaps indeed the most important factor in many instances. In contrast to their strong persistence in higher plant tissues, the chlorinated aliphatic acid herbicides, in general, are readily subject to soil microbiological decomposition. Virtually no problems from soil residues of these herbicides have been detected after their wide-scale commercial use over a period of more than 15 years.

1. Physical Properties of the Herbicides That Influence Degradation

Little work has been done to determine the effect of molecular structure of aliphatic acids on the ease of decomposition by soil microorganisms. As a rule, but with distinct exceptions, short-chain aliphatic hydrocarbons are not as readily metabolized as those of higher molecular weight, as shown by a number of research workers [58-60] . Also, unsaturated aliphatics tend to be more readily attacked than the corresponding saturated acids [61] .

A few scattered reports describe the relative persistence of the various chlorinated aliphatic acid herbicides in soils. As reviewed by Kearney et al. [16] , Jensen [62] found that the number of chlorine substituents appeared to determine the rate of decomposition. In pure culture studies, he observed that a group of Pseudomonad-like bacteria readily decomposed both monochloroacetate and monobromoacetate. The same organisms only moderately decomposed 2-chloropropionate, but had little effect on dalapon or dichloroacetate and no effect on TCA. Another group of bacteria (probably Agrobacterium) readily decomposed dalapon and dichloroacetate, but was only partially effective on 2-chloropropionate, and was ineffective on mono- or trichloroacetate. Other genera of soil microorganisms manifested still different specificities.

Hirsch and Alexander [63] also investigated the decomposition of propionic and acetic acids having various numbers of chlorine, bromine, fluorine, and iodine substitutions by a Pseudomonas sp. and a Nocardia sp. Significant differences existed between isolates in the types of halogen-containing aliphatic acids utilized, as well as the effect of halogen number and position. Both microorganisms grew well on yeast extract in the presence of fluorinated compounds but were unable to degrade these compounds. Although both microorganisms were effective on dalapon and 2-chloro-, 2-bromo-, and 2-iodopropionate, they were more effective on the 2,2-dichloroaliphatic acids than the corresponding 2-chloro substituted compound. Also, in each case, the organisms were more effective on the α-halogenated propionate than the corresponding acetate. The inability of either strain to dehalogenate

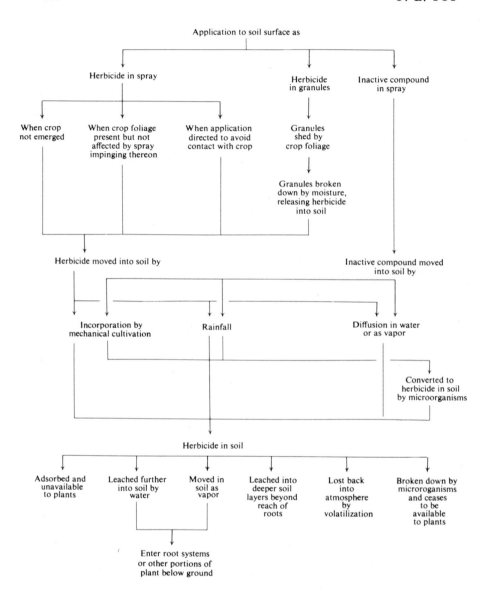

FIG. 4. Scheme showing the sequence of events following application of herbicides acting through the soil. (Reprinted from Holly [57, p. 453], by permission of Academic Press, London and New York.)

2,3-dibromo- or 2,2,3-trichloropropionate suggests that the β-substituted forms are less susceptible to degradation.

In contrast, Kaufman [64] (cited by Kearney et al. [16]) observed that eight of nine isolates which effectively degraded dalapon were more effective on 3-chloropropionic than on 2-chloropropionic acid. Two of these organisms were also more effective on 2,3-dichloroisobutyric than on 2,2-dichloro-butyric acid. The ninth isolate was about equally effective on 2-chloropropionic acid and dalapon, but it was less effective on 3-chloropropionic acid than on 2-chloropropionic acid. This same isolate was apparently effective in varying degrees on 10 chlorinated acetates, propionates, and butyrates examined. The effectiveness of all microorganisms decreased with each increase in the number of chlorine substituents. Kaufman [65] observed that the chlorinated propionates were decomposed more rapidly than the chlorinated acetates. One isolate, however, was more effective on the chlorinated butyrates than on either the propionates or acetates.

Kaufman [65], in using a soil-enrichment technique, found that the rate of microbial decomposition of halogenated acetic and propionic acids decreased as the number of halogens on the molecule increased. Microbial decomposition of the β-substituted aliphatic acids occurred more slowly than the corresponding α-substituted compounds. Halogenated propionic acids were decomposed more rapidly than the corresponding acetic and butyric acids.

2.   Degradation of the Herbicides in the Soil by Microorganisms

General soil studies indicate that both dalapon and TCA are subject to microbial decomposition, but that TCA is degraded more slowly than dalapon. Loustalot and Ferrer [66] reported that the disappearance of TCA from soil was favored by warm, moist conditions. Ogle and Warren [67] also found that breakdown of TCA in soils was most rapid under conditions conducive to high microbiological activity. TCA breakdown was low in sandy soils. Similar results have been obtained with dalapon [68, 69]. Thiegs [68] also found that dalapon degradation in soils was most rapid under warm, moist conditions. He further reported that subsequent additions of dalapon to soil were decomposed more rapidly than the initial application. Holstun and Loomis [69] found that the decomposition of dalapon is primarily a function of an undetermined fraction of the soil microorganisms. Sterilization of the soil essentially stopped all degradation, which began again after recontamination of the soil.

In other studies using both [1-^{14}C]dalapon and [2-^{14}C]dalapon, Thiegs [70] reported the recovery of $^{14}CO_2$ in both instances and concluded that the decomposition of dalapon in soil was complete to CO_2. The degradation of 2,2-dichlorobutyric acid in soil apparently follows the same pattern as dalapon and TCA, but proceeds much more slowly [8].

TABLE 5

References to Soil Microorganisms That Decompose Chlorinated Aliphatic Acids [16][a]

Organism	Mono-chloro-acetic acid	Di-chloro-acetic acid	Tri-chloro-acetic acid (TCA)	2-Chloro-propionic acid	3-Chloro-propionic acid	2,2-Di-chloro-propionic acid (dalapon)	2,2,3-Tri-chloro-propionic acid	2,2-Di-chloro-butyric acid	2,2-Di-chloro-isobutyric acid	2,3,3-Tri-chloro-isobutyric acid
Bacteria										
Agrobacterium sp.	—	64, 72, 76	—	64	64	64, 71, 72, 76, 84	—	64	64	64
Pseudomonas sp.	—	63, 64	64	63, 64	63, 64	63, 71, 84, 85, 86	64	64	64	64
Arthrobacter sp.	64	64	64, 76, 87	64	64	63, 64, 75	64	64	64	64
Micrococcus sp.	—	—	—	—	—	86	—	—	—	—
Alcaligenes sp.	64	—	—	64	64	64, 75	—	64	64	64
Bacillus sp.	—	64	—	64	64	64, 75	—	64	64	64
Pseudomonas dehalogens	—	—	77, 88	—	—	77, 88	—	—	—	—
Flavobacterium sp.	—	—	—	—	—	70	—	—	—	—
Fungi										
Trichoderma viride	72, 76, 89	72, 76, 87	72, 76	89	—	75	—	—	—	—
Clonostachys sp.	72, 89	89	—	89	—	—	—	—	—	—
Penicillium	—	—	—	—	—	63, 75	—	—	—	—
Aspergillus sp.	—	89	—	—	—	63	—	—	—	—
Acrostalagmus sp.	89	89	—	89	—	—	—	—	—	—
Penicillium lilacinum	—	—	—	—	—	75	—	—	—	—
Penicillium rouqueforti	72	—	—	—	—	—	—	—	—	—
Actinomycetes										
Nocardia sp.	63, 64	63	—	63, 64	63, 64	63, 64, 75	—	64	64	64
Streptomyces sp.	—	—	—	—	—	63, 75	—	—	—	—

[a]Reproduced by permission of John Wiley & Sons, Inc.

The final proof of the dominating role of bacteria or other microorganisms in detoxication processes in the soil is the isolation of the responsible organism, its growth in pure culture using the herbicide as a source of carbon or nitrogen, and the demonstration that it can inactivate the herbicide. Most bacterial genera seem to confine their attack to one specific group of herbicides, but some seem to be capable of decomposing chemical compounds of widely divergent structures. The genera Nocardia and Arthrobacter, for example, attack both the phenoxy and the chlorinated aliphatic herbicides.

Hirsch and Alexander [63] isolated and characterized strains of Pseudomonas and five strains of Nocardia which decomposed dalapon, liberating 90-100% of the halogen within three weeks. Other workers have reported additional species of bacteria, fungi, and actinomycetes that have decomposed chlorinated aliphatic acids. For example, Magee and Colmer [71] reported Agrobacterium, Thiegs [70] reported Flavobacterium, and Jensen [62, 72] added Penicillium, Trichoderma, Clonostachys, and Arthrobacter.

Numerous authors [62, 63, 70, 72-78] have reported that soil microorganisms can utilize chlorinated aliphatic acids as a source of energy. The current literature records a wide variety of microorganisms that have been isolated and proved effective in degrading various chlorinated aliphatic acid compounds (Table 5). Nearly all these organisms have been isolated by means of an "enrichment technique." This technique, in its several forms, consists mainly of exposing several grams of soil to an aqueous solution of the herbicide in question. The solution is then periodically analyzed for breakdown products of the herbicide. A soil-perfusion technique similar to the one described by Lees and Quastel [79] has proved to be useful as an enrichment technique for studying pesticide degradation processes [16] . Newman and Downing [80] stated that it appears that the disappearance of TCA and dalapon results from removal by leaching and microbial decomposition. The relative importance of these two means depends on soil and environment.

Kaufman [81] found that the microbial degradation of dalapon was inhibited in the presence of amitrole. Phytotoxic residues of the herbicides persisted longer in the soils when the herbicides were applied in combination than when each was used individually.

Dehalogenation of TCA by soil microorganisms has been reported by several investigators [62, 63, 74, 78, 82] . Most of the isolated microorganisms grow feebly on TCA as a sole source of carbon. Jensen [78] reported that TCA and its theoretical dehalogenation product, oxalate, could serve as a carbon source for two species of Arthrobacter.

3. Degradation of the Herbicides in Culture Solution by Microorganisms

A range of halogenated fatty acids can be metabolized by a number of soil microorganisms in culture [42, 63, 74] . The attack of microorganisms

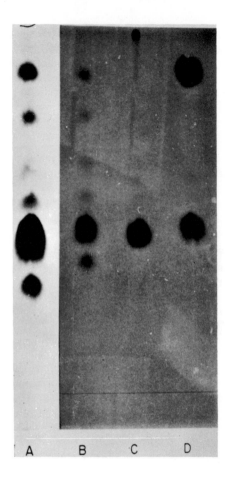

FIG. 5.   Radioautograms showing degradation products after decomposition of pure [^{36}Cl]dalapon by some species of microorganism, possibly Alternaria sp. Point of drop application on the chromatogram is at extreme top edge. (A) Ten-microliter drop, after contamination; (B) 5-μl drop, after contamination; (C) pure dalapon; (D) impure dalapon, the remainder of the activity being represented as inorganic chloride (Li^{36}Cl). (Reprinted from Foy [83] with permission.)

on the herbicide has been followed by measuring the release of ionic chlorine into the culture medium.

Foy [42, 50] found that pure [^{36}C]dalapon in aqueous stock solution was metabolized over a period of several months in the refrigerator by some species of microorganism(s), possibly Alternaria sp., which produced five new ^{36}Cl-labeled substances (Fig. 5). Two of the substances were tentatively identified as inorganic ^{36}Cl and [^{36}Cl]monochloropropionate. No attempt was made to identify the other three labeled compounds. Although serendipitous, this early (1957) observation offered positive proof that a microorganism was capable of using dalapon as sole substrate to synthesize new labeled compounds. He postulated that similar conversions probably occur in soils, accounting for the disappearance of dalapon toxicity under favorable conditions.

In Jensen's work [74] three species showed quite different toxicity in relation to the chloroacetic and propionic acids fed to them. Thus, a Pseudomonas sp. most vigorously attacked mono- and dichloroacetic acids. An Arthorobacter sp. was most active on dichloroacetic acid and dalapon. The experiments of Hirsch and Alexander [63] were rather more ambitious in that they employed a greater range of substrates, some labeled with ^{14}C to follow independently the release of chlorine and the utilization of carbon. Species of Pseudomonas and Nocardia were shown to utilize 2-monosubstituted propionic acids with great ease, irrespective of the halogen. The metabolism of the corresponding β-substituted acids was found to be much more difficult. The authors also found a difference in behavior of the two organisms in relation to Cl, Br, and I. On the strength of this, they suggested that the responsible enzyme could not be a nonspecific dehalogenase which removes a halogen from a utilizable substrate. Both organisms decomposed dalapon with the utmost vigor but had no effect at all on 2,2,3-trichloropropionate. Behavior with the acetic acids was equally complex. Two features of particular importance were that TCA was attacked by Pseudomonas but not by Nocardia, whereas the opposite behavior was shown with iodoacetic acid.

Jensen [74] found dehalogenase induction in a Pseudomonas sp. (Table 6), which he explained by assuming that the bacteria can form two different chloride-liberating enzymes, one of which is induced by monochloroacetic acid and is active toward that compound only, while the other is induced by dichloroacetic acid and can dechlorinate several compounds.

4.   Mechanism Through Which Microbes Develop the Ability to Degrade the Herbicides

The mechanism through which a soil microbial population develops the capacity to degrade a pesticide is not completely understood. Audus [90] has suggested two major possibilities that would involve either (a) chance mutation or (b) adaptive enzymes. According to Audus [90], the first theory is supported by the random manner in which a wide variety of microbial

TABLE 6

Dehalogenase Induction in _Pseudomonas dehalogenans_ [74]

P. dehalogenans group	Inducing molecule	Simultaneous adaptation to:			
		Monochloro-acetic	Dichloro-acetic	Trichloroacetic (TCA)	2, 2-Dichloropropionic (dalapon)
I–II	Monochloroacetic	+	+	–	–
III	Monochloroacetic	+	+	–	+
III	Dichloroacetic	+	+	–	+
III	Trichloroacetic	+	+	+	+
III	2-Monochloropropionic	+	+	+	+
III	2,2-Dichloropropionic	+	+	+	+

genera decompose certain herbicides. However, the tendency of several samples of a single soil to develop similar effective populations of microorganisms repeatedly, when independently enriched, conflicts with this theory.

The second theory, which proposes the induction of adaptive enzymes in certain responsive microbial genera, is in agreement with the views of induced microbial changes presented by Cohn and Monod [91]. This theory has been criticized, however, by Walker and Newman [92] on the basis that effective populations would not persist in soils after complete detoxication and that subsequent herbicide additions would also require a lag phase. Conversely, as discussed by Kearney et al. [16], mutants would retain their effective characteristics in the absence of the herbicide substrate.

The finding that pure cultures quickly lose their ability to utilize certain herbicides when supplied with a less complex substrate like glucose [63] lends support to the induction or adaptive enzyme theory. The results of investigations conducted more recently [74, 93] with pure cultures of microorganisms lend support to the adaptation theory.

Although the mechanism by which an organism becomes effective is unknown, proliferation is an important feature of this phenomenon, as pointed out by Kearney et al. [16]. Dramatic demonstration of the importance of proliferation was shown by Leasure [8], who treated potted soil with dalapon at 56 kg/ha at weekly intervals for a period of six weeks. After six weeks, the pots were seeded to Japanese millet, wild oats, radishes, and cranberry beans, and immediately treated again with a preemergence application of dalapon (56 kg/ha). All seedlings grew normally in pots which had received a total of 392 kg of dalapon per hectare within a period of less than seven weeks. Drastic effects of the herbicide were noted on all four species planted in pots treated with dalapon for the first time.

5. Effects of Herbicides on Soil Microorganisms

The literature contains many conflicting viewpoints of the effects of herbicides on soil microorganisms. Both stimulatory and inhibitory effects on soil microorganisms in response to applications of TCA or dalapon to soils have been observed. In a few instances initial reductions in growth have been followed by increases in activity. Increases may be due either to adaptation and proliferation of "effective" microorganisms or to proliferation of organisms resistant to the effects of the herbicide. Herbicide concentrations found to be inhibitory in the laboratory do not necessarily occur in the soil under ordinary field conditions. Herbicides are effective to the extent to which they are dissolved in soil water [94]. As pointed out by Kearney et al. [16], application rates must be distinguished from concentrations which may result in small areas in the soil. They cite the following example: An application rate of 11.2 kg/ha is equivalent to 5 ppm, assuming even distribution in 2.24 million kg of soil (15.24 cm, plow depth). If complete solubilization and no adsorption occurs, the concentration in the soil solution is 25 ppm in a soil with 20% water at field capacity. Since field soils are seldom maintained at field capacity, these concentrations are often even greater. Although some microorganisms may be inhibited in areas of relatively high concentrations, adjacent areas in the soil may be essentially free of herbicides and therefore represent a source of repopulation.

Nitrifying organisms are among the most sensitive microorganisms to herbicides [16]. Otten et al. [95] observed that nitrification in soil was inhibited by both TCA and dalapon. Douros [9] also observed an inhibition of nitrification by dalapon, but according to Worsham and Giddens [97] depression was only temporary. Mayeux and Colmer [98] observed that dalapon had little effect on nitrite oxidation. Since these compounds tend to persist in most soils from a couple of days to months, it seems likely that the nitrifying organisms may slowly acquire a tolerance to them.

Azotobacter bacteria appeared to be resistant to both dalapon and TCA [80, 94, 99-101]. Azotobacter is very resistant to the chlorinated aliphatic herbicides, concentrations of 1000-20,000 ppm being required to produce substantial inhibitions of growth and respiration in the three species tested by Colmer [101] and Magee and Colmer [102]. The species showed markedly different sensitivities to TCA, which decreased in the order of A. chroococcum, A. agile, and A. vinelandii. Resistance to dalapon is of the same order as that to TCA, but all three species seem equally sensitive [102]. There is little or no threat to nitrogen fixation by Azotobacter by this group of weedkillers when used at normal rates [80].

Little is known about the effect of TCA or dalapon on symbiotic nitrogen-fixing bacteria, although Worsham and Giddens [97] have reported that dalapon had no effect on soybean nodulation at application rates up to 76.16 kg/ha.

Several investigations have been conducted to determine the effects of halogenated aliphatic acids on soil microbial populations in general. Kratochvil [103] observed that TCA significantly reduced the soil microbial activity of treated soils. In contrast, Hoover and Colmer [104] found that rates of TCA higher than normally used in field application had no deleterious effect on bacteria, actinomycetes, and fungi in sugarcane fields. The methods used to investigate this action by the different investigators may account for the different results. Wehr and Klein [105] reported that dalapon and TCA did not inhibit the growth of a Bdellovibris strain or its Pseudomonas sp. host.

Magee [106] observed that dalapon stimulated the multiplication of soil bacteria, actinomycetes, and fungi. Worsham and Giddens [97] also observed an increase in numbers of actinomycetes and bacteria present in treated soil. Elkan et al. [107, 108] observed that sodium propionate greatly increased both respiration and total numbers of fungi. Hale et al. [109] found that low concentrations (50-150 ppm) of dalapon increased oxygen uptake by soil microorganisms, whereas higher concentrations (600 ppm) slightly inhibited oxygen uptake.

Dalapon and TCA also have an effect on certain plant diseases, as observed by Richardson [110, 111]; however, neither compound had an appreciable effect on the causative organism under pure culture conditions. Most other references in the literature concerning the effects of chlorinated aliphatic acids on soil microorganisms are negative.

6. Influence of Certain Physical Soil Constants on Degradation

It is well known that clays, and specifically montmorillonite clays, adsorb organic complexes on their surfaces. The clay hindrance of degradation may result from adsorption not only of the substrate but also of the enzyme. The latter type of adsorption is likely of prime or sole importance to the activity of extracellular enzymes. The degree of substrate binding varies with the nature of the compound, its molecular weight, the pH of the system, and the individual clay mineral. The vast potential for binding is suggested by the enormous surface exposed by clays.

The chemical structure of chlorinated aliphatic acids suggests little or no adsorption of these compounds on the soil colloid. This has been demonstrated experimentally [112, 113] and is illustrated in Table 7.

Dalapon exhibited a high degree of mobility in several soils studied by Holstun and Loom [69]. Warren [114] reported that in a group of 17 herbicides, dalapon and TCA were the most mobile. Kearney et al. [16] presented further evidence that the physical interactions between soil particles and dalapon are relatively unimportant. They concluded that the high initial con-

TABLE 7

Soil Properties and Adsorption Percentages for Dalapon [16][a]

Soil type	pH	Clay (%)	Organic matter (%)	CEC[b] (meq/100 g)	Adsorption dalapon (%)
Barnes clay loam	7.4	34.4	6.90	33.8	0
Hagerstown silty clay loam	5.5	30.0	4.31	12.5	2
Sharkey clay	6.2	67.1	3.90	40.2	0
Celeryville muck	5.0	—	—	142.0	20

[a]Reproduced by permission of John Wiley & Sons, Inc.

[b]Cation exchange capacity.

centration of TCA and dalapon in the soil and the short period of action reduce the significance of leachability as an important factor that might produce inconsistent performance as a herbicide.

At low rates of application, phytotoxic residues of dalapon disappeared from several soils within two to four weeks [114-116]; at high rates of application, dalapon persisted for several months [115]. Residual activity of TCA usually persisted longer than that of dalapon in soils [114, 117], but conflicting data have been reported in at least one case [115].

In their studies Day et al. [118] concluded that "the capacity of the soil to decompose dalapon was essentially random with respect to soil series, texture, C.E.C., total organic matter, and geographical source." Other researchers [62, 68, 75, 115, 116] have also reported differences among soils.

In 1955 Thiegs [68] demonstrated that temperature and moisture were factors affecting the rate of breakdown of sodium dichloropropionate in soil. Dalapon was decomposed most rapidly in warm, moist soil but was relatively stable in moist soil at 40°F and dry soil at 100°F. Thiegs found that fresh additions of dalapon to soil were decomposed more rapidly than the initial application. He also reported that the addition of organic matter increased the rate of disappearance of dalapon and that dalapon did not disappear from sterilized soil.

Holstun and Loomis [69] stated that dalapon decomposition was adversely affected by low soil moisture, low pH, temperatures below 20°-25°C, and large additions of organic matter, and they therefore concluded that it was a function of microbiological activity. Jensen [73] observed that dalapon was decomposed only feebly at pH levels below 5. Kaufman [75] found with five soils in greenhouse and laboratory studies that dalapon degradation by effective microorganisms was affected by organic matter level, pH, cation-exchange capacity, and aeration.

Corbin and Upchurch [119] reported that dalapon was almost 100% detoxified within a two-week incubation period when high-organic-matter soils were buffered to pH 6.5. Soils buffered to pH 6.5 provided optimum conditions for microbial adaptation and herbicide degradation. Inactivation rate was slower at pH 7.5 and 5.3 and negative detoxication occurred at pH 4.3. Dalapon phytotoxicity in high organic soils increased as pH decreased and reached a maximum at pH 4.3; a change of one pH unit decreased the phytotoxicity [120].

Phytotoxic residues of TCA usually disappear from soil within 30-90 days. However, TCA may persist longer, especially at high rates of application [66, 121]. As we have seen, this herbicide is readily moved downward with water [67, 114]; the absence of leaching, especially in greenhouse and laboratory experiments, may account at least in part for the extended phytotoxicity.

Under the conditions imposed by Ogle and Warren [67] , the residual activity of TCA was greatest in a fine sandy soil and least in a muck. Similarly, Crafts and Drever [115] , in a greenhouse experiment, and Barrons and Hummer [122] reported more inactivation of TCA in high-organic-matter soils than in low-organic-matter soils. Rai and Hammer [121] , on the other hand, found that toxic residues had a greater persistence in the organic soil than in the sandy soil. They suggested greater retention by the clay in soil levels where seeds germinate. However, they had worked with metal cans where leaching was restricted. This may account, in part, for the results. The rate of disappearance of TCA varied with temperature [66, 67, 117] and was usually higher with higher temperatures, the optimum being at about 45°C.

General soil persistence of chlorinated aliphatic acids other than TCA and dalapon has not been thoroughly investigated. In tests conducted at two locations in Tennessee, initial and residual activities of sodium 2, 2, 3-trichloropropionate in soil were approximately equal to those of dalapon [116] .

One additional herbicide, erbon (the 2, 4, 5-trichlorophenoxyethyl ester of dalapon), deserves mention. This potent nonselective soil sterilant may persist in soils for several months to one year or more, depending on dosage and environmental conditions. The residual activity of erbon is probably due to a slow hydrolysis which yields dalapon and 2,4,5-trichlorophenoxy ethanol. The chlorophenoxy alcohols are known to be oxidized to their corresponding acids [84, 87]; thus, 2, 4, 5-trichlorophenoxy ethanol would be converted to the potent and relatively persistent herbicide 2, 4, 5-trichlorophenoxy acetate (2, 4, 5-T).

7. Pathways of Degradation in Soil

Abundant evidence exists that soil microorganisms can dehalogenate chlorinated aliphatic acid herbicides, particularly dalapon and TCA, and use the carbon as a sole source of energy [62, 63, 74, 76, 78] . At least eight species of soil bacteria, seven or eight species of fungi, and two species of actinomycetes have been shown to be effective in degrading certain chlorinated aliphatic acids [16] (see Table 5).

The initial reactions associated with fragmentation of the halogenated aliphatic acids have been ascribed to enzymic catalysis. Foy's [42, 50] early observation of the microbial decomposition of [^{36}Cl]dalapon in vitro could be explained on this basis. The results of Jensen's work [62] suggested that degradation of the chlorinated organic acids occurred by a dechlorination process involving substrate-induced enzyme. These observations were supported by the findings of Magee and Colmer [71] . Hirsch and Alexander [63] concluded that it was unlikely that the responsible enzyme was a nonspecific dehalogenase which removed any halogen from metabolizable halogenated fatty acids.

Jensen has made a very extensive study of adaptation phenomena in various organisms which can degrade the chlorinated aliphatic acids. He used the accumulation of ionic chlorine as a measure of the progress of degradation. Although Jensen has not done any experimental studies to determine the fate of the intermediates produced, he suggested [62] that the first step probably is a hydroclastic removal of the chlorine atom and the substitution of a hydroxyl group, as shown in Eq. (2):

$$\underset{\substack{| \\ Cl}}{R\text{-}COO^-} + H_2O \rightarrow \underset{\substack{| \\ OH}}{R\text{-}COO^-} + H^+ + Cl^- \tag{2}$$

The products resulting from this general scheme of metabolism would then be the corresponding hydroxy or keto acids [Eqs. (3)-(7)].

$$CH_2ClCOOH + H_2O \rightarrow CH_2OHCOOH + H^+ + Cl^- \tag{3}$$

$$CHCl_2COOH + H_2O \rightarrow CHOCOOH + 2H^+ + 2Cl^- \tag{4}$$

$$CCl_3COOH + 2H_2O \rightarrow (COOH)_2 + 3H^+ + 3Cl^- \tag{5}$$

$$CH_3CHClCOOH + H_2O \rightarrow CH_3CHOHCOOH + H^+ + Cl^- \tag{6}$$

$$CH_3CCl_2COOH + H_2O \rightarrow CH_3COCOOH + 2H^+ + 2Cl^- \tag{7}$$

The products resulting from hydrolysis are all common metabolites found in many microbial systems. The authenticity of many of these reactions in biological systems has yet to be established.

Hirsch and Alexander [63] reported a unique case of utilization of dalapon without halide liberation. A Streptomycete sp., isolated from soil, decomposed 11.6% of the dalapon without significant accumulation of ionic chloride. The most plausible explanation offered for this phenomenon was that the organically bound chlorine was never cleaved from the molecule, but remained in solution while a portion of the carbon chain served as a carbon source. An alternative suggestion was that the chlorine was released in a volatile organic form from solutions.

The metabolism of [^{14}C]dalapon has been studied in an organic soil, in mixed bacterial populations (Arthrobacter sp.) from the organic soil, and in pure cultures [124]. A comparison of [1-^{14}C]dalapon and [2-^{14}C]dalapon metabolism in the system above showed a rapid evolution of $^{14}CO_2$ from the carboxyl-labeled dalapon, whereas the labeled carbon from the 2 position of dalapon was found primarily in the lipid, nucleic acid, protein, and cold TCA-soluble fractions of the organisms. A study of the soluble labeled products extracted from microorganisms incubated with [^{14}C]dalapon showed activity in the amino acids alanine and glutamic acid [16]. These observations would be consistent with a pathway involving pyruvate as one of the early products in the metabolic degradation of dalapon. Early labeled metabolic degradation products of [2-^{14}C]dalapon in the presence of pure cultures of Arthrobacter sp. grown under aerobic conditions were pyruvate and alanine [124].

Recently Kearney et al. [93] reported the isolation and partial purification of an enzyme from an Arthrobacter sp. that removed the organically bound chlorine from dalapon. The enzyme had an optimum pH of 8.0, was rapidly inactivated at high temperatures, and was fairly specific for dalapon. The partially purified enzyme showed no metal ion requirement and was not enhanced by reducing conditions. It is difficult, however, to determine a requirement for reducing conditions since many of these reagents have copious quantities of halide ion. The product resulting from enzymic dehalogenation of dalapon was pyruvate. Several mechanisms can be proposed for such a transformation. The immediate precursor of pyruvate in this system is probably 2-chloro-2-hydroxypropionate [125]. Kearney [125] suggested that one reaction system by which the enzyme could form 2-chloro-2-hydroxypropionate from dalapon would involve a direct substitution reaction:

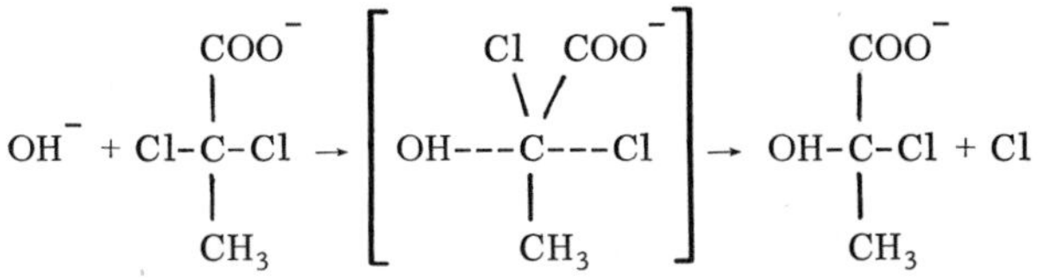

$$
OH^- +
\begin{matrix} COO^- \\ | \\ Cl\text{-}C\text{-}Cl \\ | \\ CH_3 \end{matrix}
\rightarrow
\left[
\begin{matrix} Cl\ \ COO^- \\ \backslash / \\ OH\text{---}C\text{---}Cl \\ | \\ CH_3 \end{matrix}
\right]
\rightarrow
\begin{matrix} COO^- \\ | \\ OH\text{-}C\text{-}Cl \\ | \\ CH_3 \end{matrix}
+ Cl \tag{8}
$$

In this case there would be a direct nucleophilic attack on 2-carbon, led by a hydroxyl group to form the desired product. A second reaction would involve β elimination [125], i.e., some basic group on the enzyme surface could abstract a proton from the β-carbon of dalapon to form 2-chloroacrylate or the 1-chloropropene:

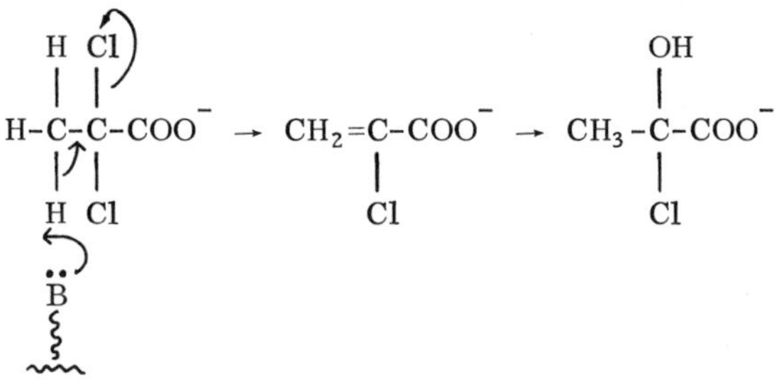

$$
\begin{matrix} H\ \ Cl \\ |\ \ | \\ H\text{-}C\text{-}C\text{-}COO^- \\ |\ \ | \\ H\ \ Cl \end{matrix}
\rightarrow
\begin{matrix} \\ CH_2{=}C\text{-}COO^- \\ | \\ Cl \end{matrix}
\rightarrow
\begin{matrix} OH \\ | \\ CH_3\text{-}C\text{-}COO^- \\ | \\ Cl \end{matrix}
\tag{9}
$$

Energetically, reaction (9) would be far simpler for the enzyme to carry out.

Smith (cited by Kearney et al. [16]) prepared this 2-chloro-2-hydroxypropionate compound and found that it rapidly went to pyruvate. This reaction is probably chemical. As seen above, two mechanisms, one involving direct

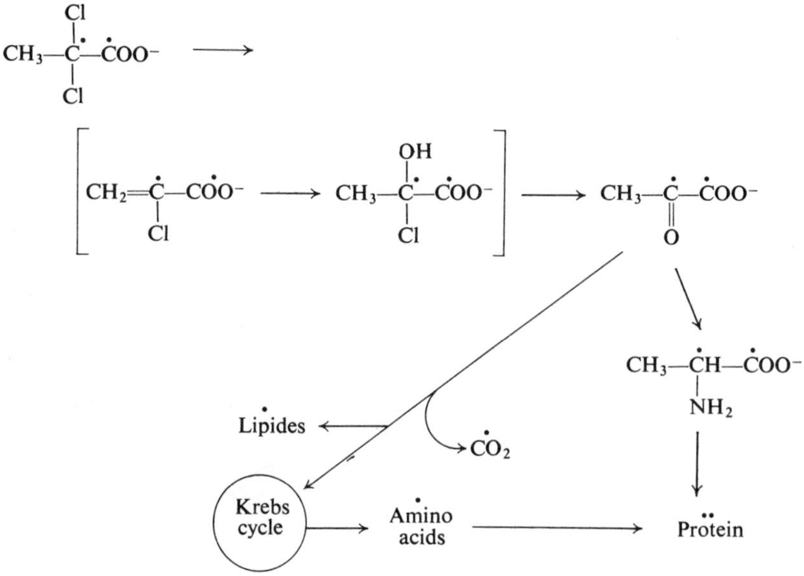

FIG. 6. Proposed pathway of metabolism of dalapon by an Arthrobacter sp. isolated from soils. The dots designate the location of ^{14}C in dalapon. The brackets enclose hypothetical intermediates involved in the enzymic conversion of dalapon to pyruvate. (Reprinted from Kearney et al. [16, p. 26] by permission of John Wiley & Sons, Inc.)

substitution and the other β elimination, have been proposed for the enzymic formation of the 2-chloro-2-hydroxypropionate. Evidence thus far collected for the metabolism of dalapon by the Arthrobacter sp., used in the studies of Beall et al. [124] and Kearney and co-workers [93, 126], suggested the sequence of reactions shown in Fig. 6.

In preliminary studies using cell-free extracts, Kearney et al. [16] found that chloride liberation from TCA was slower than the enzymic dehalo-genation of dalapon. The substrate specificity of their crude system was very broad, attacking both mono- and dichlorinated acetates and propionates. Another enzyme isolated by research workers at Cornell University, however, is apparently more specific in that it will hydrolyze 3-chloropropionate but not 2,2-dichloropropionate [127].

Thiegs [70], using both [1-^{14}C]dalapon and [2-^{14}C]dalapon, reported the recovery of labeled CO_2 in both experiments and concluded that the decomposition of dalapon in soil was complete to CO_2.

Metabolic studies with an unidentified soil microorganism incubated with [1-^{14}C]TCA and [2-^{14}C]TCA indicate rapid evolution of $^{14}CO_2$ from both forms of labeled TCA [128]. Coinciding with $^{14}CO_2$ evolution is the release of Cl into the solution. Although growth is limited on TCA alone, radioactivity from [1-^{14}C]TCA and [2-^{14}C]TCA was incorporated into all cellular components, namely, transient intermediates, lipids, nucleic acid, and proteins. One of the early products detected in TCA metabolism was the amino acid serine.

The degradation of 2,2-dichlorobutyric acid in soil reportedly follows the same pattern as dalapon and TCA, but proceeds much more slowly.

From the foregoing discussion, it is obvious that microbial degradation is an important factor affecting the persistence of the chlorinated aliphatic acids. Experimental evidence at present is not sufficient to explain precisely all the physical, chemical, and biological mechanisms involved. However, the several major factors that probably affect the biodegradability or decomposition rate of organic chemicals (including the chlorinated aliphatic herbicides) have been well summarized by Alexander [129] as follows:

1. Inaccessibility of the substrate. The compound may be deposited in a microenvironment which precludes microbial approach; it may be adsorbed to clay or other colloidal matter; or it may become entrapped or embedded within a nonmetabolizable or slowly degraded substance that prevents the organism or its enzymes from reaching the substrate.

2. Absence of some factor essential for growth. For example, no activity would occur should water, nitrogen, or a biologically utilizable terminal electron acceptor be unavailable.

3. Toxicity of the environment. This could result from biologically generated organic inhibitors, microbially formed inorganic toxins, high salt concentrations, extremes of temperature, acidity, or some other environmental condition outside of the range suitable for microbial proliferation.

4. Inactivation of the requisite enzymes. Enzymes may lose activity by adsorption to clay minerals or other colloids, as they may be inhibited by their substrates or products.

5. A structural characteristic of the molecule which prevents the enzyme from acting. An example cited by Alexander [129] was terminal quaternary groups, nonalkyl substituents, or extensive branching on an aliphatic moiety which might markedly affect the microbial decomposition of alkyl benzene derivatives, particularly where the degradation must be initiated by α or β oxidation of the alkyl portion of the molecule. Similarly, Alexander points out that substituents

that do not permit enzyme approach to the site upon which the enzyme acts could delay or prevent decomposition. Some evidence summarized above suggests that the introduction of halogens, methyl, or other substituents imparts resistance to certain herbicidal compounds.

6.   Inability of the community of microorganisms to metabolize the compound because of some physiological inadequacy. An enzyme capable of degrading the compound may simply not exist. Alternatively, the substrate may not be able to penetrate into cells which have the appropriate enzymic composition.

Two recent studies [130, 131] are pertinent to this discussion. Diaz and Alexander [130] found that sewage microorganisms readily degraded unsubstituted aliphatic acids, but the rate of decomposition was much slower with substituted acids as substrates. The type, number, and position of the substituents governed the rate of the oxidation. A single halogen, particularly if on the α-carbon, decreased the rate of biodegradation, but the dihalogenated compounds tested were especially resistant. Dimethyl-substituted aliphatic acids and alcohols were poorly utilized. Bacteria unable to grow on certain brominated fatty acids were capable of oxidizing and dehalogenating ω- but not α-bromoaliphatic acids.

In another study, Bollag and Alexander [131] reported that an enzyme preparation derived from a soil microorganism, identified as a strain of Micrococcus denitrificans, acted upon several chlorinated aliphatic acids. The enzyme system was induced by growth of the bacterium on 3-chloropropionic or 3-chlorobutyric acid. The responsible enzyme removed the chlorine present on a number of the 3- and 4-carbon alkanoic acids if the halogen was on the β but not on the α position. No evidence was obtained for 3-hydroxypropionic acid serving as an intermediate in the metabolism of β-chloropropionic acid.

D.   Volatilization of the Herbicides

The volatility of the chlorinated aliphatic acids from soil would depend on the chemical form and the presence and magnitude of physical interactions between the molecules and soil particles [16]. Since there is little interaction with soil particles, as previously noted, volatility at any given temperature is then primarily a function of the chemical form of the molecule and air movement.

Kutschinski [132] reported high dalapon losses at high temperatures when dalapon was applied to the soil as an acid. Day [133] and Richardson [110] suggest a rapid enough loss by volatilization to eliminate significant soil residues in their studies with some esters of dalapon. Foy [123] found that

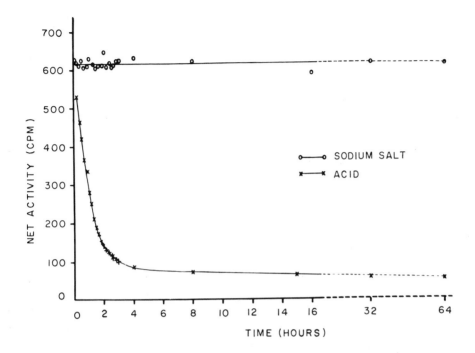

FIG. 7.   Comparison of volatilization losses of the acid and the sodium salt of [^{36}Cl]dalapon from an inert surface. Counts were taken continuously or at intervals in a laboratory hood, under a gentle fan, without change in geometry. (Reprinted from Foy [123] , with permission.)

the dalapon acid volatilized rapidly from an aluminum surface at room temperature, whereas negligible amounts of the sodium salt form disappeared in 64 hr under the same conditions (Fig. 7). Significant loss of the sodium salt of dalapon under normal field use conditions appears unlikely.

In general the same relationship between the acid and sodium salt forms would be expected to exist for the other chlorinated aliphatic acids as well.

E.   Movement and Persistence in Water

Frank et al. [134] reported that negligible concentrations of dalapon and TCA would remain in water after the water traveled a distance of 32. 2-40. 2 km, but that the low concentrations likely would not be hazardous to crops or animals. Maximum concentrations of TCA were detected within the first hour of water flow, and no TCA was detected after 48 hr at 14.5 km below the treatment site [135].

F. Dissipation or Persistence Under Field Conditions

Dalapon is used for the control of annual and perennial grasses in sugarcane, sugarbeets, corn, potatoes, asparagus, grapes, flax, new legume spring seedlings, citrus and deciduous fruit, coffee, certain stone fruits, and nut trees [19]. Scholl [136] reported that dalapon residues in seedling alfalfa and birdsfoot trefoil decreased rapidly and appeared to be insignificant eight weeks following treatment. Dalapon mixed with pyrazon retarded top and root growth of sugarbeets initially, but had no effect on gross weight yield, yield of sucrose, or sucrose content at harvest [137]. Under high nitrogen levels in the soil, dalapon applied 18-22 weeks after planting sugarbeets appeared to be helpful in increasing percent sucrose, yield, or both [138]. Dalapon applied for barley grass control in dryland lucerne showed no residual effects on barley grass seedlings, but reduced the stand of clover seedlings [137]. No residual effect occurred on cereal crops planted the year after dalapon treatments for quackgrass control were made [140]. Dalapon, TCA, and other herbicides controlled johnsongrass on drainage ditchbanks in sugarcane, but also suppressed vegetation with desirable species to the point where erosion could become a problem [141]. Hyder and Everson [142] reported that dalapon reduced undesirable perennial grass stands in abandoned croplands, but followup treatments were needed to maintain control of competing vegetation while seeded species became established. Success of seeding crested wheatgrass was slightly better after mechanical fallow than chemical fallow with dalapon in abandoned croplands, but the yield was about the same the year after planting for both methods [143].

TCA is used for control of grass seedlings and certain established perennial grasses and cattails. TCA is also useful in certain crops as spot treatment, row treatment, or as an overall treatment. It is used also on noncropland including plant sites and road shoulders for vegetation control [19]. Fenuron TCA or monuron TCA and other soil sterilant herbicides controlled herbaceous vegetation for one year or less [119].

III. MODE OF ACTION

Although the effects of higher plants on the chlorinated aliphatic acids are very slight, these compounds exhibit a wide range of effects on plant systems. The precise mechanisms of action of these compounds at the cell level in higher plants are unknown. However, the following discussion should prove useful in promoting further research in this area of plant physiology and biochemistry.

The phytotoxic effects of the chlorinated aliphatic acid herbicides as a group are very similar in some respects: All cause leaf chlorosis and abnormal growth responses typical of growth regulators [144, 145] . TCA, absorbed by grass roots, causes formative effects which indicate profound physiological disturbances. Dalapon causes similar formative effects through either foliar or root absorption. In addition to these systemic effects, the mono-, di-, and trichloroacetic acids cause a pronounced contact toxicity which is associated with an inability to be translocated from leaves. An acute toxicity of dalapon similar to that of TCA is evident at higher rates and may also be manifested under special circumstances favoring rapid penetration [18, 40] .

Trichloroacetic acid is widely used as a general protein precipitant. The acid is comparable in strength to many of the mineral acids. Redemann and Hamaker [146] have shown that dalapon is almost as effective as TCA in precipitating egg yolk and egg albumen. Whether this is important in the phytotoxic action of these two compounds is debatable, since the acid or undissociated form is reasoned to be encountered rarely under physiological conditions. It should be noted, however, that halogenated acetates and propionates are theoretically able to alkylate the sulfhydryl or amino groups in enzymes.

Dalapon characteristically exhibits two physiologically distinct types of action: acute toxicity (on occasion) and slower growth inhibition. Presumably the acute toxicity, defined as immediate and localized, is due to its action as an acid and protein precipitant [146, 147] , which causes drastic permeability changes in the plasma membranes and nonselective, localized destruction of cellular constituents. Wilkinson [148] has suggested that the toxic effects of dalapon at high rates are typical of those of strong acids, disrupting lipoidal membranes. This destruction of the plasma membrane might result equally well, however, from an attack on protein, since the ectoplast is visualized as consisting of two monolayers of lipids surrounded on either side by a monolayer of proteins [149] .

Additional studies have confirmed that, in relative terms, dalapon is readily absorbed and translocated following application to foliage. For example, couchgrass (quackgrass) plants treated with [14C]dalapon revealed that dalapon moved in the symplast and apoplast, but was less mobile in the symplast [150] . Nodes of treated shoots appeared to act as barriers to translocation. When couchgrass stubble was treated, distribution was restricted to treated shoots. Johnsongrass plants defoliated three times prior to treatment showed no increase in basipetal translocation [151] . Translocation into rhizomes increased as plants matured to the flowering stage. During the preroot stage, 14C movement was restricted to rhizome apices. Absorption of [14C]pyrazon by mustard and sugarbeet leaves was greater when dalapon plus pyrazon plus surfactant was applied to the adaxial surface of the third true leaf of seedlings [152] .

Acute toxicity produced by dalapon solutions may be brought about by a high concentration of the herbicide, of a toxic spray additive, or of hydrogen ion, or by other factors such as temperature, relative humidity, stomatal behavior, etc., which may react indirectly with these and thus favor rapid penetration.

Prasad et al. [153] reported that greater amounts of dalapon were absorbed and translocated in barley, bean, zebrina, coleus, and nasturtium under high relative humidity conditions, and greater uptake and translocation occurred in bean at higher posttreatment temperatures. High relative humidity also favored stomatal opening in zebrina, and uptake was greater through the abaxial than adaxial surface of leaf discs of zebrina, coleus, and nasturtium. A nonionic surfactant (X-77) enhanced the uptake of dalapon in bean. Dalapon activity on quackgrass was enhanced when dalapon was applied in an isoparaffinic oil carrier [154].

TCA at lower rates curtailed seedhead production in green foxtail when plants were grown under high light intensities and at moderate temperatures [155]. Rate, plant size, and availability of water for moving TCA in soil were important factors in controlling green foxtail.

Toxicity at the point of spray application can reduce or even prevent translocation and subsequent expression of systemic plant growth-regulating action [20, 41]. This factor may be much more important in herbicidal efficacy than has been commonly supposed. Whereas acute toxicity may be considered disadvantageous when translocation and systemic action are desired, it is conceivable that a slight amount of local injury to plant cells may be advantageous under certain circumstances, by weakening membranes and so increasing permeability.

The more subtle, delayed response to solutions with lower initial concentrations is manifested in the meristematic regions of a plant, following the transport of dalapon with food materials, as discussed by Foy [41]. In this case dalapon may accumulate to toxic levels in meristematic cells and act against one or more enzymes or, perhaps, against the membranes of organelles. Dalapon is known to inhibit mitotic activity in meristems [156]. To accomplish systemic toxicity, the herbicide must move across cell membranes without destruction of the cellular contents and accumulate to toxic concentrations in remote tissues.

There are as yet no absolute criteria for determining relationships between the chemical and the biological behavior of a compound such as dalapon. However, both ionic and polar properties are significant.

Ions and molecules behave differently in at least three properties, chemical reactivity, adsorption at surfaces, and penetration of membranes, and some of the selectively toxic agents are active only in the nonionized state. The ease with which some acids are absorbed by plants is inversely proportional to their degree of ionization, except at the point where pH $\leq$ pK. Dalapon showed this trend in studies with red-beet-root slices, johnsongrass foliage, corn coleoptiles, and corn foliage [20].

Nondissociated dalapon molecules, in aqueous solution at low pH, penetrated corn leaves more readily than did the anions in alkaline solutions. Systemic herbicidal response was most pronounced at pH 5-7, however, because acute toxicity at lower pH levels prevented translocation of the herbicide along with the products of photosynthesis.

Acetic and propionic acids, weaker members than their chlorinated analogs in the two series, are usually considered rather nontoxic, but in the nondissociated state either may be highly toxic to plant cells [157]. Thus, their highest activity occurs in oil, an apolar solvent; in aqueous solutions they are active only at low pH.

The distinction between polar and nonpolar (apolar) compounds is often stressed in literature on the action of herbicides. Nonpolar compounds are generally regarded as oillike, hence more lipophilic; polar compounds, as more waterlike, or hydrophilic. Daniels [158] refers to a relationship between the ionic character of a compound and its relative polarity: Polar compounds exhibit chiefly electrostatic attraction, which results in the formation of heteropolar bonds (ionic linkages). Nonpolar (homopolar or electron-pair) bonds involve an "exchange energy binding," understandable on the basis of quantum mechanics. However, the two types of bonds are not mutually exclusive, and both are operative in most linkages between atoms. Also, there are gradations between. Therefore, although dalapon may be regarded as relatively ionic, hydrophilic, and polar, it becomes permissible to speak also of its more penetrative, lipophilic tendencies when it is in the nondissociated state.

Chemical manipulation also, such as the addition of a suitable surfactant to a polar, aqueous solution of dalapon, may make it more compatible with plant waxes and possibly even with plasma membranes. Another possible factor is that surfactants (amphiphatic molecules) may accumulate or become oriented at the plasma membrane and exert a kind of narcotic action. However, considering the broad array of possible solution additives, there is at present no clear-cut relation between surface activity and dalapon toxicity. Little understood but specific cation-competition effects also apparently exist.

Robbins et al. [159] stated: "From consideration of the toxicity of hydrocarbons . . ., it is apparent that increase in polarity enhances the inherent reactivity of a molecule whereas increase in its oil-like properties promotes penetration." As the toxicity of herbicidal solutions and their penetration seem to be opposite processes, these authors concluded that "there must be an optimum point in the balance between them, and this in reality represents a compromise between toxicity resulting from polarity of the molecule and compatibility with the cuticle resulting from oil-like properties. " This line of reasoning may be helpful in interpreting the penetrating ability of systemic growth regulators such as dalapon. Surfactants are probably important sta-bilizers at the solution-plant surface interfaces, in effect achieving a more nearly optimum balance between polar and oillike properties within the same herbicide molecule.

Both passive and active (metabolic) phases are involved in the uptake of dalapon by roots and fronds of Lemna minor, a free-floating aquatic plant [51] . The metabolic rate of uptake could be partially inhibited by the addition of a structurally similar compound, pyruvic acid, or a range of respiratory inhibitors such as dinitrophenol, sodium azide, sodium arsenite, iodoacetic acid, and phenylmercuric nitrate. The uptake mechanism was most sensitive to inhibitors affecting the sulfhydryl groups, and this inhibition was partially reversed by cysteine and glutathione.

A number of growth responses have been ascribed to the chlorinated aliphatic acid herbicides. However, their exact role and biochemical level remain obscure.

Fawcett et al. [160] reported that certain concentrations of various chlorinated aliphatic acids stimulated growth in the wheat cylinder and oat coleoptile bioassays, but that the results were not typical of auxin responses. He suggested that they might be due to changes in membrane permeability. Ingle and Rogers [161] showed that although the chlorinated aliphatic acids did cause slight elongation in the wheat cylinder tests, they were inactive in the pea stem test and had little effect on oxygen uptake by mitochondria. They concluded that these compounds do not interfere with the production of meta-bolic energy but rather with its utilization.

In a later study [162] , Ingle compared the growth-regulating properties of several halogenated aliphatic acids. Wheat shoot growth was inhibited more than wheat root growth, and tomato root growth was inhibited more than tomato shoot growth by 2, 2-dichloropropionic acid, 2, 2-dichlorobutyric acid, 2, 3-dichlorobutyric acid, dichloroacetic acid, dibromoacetic acid, tribromo-acetic acid, and trifluoroacetic acid [162] . Propionic and butyric acid derivatives were more inhibitory than acetic acid derivatives with the excep-tion of tribromoacetic acid on tomato growth. Trifluoroacetic acid gave a slight stimulation of wheat root growth, but had little effect on tomato roots. It is suggested that the relative concentration of molecules and ionic species may be related to biological activity.

Other researchers [163] have shown that oat root or shoot growth and cucumber root growth were stimulated by sublethal doses of dalapon. Studies on oat seedling respiration, photosynthesis, protein content, free amino acid content, and total available soluble carbohydrates failed to provide any consistent explanation for stimulation.

Wilkinson [148] reasoned that since auxin-type growth stimulation seemed to require both a ring structure and an acidic side chain and the chlorinated aliphatic acids are definitely growth regulators but not auxin-type stimulators, the ring portion of auxinlike compounds may be required for stimulation but not for all growth-regulator responses.

The composition of plant parts can be altered by applications of these acids even though no formative effects or growth suppression are necessarily noted. Corns and Miller [164, 165] showed that both TCA and dalapon decreased the moisture content of sugarbeet seedlings and increased both sugar and dry-matter content, but had no effect on either water-soluble nitrogen or total nitrogen content.

One striking effect was the increased cold resistance of treated seedlings which can be either a direct effect or an indirect result of the changes in leaf composition. In similar tests, isopropyl m-chlorocarbanilate (chlorpropham), 2, 3, 6-trichlorobenzoic acid (2, 3, 6-TBA), and sodium chloride had no effect.

Both TCA and dalapon cause a marked reduction in the amount of surface wax produced on the leaves of various plants [166, 167], but no mechanism for this response was suggested, other than a general interference with the physiology and biochemistry of the plant. The cuticle is actually a complex mixture of products of cellular metabolism that have migrated to, and outward from, the external surfaces of epidermal cells. Disturbed metabolism in some way interferes with the normal deposition of cuticular components. This phenomenon, in turn, apparently exerts important secondary influences as well. In the case of TCA, for example, Dewey et al. [168] showed that Polygonum aviculare is much more sensitive to postemergence applications of dinitrophenol herbicides if the soil is pretreated with TCA. Further experiments of Pfeiffer et al. [169] revealed an inhibiting action of TCA on the formation of a normal cuticle. Any pretreatment that interferes with or alters the deposition of wax or other cuticular components may drastically affect herbicidal effectiveness and selectivity based on differential wetting. Also, loss or alteration of the cuticle by TCA-treated plants could conceivably lead to a higher rate of transpiration and thereby contribute to an earlier death by wilting. Interestingly, Prasad and Blackman [170] observed that when Salvinia natans is treated with dalapon, a considerable portion of the floating leaves became submerged in the culture solution. This was attributed, at least in part, to a reduced density of epidermal hairs. Judging from the work of Juniper and Bradley [167], the waxy component of the cuticle was also probably altered, affecting wettability.

Foy [41] concluded that neither penetrability, translocatability, nor metabolic activation appeared to play a major role in determining species specificities in the phytotoxic action of dalapon against cotton and sorghum. The key to its moderate selectivity still seems unquestionably to reside in the protoplasm, i.e., at the biochemical level.

Most studies on the site of action of 2,2-dichloropropionate and other chlorinated aliphatic acids have been done with microorganisms. Ram and Rustagi [171] observed recently, for example, that the inhibition of yeast growth on chlorosubstituted isobutyric acid, propionic acid, and acetic acids could be protected by an exogenous supply of β-alanine. The conclusions from such research may or may not apply equally well to common soil microorganisms and may not be important to interpreting the mechanism(s) of action in higher plants.

Anderson et al. [39], studying yellow foxtail and sugarbeets, had suggested that an indirect detoxication reaction could be in operation, accounting for selective toxicity. They observed that dalapon caused the degradation of protein to ammonium compounds, and even all the way to ammonia, in both tolerant and susceptible species. It would seem then that the abnormal protein metabolism accompanied by unusual accumulations of metabolites might be the cause of dalapon's toxicity as well as its selectivity. The capability to detoxify the breakdown products of proteins of certain plants could be the basis of their tolerance [172].

As pointed out by Anderson et al. [39], however, a change in nitrogen metabolism as a biochemical expression of herbicide injury is not unique with dalapon. For example, Rebstock et al. [173] found more protein (N × 6.25) and arginine in the shoots of wheat growing in soil treated with TCA than in the control plants. Levels of other amino acids were generally decreased by the treatment. However, the significance of these long-term changes in nitrogen metabolism may be somewhat questionable in relation to determining primary sites of action. Numerous workers have shown that under a variety of stress conditions with various organisms, free amino acids have a tendency to accumulate.

Although Hilton et al. [174, 175] have shown interference with pantothenic acid metabolism in microorganisms, the same phenomenon could not be demonstrated readily in higher plants. Prasad and Blackman [51] were able to offset the dalapon-induced growth inhibition of Lemna minor only partially with calcium pantothenate and then over a narrow range of concentrations only. β-Alanine was completely ineffective. Their results, therefore, do not support a general view that the primary action of dalapon is to interfere with the biosynthesis of coenzyme A. This is largely in agreement with the findings of Ingle and Rogers [161], who observed that inhibition of cucumber root growth by dalapon could be reversed only to a minor extent with pantothenic acid, 1-pantoic acid, and β-alanine. The latter work [161] also suggests that species factors are involved even in higher plant responses to dalapon. They found partial reversal of dalapon inhibition by these metabolites in cucumber but not in corn.

Hilton et al. [176], in their review of herbicidal action, concluded that in higher plants the protective effect of pantothenate and its precursors against the inhibition of growth by dalapon is small and the antagonisms are not readily apparent.

While there is a general agreement that TCA increases the respiration of a number of plant species [177], dalapon has no effect on the oxygen uptake of maize roots or of soybean mitochondria [178]. Foy and Penner [179], using isolated cucumber mitochondria, found O_2 uptake inhibited by TCA and dalapon, but only at high $(10^{-2}-10^{-4}$ M) concentrations. The only pronounced effect of dalapon found by Ingle and Rogers was that of a 50% reduction in the uptake of the phosphate ion $(^{32}PO_4^{3-})$ into corn seedling roots. On the basis of these results, they suggest that dalapon does not interfere with the respiration or the production of metabolic energy, but rather with the utilization of the latter.

Several recently observed biochemical effects are also persistent [180-183]. Lotlikar et al. [180] found that dalapon and 2,2,3-TPA at concentrations as high as 1×10^{-2} M did not have a large effect on oxidative phosphorylation in cabbage mitochondria. Also, Ashton et al. [181] observed that dalapon did not appreciably reduce proteolytic activity in the cotyledons of 3-day-old squash seedlings. On the other hand, Jones and Foy [182] showed that the activity of α-amylase in barley half-seeds, induced by exogenous gibberellic acid, was moderately inhibited by 5×10^{-4} M commercial formulation of dalapon. Dalapon reportedly stimulates dipeptidase activity of the cotyledons of squash [183].

An extremely interesting effect of certain chlorinated aliphatic acids is their action as somewhat selective gametocides in certain plants [13, 14]. Scott [184], working with cotton, reported that both dalapon and 2,3-dichloro-isobutyric acid induced nondehiscent anthers when applied at rates of 100-500 μg per plant. Plants thus affected were made sterile and readily produced hybrid cottonseed when cross-fertilized. No other effects of treatment were noted, either on the treated plants or in their hybrid progeny. The application of either pantothenic acid or D-ribose partially reversed the effect of the chlorinated acids.

Susceptibility of plants to dalapon may be under partial genetic control, a fact not altogether unexpected in view of its growth-regulating activity in meristematic regions. Scott reported rather widely differing results in his gametocide experiments with several varieties of cotton. Funderburk and Davis [185] reported that hybrid varieties of corn differed in their susceptibility to dalapon. Foy [186] and Behrens [187] also noted a widely different tolerance to dalapon in a number of inbred lines of corn.

Roché and Muzik [188], in studies involving 12 selections of barnyardgrass, found that susceptibility to dalapon varied with the stage of development of the plant. Furthermore, significantly different responses to dalapon occurred among the biotypes tested.

Wheat seedlings absorbed and retained more [2-^{14}C]TCA from nutrient solution than did oats [189]. KCN and 2,4-D decreased [2-^{14}C]TCA absorption by wheat more than by oats. MCPA, 2,4,5-T, and silvex also exhibited a strong inhibition of the accumulation of TCA in wheat [190]. TCA caused an inhibition of 2,4-D accumulation in oats, but not in wheat.

Obviously a number of plant processes are affected by the chlorinated aliphatic acid herbicides, and it is likely that more than a single pathway is inhibited. The evidence points toward multiple pathways and to more than one site of action.

Disturbance of energy metabolism, particularly the utilization of metabolic energy, could logically lead to the observed growth-regulating effects of the chlorinated aliphatic acid. All anatomical and morphological changes in plants are preceded by biochemical changes. As inferred earlier, no clear-cut relationship exists between formative effects and lethality. Attempts to explain mechanisms of action of a substance which produces formative effects must take cognizance of the influence of such compounds upon the normal processes of growth and differentiation, which are themselves incompletely understood. The principal factor responsible for overall inhibition of growth at the cellular level is apparently a reduction in meristematic activity. Most observed effects reported in the literature, particularly long-term effects, are probably secondary and do not operate directly on cell division [51].

Enzymatic studies using relatively simpler biological systems than those of higher plants can conceivably be useful in elucidating the biochemical mechanisms of action and selectivity of the chlorinated aliphatic acids. As Foy pointed out [42], the specific point of attack of a toxicant might be on the production of an essential substrate, on enzymes that bring about the release of energy from substrate, or on enzymes required in the essential utilization of this energy. Several sites have been suggested. The final toxic result, however, may actually be produced by a complex series of sequential and consequential reactions. This may account for the fact that the mechanism(s) of action is known for so few growth regulators, if indeed any are known with certainty.

Although far from satisfactory as a general explanation of the phytotoxicity of the chlorinated aliphatic acids in higher plants, one of the most plausible areas of biochemical sensitivity in microorganisms seems to revolve around pyruvate metabolism, which occupies a key position in relation to other metabolic crossroads in higher organisms.

Redemann and Meikle [191] studied dalapon as a competitive inhibitor of an enzyme system involving pyruvate as a substrate. The three enzymes selected for their study were pyruvate oxidase from Streptococcus faecalis, pyruvate oxidase from Proteus vulgaris, and carboxylase from yeast. Interpretation of double reciprocal plots of 1/S versus 1/V led to the conclusion that inhibition of pyruvate oxidase from S. faecalis most closely resembled noncompetitive inhibition, while yeast carboxylase and pyruvate oxidase from P. vulgaris both appeared to be competitively as well as noncompetitively inhibited. The noncompetitive inhibition occurred only at dalapon concentrations greater than 7×10^{-4} M with pyruvate oxidase from precipitation of the enzyme substrate complex. Based on calculations using expected pyruvate concentrations in plant tissues, these workers concluded that if such an inhibition is not responsible for the herbicidal action of dalapon, it certainly could contribute to the stunting effect following translocation. It should be pointed out, however, that the bacteria selected are known to contain highly active pyruvate-oxidase systems and that depression in oxygen consumption was employed as a measure of interference by dalapon in the presence and absence of pyruvic acid. On the other hand, in studies of higher plants, only at very high concentration is the rate of oxygen consumption depressed [51, 161, 178, 179].

Perhaps an equally (or more) important primary site of action, still involving pyruvate indirectly, is the competitive inhibition of pantothenic acid synthesis [171]. Hilton et al. [174, 175, 192] showed that the growth of yeast (Saccharomyces cerevisiae) was inhibited by chlorosubstituted isobutyric, propionic, and acetic acids, but this inhibition was partially reversed by exogenous additions of calcium pantothenate. Hilton et al. [175] reasoned that because of the structural similarity of the chlorinated aliphatic acids to propionic and acetic acids, it seemed probable that their phytotoxicity might result, in part, from interference with pantothenic acid synthesis. Such an effect had been demonstrated with the two unchlorinated acids by King and Cheldelin [193] where they used yeast as a test organism. Their results led to the conclusion that "propionate inhibits growth by competing with β-alanine for attachment within the yeast cell, thereby preventing the coupling of the pantothenic acid moieties."

First indications were that dalapon competed with β-alanine, but later results with pure enzyme preparations suggested that competition is with pantoic acid rather than β-alanine. In using a fresh sample of dalapon that was 99% pure in comparison with propionic acid, Hilton et al. [175] showed that although β-alanine offered some protection against this herbicide, the effect was not so pronounced as that observed with the unchlorinated parent acid. They suggested that chlorination of propionic acid obviously decreased the ability of the molecule to compete with β-alanine. Is chlorine substituted off slowly before the toxic system manifests itself? This agrees with the slowness to react, but there it no absolute evidence for this theory. The remote possibility that initial dehalogenation of dalapon in tissue could, instead of constituting a detoxication mechanism, actually tend to decrease its potentialities as an inhibitor of pantothenate synthesis has been discussed by Foy [83]. This hypothesis is of improbable herbicidal significance, however, in view of the resistance of dalapon to metabolic decomposition in both tolerant and susceptible species.

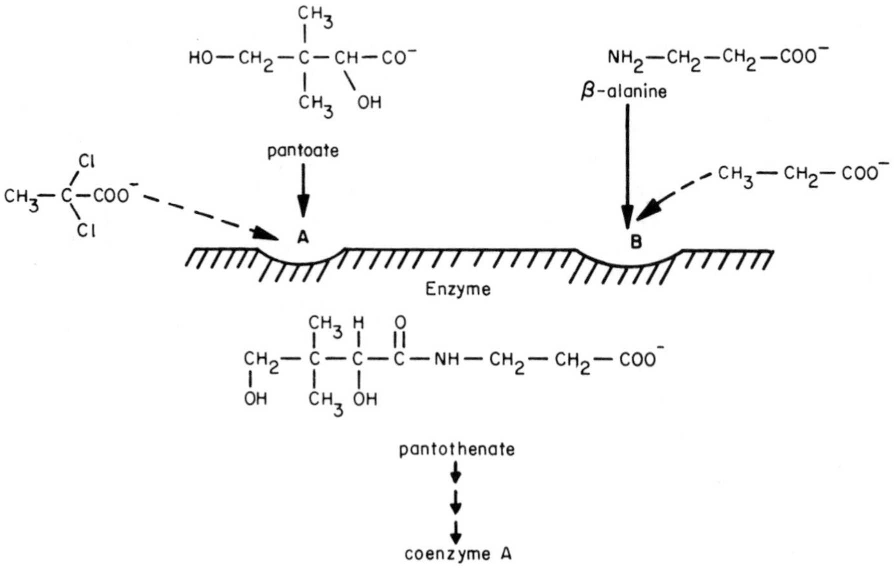

FIG. 8.   Schematic diagram of pantothenate enzyme surface, showing site A where dalapon competes with pantoate and site B where propionate competes with β-alanine. (Reprinted from Kearney et al. [16, p. 24] by permission of John Wiley & Sons, Inc.)

Van Oorschot and Hilton [194] used the pantothenate-synthesizing enzyme and growing cultures of Escherichia coli to study antimetabolite relations involved in inhibitions by salts of aliphatic acids and their chlorine-substituted derivatives. Protection of the enzyme against these inhibitors followed two different patterns. β-Alanine protected the enzyme against inhibition by unchlorinated aliphatic acids, monochloroacetic acid, and β-chloropropionic acid. Pantoate protected the enzyme against inhibition by α-chloropropionic acid and di- or trichlorosubstituted acids of the acetic and propionic series. Chlorosubstituted compounds in the first group were less inhibitory to the enzyme than the unchlorinated compounds. However, toxicity to the enzyme increased with additional chlorine substitutions on chlorosubstituted compounds antagonized by pantoate.

The kinetic data thus suggested that the enzyme has two sites on which pantoate and β-alanine react to yield pantothenate [16]. The two sites can be called A and B, as illustrated in Fig. 8. Dalapon and TCA compete with pantoate for site A on the enzyme surface. Conversely, acetate and propionate are competitive with β-alanine for site B and uncompetitive with pantoate. The uncompetitive relationship implies that propionate inhibits the enzyme-

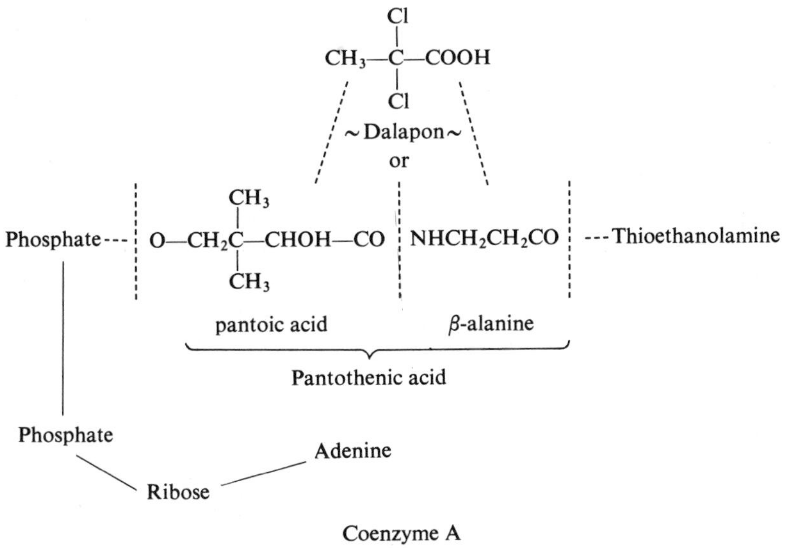

FIG. 9. Structural relationships of coenzyme A, depicting how dalapon may disturb normal metabolic functions. Competitive inhibition of pantothenic acid synthesis is illustrated by the replacement of either moiety of pantothenate by dalapon. (Reprinted from Foy [83], with permission.)

pantoate complex [14] . Whatever the mechanism, as discussed by Foy [42] and Leasure [8] , if dalapon disrupts pyruvate metabolism or competes with either β-alanine or pantoic acid, it is reasonable to assume that the synthesis of pantothenic acid would be disrupted and the supply of functional coenzyme A would be impaired. Figure 9 shows the structural relationships of coenzyme A, depicting how dalapon might disturb normal metabolic functions. Competitive inhibition of pantothenic acid synthesis is illustrated by the replacement of either moiety of pantothenate by dalapon.

Most of the pantothenic acid in animals and microorganisms is reportedly present as coenzyme A, although it is also found in nature in other combined forms such as pantetheine and pantethine. Less is known of its activity and occurrence in higher plants; however, several edible reproductive structures are known sources of the vitamin. It is probably synthesized in the leaves and transported into these storage organs during periods of rapid development. The specific effects of pantothenic acid or coenzyme A upon cell differentiation are little known.

The improper functioning of coenzyme A in plants could lead to drastic changes in plant growth and development. For example, some of the important processes believed to be mediated by coenzyme A are pyruvate oxidation, citrate synthesis, α-ketoglutarate oxidation in the citric acid cycle, fatty acid and steroid synthesis and breakdown, and perhaps auxin action [auxins may act through an ester with coenzyme A (auxinyl coenzyme A)]. Hence, coenzyme A is a key compound in plant metabolism and growth through its control of the energy transfers in carbohydrate, nitrogen, and fat metabolism. Interference of dalapon with the citric acid cycle in any way, e.g., by competing with pyruvate for enzyme attachment of with β-alanine for attachment to another moiety of pantothenic acid, could indirectly cause disturbances in ancillary processes, such as nitrogen metabolism. (The dark green appearance of dalapon-treated plants, delayed maturation, and prolongation of vegetative growth are characteristic of plants having high levels of available nitrogen.)

It is entirely probable that dalapon exhibits more than one primary site of action. Note that β-alanine yields pyruvic acid on deamination. It is possible that dalapon and related compounds (perhaps increasingly with decreased chlorination) are able to compete with several metabolites that are structurally similar, e.g., β-alanine, pantoic acid, pyruvic acid, variably among plant species.

The low response in roots cannot always be attributed to a lack of accumulation of dalapon, as Foy has shown [42]. It seems unlikely that the principal action is in competition with pyruvate, because this substance certainly occurs, albeit fleetingly, in all regions of high respiratory activity, which would include root tips as well as shoot apical meristems. If the principal mechanism is the interference with pantothenic acid synthesis, one possible explanation is that synthesis occurs in the shoots, requiring the products of photosynthesis, and if the process should be altered, the meristematic areas of the shoots (by virtue of their closer proximity) would show the deficiency most readily. Abnormalities can occur in roots under certain conditions as shown by the work of Grigsby et al. [195] and Prasad and Blackman [156]. This suggests the indirect involvement of light, which is consistent with other observations. Also, it seems safe to assume that light (and elevated temperatures) may indirectly exert an effect by causing an increase in accumulation of dalapon in the tops from soil or nutrient solution through an increase in transpiration.

Differences in susceptibility to dalapon among species and among tissues of a given plant are seemingly dependent on the presence or absence of key enzymes or enzyme precursors. Further experimentation into the mechanisms of herbicidal selectivity of the chlorinated aliphatic acid herbicides should emphasize biochemical distinctions between susceptible and resistant species.

ACKNOWLEDGMENTS

The assistance of Peter Gous, Donald Jones, and Harold Witt in conducting portions of the literature search is gratefully acknowledged.

REFERENCES

1. J. Mukula and E. Ruuttunen, Ann. Agr. Fenn., 8, Suppl. 1, 10 (1969).

2. E. W. Bousquet, U.S. Pat. 2,393,086 (1944).

3. A. I. Virtanen, Finn. Pat. 22,562; Swed. Pat. 122,159 (1948).

4. K. C. Barrons, U.S. Pat. 2,642,354 (1951).

5. K. C. Barrons, U.S. Pat. 2,807,530 (1957).

6. B. V. Toornman, U.S. Pat. 2,880,082 (1959).

7. A. G. Norman, C. E. Minarik, and R. L. Weintraub, Ann. Rev. Plant Physiol., 1, 141 (1950).

8. J. K. Leasure, J. Agr. Food Chem., 12, 40 (1964).

9. P. W. Zimmerman and A. E. Hitchcock, Contrib. Boyce Thompson Inst., 16, 209 (1951).

10. G. L. McCall and J. W. Zahnley, Kansas State Coll. Agr. Exp. Sta. Circ. No. 255, 1949.

11. Dow Chemical Company, Dalapon Bull. No. 2, 1953.

12. Environmental Protection Agency, Summary of Registered Agricultural Pesticide Chemical Uses, 1973, pp. I-D-1.1, I-S-8.

13. Rohm and Haas Company, Progress Report on FW-450 Chemical Gametocide, 1959.

14. For several pertinent references, see reports of the Cotton Gametocide Symposium, Proc. Cotton Improvement Conf., National Cotton Council, Memphis, Tenn., 1958, pp. 57-101.

15. L. F. Fieser and M. Fieser, Advanced Organic Chemistry, Reinhold, New York, 1961, p. 360.

16. P. C. Kearney, C. L. Harris, D. D. Kaufman, and T. J. Sheets, Adv. Pest Control Res., 6, 1 (1965).

17. V. H. Freed, Mimeograph for Project, 41-47, Oregon State University, Corvallis, 1956.

18. V. H. Freed and M. Montgomery, Res. Prog. Rep. West. Weed Control Conf., 1956, 98.

19. Weed Science Society of America, Herbicide Handbook, 2nd ed., 1970, pp. 180, 319.

20. C. L. Foy, Hilgardia, 35, 125 (1963).

21. C. L. Foy, Weeds, 10, 35 (1962).

22. C. L. Foy, Weeds, 10, 97 (1962).

23. C. L. Foy and L. W. Smith, Weeds, 13, 15 (1965).

24. L. W. Smith, C. L. Foy, and D. E. Bayer, Weed Res., 6, 233 (1966).

25. L. W. Smith, C. L. Foy, and D. E. Bayer, Weeds, 15, 87 (1967).

26. L. L. Jansen, Weeds, 13, 117 (1965).

27. L. L. Jansen, J. Agr. Food Chem., 12, 223 (1964).

28. C. G. McWhorter, Weeds, 11, 83 (1963).

29. V. H. Freed, K. McKennon, and M. Montgomery, Res. Prog. Rep. West. Weed Control Conf., 1955, 81.

30. V. H. Freed and M. Montgomery, Res. Prog. Rep. West. Weed Control Conf., 1956, 96.

31. J. B. Neilands and P. K. Stumpf, Outlines of Enzyme Chemistry, 2nd ed., Wiley, New York, 1958, p. 118.

32. American Cyanamid Company, Tech. Data Experimental Herbicide 6249, 1955.

33. B. A. Kratky and G. F. Warren, Weed Res., 11, 257 (1971).

34. W. S. Hardcastle and R. E. Wilkinson, Weed Sci., 18, 336 (1970).

35. R. C. Hiltibran, Trans. Ann. Fish. Soc., 96, 414 (1967).

36. Water Quality Criteria, Report of the National Tech. Adm. Comm. to Sec. of the Interior, Fed. Water Pollution Centr. Adm. USDI, 1968, 234 pp.

37. H. O. Sander and O. B. Cope, Limnol. Oceanogr., 13, 112 (1968).

38. R. G. Heath et al., U. S. Bur. Sport. Fish. Wildlife, Patuxent Wildlife Res. Center, unpublished work, 1970.

39. R. N. Anderson, R. Behrens, and A. J. Linck, Weeds, 10, 4 (1962).

40. F. A. Blanchard, W. W. Muelder, and G. N. Smith, J. Agr. Food Chem., 8, 124 (1960).

41. C. L. Foy, Plant Physiol., 36, 688 (1961).

42. C. L. Foy, Plant Physiol., 36, 698 (1961).

43. G. N. Smith and D. L. Dyer, J. Agr. Food Chem., 9, 155 (1961).

44. F. A. Blanchard, Weeds, 3, 274 (1954).

45. F. Mayer, Biochem. Z., 328, 433 (1957).

46. K. C. Barrons and R. W. Hummer, Agr. Chem., 6, 48 (1951).

47. T. W. Tibbetts and L. G. Holm, Weeds, 6, 146 (1954).

48. R. N. Anderson, A. J. Linck, and R. Behrens, Weeds, 10, 1 (1962).

49. A. S. Crafts and C. L. Foy, Down to Earth, 14, 1 (1959).

50. C. L. Foy, Ninth Int. Bot. Congr. Proc. Abstr., 2, 121 (1959).

51. R. Prasad and G. E. Blackman, J. Exp. Bot., 16, 545 (1965).

52. C. L. Foy and J. H. Miller, Weeds, 11, 31 (1963).

53. J. Giovanelli and P. K. Stumpf, J. Am. Chem. Soc., 79, 2652 (1957).

54. M. M. Schreiber, J. Agr. Food Chem., 7, 427 (1959).

55. G. I. McIntyre, Weed Res., 2, 165 (1962).

56. A. H. Kutchinski, J. Agr. Food Chem., 9, 365 (1961).

57. K. Holly, in Plant Physiology and Biochemistry of Herbicides (L. J. Audus, ed.), Academic Press, New York, 1964, p. 453.

58. W. R. Fennerty, E. Hawtrey, and R. E. Kallis, Z. Allg. Mikrobiol., 2, 169 (1962).

59. E. N. Bokova, Mikrobiologiya, 23, 15 (1954).

60. J. N. Ladd, Aust. J. Biol. Sci., 9, 92 (1956).

61. C. E. Zobell, Adv. Enzymol., 10, 433 (1950).

62. H. L. Jensen, Can. J. Microbiol., 3, 151 (1957).

63. P. Hirsch and M. Alexander, Can. J. Microbiol., 6, 241 (1960).

64. D. D. Kaufman, unpublished data, USDA, ARS, CRD, Beltsville, Md., 1963.

65. D. D. Kaufman, Am. Soc. Agron. Abstr., Soil Sci. Div., p. 85 (1965).

66. A. J. Loustalot and R. Ferrer, Agron. J., 42, 323 (1950).

67. R. E. Ogle and G. F. Warren, Weeds, 3, 257 (1954).

68. B. J. Thiegs, Down to Earth, 11, 2 (1955).

69. J. T. Holstun and W. E. Loomis, Weeds, 4, 205 (1956).

70. B. J. Thiegs, Down to Earth, 18, 7 (1962).

71. L. A. Magee and A. R. Colmer, Can. J. Microbiol., 5, 255 (1959).

72. H. L. Jensen, Nature, 180, 1416 (1957).

73. H. L. Jensen, Soils Fert., 23, 60 (1960).

74. H. L. Jensen, Acta Agr. Scand., 10, 83 (1960).

75. D. D. Kaufman, Can. J. Microbiol., 10, 843 (1964).

76. H. L. Jensen, Tidsskr. Planteavl, 63, 470 (1959).

77. P. Hirsch and R. Stellmach-Hellwig, Zentr. Bakteriol. Parasitenk., 2, 683 (1961).

78. H. L. Jensen, Acta Agr. Scand., 13, 404 (1963).

79. H. Lees and J. H. Quastel, Biochem. J., 40, 803 (1966).

80. A. S. Newman and C. R. Downing, J. Agr. Food Chem., 6, 352 (1958).

81. D. D. Kaufman, Weeds, 14, 130 (1966).

82. C. G. Gemmell and H. L. Jensen, Arch. Mikrobiol., 48, 386 (1964).

83. C. L. Foy, Ph.D. Thesis, University of California, Davis, 1958.

84. R. B. Carroll, Contrib. Boyce Thompson Inst., 16, 409 (1952).

85. K. C. Barrons, Down to Earth, 6, 8 (1951).

86. A. N. McGregor, J. Gen. Microbiol., 30, 497 (1963).

87. R. B. Carroll, Proc. South. Weed Conf., 4, 13 (1951).

88. P. Hirsch, Weed Soc. Am. Abstr., 11, 547 (1962).

89. H. L. Jensen, Acta Agr. Scand., 9, 421 (1959).

90. L. J. Audus, in Herbicides and the Soil (E. K. Woodford and G. R. Sagar, eds.), Blackwell, Oxford, 1960, p. 1.

91. M. Cohn and J. Monod, Symp. Soc. Gen. Microbiol., 3, 132 (1953).

92. R. L. Walker and A. S. Newman, Appl. Microbiol., 4, 201 (1956).

93. P. C. Kearney, D. D. Kaufman, and M. L. Beall, Biochem. Biophys. Res. Commun., 14, 29 (1964).

94. W. W. Fletcher, in Herbicides and the Soil (E. K. Woodford and G. R. Sagar, eds.), Blackwell, Oxford, 1960, p. 20.

95. R. T. Otten, J. E. Dawson, and M. M. Schreiber, Proc. Northeast Weed Control Conf., 11, 120 (1957).

96. J. D. Douros, Diss. Abstr., 19, 19 (1958).

97. A. D. Worsham and J. Giddens, Weeds, 5, 316 (1957).

98. J. V. Mayeux and A. R. Colmer, Appl. Microbiol., 10, 206 (1962).

99. A. R. Colmer, Bacteriol. Proc., 53, 16 (1953).

100. A. R. Colmer, Appl. Microbiol., 1, 184 (1953).

101. A. R. Colmer, Proc. South. Weed Conf., 7, 237 (1954).

102. L. A. Magee and A. R. Colmer, Appl. Microbiol., 3, 288 (1955).

103. D. E. Kratochvil, Weeds, 1, 25 (1951).

104. M. E. Hoover and A. R. Colmer, Proc. Natl. Acad. Sci. (U. S.), 16, 21 (1953).

105. U. B. Wehr and D. A. Klein, Soil Biol. Biochem., 3, 143 (1971).

106. L. A. Magee, Diss. Abstr., 19, 413 (1958).

107. G. H. Elkan, K. W. King, and W. E. C. Moore, Bacteriol. Proc., 58, 7 (1958).

108. G. H. Elkan and W. E. C. Moore, Can. J. Microbiol., 6, 339 (1960).

109. M. G. Hale, F. H. Hulcher, and W. E. Chappell, Weeds, 5, 331 (1957).

110. L. T. Richardson, Can. J. Plant Sci., 37, 196 (1957).

111. L. T. Richardson, Can. J. Plant Sci., 39, 30 (1959).

112. A. C. Leopold, P. van Schaik, and M. Neal, Weeds, 8, 48 (1960).

113. G. F. Warren, Proc. North Central Weed Control Conf., 13, 5 (1956).

114. G. F. Warren, Proc. North Central Weed Control Conf., 11, 5 (1954).

115. A. S. Crafts and H. Drever, Weeds, 8, 12 (1960).

116. R. F. Richards, Proc. South. Weed Conf., 9, 154 (1956).

117. H. Beinhauer, Int. Congr. Plant Protection Proc., 1, 527 (1957).

118. B. E. Day, L. S. Jordan, and R. C. Russell, Soil Sci., 95, 326 (1963).

119. F. T. Corbin and R. P. Upchurch, Weed Sci., 15, 370 (1967).

120. F. T. Corbin, R. P. Upchurch, and F. L. Selman, Weed Sci., 19, 233 (1971).

121. G. S. Rai and C. L. Hammer, Weeds, 2, 271 (1953).

122. K. C. Barrons and R. W. Hummer, Proc. South. Weed Conf., 4, 3 (1951).

123. C. L. Foy, Hilgardia, 30, 153 (1960).

124. M. L. Beall, P. C. Kearney, and D. D. Kaufman, Weed Soc. Am. Abstr., 1964, 11.

125. P. C. Kearney, Adv. Chem. Ser., 60, 257 (1966).

450 C. L. FOY

126. P. C. Kearney, D. D. Kaufman, and M. L. Beall, Weed Soc. Am. Abstr., 1964, 13.

127. P. C. Kearney, private communication, 1967.

128. P. C. Kearney and D. D. Kaufman, American Chemical Society, 150th Meeting, 1965, p. 16A.

129. M. Alexander, Adv. Appl. Microbiol., 7, 35 (1965).

130. F. F. Diaz and M. Alexander, Appl. Microbiol., 22, 1114 (1971).

131. J. M. Bollag and M. Alexander, Soil Biol. Biochem., 3, 91 (1971).

132. A. H. Kutschinski, Down to Earth, 10, 14 (1954).

133. B. E. Day, Weed Res., 1, 177 (1961).

134. P. A. Frank, R. J. Demint, and R. D. Comes, Weed Sci., 18, 687 (1970).

135. R. D. Comes, P. A. Frank, and R. J. Demint, Weed Sci. Soc. Am. Abstr., 1972, 36.

136. J. M. Scholl, Down to Earth, 25 (3), 15 (1969).

137. E. E. Schweizer, Weed Sci., 18, 386 (1970).

138. F. G. Andriessen, Down to Earth, 23 (3), 19 (1967).

139. O. R. Southwood, Weed Res., 11, 231 (1971).

140. A. C. Carder, Weed Sci., 15, 201 (1967).

141. R. W. Millhollon, Weed Sci., 17, 370 (1969).

142. D. N. Hyder and A. C. Everson, Weed Sci., 16, 531 (1968).

143. A. C. Everson, D. N. Hyder, H. R. Gardener, and R. E. Bement, Weed Sci., 17, 548 (1969).

144. E. K. Woodford, K. Holly, and C. C. McCready, Ann. Rev. Plant Physiol., 9, 311 (1958).

145. J. van Overbeek, in The Physiology and Biochemistry of Herbicides (L. J. Audus, ed.), Academic Press, New York, 1964, p. 392.

146. C. T. Redemann and J. Hamaker, Weeds, 3, 387 (1954).

147. E. A. Olsson, Jr., M.S. Thesis, Colorado State University, Fort Collins, 1957.

148. R. E. Wilkinson, Ph.D. Thesis, University of California, Davis, 1956.

149. H. Davson and J. F. Danielli, The Permeability of Natural Membranes, Macmillan, New York, 1943, p. 361.

150. K. Lund-Hoie and A. Bylterud, Weed Res., 9, 205 (1969).

151. R. J. Hull, Weed Sci., 17, 314 (1969).

152. W. S. Belles, Weed Sci. Soc. Am. Abstr., 1971, 56.

153. R. Prasad, C. L. Foy, and A. S. Crafts, Weed Sci., 15, 149 (1967).

154. R. J. Burr and G. F. Warren, Weed Sci., 19, 701 (1971).

155. P. N. P. Chow, Weed Sci., 20, 172 (1972).

156. R. Prasad and G. E. Blackman, J. Exp. Bot., 15, 48 (1964).

157. J. van Overbeek and R. Blondeau, Weeds, 3, 55 (1954).

158. F. Daniels, Outlines of Physical Chemistry, Wiley, New York, 1953.

159. W. W. Robbins, A. S. Crafts, and R. N. Raynor, Weed Control, 2nd ed., McGraw-Hill, New York, 1952.

160. C. H. Fawcett, R. L. Wain, and F. Wightman, Nature, 178, 972 (1958).

161. M. Ingle and B. J. Rogers, Weeds, 9, 264 (1961).

162. L. M. Ingle, Proc. West Va. Acad. Sci., 40, 1 (1968).

163. S. J. Wiedman and A. P. Appleby, Weed Res., 12, 65 (1972).

164. W. G. Corns, Can. J. Bot., 34, 154 (1956).

165. S. R. Miller and W. G. Corns, Can. J. Microbiol., 35, 5 (1957).

166. B. E. Juniper, New Phytol., 58, 1 (1959).

167. B. E. Juniper and D. R. Bradley, Ultrastruct. Res., 2, 16 (1958).

168. O. R. Dewey, P. Gregory, and R. K. Pfeiffer, Proc. Brit. Weed Control Conf., 3rd Blackpool, 1, 313 (1956).

169. R. K. Pfeiffer, O. R. Dewey, and R. T. Brunskill, Proc. 4th Int. Congr. Plant Protection, Hamburg, 1957, 523.

170. R. Prasad and G. E. Blackman, J. Exp. Bot., 16, 86 (1965).

171. H. Y. Ram and P. N. Rustagi, Sci. Cult. (Calcutta), 32, 286 (1966).

172. T. R. Richmond, Crop Sci., 2, 58 (1960).

173. T. L. Rebstock, C. L. Hammer, R. W. Luecke, and H. M. Sell, Plant Physiol., 28, 437 (1953).

174. J. L. Hilton, J. S. Ard, L. L. Ard, L. L. Jansen, and W. A. Gentner, Weeds, 7, 381 (1959).

175. J. L. Hilton, L. L. Jansen, and W. A. Gentner, Plant Physiol., 33, 43 (1958).

176. J. L. Hilton, L. L. Jansen, and H. Hull, Ann. Rev. Plant Physiol., 14, 353 (1963).

177. G. S. Rai and C. L. Hammer, Quart. Bull. Mich. Agr. Exp. Sta., 38, 555 (1956).

178. C. M. Switzer, Plant Physiol., 32, 42 (1954).

179. C. L. Foy and D. Penner, Weeds, 13, 226 (1965).

180. P. D. Lotlikar, L. F. Remmert, and V. H. Freed, Weed Sci., 16, 161 (1968).

181. F. M. Ashton, D. Penner, and S. Hoffman, Weed Sci., 16, 169 (1968).

182. D. W. Jones and C. L. Foy, Weed Sci., 19, 595 (1971).

183. R. Tsay and F. M. Ashton, Weed Sci., 19, 682 (1971).

184. R. A. Scott, Plant Physiol., 36, 529 (1961).

185. H. H. Funderburk and D. E. Davis, Weeds, 8, 6 (1960).

186. C. L. Foy, Res. Prog. Rep. West. Weed Control Conf., 1964, 108.

187. R. Behrens, in Summary of 1962 Weed Control Trials in Field Crops, Minn. Agr. Exp. Sta., 1962.

188. B. F. Roché, Jr., and T. J. Muzik, Agron. J., 56, 155 (1964).

189. P. N. P. Chow, Weed Sci., 18, 492 (1970).

190. P. N. P. Chow, Weed Sci. Soc. Am. Abstr., 1972, 90.

191. C. T. Redemann and R. W. Meikle, Arch. Biochem. Biophys., 59, 106 (1955).

192. J. L. Hilton, Science, 128, 1509 (1959).

193. T. E. King and V. H. Cheldelin, J. Biol. Chem., 174, 273 (1948).

194. J. L. P. van Oorschot and J. L. Hilton, Arch. Biochem. Biophys., 100, 289 (1963).

195. B. H. Grigsby, T. M. Tsou, and G. B. Wilson, Proc. South. Weed Conf., 8, 279 (1955).

196. F. R. Kemp, L. P. Stoltz, J. W. Herron, and W. T. Smith, Jr., Weed Sci., 17, 444 (1969).

197. C. G. McWhorter, Weed Sci. Soc. Am. Abstr., 1971, 84.

198. G. D. Wills, Weed Sci. Soc. Am. Abstr., 1971, 84.

Chapter 9

DINITROANILINES

GERALD W. PROBST, TOMASZ GOLAB,
and WILLIAM L. WRIGHT

Lilly Research Laboratories
Division of Eli Lilly and Company
Indianapolis, Indiana

I. INTRODUCTION

An important series of selective herbicides was introduced in agriculture in the 1960s. Trifluralin (α, α, α-trifluoro-2,6-dinitro-N,N-dipropyl-p-toluidine), the most prominent member of the series, was first registered for use on a food crop in October 1964. Nitralin (4-methylsulfonyl-2,6-dinitro-N,N-dipropylaniline) and benefin (N-butyl-N-ethyl-α, α, α-trifluoro-2,6-dinitro-p-toluidine) have also received widespread usage for the selective preemergence control of a wide variety of grasses and broadleaf weeds. The 1970s have been an era for the commercial introduction of other dinitroanilines that have the same general herbicidal properties. Isopropalin (2,6-dinitro-N,N-dipropylcumidine) was registered by the Environmental Protection Agency for use on direct-seeded and transplant tomatoes in 1971 and on tobacco in 1972. Additional candidates include dinitramine (N^4, N^4-diethyl-α, α, α-trifluoro-3,5-dinitrotoluene-2,4-diamine), oryzalin (3,5-dinitro-N^4, N^4-dipropylsulfanilamide), fluchloralin [N-(2-chloroethyl)-2,6-dinitro-N-propyl-4-trifluoromethylaniline] , profluralin (N-cyclopropyl-methyl-α, α, α-trifluoro-2,6-dinitro-N-propyl-p-toluidine), butralin [N-(sec-butyl)-4-(t-butyl)-2,6-dinitroaniline] , and penoxalin [N-(1-ethylpropyl)-2,6-dinitro-3,4-xylidine] . The acceptance of the dinitroanilines as effective, dependable herbicides and their extensive use in many important crops have stimulated interest in studying their degradation in plants and soil.

A. History

Dinitroanilines have been recognized as dye intermediates for several decades. The fungicidal activity of substituted dinitroanilines has been described [1–3] . Phytotoxic studies on bean plants were initially reported for 2,4-dinitroaniline in 1955 [4] . Of prime interest was the report by Alder et al. [5] and Soper et al. [6] that 2,6-dinitroaniline possessed a marked general herbicidal activity as compared to the 2,4- or 2,3-dinitroanilines. The 2,6-dinitroaniline exhibited both contact and preemergence herbicidal activity toward multiple plant species. Selective herbicidal activity was obtained by substitutions on the amino group of the 2,6-dinitroaniline molecule. Selectivity was directed toward grasses rather than broadleaf weeds and preemergence action rather than foliar contact.

Substitution at the 3 and/or 4 position of the ring modifies the degree but not the type of herbicidal activity. For example, herbicidal potency is modified by 4-position substitution in the order $CF_3 > CH_3 > Cl > H$. The compounds substituted at both the 3 and 4 positions, such as 3-chloro-4-methyl-, also possess herbicidal activity within an acceptable range. A very wide variation in molecular structure of 2,6-dinitroanilines is possible; and, as might be expected, other analogs possess herbicidal properties to various degrees (Fig. 1).

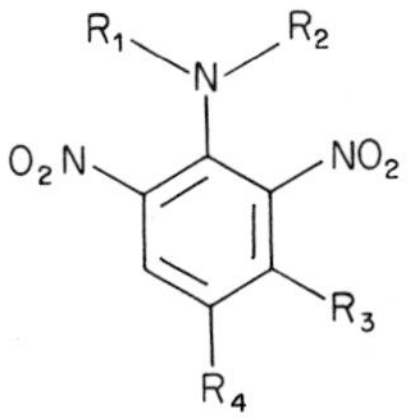

R_1 = alkyl, haloalkyl, cycloalkyl, etc.

R_2 = alkyl, haloalkyl, cycloalkyl, H, etc.

R_3 = NH_2, Cl, H, CH_3

R_4 = CF_3, CH_3, Cl, SO_2CH_3, SO_2NH_2, C_2-C_4 alkyl, etc.

FIG. 1. Substitution patterns on 2,6-dinitroanilines in use or under consideration for development.

Alder and Bevington [7] reported that both the trifluoromethyl- and methyl-substituted N,N-dipropyl-2,6-dinitroanilines were preemergence herbicides, with trifluralin (1) being more active than dipropalin (2). Dipropalin (2,6-dinitro-N,N-dipropyl-p-toluidine) exhibited some foliar contact phytotoxicity, whereas trifluralin had less. This result was confirmed by Gentner [8]. Benefin [9] and nitralin [10] were commercialized in the late 1960s.

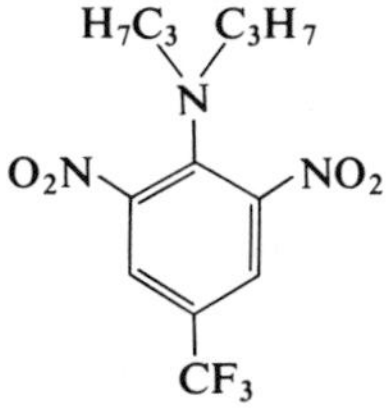

Trifluralin (1) 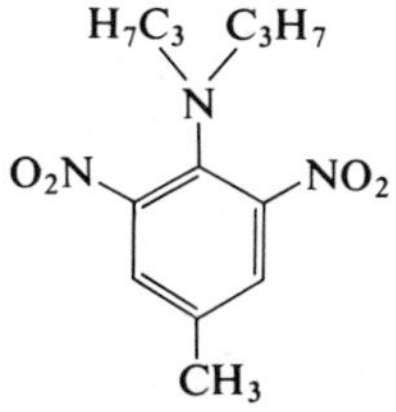   Dipropalin (2)

Several new herbicides in the dinitroaniline class have been introduced. Isopropalin has demonstrated effective control of grasses and a variety of broadleaf weeds in solanaceous crops [11]. Oryzalin was developed as a selective, surface-applied herbicide for soybeans [12]. Dinitramine shows a weed and crop tolerance spectrum similar to that of trifluralin [13].

Various reports indicate that several other dinitroaniline herbicides are under various stages of development for commercial use: fluchloralin, profluralin, butralin, and penoxalin. In addition, ethalfluralin, N-ethyl-N-(2-methyl-2-propenyl)-2,6-dinitro-4-(trifluoromethyl)benzenamine, and prosulfalin, N-[[4-(dipropylamino)-3,5-dinitrophenyl]sulfonyl]-S,S-dimethyl-sulfilimine, are being developed for weed control in cotton and turf, respectively. These compounds are reported to have herbicidal characteristics similar to other members of the class.

B. Chemical and Physical Properties

The physicochemical properties of the more important dinitroaniline herbicides are listed in Table 1.

Extensive toxicological data obtained on trifluralin [14] indicate that it constitutes no hazard to man or animals when used as directed. The acute oral LD_{50} of trifluralin for adult rats is greater than 10 g/kg. Exposure of rats to a mist containing 2.8 mg of trifluralin per liter caused no adverse effects. No skin irritation was observed in rabbits treated dermally with 2.5 g/kg. Rats were fed 2000 ppm of trifluralin in their diet for a two-year period without adverse effect and with no change in reproduction or fertility through three generations. Trifluralin was given daily to dogs at 1000 ppm in their diet for two years without adverse effect.

The LC_{50} of trifluralin as an emulsifiable concentrate in static fish ponds is 0.058, 0.094, and 0.560 ppm for bluegills, fathead minnows, and gold-fish; respectively [15]. If trifluralin is first sprayed on soil, then added to static water, the LC_{50} value for bluegills is 2.8 ppm on Princeton fine sand and 13.2 ppm on Brookston silty clay loam.

For benefin, the acute oral LD_{50} for adult rats is greater than 10 g/kg, and the LD_{50} for newborn rats is 0.8 g/kg [16]. The acute oral LD_{50} for dog, chicken, and rabbit is greater than 2 g/kg in each species, which is the same value reported for nitralin [10] in mice and rats. Five freshwater fish species tolerated a suspension of nitralin at 20 ppm for 48 hr.

The acute oral LD_{50} of isopropalin for adult rats is greater than 5 g/kg [11]; for oryzalin, it is greater than 10 g/kg [12]; and for dinitramine, the LD_{50} is 3 g/kg [13]. Acute dermal, eye irritation, and subacute feeding studies indicate that isopropalin, oryzalin, and dinitramine cause few to no adverse effects.

Generally, the reported values for acute oral toxicity, dermal, and inhalation toxicities reveal the similarity of the various dinitroanilines for little toxicity to mammalian species. The notable point of toxicity is that related to fish. Nitralin and oryzalin exhibit the least toxicity to fish, which is probably associated with the 4-position substitution of methylsulfonyl and the sulfanilamide group in the molecule. The other dinitroanilines having a trifluoromethyl, methyl, isopropyl, or tert-butyl group in the 4 position exhibit marked fish toxicity; but these compounds similarly exhibit low water solubility and a tendency toward tenacious soil adsorption, an important characteristic for preemergence, soil-incorporated dinitroaniline herbicides.

TABLE 1

Chemical and Physical Properties of Some Dinitroaniline Herbicides

Common name	Chemical name and structure	Properties
Trifluralin	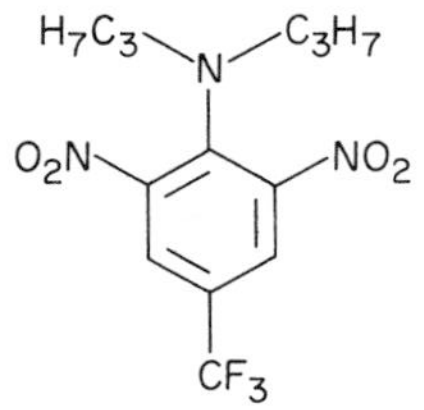 H_7C_3—N—C_3H_7 ... NO_2 ... O_2N ... CF_3 α,α,α-Trifluoro-2,6-dinitro-N,N-dipropyl-p-toluidine (1)	Yellow–orange, crystalline solid Melting range: 48.5–49°C Mol wt: 335 Boiling range: 96–97°C at 0.18 mm Hg Vapor pressure: 1.99×10^{-4} mm Hg at 29.5°C Solubility: Soluble in acetone and xylene; slightly soluble in ethanol; water solubility 0.1–0.3 ppm at 27°C
Benefin	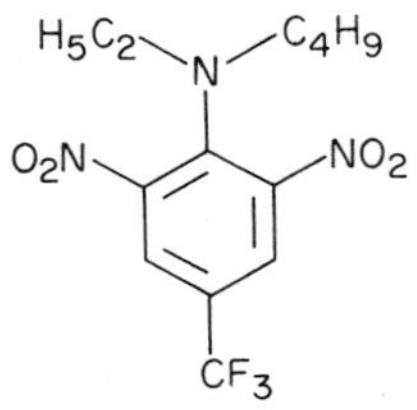 H_5C_2—N—C_4H_9 ... NO_2 ... O_2N ... CF_3 N-Butyl-N-ethyl-α,α,α-trifluoro-2,6-dinitro-p-toluidine (3)	Yellow–orange, crystalline solid Melting range: 65–66.5°C Mol wt: 335 Boiling range: 121–122°C at 0.5 mm Hg Vapor pressure: 3.89×10^{-5} mm Hg at 30°C Solubility: Soluble in acetone and xylene; slightly soluble in ethanol; water solubility 0.2–0.5 ppm at 25°C

(continued)

TABLE 1 (continued)

Common name	Chemical name and structure	Properties
Nitralin	 4-Methylsulfonyl-2, 6-dinitro- N, N-dipropylaniline (4)	Golden-orange, crystalline solid Melting range: 151–152°C Mol wt: 345 Vapor pressure: 1.5×10^{-6} mm Hg at 25°C Solubility: Soluble in acetone and slightly soluble in ethanol and xylene; water solu- bility 0.6 ppm at 25°C
Isopropalin	 2, 6-Dinitro-N, N-dipropyl- cumidine (5)	Red-orange oil Mol wt: 309 Vapor pressure: 1.45×10^{-5} mm Hg at 30°C Solubility: Soluble in most organic solvents; water solubility < 0.5 ppm at 25°C

Oryzalin

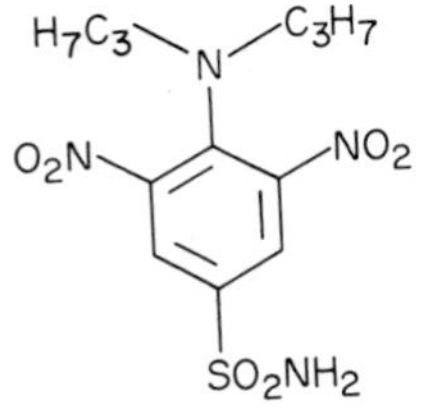

3,5-Dinitro-N^4,N^4-dipropyl-
sulfanilamide (6)

Yellow-orange, crystalline solid
Melting range: 141-142°C
Mol wt: 346
Vapor pressure: $< 1 \times 10^{-7}$ mm Hg at 30°C
Solubility: Soluble in acetone, methanol, ethanol,
and acetonitrile; slightly soluble in benzene
and xylene; insoluble in hexane; water solu-
bility 2.5 ppm at 25°C

Dinitramine

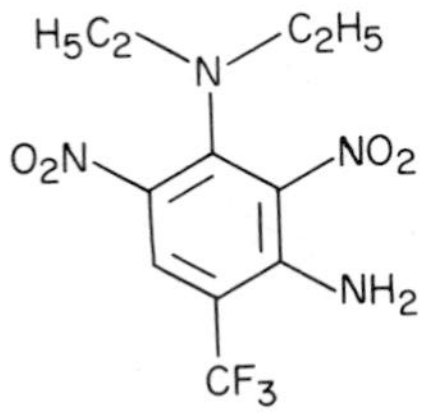

N^4,N^4-Diethyl-α,α,α-trifluoro-
3,5-dinitrotoluene-2,4-diamine (7)

Yellow, crystalline solid
Melting range: 98-99°C
Mol wt: 322
Vapor pressure: 2.6×10^{-6} mm Hg at 25°C
Solubility: Soluble to 10% in ethanol and
57% in acetone at 25°; water solubility
1 ppm at 25°C

(continued)

TABLE 1 (continued)

Common name	Chemical name and structure	Properties
Fluchloralin	 N-(2-Chloroethyl)-2,6-dinitro-N-propyl-4-trifluoromethylaniline (8)	Orange-yellow crystals Melting range: 42–43°C Mol wt: 356 Vapor pressure: 6×10^{-6} mm Hg at 20°C Solubility: Soluble in acetone, ether, benzene, chloroform, ethyl acetate; slightly soluble in cyclohexane and ethanol; water solubility 1 ppm at 20°C
Profluralin	 N-Cyclopropylmethyl-α,α,α-trifluoro-2,6-dinitro-N-propyl-p-toluidine (9)	Yellow-orange crystals or deep orange liquid Melting range: 32.1–32.5°C Mol wt: 347 Vapor pressure: 6.9×10^{-5} mm Hg at 25°C Solubility: Soluble in acetone, xylene, ethanol, n-hexane; water solubility 0.1 ppm at 27°C

Butralin

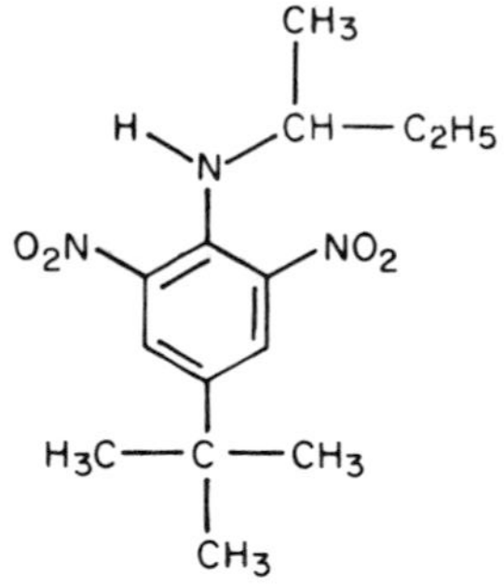

N-(sec-Butyl)-4-(t-butyl)-
2,6-dinitroaniline (10)

Yellow-orange crystals
Melting range: 60–61°C
Mol wt: 295
Vapor pressure: Not reported
Solubility: Soluble in benzene, xylene, acetone,
methyl ethyl ketone, dioxane; slightly soluble
in carbon tetrachloride, ethanol, isopropanol;
water solubility 1.0 ppm at 24°C

Penoxalin

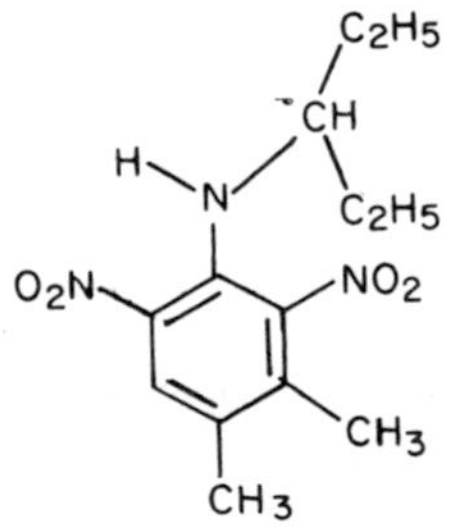

N-(1-ethylpropyl)-2,6-dinitro-
3,4-xylidine (11)

Orange-yellow crystals
Melting range: 56–57°C
Mol wt: 281
Vapor pressure: Not reported
Solubility: Soluble in chlorinated hydrocarbon
and aromatic solvents; slightly soluble in
paraffinic hydrocarbon solvents

C.   Synthesis

Trifluralin is prepared by reacting 4-chloro-3,5-dinitro-α,α,α-trifluorotoluene (12) with dipropylamine, whereas benefin is prepared with butylethylamine [17]. Reacting dipropylamine with 4-chloro-3,5-dinitro-

(12)   (1)

phenylmethylsulfone yields nitralin [18]. Isopropalin is prepared by reacting 4-chloro-3,5-dinitrocumene with dipropylamine [17]. Reacting ammonia with 3,5-dinitro-N,N-dipropylsulfanilyl chloride yields oryzalin [19]. Dinitramine may be prepared by reacting ethanolic ammonia with 3-chloro-N,N-diethyl-α,α,α-trifluoro-2,6-dinitro-p-toluidine [20].

The previous illustrations serve as a general pattern for the synthesis of the dinitroaniline herbicides. Generally, the synthetic pathway is based on a precursor molecule containing the three or four functional groups of the molecule with a chlorine substituted in the 1 position. This essentially allows for directed nitration in the 2,6 positions and subsequent substitution of the chlorine with the appropriate amine.

D.   Formulations

Trifluralin is commercially available as a 4 lb/gal emulsifiable concentrate and as a 5% active granular. Benefin is available as a 1.5 lb/gal liquid concentrate and as a 2.5 and 5% granular. Nitralin is formulated as a 50% wettable powder and as a liquid dispersion containing 4 lb/gal active ingredient. Isopropalin is available as a 6 lb/gal emulsifiable concentrate. Oryzalin is formulated as a 75% wettable powder; dinitramine is a 2 lb/gal emulsifiable concentrate; fluchloralin, profluralin, and butralin are 4 lb/gal emulsifiable concentrates, whereas penoxalin is formulated as a 3 lb/gal emulsifiable concentrate.

The selective herbicidal properties of members of the dinitroaniline series appear to be greatly similar with respect to the weed spectrum controlled at recommended rates. Application rates for effective weed control are slightly greater for nitralin [10], benefin [16], isopropalin [11], and oryzalin [12] than for trifluralin [14] and dinitramine [13]. Crop tolerance varies slightly among these herbicides. Harvey [21] compared the relative

phytotoxicity of 12 substituted dinitroaniline herbicides to soybeans, velvetleaf, giant foxtail, and foxtail millet under greenhouse conditions. Dinitramine
was generally most phytotoxic to these species. All of the herbicides examined inhibited root growth. Oryzalin, dinitramine, and fluchloralin were
most inhibitory to soybean root growth. Similarly, oryzalin, dinitramine,
chlornidine [2, 6-dinitro-N, N-di-(2-dichloroethyl)-p-toluidine], nitralin,
and CGA-14397 (N-propyl-N-tetrahydrofurfuryl-4-trifluoromethyl-2, 6-
dinitroaniline) reduced velvetleaf root growth. It must be emphasized that
the differences noted among members of the dinitroaniline herbicides are
related primarily to degree of activity rather than to significant qualitative
differences in selectivity.

II. DEGRADATION PATHWAYS

The effectiveness of trifluralin as a herbicide stimulated interest in its
degradation in soil [22, 23] as well as its metabolism in plants [23, 24] and
in animals [25-28]. Originally the degradation of trifluralin [23, 24] and
benefin [29] in soils, plants, and animals was thought to proceed in two
distinct patterns, essentially in a sequence of oxidation and reduction reactions. Under aerobic conditions, oxidative dealkylation predominated;
whereas, under anaerobic conditions, reduction of nitro groups predominated.
As methodology and sophistication in identification have progressed, other
degradation products have been elucidated for trifluralin and other dinitroanilines.

Crosby and Moilanen [30] using collimated uv light with trifluralin,
Smith et al. [31] with dinitramine in soil, Laanio et al. [32] with dinitramine
in microbial culture, and Crosby and Leitis [33] with trifluralin photodecomposition in water almost simultaneously reported the formation of
benzimidazole-type compounds and reopened investigations as extensions of
those performed earlier.

Golab [34] detected a benzimidazole compound formed from trifluralin
exposed to laboratory fluorescent light on silica gel plates or in common
organic solvent. Short exposure of trifluralin resulted in the formation of
three major photodegradation products, namely, the monodealkylated derivative of trifluralin (α, α, α-trifluoro-2, 6-dinitro-N-propyl-p-toluidine), the
didealkylated derivative (α, α, α-trifluoro-2, 6-dinitro-p-toluidine), and a
benzimidazole derivative (2-ethyl-7-nitro-1-propyl-5-trifluoromethylbenz-
imidazole).

Analogous compounds have been identified in photodecomposition of
trifluralin [35]; in soil degradation studies with trifluralin, isopropalin, and
oryzalin [36]; as well as for trifluralin, profluralin, butralin, dinitromine,
fluchloralin, and chlornidine [37]. From these studies it is apparent that a
variety of reactions enter into the degradation of the dinitroanilines, namely,

dealkylation, reduction, oxidation, hydrolysis, cyclization, and combinations of these reactions.

Analytical methods used in investigations on the mode of degradation of trifluralin were thin-layer chromatography [38], gas chromatography [39], and radiochemical methods, including autoradiography and reverse isotope dilution techniques. Early investigations have been furthered by thin-layer chromatography, gas-liquid chromatography, and by the use of trifluralin synthesized with ^{14}C in various positions [40], namely, in the propyl groups, the trifluoromethyl group, and mixed labeling of the benzine ring portion and the trifluoromethyl group. Since these original studies, other dinitroaniline investigations have been conducted with ^{14}C-ring-labeled compounds. Use of sensitive mass spectrometry alone and coupled with gas-liquid chromatography has aided in detection and elucidation of chemical structures of new degradation products. Methods were adapted and modified as necessary to permit analysis of these products. Generally, the same type of methodology and radioactive labeling is used to investigate other dinitroanilines to fulfill the requirements of the Environmental Protection Agency for registrations.

With few exceptions, the dinitroaniline herbicides degrade in a similar pattern; minor differences are observed as in-depth investigations are extended.

A. Dissipation by Volatilization

The apparent enhancement of trifluralin herbicidal activity by soil incorporation as compared to surface application [41] suggests possible loss by either volatilization or photodecomposition. The vapor pressure of trifluralin suggests that volatility can be a factor under certain field conditions. Label directions for maximum effectiveness require soil incorporation of many dinitroanilines.

Spencer and Claith [42] reported that trifluralin volatilized more rapidly when surface applied than when incorporated into the soil. Vapor density was markedly less in drier soils. When the herbicide was applied at 1-10 kg/ha to wet soil surfaces, up to 4 kg/ha were lost per day. The rate was a function of temperature. Parochetti et al. [43] examined the volatility of eleven dinitroanilines. Vapor loss measurements from moist Lakeland sand at 50°C, with an air flow rate of 50 ml/min for 3 hr, revealed the following: no vapor loss for nitralin and oryzalin and up to 25% for benefin, profluralin, and trifluralin, with the others ranging from 2 to 13%.

In conjunction with a study of the degradation of trifluralin under laboratory conditions and soil anaerobiosis, Paar and Smith [44] continuously scrubbed the effluent gases to determine the rate and amount of trifluralin volatilized from moist soil incubated aerobically or anaerobically in either a moist or flooded state. The greatest amount of trifluralin volatilized from the moist aerobic environment, with a cumulative value of 2.5% of the applied trifluralin in 20 days. In the Paar and Smith experiment, the volatilization rate reached a maximum during the first 3 to 4 days, after which it became progressively slower. Volatilization from the moist anaerobic systems reached a maximum rate after several days and then decreased rapidly. Less than 1% of the applied trifluralin volatilized after 20 days from the moist anaerobic systems. Flooding appeared to greatly retard the volatilization of soil-applied trifluralin. Less than 0.1% of the trifluralin applied was volatilized during the experiment.

B. Degradation by Photodecomposition

Photochemical decomposition is characteristic of substituted aromatic nitro compounds [45]. Wright and Warren [46] exposed trifluralin, as a thin film on glass or sprayed on the soil surface, to sunlight and artificial light from a mercury vapor source. Changes in the absorption spectrum occurred within 2 hr of exposure. Marked alteration in absorbance and a decrease in the inhibitory effect on millet growth were observed after 4-6 hr of exposure. Photodecomposition also occurred on the soil surface but to a lesser degree.

Trifluralin and its related compounds in methanol and heptane solutions decompose extensively when exposed to ultraviolet radiation. Day [47] detected at least 10 trifluralin-related compounds by gas chromatography after exposure of anhydrous methanol solutions containing trifluralin. Two compounds matched reference models, namely, α,α,α-trifluoro-2,6-dinitro-N-propyl-p-toluidine (13) and α,α,α-trifluoro-2,6-dinitro-p-toluidine (14). The concentration of (13) diminished rapidly after 1 hr, but that of (14) remained at a high level for up to 5 hr. Two other major products were formed in the first hour, but they were not identified. All these products gradually disappeared on further irradiation.

The model system of the photochemical demethylation of α,α,α-trifluoro-2,6-dinitro-N-methyl-p-toluidine in n-heptane with an ultraviolet source resulted in the formation of formaldehyde and (15) α,α,α-trifluoro-6-nitro-2-nitroso-p-toluidine [48]. Propionaldehyde was identified in a similar irradiation of a monopropyl derivative of trifluralin (α,α,α-trifluoro-2,6-dinitro-N-propyl-p-toluidine) exposed to intense ultraviolet radiation; the absorption maxima shifted from 275 and 376 nm to 231 and 434 nm. The latter absorption maxima correspond to those exhibited by compound (15) and suggest it to be a minor intermediate in the photodecomposition sequence under these conditions.

Crosby and Moilanen [30] have developed a laboratory apparatus that provides a beam of collimated uv light in a suitably designed reaction chamber. The pesticide is vaporized within the chamber and simultaneously irradiated

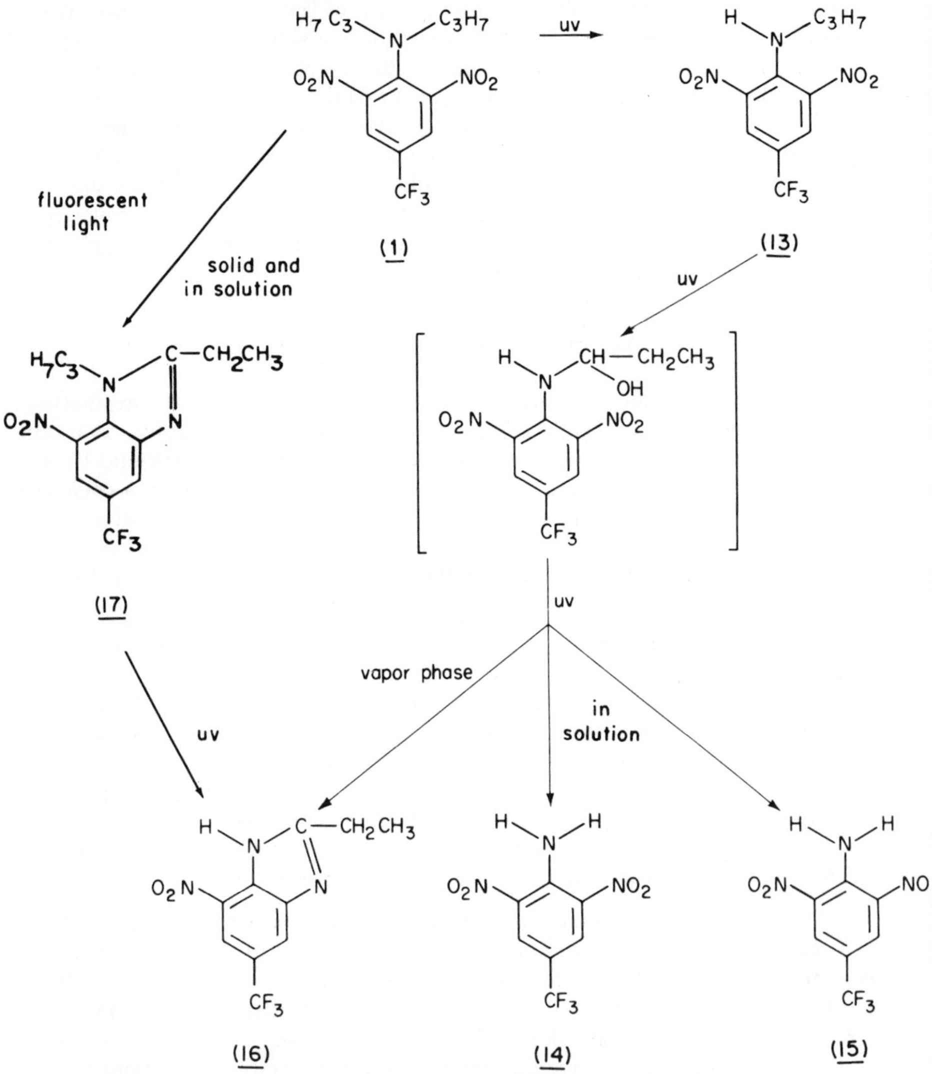

FIG. 2.   Possible reaction sequences in the photodecomposition of trifluralin based on observed degradation products.

with uv light. Trifluralin was subjected to vapor-phase photolysis for 4 days at 35°C. Although the amount of trifluralin exposed to the collimated uv light source was estimated as 2.34% of the total, several degradation products were observed in the gas-chromatographic analysis of the exposed reaction mixture. Two photoproducts were identified in that time. The major photo-lyzed product was the monodealkylated product, compound (13), identified by mass spectrometry with an authentic model compound. The second photoproduct has m/e 259 and is a benzimidazole, 2-ethyl-7-nitro-5-trifluoro-methylbenzimidazole (16), which results from the cyclization of the nitro group with the α-carbon of the propyl group of compound (13), the mono-dealkylated derivative.

The cyclization has also been observed by Golab [34] from exposing trifluralin to laboratory fluorescent light on thin-layer chromatographic plates and in common organic solvents. In addition to mono- and didealkyl-ated photoproducts, a benzimidazole with m/e 301 was isolated and identified as 2-ethyl-7-nitro-1-propyl-5-trifluoromethylbenzimidazole (17).

Photodecomposition studies with trifluralin have been extended by Crosby and Leitis [35, 49]. Their in-depth investigations reveal a multitude of products formed by sunlight wavelengths. Dealkylated intermediates appear in minor amounts, but under acidic conditions, 2-amino-6-nitro-α, α, α-trifluoro-p-toluidine is a principal product, indicating a nitro group reduc-tion. Under alkaline conditions, 80% of the photolysis products was 2-ethyl-7-nitro-5-trifluoromethylbenzimidazole. In addition, 2-ethyl-2,3-dihydroxy-7-nitro-1-propyl-5-trifluoromethylbenzimidazoline and 2-ethyl-7-nitro-5-trifluoromethylbenzimidazole 3-oxide were readily detectable. Soderquist et al. [50] studied the occurrence of trifluralin and its photoproducts in air above both surface-treated and soil-incorporated fields. Vapor traps with air flow rate greater than 1 meter^3/min were mounted at 0.5 and 1.8 meters above ground. Collected air samples contained mostly trifluralin, traces of dealkylation products (13) and (14), and traces of both benzimidazoles (16) and (17).

On the basis of these studies, a simplistic reaction sequence of tri-fluralin photodecomposition is postulated in Fig. 2. The figures provide a view of cyclization, dealkylation, oxidation, and reduction. A detailed description of this photodecomposition is presented in Chapter 18, Volume 2.

A detailed description by Newsom and Woods [51] of dinitramine photodecomposition reveals the cyclization of a nitro group and an adjacent N-ethyl group to yield benzimidazole products. The cyclization reaction has been observed with trifluralin, isopropalin, and oryzalin [36] in field soils and with fluchloralin [52] after exposure to sunlight and artificial uv light. In the case of fluchloralin, a compound having an unsymmetrical side chain containing halogen, a diverse number of cyclized and dealkylated products were observed by Nilles and Zabik [52], including a cyclized quinoxaline and benzodiazepine. Kearney et al. [37] detected the formation of 4-methyl-2,6-dinitrophenylmorpholine from chlornidine.

Butralin was irradiated in saturated aqueous solution and methanolic solution with sunlight, mercury vapor lamp, and a QE sunlamp. In this study, Plimmer and Klingebiel [53, 54] observed that butralin was more photostable than trifluralin or dinitramine, but 4-tert-butyl-2-nitro-6-nitrosoaniline was produced as a major product. In addition, the mixture contained the dealkylated product 4-tert-butyl-2,6-dinitroaniline. Analysis of minor components suggests two pathways of photodecomposition which can operate simultaneously. The major pathway involves excitation of the nitro group with subsequent reductions, followed by oxidation of other reactants. A minor pathway consists of displacement of a nitro group by a hydroxyl group. Cyclization to a benzimidazole, as occurs with trifluralin and dinitramine, is blocked by the branching effect of the N-sec-butyl group in the parent molecule and represents an exception in this series of compounds reported to date.

C. Degradation in Plants

The study of the degradation of the dinitroaniline herbicides in or by plants has been directed primarily toward the detection of suspected compounds resulting from oxidative and reductive processes on the basic molecular structure. In most instances, the radioactive-labeled herbicide is incorporated in soil or other medium; then, test plants are grown, harvested, extracted, and suitably analyzed for the particular herbicide and its metabolites. Under these conditions, present evidence is inadequate to claim active metabolism by the plant. As a chemical class, the dinitroaniline herbicides, including their recognizable degradation products, do not appear to be readily translocated from soil through the roots to the upper portion of growing plants. Those detectable degradation products that may be observed frequently tend to reflect the same residues observed in soil but to a much lesser degree. This premise is supported by the microradioautographic investigations of Strang and Rogers [55] .

Residue analysis with a sensitivity of 5-10 ppb has indicated that trifluralin and benefin or their degradation products were not incorporated in the leaves, seeds, or fruit of a wide variety of plants. Roots from plants grown in soil containing these herbicides exhibit a residue, but only in that region of contact with the herbicide. Although no residue has been encountered in most tolerant crops, studies were undertaken to determine the distribution of residual radioactivity in plants grown in soil containing ^{14}C-labeled trifluralin and benefin [23, 24, 29] , isopropalin, and oryzalin [56, 57].

Analyses of soybean and cotton plants grown in soil containing ^{14}C]trifluralin labeled in the propyl group or in the trifluoromethyl group revealed

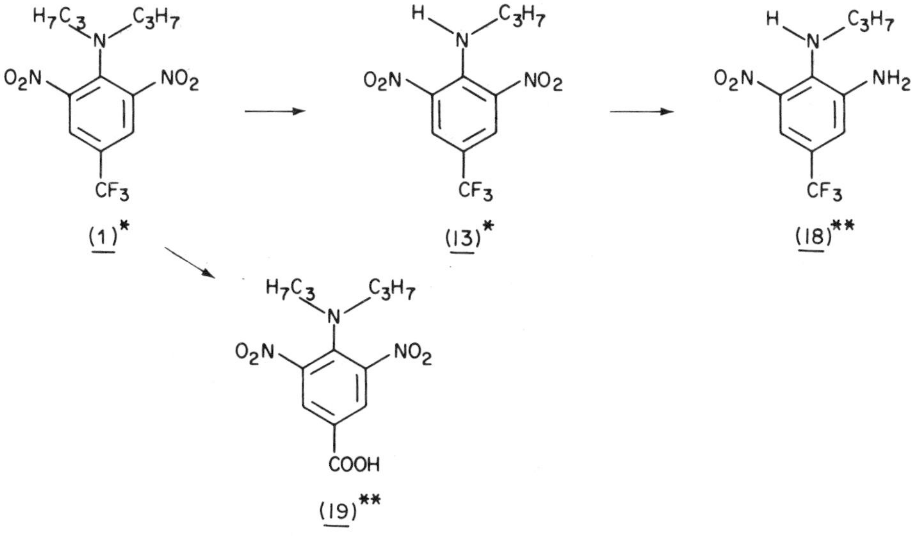

FIG. 3.   Reaction sequence of trifluralin degradation in carrot root based on identified (*) and detected (**) products.

the presence of radioactivity in the lipids, glycosides, hydrolysis products, proteins, and cellular fractions. The universal distribution of the radioactivity without definite identification of trifluralin or recognizable metabolites suggests nondescript incorporation.

Residues of trifluralin were found on the outside layer or peel of some root crops. The amount of trifluralin found in carrot root depends on the age and size of the root as well as the rate and depth of trifluralin incorporation in the soil.

Carrots were grown in soil containing ^{14}C-trifluoromethyl-labeled trifluralin [24]. The average total radioactivity found in carrot roots was 0.65 ppm, calculated as trifluralin, distributed principally in the peel (69%) and at the approximate junction of the phloem and the xylem (10%). Trifluralin constituted 84% of the radioactivity in the carrot extract, with the major metabolic product (13) representing approximately 4.3% of the total radioactivity. Trace amounts of α, α, α-trifluoro-5-nitro-N^4-propyltoluene-3,4-diamine (18) and 4-(dipropylamino)-3,5-dinitrobenzoic acid (19) were detected by thin-layer chromatography with one trace zone of radioactivity not identified. Less than 5% of the radioactivity present was in the unidentified polar product fraction. Figure 3 shows the pathway of trifluralin metabolism in carrot root based on identifiable metabolites only. It was not determined if the compounds found in carrot root were the result of degradation in soil or the result of biological conversion by carrot tissue. The presence of compound (19), in which the trifluoromethyl group was converted to a carboxyl group, does suggest that this compound was not absorbed from soil, as it has been found only in the plant tissue. A trace amount of trifluralin was the only identifiable product in carrot tops.

Biswas and Hamilton [58] exposed bare-rooted sweet potato and peanut plants to ^{14}C-trifluoromethyl-labeled trifluralin at 7 ppm aqueous concentration for 72 hr. In another study, crude leaf extracts of these plants were incubated with ^{14}C-labeled trifluralin at 32°C for intervals ranging from 1/3 to 50 days. Identification of the degradation products as a result of the treatments were made by comparing thin-layer chromatographic data with those previously reported by Golab [38]. Without the aid of suitable model compounds for verification, the study suggested that the crude leaf extracts degraded trifluralin differently. The initial degradation product in crude peanut leaf extracts was the monodealkylated compound (13), whereas the initial degradation product formed in crude sweet potato leaf extract was the mononitro-reduced compound, N^2C, N^2-di-n-propyl-3-nitro-5-trifluoro-methyl-o-phenylenediamine (20). Additional data indicate extensive degradation by both oxidative and reductive pathways. Preliminary infrared analysis seems to indicate the formation of a benzoic acid or phenolic derivative. The radioactivity remaining at the origin of thin-layer chromatograms prepared from extracts of the intact plants ranged from 64 to 86%, indicating rapid degradation into unidentifiable polar products.

Golab et al. [29] investigated the metabolism of ^{14}C-ring-labeled benefin in peanut plants and alfalfa. Mature peanut plants grown under growth room conditions were harvested and divided into leaves, stems, roots, hulls, and meat. The distribution of total radioactivity, benefin, and its metabolites was determined by radiochemical methods in conjunction with thin-layer chromatographic methods [59].

A striking feature of the results is a comparison of the portion of total radioactivity associated with recognizable compounds. More than 93% of the radioactivity, like that associated with trifluralin in soybeans and cotton plants, was nondescript and is viewed as extensively degraded benefin. In peanut plant parts, benefin was the principal compound detected, representing approximately 3% of the total radioactivity. Of the total benefin, leaves and stems account for < 0.005 ppm, and meat accounts for < 0.001 ppm. The detectable degradation products, which matched the cochromatographed model compounds on radioautographs of thin-layer chromatograms, constituted a smaller amount of radioactivity than that reported for benefin. The maximum concentration of the major degradation product in roots, 2,6-dinitro-α,α,α-trifluoro-p-cresol, was calculated to be 0.057 ppm. Examination of the degradation products in the various plant parts reveals a similarity of products in roots and hulls with those recognized in soil. In order, stems, leaves, and meats contain fewer products in lesser amounts. These differences suggest the absorption of benefin and its degradation products from soil and translocation in the plant rather than the absorption of benefin and its metabolism in the plant tissue. If benefin were metabolized directly in the plant, the recognizable degradation products should be distributed more uniformly. This view is further supported by observations with alfalfa grown in soil treated with labeled benefin at the same rate [29]'.

Six cuttings of alfalfa were harvested over a period of 227 days. Although the amount of radioactivity found in the different cuttings gradually declined over the experimental period, the total radioactivity calculated as benefin ranged from 0.34 to 1 ppm. Benefin constituted 1.5% of the total radioactivity in the alfalfa plants. Recognizable degradation products, α, α, α-trifluorotoluene-3,4,5-triamine (21), α, α, α-trifluoro-2,6-dinitro-p-cresol, and α, α, α-trifluoro-5-nitrotoluene-3,4-diamine (22), were detectable only in trace amounts. Again, the majority of radioactivity, approximately 94%, resided in the fractions described as extractable polar products and nonextractable products. The radioactive compounds in the alfalfa plant are similar to the degradation products found in soil but are present in lesser amounts.

Nitralin, as the ^{14}C-ring-labeled compound, has been studied in soybeans, peanuts, peas, and carrots grown under modified field conditions in which the herbicide was soil incorporated at 0.6, 0.75, and 1.0 lb/acre, respectively [60]. Foliage, stalks, and the raw agricultural commodity were examined for total radioactivity calculated as nitralin. In soybeans, the whole plant contained 0.02 ppm radioactivity calculated as nitralin. The seeds contained less than 0.01 ppm radioactivity. Similarly, whole plant parts from peanuts and peas contained an equivalent of 0.03 ppm radioactivity calculated as nitralin, with the seeds in a range similar to that of soybeans.

Carrots were grown as a rotational crop under greenhouse conditions 11 weeks after the initial incorporation study with soybeans [60]. Six weeks after planting, carrots were harvested and the roots were divided into peel and cores for analysis. The roots contained 0.066 ppm radioactivity calculated as nitralin with 78 and 22% of the radioactivity residing in the peel and cores, respectively. Chromatography of the carrot extracts on microgranular cellulose developed with a multigradient elution system resulted in the detection of at least 11 chromatographic peaks. Concentrations were insufficient for identification purposes.

The distribution of uniformly ^{14}C-ring-labeled isopropalin in tomato plants and fruit was investigated by Golab and Althaus [56]. Direct-seeded tomatoes were grown under field conditions in soil treated with [^{14}C]isopropalin at 1.68 and 3.36 kg/ha. At harvest, the tomato fruit contained less than 0.002 ppm of radioactivity calculated as isopropalin, with a detection sensitivity of 0.002 ppm. Leaves and stalks of tomato plants contained less than 0.030 ppm at 1.68 kg/ha and 0.060 ppm at 3.36 kg/ha of isopropalin. Similar results were obtained with direct-seeded peppers. Gas-liquid chromatographic and thin-layer radioautographic analyses of the plant extracts indicated the absence of any detectable residue of isopropalin. No identifiable transformation products were observed on the radioautographs.

Dry tobacco leaves contained 0.126-0.324 ppm radioactivity calculated as isopropalin at an application rate of 1.68 kg/ha and approximately twice this amount at a 3.36 kg/ha rate [56]. Traces of isopropalin were detected in radioaudiographs prepared from extracts of eight-week-old tobacco leaves, but none was found in the mature dry leaves. Attempts to characterize the radioactivity in immature leaves indicated that the residual radioactivity was distributed randomly throughout the leaves. Isopropalin or its recognizable degradation products could not be identified.

At the rates described above only negligible amounts of radioactivity were detected in mature wheat grain grown as a rotational crop on the radioactive plots. Solvent fractionation of the radioactivity from wheat straw indicated a random distribution of radioactivity throughout the plant tissue. This distribution probably occurs through the assimilation of simple degradation products of the degraded dinitroaniline carbon moiety.

Smith et al. [61] incorporated dinitramine labeled with ^{14}C in the trifluoromethyl group as well as the uniformly ring-labeled compound in Anaheim silty loam soil. The degradation in soil and the radioactivity in soybeans planted in the soil under greenhouse conditions were investigated. No differences were observed in soil cropped to soybeans as compared to uncropped soil. The fraction of incorporated radioactivity which accumulated in the plant tissues was negligible.

Investigations with oryzalin by Golab et al. [57] further indicate that no significant radioactive residues were detectable in either seed or forage of soybean and wheat plants. No specific metabolites of oryzalin were identified in soybean plants. Trace amounts of radioactivity found in plant tissue appeared to be associated with the various plant constituents.

The use of radioactive-labeled dinitroaniline herbicides in a wide variety of plant species indicates that the parent compounds or their degradation products are not transported in plant tissues in significant amounts. The small amounts of total radioactivity found in plants and in the raw agricultural commodity could be extensively degraded fragments incorporated as carbon sources of normal metabolic consequence.

The extensive studies of trifluralin, benefin, nitralin, dinitramine, isopropalin, and oryzalin reported to date reveal conclusively that these dinitroanilines and their degradation products observed in soil do not accumulate in the edible portions of crops tolerant to the action of these herbicides. The literature on these compounds supports the thesis that there is little translocation of the parent compounds or their degradation products into the raw agricultural commodity, with the exception of certain root crops. It is apparent that root crops such as carrots incorporate a minor amount of the parent compound and those degradation products observed in soil, but none in sufficient quantity to approach a significant residue. The description of the degradation of dinitro-aniline herbicides in soil in the subsequent section provides a wealth of interesting biochemical phenomena. The degradation pathways reveal inter-esting chemistry but have little bearing on the accumulation in edible plant tissue or safety of these herbicides in crop production. With the exception of possible accidental contamination of aquatic systems, soil-incorporated or surface-applied dinitroaniline herbicides for weed control in crops provide a pesticide of maximum safety to the environment.

D. Degradation in Soil

Evidence from studies with trifluralin [23] and benefin [29] clearly demonstrates that degradation of the dinitroanilines in soil is affected by aerobic and anaerobic conditions. The aerobic pathway proceeds by a series of oxidative dealkylation steps; the anaerobic pathway is initiated via a sequential reduction of the nitro groups. Investigations with nitralin [60] , isopropalin [56] , oryzalin [57] , and dinitramine [61] tend to support the dual pathway. Cyclization involves both oxidation and reduction. The difference in the nature of the degradation products formed by aerobic versus anaerobic pathways is more quantitative than qualitative; thus, reduction also occurs under aerobic conditions and oxidative dealkylation under anaerobic conditions.

A number of factors influence the degradation or disappearance of the dinitroaniline herbicides in soil, including volatility, photodecomposition, and metabolism. The rate of degradation is also a function of soil type, moisture content, temperature, and rate and method of incorporation. Experimental conditions can be created, and field conditions naturally occur, which distinguish the predominance of oxidative or reductive degradation. Seasonal variation provides conditions for both forms of degradation to operate simultaneously.

1. Aerobic Degradation

Long-term degradation studies under natural field conditions have been performed on trifluralin, oryzalin, and isopropalin by Golab and Amundson [36] , Golab et al. [57] , and Golab and Althaus [56] , respectively. Prelim-inary data obtained on trifluralin over a 28-month observation period and complete data obtained on both oryzalin and isopropalin during 36 months of observation indicate a similar general pattern of degradation for these three herbicides. Isolation and identification of more and/or different degradation products for one dinitroaniline herbicide than for another lies mostly in the depth of study with each compound.

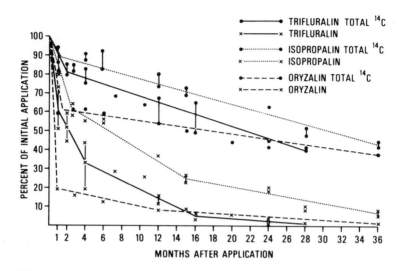

FIG. 4.   Dissipation of trifluralin, oryzalin, and isopropalin in soil.

Some of the minor differences in the chemical structure of the parent compounds might be responsible for the absence of some degradation products. In these studies, mixed ^{14}C-labeled trifluralin (85% in CF_3 group, 15% in benzene ring) and uniformly ^{14}C-ring-labeled trifluralin, oryzalin, and isopropalin were used. The [^{14}C]trifluralin and [^{14}C]isopropalin were incorporated in the top three inches of soil, and [^{14}C]oryzalin was sprayed on the soil surface. Application rates of each compound included the recommended rate for the soil type and twice that rate. Appropriate crop species (trifluralin: soybean; oryzalin: soybean; isopropalin: tomato, pepper, and transplant tobacco) were seeded or transplanted into the plots immediately after herbicide treatment. Soil samples were analyzed for total radioactivity before and after extractions, for the parent herbicide and for transformation or degradation products.

The extent of dissipation of trifluralin, oryzalin, and isopropalin is shown in Fig. 4. The observed rates of degradation were found to be independent of amount of herbicide applied. The rates of degradation of trifluralin and oryzalin were quite similar, while that of isopropalin was considerably slower over the 28-month period. At that time 1% of the applied trifluralin, 2% oryzalin, and 8% isopropalin could be detected in soil plots. The loss of total radioactivity from the isopropalin plots was also slower than with the other herbicides [36].

Extractable radioactivity (aqueous methanol or aqueous acetone) decreases with time with each herbicide, while nonextractable or "soil-bound" radioactivity increases with time. Twelve months after application of the herbicides the amount of "soil-bound" radioactivity as percentage of initial application was 27-41, 30, and 15 for trifluralin, oryzalin, and isopropalin, respectively. These significant quantities of nonextractable or "soil-bound" radioactivity could be released by extraction with 0.5 N alkali (NaOH or KOH). Radioactivity occurred in the humin fraction and in the humic and fulvic acid fractions of the soil. The process of extraction and separation of the humin, humic, and fulvic acid fractions was performed in a manner similar to that described by Marshall [62].

The degradation pathways observed with these three dinitroanilines are shown in Fig. 5 and are based on identification of the extractable transformation products. It is readily apparent that a variety of mechanisms, i.e., dealkylation, reduction, oxidation, hydrolysis, cyclization, and combinations of these, are involved in the transformation of dinitroanilines in soil. There is no indication of a buildup of any of the transformation products in soil, and none has ever been found at a concentration greater than 4% of the original amount of herbicide applied. Radiochemical methods were used for quantitation. Confirmation and identification of the degradation products were accomplished by comparative chromatography, tlc, glc, and glc-mass spectrometry with authentic synthetic model compounds. The aerobic degradation pathway of trifluralin, oryzalin, and isopropalin as proposed in Fig. 5 supplements the degradation patterns proposed previously by Probst et al. for trifluralin [23] by the addition of new degradation products, namely, dihydroxybenzimidazolines, benzimidazole N-oxides, and benzimidazoles.

The recently proposed degradation patterns of various dinitroaniline herbicides are not likely to be final. As new investigations proceed with more sophisticated instruments and methodology, the status of degradation products formed under aerobic and anaerobic conditions will be extended to supplement present knowledge.

Portions of the material designated in the previous studies as extractable polar products of trifluralin [23], formed under aerobic conditions and located on or close to the origin of the tlc plates, contained mostly polar benzimidazole-type degradation products, according to a recent examination [36]. The nature of the remaining small portion of this extractable polar material in the region of the origin on the tlc plates remains to be determined. Distribution of similar polar products derived from isopropalin and oryzalin has been described [56, 57].

The degradation pattern of benefin under aerobic conditions [29] has not been reexamined. Considering new observations made on other dinitroaniline

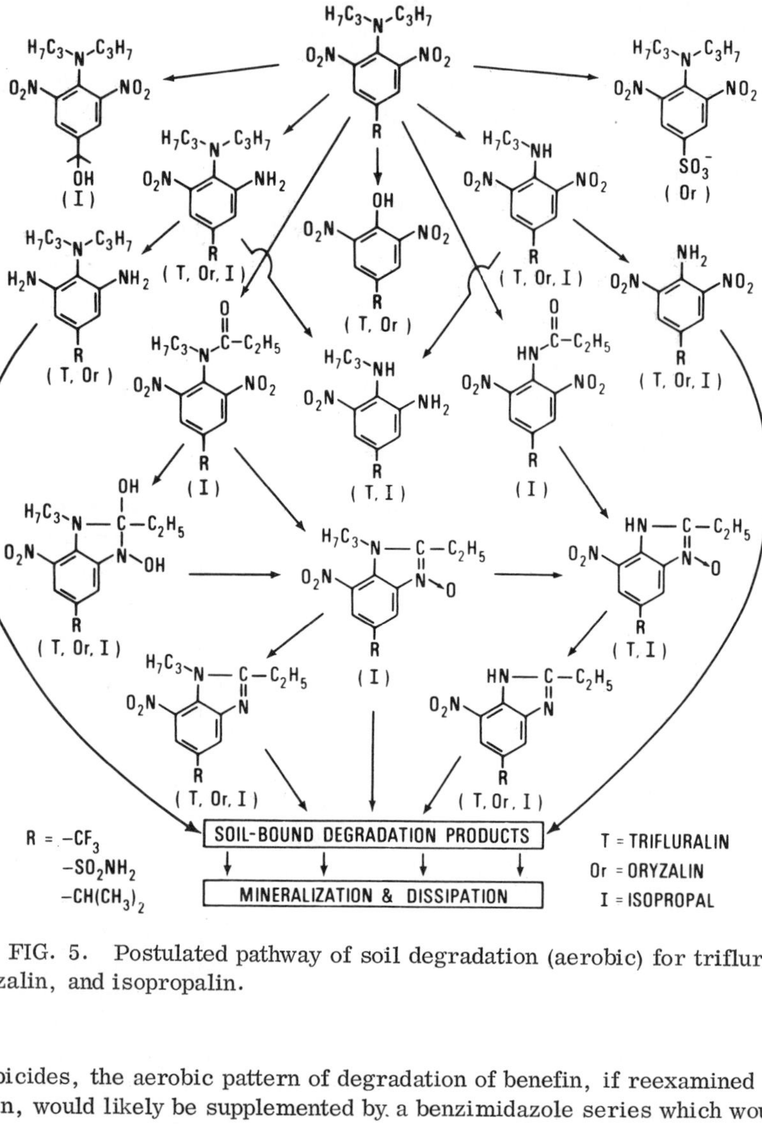

FIG. 5. Postulated pathway of soil degradation (aerobic) for trifluralin, oryzalin, and isopropalin.

herbicides, the aerobic pattern of degradation of benefin, if reexamined again, would likely be supplemented by a benzimidazole series which would correspond to the structure of benefin.

Bode and Gebhardt [63] determined the diffusion coefficients for trifluralin under laboratory conditions. Using two half-cells, one containing ^{14}C-labeled trifluralin, the rate of diffusion was determined at 14- and 28-day intervals. The rate of diffusion was more rapid in moist soil than in dry soil. For soil containing 20% water, the average diffusion coefficient was 0.9001×10^{-7} cm^2/sec; whereas, in air-dried soil, the value was 0.2828×10^{-8} cm^2/sec.

Chemical assays and bioassays indicate that trifluralin does not readily leach in soil. Repeated applications of trifluralin at recommended rates in soil do not result in a buildup of trifluralin with time, according to Parka and Tepe [64]. In a recent report by Miller et al. [65], soil persistence of trifluralin, benefin, and isopropalin further confirmed that these herbicides do not leach or accumulate with repetitive applications.

To determine if microorganisms play an important role, trifluralin was incorporated in both autoclaved and nonautoclaved soil at 8 ppmw [23]. This soil was maintained at 75% field moisture capacity and incubated at 80°F. Using the inhibition of crabgrass as a bioassay, the study revealed that trifluralin degradation proceeds slightly more rapidly in nonautoclaved than in autoclaved soil. Examination of several trifluralin-treated soils failed to show specific microorganisms that caused trifluralin degradation.

Of special interest was the detection of α,α,α-trifluoro-2,6-dinitro-p-cresol derived from trifluralin and as observed previously with benefin [29]. The corresponding compound, 3,5-dinitro-4-hydroxybenzenesulfonamide, was formed from oryzalin under the same aerobic conditions [57]. This displacement of the dipropylamino group by a hydroxyl group was not observed with isopropalin. However, hydroxylation of isopropalin was observed in the isopropyl group resulting in the formation of a new degradation product, α,α-dimethyl-3,5-dinitro-4-dipropylaminobenzyl alcohol [56].

Dinitramine, both ^{14}CF$_3$-labeled and uniformly ^{14}C-ring-labeled, was incorporated in Anaheim soil at 0.5 lb/acre (0.6 ppm). According to Smith et al. [60], the methanol- and acetonitrile-extractable radioactivity decreased rapidly in the initial 60 days and decreased to 20% of the original in 244 days. After 100 days, the concentration of dinitramine decreased to 0.05 ppm. Thin-layer chromatography indicated four compounds present in the extract: dinitramine (7); 3,4-dinitro-N^3-ethyl-6-trifluoromethyl-m-phenylenediamine (23), at 0.01 ppm; 6-amino-1-ethyl-2-methyl-7-nitro-5-trifluoromethylbenzimidazole (24), at 0.06 ppm; and 6-amino-2-methyl-7-.nitro-5-trifluoromethylbenzimidazole (25), at 0.02 ppm. The degradation products were characterized with authentic synthetic model compounds. The relationship of these degradation products in soil is shown in Fig. 6. One-half of the ^{14}C radioactivity remaining after solvent extraction was recovered by basic extraction and constituted the polar fraction.

Funderburk et al. [66] observed that trifluralin was the only radioactive compound present in extracts of ^{14}C-labeled trifluralin-enriched soil. They also examined four species of fungi (Sclerotium rolfsii, Aspergillus niger, Fusarium sp., and Trichoderma sp.) grown in liquid medium containing trifluralin. Extracts of the solutions, assayed by electron-capture gas chromatography, showed little change in trifluralin concentration. However,

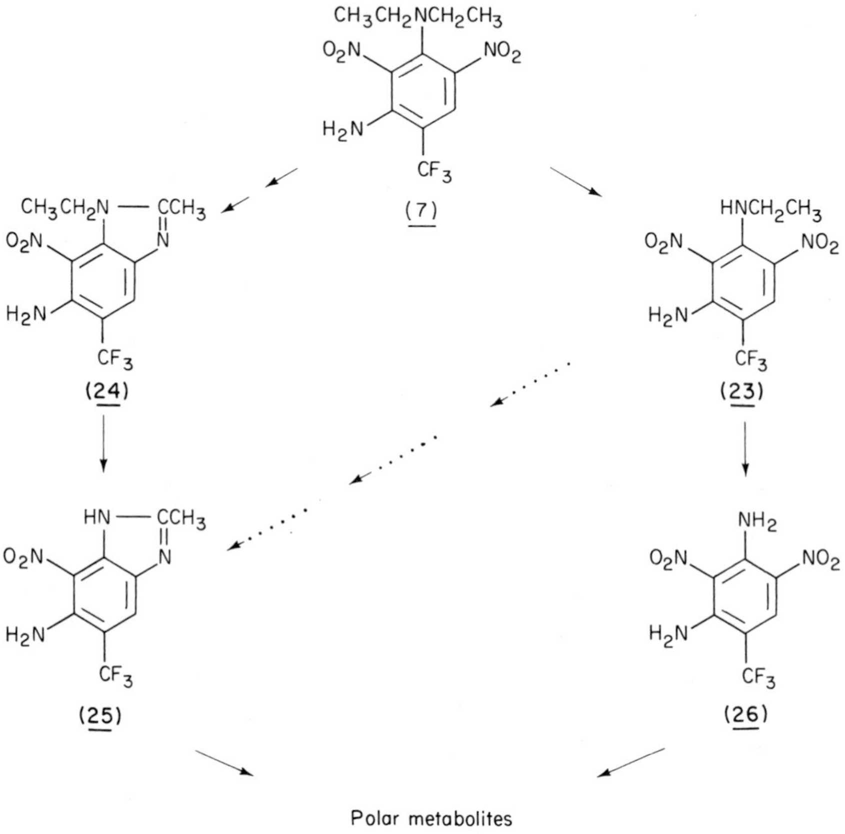

FIG. 6. Proposed pathway of dinitramine degradation in soil and by fungal cells in vitro.

the extract from Aspergillus niger, in addition to trifluralin, contained a small quantity of α,α,α-trifluoro-2,6-dinitro-N-propyl-p-toluidine (13).

Addison [67] demonstrated that mass inoculation of soil with ten native microorganisms (six actinomycetes, three fungi, and one bacterium) was of little or no value in promoting the degradation of trifluralin. Similar results were obtained using five facultative anaerobes. Hamdi and Tewik [68] isolated a bacterium from cotton field soil previously treated with trifluralin which was capable of decomposing trifluralin in the presence of glutamate, lactate, acetate, or yeast extract. The microorganism Pseudomonas sp., group III, degraded 95% of a 0.01% concentration of trifluralin incorporated in the medium in 21 days at a pH optimum of 7.4. The degradation pathway was not established.

The investigation of Messersmith et al. [69] supports the contention that the breakdown of trifluralin by soil microorganisms accounts for only a small fraction of the total dissipation. Using ^{14}C-propyl-labeled trifluralin at 1 and 100 ppm in Sharpsburg silty clay loam and Anselmo sandy loam, the rate of dissipation was greater at a field capacity of 1.6 as compared to 0.8, with the lower concentration degrading more rapidly. The amount of $^{14}CO_2$ liberated from the soils was 5 and 3%, respectively. In soils of this moisture content, oxidative dealkylation is the predominant route of degradation. Hence, with ^{14}C-propyl-labeled trifluralin, information on the initial degradation steps can be measured. The results tend to support the thesis that biological breakdown accounts for only a small fraction of trifluralin degradation. Thus, microorganisms may contribute to the destruction of trifluralin to simpler compounds, but the evidence does not suggest that this is the major mode of degradation.

The metabolism of four structurally related dinitroaniline herbicides by various species of fungi was recently reported by Laanio et al. [70]. The preliminary report emphasized the metabolism of dinitramine (7) by Aspergillus fumigatus, Fusarium oxysporum Schlect, and a Paecilomyces species. Aspergillus fumigatus was cultured for 12 days on a minimal nutrient medium containing 50 ppm of ^{14}C-labeled dinitramine. The fungal cells and nutrient medium were separated and extracted with acetone and benzene, respectively. Most of the radioactivity was found in the acetone extract, indicating an absorption affinity by the fungal cells. The extract was separated by thin-layer chromatography. The medium polar metabolites were characterized by mass spectrometry and/or by cochromatography with reference model compounds. Three metabolites were characterized from the Aspergillus culture extracts, and a fourth structure tentatively proposed by mass spectrometry was isolated from the Paecilomyces culture.

The proposed pathway of fungal metabolism of dinitramine is shown in Fig. 6. Dinitramine is degraded into the corresponding mono- and didealkylated derivatives, compound (23) and 2,4-dinitro-6-trifluoromethyl-m-phenylenediamine (26). The cyclized product (24) was identified and compound (25) was tentatively identified by Smith et al. [61].

Kearney et al. [71] examined solution cultures of Paecilomyces sp. enriched with butralin. In addition to trace amounts of the dealkylated product, a polar compound was isolated which, according to mass spectral data, had a m/e of 311 and was shown to be $C_{14}H_{21}N_3O_5$. A comparison of fragmentation pathways suggested an oxidation of the N-sec-butyl moiety. Such an oxidation product supports the proposed oxidative intermediate identified in the photodecomposition of trifluralin shown in Fig. 2.

By subjecting the different dinitroaniline herbicides to a variety of conditions, intermediates in the degradative pathway have been identified with the aid of synthetic models. Present investigations tend to verify the probable oxidative and reductive degradation products predicted from theoretical organic chemistry.

2. Anaerobic Degradation

In the course of investigating trifluralin degradation in soil with moisture contents adjusted to 0, 50, 100, and 200% of field capacity (FC), it was observed that trifluralin at 200% FC was degraded rapidly as compared to other moisture contents [22, 23]. At 200% FC, 50% of the added trifluralin disappeared in 10 days and 84% in 24 days. Rapid degradation of trifluralin under these anaerobic conditions could not be associated with anaerobic microorganisms or soil type.

This type of anaerobic state may exist for a short time under conditions of excessive rainfall coupled with poor drainage. Soil under water provides a unique condition for monitoring the degradation of trifluralin, benefin, and other dinitroanilines incorporated in soil.

Separate experiments were performed in which both ^{14}C-labeled trifluralin and benefin were incorporated in soil at 4 ppmw; then the respective soils were flooded with water and incubated at 24.5° C in the growth room [23, 29]. Figure 7 compares the rapid degradation of trifluralin and benefin and the sequential formation of their respective degradation products. In both cases, the major reduction products (α,α,α-N^4,N^4-dipropyl-5-nitro-toluene-3,4-diamine (20) and α,α,α-trifluoro-N-butyl-N-ethyl-5-nitro-toluene-3,4-diamine (27), trifluralin and benefin with one nitro group reduced to an amino group, respectively) appeared maximal on the fifth to the seventh day and then gradually declined. With the decline of compound (20), in the trifluralin study, there is a simultaneous rise in α,α,α-trifluoro-N^4,N^4-dipropyltoluidine-3,4,5-triamine (28), as well as extractable polar products. A small amount of radioactivity was associated with the chromatographic position of the model compound (21), the dealkylated triamine, indicating a final recognizable metabolite of the reduction and dealkylation processes.

As shown in Fig. 7, benefin was degraded in a pattern similar to that of trifluralin with the exception that extractable polar products formed more slowly. The first recognizable reaction in the anaerobic degradation of both trifluralin and benefin appears to be the formation of the monoamino product. However, as noted in the figure, both extractable polar products and nonextractable products are formed prior to the detection of the reduced compounds (20) and (27). This suggests that formation of polar products is not restricted to the identified sequence.

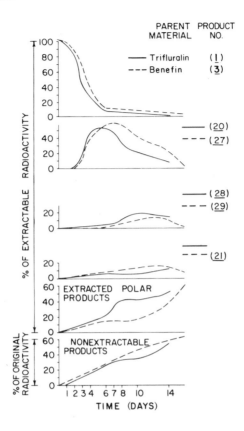

FIG. 7.   Degradation of [^{14}C]trifluralin and [^{14}C]benefin in soil under water.

The use of thin-layer chromatography coupled with radioautography provided evidence for the occurrence of lesser metabolites in the soils. Figures 8 and 9 show the postulated pathways of anaerobic trifluralin and benefin degradation in soil. All compounds shown in Fig. 8, except compound (30), were compared in thin-layer radioautograms with available synthetic model compounds. Williams and Feil [72] synthesized α, α, α-trifluoro-N^4-propyltoluene-3,4,5-triamine (30) and identified it as a trifluralin degradation product from the action of rumen bacteria. An authentic sample from Williams confirmed the identity of previously postulated compound (30) formed in soil under water [23] . $\alpha, \alpha,\ \alpha$-Trifluoro-N^4-butyltoluene-3,4,5-triamine (31), a similar analog in the benefin series, was observed in the anaerobic degradation of benefin in soil [29] (Fig. 9).

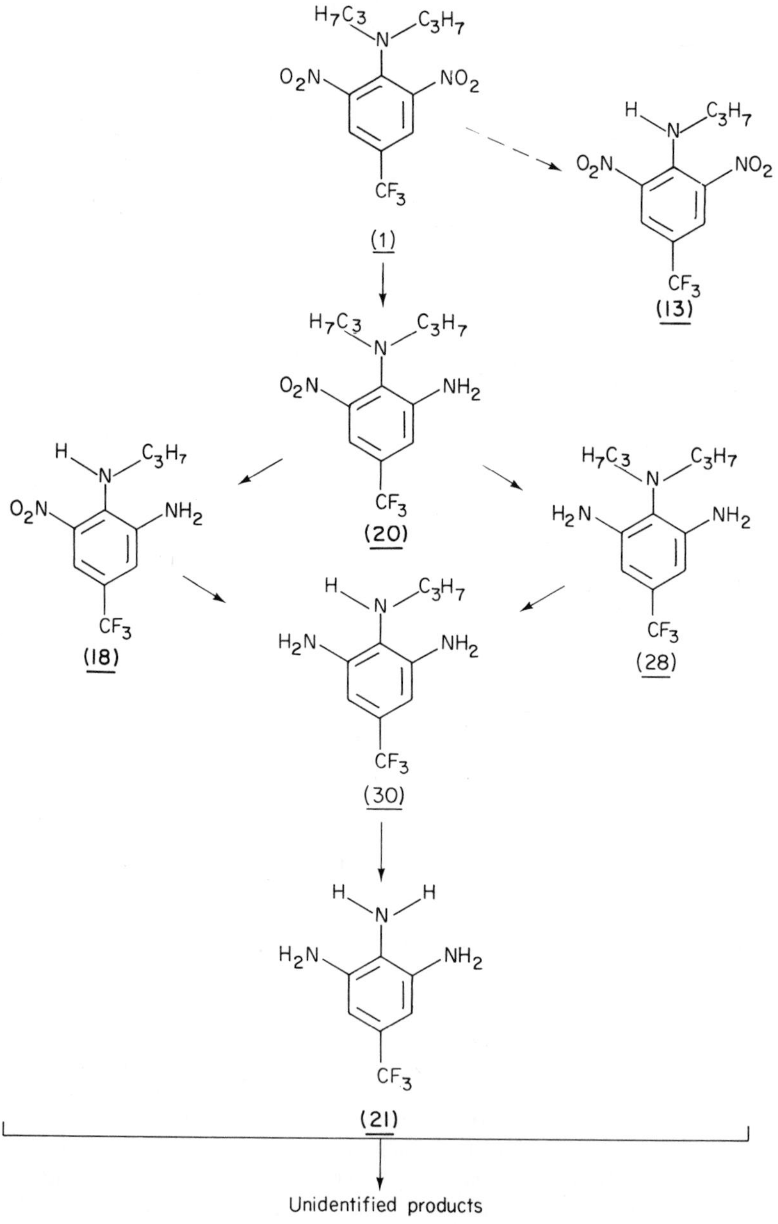

FIG. 8. Postulated anaerobic pathway of trifluralin degradation in soil.

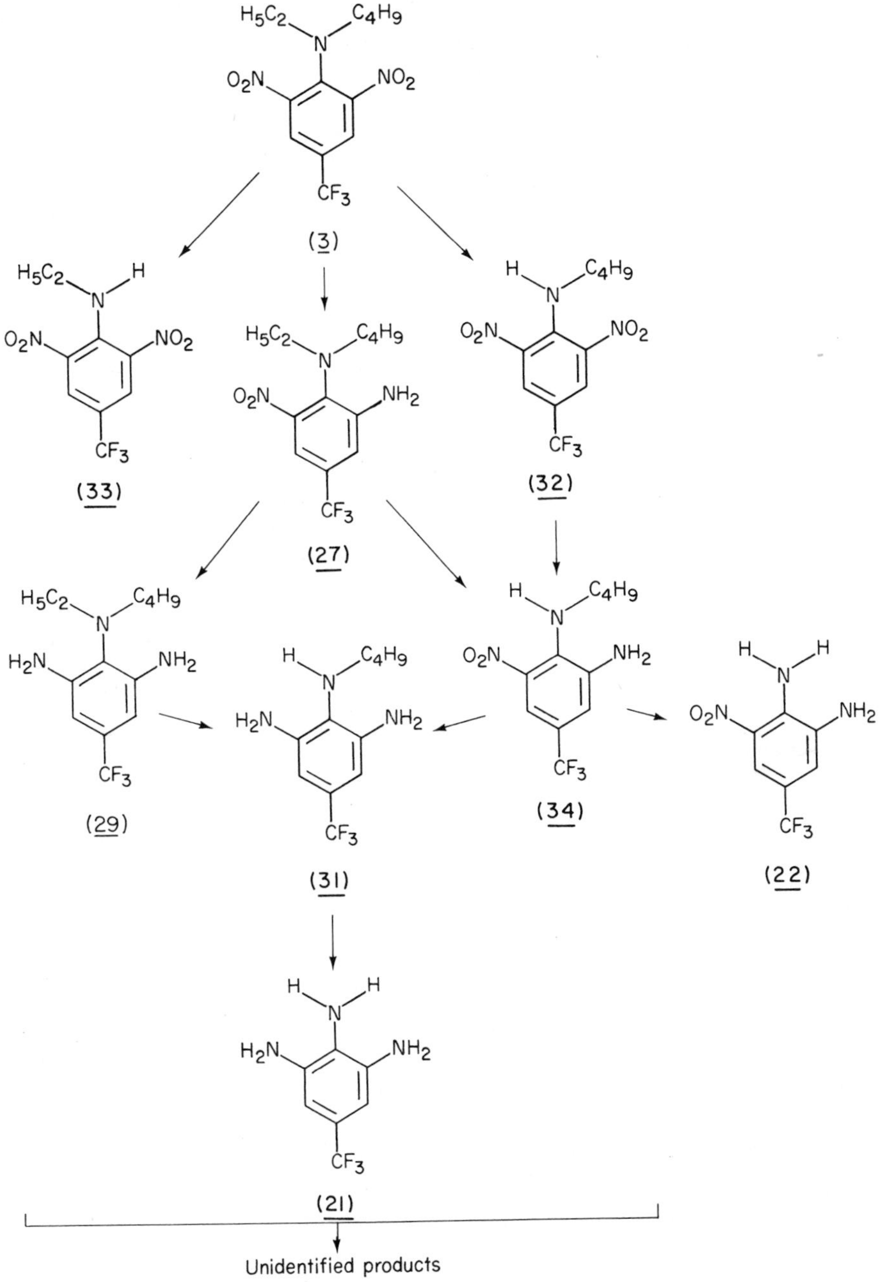

FIG. 9.   Postulated anaerobic pathway of benefin degradation in soil.

Experiments with trifluralin and benefin in soil under aerobic and anaerobic conditions had been performed in the late 1960s [23, 29]. Since that time similar research has not been performed with benefin; thus, no additional degradation products have been detected to supplement the degradation pathway proposed previously [29] and as shown in Fig. 9. Conversely, trifluralin has been studied by many investigators under a variety of physical conditions, mostly aerobic. Therefore, the degradation pathway described by Probst et al. [23] can be integrated by benzimidazoles, as indicated in Fig. 5.

There are few additional reports on the degradation of trifluralin in soil under anaerobic conditions since the original report of Probst et al. [23]. Paar and Smith [44] reported on anaerobic degradation of trifluralin in soil, but they did not identify new degradation products. They again recognized the two principal pathways of trifluralin degradation: the aerobic pathway involving sequential dealkylation of propyl groups and the anaerobic pathway involving initial reduction of the nitro groups. Degradation of trifluralin was more rapid and extensive in soil under anaerobic (N_2) conditions compared with well-aerated systems and followed the order of moist anaerobic > flooded anaerobic > moist aerobic. Degradation in these environments after 20 days was 99, 45, and 15%, respectively. According to their investigation, the evidence for microbial involvement in the initial degradation of trifluralin under soil anaerobiosis was obtained. They reported (a) enhanced degradation of trifluralin in the presence of an organic substrate, (b) lack of trifluralin degradation in moist anaerobic environments after autoclaving, (c) resumption of respiratory activity in autoclaved systems after reinoculation, which corresponded with increased degradation activity, and (d) a temporary lag period or suppression of respiratory activity and trifluralin degradation from the presence of KN_3 as a biological inhibitor, with simultaneous resumption of respiration and degradation soon after chemical dissipation of KN_3. This kind of evidence of microbial involvement in the degradation of a pesticide is not shared by all investigators. According to Kaufman [73] and Kaufman et al. [74], the dividing line between purely chemical and biochemical mechanisms relative to pesticide degradation is not always clear, although the inability of autoclaved soil to degrade a pesticide is often cited as direct evidence in favor of microbial involvement.

The organic matter in soil could provide a ready source of protons for the reductive degradation of trifluralin-type compounds under the defined anaerobic conditions. The reductive pathway might also result from proton donation by the alkyl group produced by an initial oxidative dealkylation.

It is generally recognized that under anaerobic conditions trifluralin undergoes initial reduction of the nitro groups, although oxidative dealkylation is also observed to a minor extent. Therefore, the combination of oxidation and reduction reactions is to be expected. Formation of cyclic degradation products, i.e., benzimidazole-type compounds, was observed

by Golab [75] while comparing degradation rates of several new dinitroaniline herbicides with trifluralin in soil flooded with water; 2-ethyl-7-
nitro-5-trifluoromethylbenzimidazole (16) and 2-ethyl-7-nitro-1-propyl-5-
trifluoromethylbenzimidazole (17) were tentatively identified. If both these
benzimidazoles were formed via a 2,3-dihydroxybenzimidazolidine, then
such formation was indeed an internal oxidation-reduction pathway. Inte-
gration of new degradation products in the postulated anaerobic pathway of
trifluralin degradation in soil (Fig. 8) and of other dinitroaniline herbicides
is only a question of time.

E.   Degradation in Animals

The products identified in the metabolism of trifluralin in ruminant
animals, artificial rumen fluid, and rumen bacteria indicate a reductive
degradation pathway. In monogastric animals trifluralin undergoes both
oxidative and reductive pathways of degradation.

Emmerson and Anderson [25] administered a single oral dose of ^{14}C-
trifluoromethyl-labeled trifluralin to rats. Eighty percent of the radioactivity
was excreted in the feces while the remaining portion appeared in the urine.
Trifluralin and the amino derivative of trifluralin (20) were isolated and
identified from feces. Three urinary metabolites were isolated and identified
as the dealkylated product (14), the diamine (22), and the reduced monodealkylated derivative (18). Other minor metabolites were detected on radioautographs of thin-layer chromatograms. Figure 10 shows the dealkylative
and reductive pathways of trifluralin degradation in the rat based only on
the identified metabolites in urine and feces. The monopropyl derivative
(13) shown in the figure with brackets was apparent in thin-layer chromatography but was not confirmed by other criteria. Trifluralin metabolism in
the dog is similar to that observed in the rat.

A lactating cow was fed trifluralin at 1 and at 1000 ppm in its ration for
39 and 13 days, respectively, to determine the absorption, metabolism, and
excretion in ruminant animals. Examination of urine, feces, blood, and
milk revealed detectable residues only in feces after the ingestion of trifluralin at 1000 ppm. Trifluralin, compound (20) and compound (28) with
one and two nitro groups reduced, respectively, were detected in the feces
by gas chromatography [28]. The results were not adequate to determine
the fate of trifluralin in the dairy cow but indicated that it was rapidly
metabolized.

Since less than 1% of the trifluralin administered to the cow could be
accounted for, the metabolism of ^{14}C-labeled trifluralin was studied in the
goat [28]. A goat was fed nonlabeled trifluralin at 1 ppm in the diet for
11 days, labeled trifluralin for 1 day, and nonlabeled trifluralin again
for 14 days. Within 6 days, 99% of administered radioactivity was

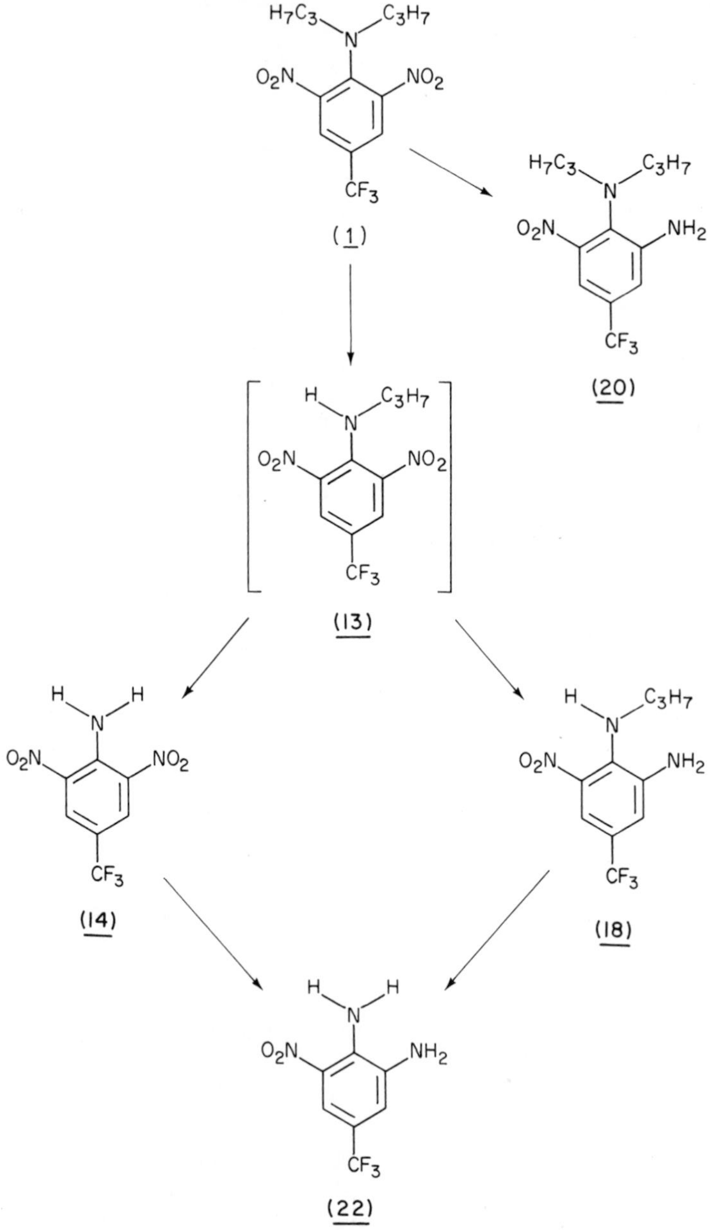

FIG. 10.   Reaction sequence of trifluralin degradation in the rat based on identified products.

accounted for in the urine (17.8%) and feces (81.2%). Radioactivity was
present in the urine for 3 days and in the feces for 6 days after administration of the labeled compound. No increase in radioactivity above normal
level was found in the milk or blood. Of the total radioactivity accounted
for in the urine and feces, only 10% was in the form of recognizable compounds. Neither urine nor feces contained trifluralin. The major metabolite
found in the urine and the only one in feces was compound (28), trifluralin
with both nitro groups reduced. Detection of several other minor metabolites
in urine supports a reductive pathway of trifluralin degradation in ruminant
animals. Similar studies conducted with labeled benefin [29] revealed the
turnover time to be essentially identical with that observed with trifluralin
[28].

^{14}C-Ring-labeled nitralin [60] fed to rats was rapidly excreted, with
52% of the radioactivity being in the urine and 47% in the feces. No $^{14}CO_2$
was observed in an experiment run in a metabolism cage. Multiple metabolic
products were observed. Guernsey cows were fed nitralin at 50 ppm in their
ration. Examination of milk revealed no detectable residues of nitralin or
its degradation products 4-methylsulfonyl-2,6-dinitro-N-propylaniline,
4-methylsulfonyl-2,6-dinitrophenol, and 4-methylsulfonyl-2,6-dinitroaniline.
Analyses of various tissues, feces, and urine showed no detectable residues
of nitralin or 4-methylsulfonyl-2,6-dinitrophenol.

Thin-layer chromatographic examination of the extracts of urine from
rats fed isopropalin [5] indicated the probable presence of the monodealkylated metabolite (2,6-dinitro-N-propylcumidine) and the didealkylated metabolite (2,6-dinitrocumidine) [76]. The presence of these compounds suggests
that a metabolic pathway similar to that of trifluralin (Fig. 11) is followed
by isopropalin. Gas chromatographic examination of the extracts of the
liver [77] from these rats verified the presence of these two compounds.
Page [76] found that liver microsomal enzymes degraded isopropalin to
2,6-dinitro-N-propylcumidine, 2,6-dinitrocumidine, and unidentified
products.

Artificial rumen fluid rapidly degrades both trifluralin and benefin
[28, 29]. Both compounds degraded to less than 1% of the amount
applied in about 12 hr, after incubation. It is interesting to note that
under the same conditions, isopropalin was degraded in about 24 hr and
oryzalin in less than 30 min [78]. Artificial rumen fluid is a unique medium
for observing the rapid formation of degradation products. In separate
experiments, labeled trifluralin and benefin were introduced into a mixture
of rumen fluid and artificial saliva. The mixture was maintained in an
atmosphere of carbon dioxide and incubated at 37°C. Figures 11 and 12
compare the rapid destruction of trifluralin and benefin as well as the
formation of major metabolites.

Trifluralin reduction products, compounds (20) and (28), constitute the
major metabolites observed during the course of the reaction (Fig. 11). A

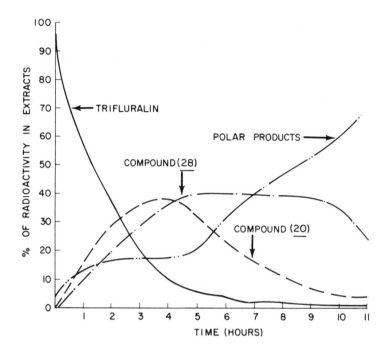

FIG. 11.   Trifluralin degradation and the formation of degradative products in artificial rumen fluid.

decline of these metabolites precedes an increase in polar products. Radio-autographs of thin-layer chromatograms indicate the presence of compounds (13) and (18) as well as an additional unknown. These latter metabolites are products of dealkylation and represent only 0.1% of the original trifluralin present in the system. Williams [79] demonstrated that anaerobic growth of Lachnospira multiparus and Bacteroides ruminicola subsp. brevis was not suppressed by trifluralin up to a concentration of 0.5 and 0.1%, respectively.

Williams and Feil [72] found that rumen bacteria were the etiological agents degrading trifluralin in rumen fluid. Only two [Bacteroides ruminicola subsp. brevis (GA-33) and Lachnospira multiparus (D-32)] of 12 character-ized rumen bacterial strains degraded trifluralin. The degradation products identified by spectral comparison with authentic compounds were α,α,α-trifluoro-N^4,N^4-dipropyltoluidine-3,4,5-triamine (28), α,α,α-trifluoro-5-nitro-N-propyltoluene-3,4-diamine (18), α,α,α-trifluoro-N-propyltoluene-3,4,5-triamine (30), α,α,α-trifluoro-2,6-dinitro-N-propyl-p-toluidine (13), and polar products. No loss of the trifluoromethyl group or cleavage

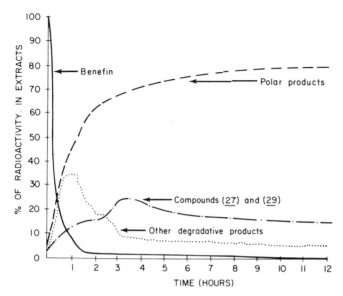

FIG. 12.   Benefin degradation and formation of degradative products
in artificial rumen fluid.

of the ring of trifluralin was observed. Trifluralin degradation by rumen
bacteria generally follows the pathway outlined for anaerobic degradation in
soil shown in Fig. 8.

   In the artificial rumen fluid system, benefin is degraded somewhat
faster than trifluralin. Figure 12 shows the combined formation of the reduction products (27) and (29). Other degradation products include the minor
metabolites associated with the anaerobic benefin degradation in soil shown
in Fig. 9. Polar products of benefin accumulate rapidly and constitute 79%
of the extracted radioactivity after 16 hr. Williams and Feil [72] attempted
to separate the polar products obtained from the degradation of trifluralin
by rumen bacteria. Eighty-eight to 93% of the original radioactivity was
extracted from the rumen microbial media. Approximately 45% of this
radioactivity was polar products. These polar products ran as a composite
on thin-layer chromatography with R_f values of 0.74 (solvent, isopropyl
alcohol) and 0.60 (solvent, diisopropyl ketone-formic acid-water, 40:15:2).
Chromatography on cellulose P and cellulose AE columns resulted in two
fractions, suggesting a mixture. The composite material was resistant to
acid and base hydrolysis and acetylation-esterification. Hydrolysis of a
polar mixture from the degradation of trifluralin in artificial rumen fluid by
Golab et al. [28] failed to change its chromatographic behavior. Reduction

of the polar mixture with tin and hydrochloric acid yielded a mixture containing a major portion of radioactivity that matched with the position of the unlabeled reference compound α, α, α-trifluorotoluene-3, 4, 5-triamine (21).

Similarly, as observed by Golab et al. [28] , the polar mixture obtained by Williams and Feil [72] could be reduced. The reduced product sublimed at 200°C. Infrared and mass spectrographic analysis indicated the presence of a trifluoromethyl group attached to the aromatic ring. None of the model compounds used by Williams and Feil matched the reduced polar products. They concluded that polar products were not likely to be azo compounds as suggested by Golab et al. [28, 29] , since they are readily degraded by rumen microorganisms, according to Katz et al. [80] .

Condensation of degradation products of dinitroaniline herbicides to azo-type compounds under anaerobic as well as aerobic conditions is possible, although no current conclusive evidence for this has been reported in the literature. Leitis and Crosby [35] have unconfirmed evidence for the formation of two azoxy compounds as minor photodecomposition products of trifluralin. To date, there are no reports on the cyclization of dinitroaniline herbicides to benzimidazole-type compounds in artificial rumen fluid, by rumen bacteria, or in ruminant animals. After a reevaluation of existing radioautograms from experiments performed on [^{14}C]oryzalin in artificial rumen fluid, Golab [78] observed that benzimidazole-type compounds may also be formed in this anaerobic medium.

III. MODE OF ACTION

Trifluralin and related compounds exert their herbicidal effect by inhibiting both root and shoot growth and development. Swelling of root tips of cotton growing in trifluralin-treated soil was first reported by Standifer et al. [81] . Fischer [82] observed inhibition of secondary root formation in young cotton plants. The inhibition of root development can be minimized by limiting the depth of soil incorporation of the herbicide. Trifluralin at 0. 01-3. 0 ppm in nutrient solution had no affect on germination but decreased root length of corn seedlings and cucumber hypocotyls, according to Schultz et al. [83] . They also observed that Avena coleoptiles were inhibited by trifluralin in the presence or absence of indoleacetic acid at 10^{-7} M concentration.

The response of 12 plant species to soil-incorporated trifluralin and benefin under greenhouse conditions was explored by Negi and Funderburk [84] . The species were grouped on the basis of inhibition of emergence and growth at different rates of application ranging up to 15 ppm. The most tolerant species were mustard, peanuts, soybeans, and cotton. Species intermediate in tolerance were morningglory, corn, cucumber, tomato, and bean. The susceptible group included oats, johnsongrass, and sorghum. Generally, trifluralin was more toxic to these species than benefin.

Histological examination of root tips grown in the presence of trifluralin provides additional insight into the physiological action of the herbicide. Talbert [85] observed that trifluralin appeared to inhibit cell division in meristematic root tissue of intact soybean roots grown in aerated nutrient solution. A noticeable increase in cell nuclei in the prophase stage of division appeared after 2 hr of treatment, as compared to controls. With continued exposure to trifluralin, numerous large polynucleated cells appeared after 24 hr. The cells of corn and cotton root tips examined by Amato et al. [86] were small and dense; many were multinucleated, with some cells containing up to five nuclei. Normal mitosis was absent. Similar results were obtained by Negi et al. [87]. They observed that the epidermis of both roots and shoots of treated corn seedlings was either disorganized or not present. It was suggested that the increase in epidermal cell size could not keep pace with the rate of increase in size of the cortex, which was also irregular in shape.

Lignowski [88] investigated the influence of trifluralin on cell division in wheat root tips. An increase in metaphase figures and a decrease in anaphase and telophase figures after 3 hr of treatment with 4×10^{-6} M trifluralin showed that mitosis was being blocked at metaphase. It was concluded that trifluralin inhibited mitosis at metaphase by interfering with the function of the spindle.

Soybean and wax bean plants treated with high rates of trifluralin displayed swelling and adventitious root generation in the hypocotyl region, according to Swann [89]. Histological studies showed that the observed swelling was the result of enlargement of cortical parenchyma cells and that the adventitious roots originated in the vascular cambium.

Strang and Rogers [55] studied the absorption and translocation of trifluralin by cotton and soybean using a microradioautographic technique. The accumulation of trifluralin in roots of both species was due largely to adsorption to the cuticle, epidermis, and cell walls. There appeared to be little active accumulation within living root cells. There was little movement out of cotton roots. The endodermis appeared to be a formidable barrier to trifluralin movement into the stele. Within the root, trifluralin accumulated in the cell walls of the xylem vessels, in cortical cell walls, and in the pericycle. Thickening of the cell walls of the xylem, radial expansion of the cortical cells, and enlargement of pericycle cells accompanied this observed accumulation of trifluralin.

Additional research on the influence of trifluralin on plant cell growth and development and on plant metabolism has been reported. Boulware and Camper [90] observed marked absorption of [^{14}C]trifluralin by isolated plant cells. Trifluralin was retained within the cell and appeared to be distributed

uniformly between the mitochondrial and the chloroplasts-nuclei fractions
of the cell organelle suspensions.

Mallory and Bayer [91] studied the influence of trifluralin on lateral
root initiation of cotton. Cotton root cells which normally produce lateral
roots were found to form unique structures, termed primordiomorphes,
when treated with trifluralin. Microscopic examination of these structures
revealed cells that contained numerous micronuclei and a greater number
of cell organelles. Hess and Bayer [92] reported an absence of microtubules in cells of lateral root meristems treated with trifluralin. Mitosis
was arrested when the chromosomes coalesced after failing to align at
metaphase.

Further work on microtubule formation has been reported by Bartels
and Hilton [93]. Trifluralin and oryzalin caused the loss of spindle and
cortical microtubules in root cells of wheat and corn. The rate of loss was
dependent on length of exposure, but spindle microtubules disappeared
more rapidly than cortical microtubules. Neither herbicide inhibited aggre-
gation and polymerization of microtubules in studies in vitro with pig brain
tissue. The authors theorized that microtubule disappearance could be the
result of inhibition of synthesis of microtubule protein or alteration of
endoplasmic reticulum membranes which are involved in synthesis or transport of microtubular protein to and from the site of polymerization. In
addition to the effect on microtubules, trifluralin and oryzalin caused
vacuolization of the meristematic cells in the root tips. Jackson and Stetler
[94] also observed loss of microtubules from dividing cells in the endosperm of African blood lily grown on a medium containing trifluralin. They
also reported the accumulation of large vesicles in the cell plate region,
which excluded microtubules and cell plate material. The authors suggest
that the decrease in number of microtubules and amount of cell plate material
leads to fusion of daughter nuclei.

In summary, morphological and histological evidence demonstrates
effects of trifluralin on plant cell division. While these effects have been
described in detail, little information is available on the biochemical processes responsible for the observed cellular alterations.

Some observations of the effect of trifluralin on plant metabolism have
been reported. Schultz and Funderburk [95] investigated the effect of trifluralin on nucleic acid and protein synthesis in corn roots. Dry weight,
total nucleic acids, DNA, RNA, and protein were determined in roots of
3-day-old corn seedlings germinated on paper toweling impregnated with
trifluralin. The dry weights of treated and untreated roots were similar,
but nucleic acid and protein per milligram of dry weight were reduced in
the treated roots. Reduction in total nucleic acid was reflected in the DNA
fraction as RNA was increased slightly in the treated plants. Concentrations
of 5 and 10 ppm of trifluralin applied to the roots of both sweet potatoes and

peanut plants grown in nutrient culture for 73 hr increased both carbohydrates and nucleic acids in the harvested plants, according to Dukes and Biswas [96] . The disparity of results in these two investigations in relation to the nucleic acid content possibly can be explained by the differences in the experimental methods employed.

Inhibition of oxygen uptake and oxidative phosphorylation in isolated mitochondria of corn, sorghum, and soybean by both trifluralin and nitralin was investigated by Negi et al. [97] . Trifluralin at 10^{-4} M inhibited the oxygen uptake by approximately 25% in corn and sorghum, with soybean inhibition being about three times greater. At the same concentration, the net phosphorus uptake was reduced by more than 50% in sorghum and corn. The phosphorus uptake in soybean was reduced by the same magnitude with trifluralin concentration at 10^{-6} M. Results with nitralin were similar, but somewhat more variable. As pointed out by Negi et al., it is not certain that the effect of trifluralin and nitralin on isolated mitochondria takes place in the intact plants, as the active agents must enter the cell and be available to the mitochondria at appropriate concentrations. There is little evidence that trifluralin enters the cell, but inhibition of these processes could be involved in the mode of action of trifluralin. However, the selective herbicidal activity of trifluralin is not explained, as the reduction of oxygen and phosphorus uptake was greater in the resistant species, soybeans, than in the susceptible species, sorghum.

In more recent studies, Moreland et al. [98] measured the effects of a series of 2, 6-dinitroanilines on electron transport and phosphorylation in isolated spinach chloroplasts and mung bean mitochondria. The 2, 6-dinitroanilines inhibited both electron transport and phosphorylation in isolated chloroplasts and mitochondria. There was no significant difference in type of activity among the twelve compounds evaluated, although some were stronger inhibitors than others. In chloroplasts, photoreduction and coupled photophosphorylation were both inhibited when water was employed as the electron donor and ferricyanide as the oxidant. Cyclic photophosphorylation was inhibited by all 2, 6-dinitroanilines in a system with phenazine methosulfate as the electron mediator under argon gas. In mitochondria, all the compounds except nitralin inhibited phosphorylating electron transport (state 3 respiration), and most were inhibitors of NAD and succinate oxidation. Some inhibited nonphosphorylating electron transport (state 4 respiration).

In studies with excised plant tissues and seedlings of spinach, mung bean, and corn, Moreland et al. [99] reported partial inhibition of photosynthesis by all compounds. Only oryzalin inhibited respiration in mung bean root tips and hypocotyls. In corn root tips, trifluralin suppressed oxygen uptake, and in corn seedlings, the compound uncoupled phosphorylation in mitochondria.

Hilton et al. [100] investigated the role of plant lipids in the selective action of trifluralin. Externally applied isoprenoid compounds and unsaturated fatty acids prevented the phytotoxic action of trifluralin to germinating seeds in the laboratory. Phytol, α-tocopherol acetate (vitamin E acetate), vitamin K, methyl palmitate, and methyl stearate were among the specific compounds found to prevent trifluralin action. This protective effect was attributed to prevention of sorption of trifluralin by seedlings, although protection by endogenous lipids is also likely. In these studies, there was a significant correlation between seedling sensitivity of eleven species of weeds and crops and their lipid content, further suggesting protection from stored lipids [101].

Interference with photosynthesis and respiration by trifluralin and other dinitroanilines has been demonstrated in both in vitro and in vivo studies. The role of plant lipids in the selective action of trifluralin has been explored. The importance of these findings as they relate to primary mode of action or to mitotic effects has not been established.

IV. SUMMARY

Interest in the selective herbicidal activity of the substituted 2,6-dinitroanilines has resulted in the development of several members as important herbicides for agronomic and horticultural crops. Substitution patterns around the 2,6-dinitroaniline molecule affect the degree of both selectivity and efficacy.

The degradation of dinitroaniline herbicides, as exemplified by extensive studies with trifluralin, oryzalin, isopropalin, and benefin, follows oxidative dealkylation and reduction pathways. Under aerobic conditions, dealkylation predominates; under anaerobic conditions, reduction of nitro groups is favored. The cyclization of 2,6-dinitroaniline herbicides to a benzimidazole-type compound also appears to be characteristic under a variety of conditions and has been recognized as a prominent degradative pathway for this class of compounds. The decomposition rate of these herbicides and the accumulation of recognizable degradation products vary in soils, plants, and animals.

The mode of action of the dinitroanilines has not been completely defined. Studies on absorption and translocation of trifluralin indicate that little is translocated from roots to aerial plant portions. Effects on cell division in roots and in germinating seedlings of susceptible plants indicate that inhibition of mitosis may be the principal means by which dinitroaniline herbicides control weeds. They are also indications of interference with photosynthesis and respiration by dinitroanilines in plants. Variation of lipid contents of various plants might be involved in the selective action of trifluralin and other dinitroaniline herbicides.

REFERENCES

1. N. G. Clark and A. F. Hams, Biochem. J., 55, 839 (1963).

2. R. F. Brookes, N. G. Clark, A. F. Hams, and H. A. Stevenson, Brit. Pat. 845, 916 (1960).

3. B. F. Malichenko, E. M. Lavchenko, N. A. Malichenko, G. I. Alyab'eva, E. A. Shomova, and L. M. Yapupol'skii, Fiziol. Aktiv. Veshchestra, 1969 (2), 75 (1969).

4. Plant Regulators, CBCC Positive Data, Series 2, National Research Council, Washington, D.C., June 1955.

5. E. F. Alder, W. L. Wright, and Q. F. Soper, Proc. North Central Weed Control Conf., 17, 23 (1960).

6. Q. F. Soper, E. F. Alder, and W. L. Wright, Proc. South. Weed Conf., 14, 86 (1961).

7. E. F. Alder and R. B. Bevington, Proc. Northeast. Weed Control Conf., 16, 505 (1962).

8. W. A. Gentner, Weeds, 14, 176 (1966).

9. L. Guse, W. Humphreys, J. Hooks, J. V. Gramlich, and W. Arnold, Proc. South. Weed Conf., 19, 121 (1966).

10. W. J. Hughes and R. H. Schieferstein, Proc. South. Weed Conf., 19, 170 (1966).

11. Technical Report on Isopropalin (EL-179), Report No. EA0045, Elanco Products Co., Indianapolis, Indiana, January 1970.

12. Technical Report on Oryzalin, Report No. EA2080, Elanco Products Co., Indianapolis, Indiana, March 1972.

13. Technical Report on USB-3584, Report No. USBR-CTDS, U.S. Borax Research Corp., Anaheim, California, 1971.

14. Technical Report on TREFLAN, Report No. EA6103, Elanco Products Co., Indianapolis, Indiana, April 1966.

15. S. J. Parka and H. M. Worth, Proc. South. Weed Conf., 18, 469 (1965).

16. Technical Report on BALAN, Report No. EA6104, Elanco Products Co., Indianapolis, Indiana, May 1966.

17. Q. F. Soper, U.S. Pat. 3,257,190 (1966).

18. S. B. Solaway and K. D. Zwahlen, U.S. Pat. 3,227,734 (1966).

19. Q. F. Soper, U.S. Pat. 3,367,949 (1968).

20. D. L. Hunter, W. G. Woods, J. D. Stone, C. W. LeFevre, U.S. Pat. 3,617,252 (1971).

21. R. G. Harvey, Weed Sci., 21, 517 (1973).

22. S. J. Parka and J. B. Tepe, Abstr. 6th Meeting Weed Soc. Am. (St. Louis), p. 40 (1966).

23. G. W. Probst, T. Golab, R. J. Herberg, F. J. Holzer, S. J. Parka, C. Van der Schans, and J. B. Tepe, J. Agr. Food Chem., 15, 592 (1967).

24. T. Golab, R. J. Herberg, S. J. Parka, and J. B. Tepe, J. Agr. Food Chem., 15, 638 (1967).

25. J. L. Emmerson and R. C. Anderson, Toxicol. Appl. Pharmacol., 9, 84 (1966).

26. R. J. Herberg, T. Golab, A. P. Raun, and F. J. Holzer, 153rd Meeting, Am. Chem. Soc., Miami Beach, Florida, 1967, p. 45.

27. T. Golab, E. W. Day, Jr., and G. W. Probst, 153rd Meeting, Am. Chem. Soc., Miami Beach, Florida, 1967, p. 44.

28. T. Golab, R. J. Herberg, E. W. Day, A. P. Raun, F. J. Holzer, and G. W. Probst, J. Agr. Food Chem., 17, 576 (1969).

29. T. Golab, R. J. Herberg, J. V. Gramlich, A. P. Raun, and G. W. Probst, J. Agr. Food Chem., 18, 838 (1970).

30. D. G. Crosby and K. W. Moilanen, Abstr. Pest. 027, 163rd National Meeting, Am. Chem. Soc., Boston, Mass., 1972.

31. R. A. Smith, W. S. Belles, Kei-Wei Shen, and W. G. Woods, Abstr. Pest. 027, 164th National Meeting, Am. Chem. Soc., New York, 1972.

32. T. L. Laanio, P. C. Kearney, and D. D. Kaufman, Abstr. Pest. 026, 164th National Meeting, Am. Chem. Soc., New York, 1972.

33. D. G. Crosby and E. Leitis, Western Regional Meeting, Am. Chem. Soc., San Francisco, 1972.

34. T. Golab, unpublished work, 1971.

35. E. Leitis and D. G. Crosby, J. Agr. Food. Chem., 22, 843 (1974).

36. T. Golab and M. E. Amundson, 3rd International Congress of Pesticide Chemistry, Helsinki, Finland, 1974.

37. P. C. Kearney, W. B. Wheeler, J. R. Plimmer, A. Kontson, and U. I. Klingebiel, 189th Meeting, Weed Sci. Soc. Am., Washington, D. C., 1975, p. 72.

38. T. Golab, J. Chromatogr., 18, 406 (1965).

39. R. E. Scroggs and J. B. Tepe, Analytical Methods for Pesticides, Plant Growth Regulators, and Food Additives, Vol. 5 (Supplemental Volumes) (G. Zweig, ed.), Academic Press, New York, 1967, p. 527.

40. F. J. Marshall, R. E. McMahon, and R. G. Jones, J. Agr. Food Chem., 14, 498 (1966).

41. S. J. Pieczarka, W. L. Wright, and E. F. Alder, Proc. South. Weed Conf., 15, 92 (1962).

42. W. F. Spencer and M. M. Claith. J. Agr. Food Chem., 22, 987 (1974).

43. J. V. Parochetti, G. W. Dec, Jr., and G. W. Burt, Weed Sci. Soc. Am. Abstr., p. 72 (1975).

44. J. F. Paar and J. Smith, Soil. Sci., 115, 55 (1973).

45. M. L. Scheinbaum, Ph.D. Thesis, Howard University, Washington, D. C., 1964.

46. W. L. Wright and G. F. Warren, Weeds, 13, 329 (1965).

47. E. W. Day, Jr., unpublished work, 1963.

48. R. E. McMahon, Tetrahedron Lett., 1966 (21), 2307 (1966).

49. D. G. Crosby and E. Leitis., Bull. Environ. Contam. Toxicol., 10, 237 (1973).

50. C. J. Soderquist, D. A. Crosby, K. W. Moilanen, J. N. Seiber, J. E. Woodrow, J. Agr. Food Chem., 23, 304 (1975).

51. H. C. Newsom and W. G. Woods, J. Agr. Food Chem., 21, 598 (1973).

52. G. P. Nilles and M. J. Zabik, J. Agr. Food Chem., 22, 684 (1974).

53. J. R. Plimmer and U. I. Klingebiel, Abstr. Pest. 024, 164th Meeting, Am. Chem. Soc., New York, 1972.

54. J. R. Plimmer and U. I. Klingebiel, J. Agr. Food Chem., 22, 689 (1974).

55. R. P. H. Strang and R. L. Rogers, Weed Sci., 19, 363 (1971).

56. T. Golab and W. A. Althaus, Weed Sci., 23, 165 (1975).

57. T. Golab, C. E. Bishop, A. L. Donoho, J. A. Manthey, and L. L. Zornes, Pest. Biochem. Physiol., 5, 196 (1975).

58. P. K. Biswas and W. Hamilton, Jr., Weed Sci., 17, 206 (1969).

59. T. Golab, J. Chromatogr., 30, 253 (1967).

60. Private communication from Shell Chemical Co., 1972.

61. R. A. Smith, W. S. Belles, K. W. Shen, and W. G. Woods, Pest. Biochem. Physiol., 3, 278 (1973).

62. C. E. Marshall, The Physical Chemistry and Minerology of Soils, Vol. I., Soil Materials, Wiley, New York, 1964.

63. L. E. Bode and M. R. Gebhardt, Abstr. 11th Meeting Weed Sci. Soc. Am., p. 198 (1971).

64. S. J. Parka and J. B. Tepe, Weed Sci., 17, 119 (1969).

65. J. H. Miller, P. E. Keeley, C. H. Carter and R. J. Thullen, Weed Sci., 23, 211 (1975).

66. H. H. Funderburk, Jr., D. P. Schultz, N. S. Negi, R. Rodriguez-Kabana, and E. A. Curl, Proc. South. Weed Conf., 20, 389 (1967).

67. D. A. Addison, Ph.D. Thesis, Clemson University, Clemson, South Carolina, 1968, p. 76.

68. A. Y. Hamdi and M. S. Tewik, Acta Microbiol. Pol. (Ser. B.), 1, 83 (1969).

69. C. G. Messersmith, O. C. Burnside, and T. L. Lavy, Weed Sci., 19, 285 (1971).

70. T. L. Laanio, P. C. Kearney, and D. D. Kaufman, Pest. Biochem. Physiol., 3, 271 (1973).

71. P. C. Kearney, J. R. Plimmer, and V. P. Williams, Abstr. Pest. 025, 164th Meeting, Am. Chem. Soc., New York, 1972.

72. P. P. Williams and V. J. Feil, J. Agr. Food Chem., 19, 1198 (1971).

73. D. D. Kaufman, Pesticides in the Soil: Ecology, Degradation and Movement, Pesticide Metabolism International Symposium, Michigan State University Press, East Lansing, Mich., 1970, pp. 73-88.

74. D. D. Kaufman, J. R. Plimmer, P. C. Kearney, J. Blake, and F. S. Guardine, Weed Sci., 16, 266 (1968).

75. T. Golab, unpublished work, 1974.

76. J. Page, unpublished work, 1971.

77. F. J. Holzer, unpublished work, 1971.

78. T. Golab, unpublished work, 1971.

79. P. P. Williams, Bucknal Proc. 67, 8 (1967).

80. S. E. Katz, C. A. Fassbender, and R. F. Strusz, J. Assoc. Offic. Anal. Chem., 52, 1213 (1969).

81. L. C. Standifer, Jr., L. W. Sloane, and M. E. Wright, Proc. South. Weed Conf., 18, 92 (1965).

82. B. B. Fischer, Aust. J. Exp. Agr. Anim. Husb., 6, 214 (1966).

83. D. P. Schultz, H. H. Funderburk, Jr., and N. S. Negi, Abstr. 6th Meeting, Weed Sci. Soc. Am. (St. Louis), p. 42 (1965).

84. N. S. Negi and H. H. Funderburk, Jr., Proc. South. Weed Conf., 20, 369 (1967).

85. R. E. Talbert, Proc. South. Weed Conf., 18, 642 (1965).

86. V. A. Amato, R. R. Hoverson, and J. Hacskaylo, Proc. Assoc. Southern Agr. Workers, 62nd Ann. Conv., 1965, p. 234.

87. N. S. Negi, H. H. Funderburk, Jr., and D. P. Schultz, Agr. Exp. Station, Auburn University, Botany and Plant Pathology, Departmental Series No. 2, Auburn, Alabama, 1967.

88. E. M. Lignowski, Ph.D. Dissertation, West Virginia University, Morgantown, 1969.

89. C. W. Swann, Ph.D. Dissertation, University of Minnesota, Minneapolis, 1969.

90. M. A. Boulware and N. D. Camper, Weed Sci., 21, 145 (1973).

91. T. Mallory and D. E. Bayer, Abstr. 14th Meeting, Weed Sci. Soc. Am. (Las Vegas), p. 87 (1974).

92. B. Hess, D. E. Bayer, and F. Ashton, Abstr. 14th Meeting, Weed Sci. Soc. Am. (Las Vegas), p. 17 (1974).

93. P. G. Bartels and J. L. Hilton, Pest. Biochem. Physiol., 3, 462 (1973).

94. W. T. Jackson and D. A. Stetler, Can. J. Bot., 51, 1513 (1973).

95. D. P. Schultz and H. H. Funderburk, Jr., Weed Soc. Am. Abstr., p. 58 (1967).

96. I. E. Dukes and P. K. Biswas, Weed Soc. Am. Abstr., p. 59 (1967).

97. N. S. Negi, H. H. Funderburk, Jr., D. P. Schultz, and D. E. Davis, Weeds, 11, 265 (1968).

98. D. E. Moreland, F. S. Farmer, and G. G. Hussey, Pest. Biochem. Physiol., 2, 352 (1972).

99. D. E. Moreland, F. S. Farmer, and G. G. Hussey, Pest. Biochem. Physiol., 2, 354 (1972).

100. J. L. Hilton, M. N. Christiansen, J. B. St. John, and K. H. Norris, Abstr. 11th Meeting, Weed Sci. Soc. Am. (Dallas), p. 109 (1971).

101. J. L. Hilton, Abstr. 13th Meeting, Weed Sci. Soc. Am. (St. Louis), p. 68 (1972).